CW01483794

North Wyke Research Station
Institute of Grassland and Environmental Research
Okehampton
Devon EX20 2SB
UK

We thank the Institute of Grassland and Environmental Research (IGER) for supporting both this book and the 12th Nitrogen Workshop. IGER is sponsored by the Biotechnology and Biological Sciences Research Council (BBSRC), who also provided additional support.

Controlling nitrogen flows and losses

Edited by:
D.J. Hatch
D.R. Chadwick
S.C. Jarvis
J.A. Roker

Wageningen Academic
P u b l i s h e r s

CIP-data Koninklijke Bibliotheek
Den Haag

ISBN 9076998434

hardcover

Subject headings:
Fertiliser
Manures
Nutrient cycling

First published, 2004

Wageningen Academic Publishers
The Netherlands, 2004

Printed in The Netherlands

PREFACE

The need to protect the environment, whilst maintaining efficient food production, is a challenge that can only be met by adopting a sustainable approach to nutrient use. The case for more efficient use of nitrogen (N) is clear: productivity should not be achieved at the expense of pollution, either of the atmosphere, or of our ground and surface waters. To achieve this goal, a tighter N budget is required and this, together with increasing legislation to limit pollution, suggested that a workshop devoted to the problems of 'controlling nitrogen flows and losses' would be timely.

This theme was adopted as the overall title for the 12th Nitrogen Workshop, held at the University of Exeter, UK (21st-24th September 2003) for an international meeting organised by the Institute of Grassland and Environmental Research, North Wyke, Research Station, Devon, UK. Previous Workshops were held in the UK from 1982 until 1992, after which they moved to mainland Europe (see list below) and have attracted an increasingly international audience. This trend has continued with the return to a UK venue, attracting over 260 delegates from more than 30 countries.

This book is a synthesis of the contributions from the 12th Workshop, drawn from both oral and poster presentations (presented previously as single page summaries in the 'Book of Abstracts'). We have edited all the expanded contributions, including the poster presentations, with the intention of providing consistency in presentation, but we have been entirely inclusive in our approach, as we wished to give all the participants an opportunity to present an up to date overview of current international research programmes and activities.

As such, we hope, this volume represents a valuable compilation of contemporary research, aimed at reconciling the environmental and economic components of N cycling within the context of a productive agricultural industry. The full papers, forming the main structure of the book, are drawn from the keynote and supporting speakers in each themed session and these have all been subjected to a refereeing process that enlisted three independent scientists to referee each paper, thus ensuring scientific rigour. We are most grateful to them for providing this service. Nevertheless, the accuracy and content of these papers remain the responsibility of the authors concerned.

The book is divided into seven main sections, which examine systematically, the nature of the problems associated with losses of N and a range of possible solutions.
- *Section 1*: 'Drivers towards sustainability-why change?' identifies the need to adopt new strategies to avoid losses to the environment.
- *Section 2*: considers the options for 'Matching supply with demand' to improve the N farm budget.
- *Section 3*: outlines the reasons for, and means of, 'Controlling losses to air'.
- *Section 4*: adopts a similar approach for 'Controlling losses to 'water'.
- *Section 5*: describes the challenges of 'Reconciling productivity with environmental considerations'.

These five Sections are each supported by abstracts from relevant posters and together, cover the main themed sessions from the Workshop. The remaining sections describe a selection of mathematical models, to assist the researcher and others (*Section 6*), which were demonstrated at the Workshop, and the final section (*Section 7*) is devoted to reports from the 'Themed Working Groups' which debated the following topical questions:

- *Organic matter: does it matter, or can technology overcome most problems related to soil fertility ?*
- *Optimising N additions: can we integrate fertilizer use and manure use?*
- *Controlling gaseous N emissions: what is achievable?*
- *Missing N: is the solution in dissolved N?*
- *Pollution problems: mitigation, or are we swapping one form of pollution for another?*
- *System studies: do we need them, or can they be replaced by desktop studies?*
- *Model answers: can we improve their level of confidence and applicability?*

We believe this book will be of value to researchers, policy makers and all those wishing to promote more efficient use of N.

The steering committee of the Workshop (editors of this book) wish to record their thanks to other colleagues who helped to organise the 12th N Workshop and, in particular, M. Alfaro, A. Bristow, L. Brown, R. Bol, P. Butler, J. Chisholm, L. Deeks, A. del Prado, H. Dennis, E. Dixon, S. Gilhespy, S. Granger, T. Harrod, J. Hawkins, L. Jewell, E. Jewkes, A. Joynes, J. Laws, G. Lewis, M. McHugh, M. Mills, T. Misselbrook, P. Murray, J. Shaw and S. Yamulki. We also thank other members of IGER, North Wyke staff for their invaluable contribution to the administration and logistics of the Workshop and preparation of this book.

The Editors, March 2004.

Previous Nitrogen Workshops

1. Rothamsted Experimental Station, UK, 20th July 1982
2. Rothamsted Experimental Station, UK, 17th July 1984
3. GRI, University of Reading, UK, 16th-17th December 1985
4. University of Aberdeen, UK, 6th-9th April 1987
5. Rothamsted Experimental Station, Silsoe, UK, 13th-14th December 1988
6. The Queen's University, Belfast, UK, 17th-19th December 1990
7. University of Edinburgh, UK, 23rd-26th June 1992
8. University of Ghent, Belgium, 5th-8th September 1994
9. Technische Universitat Braunschweig, Germany, 9th-12th September 1996
10. The Royal Veterinary and Agricultural University, Copenhagen, Denmark, 23rd-26th August 1999
11. INRA, Reims, France, 9th-12th September 2001
12. IGER, University of Exeter, UK, 21st-24th September 2003

13. PRI, Maastricht, The Netherlands, 24th-26th October 2005

Table of contents

Preface   **7**

Section 1 – Drivers towards sustainability: why change?   **27**

Drivers towards sustainability: why change?   **29**
J.J. Neeteson, J.J. Schröder and C. Jakobsson

Nitrogen balances over seven years on a mixed farm in the Cotswolds   **39**
K.A. Leach, K.W.T. Goulding, D.J. Hatch, J.S. Conway and K.D. Allingham

Watershed nitrogen modelling   **47**
N.J. Hutchings, T. Dalgaard, B.M. Rasmussen, J.F. Hansen, M. Dahl, P. Rasmussen,
L.F. Jørgensen, V. Ernstsen, F. von Platen-Hallermund and S.S. Pedersen

Mitigating non-point source pollution from dairy farms: Economic evaluation on the Don watershed   **54**
N. Turpin, G. Rotillon, P. Bontems, T. Bioteau and R. Laplana

Relationships between fertiliser nitrogen additions, crop carbon returns and soil quality   **63**
A. Bhogal, B.J. Chambers and F.A. Nicholson

Fluxes of nitrogen following clearing of Brazilian Amazonian tropical forest for pasture   **65**
M.C. Piccolo, C. Neill, C.C. Cerri and J.M. Melillo

Nitrogen losses from forage legumes and effects on succeeding cereal crops in organic farming under pannonic climate conditions in Eastern Austria   **67**
R. Farthofer, J.K. Friedel, G. Pietsch, W. Loiskandl and B. Freyer

Your farm and NVZs: A decision support system for farmers and consultants   **70**
C.P. Fawcett, M.M. Gibbons, F.A. Brown, P.M.R. Dampney and S.J. Richardson

Fate of applied nitrogen in a heathland ecosystem   **73**
E.R. Green, S.A. Power and E.M. Baggs

Effects of intensification of dairy farming in New Zealand on nitrogen efficiency, energy use and environmental emissions   **76**
S.F. Ledgard, J.D. Finlayson, M.S. Sprosen, D.M. Wheeler and N.A. Jollands

Effect of groundnut stover and weed management methods in the dry season on N cycling and maize yield 78
S. Promsakha na sakonnakhon, B. Toomsan, V. Limpinuntana, P. Vityakon, E.M. Baggs and G. Cadisch

Development of a decision support tool for global diffuse nitrogen pollution assessment 81
C. Macleod, J. Dela-Cruz, P. Haygarth, G. Glegg, D. Scholefield and L. Mee

Organic fertilisation in silvopastoralism and nitrate losses: sustainable systems 83
M.R. Mosquera-Losada, S. Rodríguez-Barreira and A. Rigueiro-Rodríguez

How nitrate vulnerable zones can reduce the adverse environmental impact of productive arable agriculture 85
J.U. Smith, P. Smith, A.G. Dailey and M.J. Glendining

Decline in species richness in acid grasslands along a gradient of nitrogen deposition 88
C.J. Stevens, N.B. Dise, D.J. Gowing and J.O. Mountford

Development of a modelling system for prediction and regulation of livestock waste pollution in the humid tropics 90
T.P. Tee, I.J. Lean, E.M. Baggs and G. Cadisch

The agricultural area survey as a tool for implementing the European Nitrates Directive in the Walloon Region of Belgium 93
C. Vandenberghe, A.C. Mohimont, T. Garot and J.M. Marcoen

Environmental economics of soil organic matter management 96
N. Wrage and O. Oenema

Section 2 – Matching supply with demand 99

Matching supply with demand 101
D.V. Murphy, E.A. Stockdale, F.C. Hoyle, J.U. Smith, I.R.P. Fillery, N. Milton, W.R. Cookson, L. Brussaard and D.L. Jones

Can N mineralisation be predicted from soil organic matter? Carbon and gross N mineralisation rates as affected by long-term additions of different organic amendments 113
A. Herrmann, E. Witter and T. Kätterer

N mineralisation from decomposition of catch crop residues under field conditions: measurement and simulation using the STICS soil-crop model 122
E. Justes and B. Mary

Gross N transformation rates and soil organic matter contents in grassland soils of different age   131
F. Accoe, P. Boeckx, O. Van Cleemput and G. Hofman

The relative efficiency of fertiliser N in NK blends and concentrated complex fertilisers   133
N.A. Akhonzada and J.S. Bailey

Nitrogen mineralisation from fish sludge as affected by plant uptake   136
M.A. Alfaro, F.J. Salazar and A. Valdebenito

Modelling of N-cycle for processes in soil using soil core experiments   138
Á. Bálint, Cs. Mészáros, Gy. Heltai, E. Nótás and K. Jung

Mineralisation of nitrogen in relation to climatic variation and soil   140
H. Björnsson

Resolving differences in N cycling between more polluted and pristine forests using ^{15}N isotope dilution   143
P. Boeckx, R. Godoy, C. Oyarzún, J. Bot and O. Van Cleemput

Effects of tillage and crop rotation on the microbial population and dynamics of soil organic matter   145
C. Carranca, A. Oliveira, A. de Varennes, M. Pampulha, M. Costa, A. Prazeres, J. Baeta, C. Neto, M.P. Andrada and M.O. Torres

The changes in microbial biomass nitrogen when different rates and forms of N were applied in a long-term experiment with maize   148
J. Černý and J. Balík

Screening of organic biological waste products for their potential to manipulate the N release from crop residues   151
B. Chaves, S. De Neve, G. Hofman, P. Boeckx and O. Van Cleemput

Nitrogen dynamics in soil with mulch and incorporated crop residues   153
F. Coppens, S. Recous, P. Garnier and R. Merckx

Prediction of nitrogen mineralisation from organic residues and supply to ryegrass   156
C.M.d.S. Cordovil, J. Coutinho and F. Cabral

Reliability of a chemical method to assess nitrogen uptake by winter wheat   158
C.M.d.S. Cordovil, J. Coutinho and F. Cabral

Rhizodeposition and symbiotic N$_2$ fixation in trimmed and untrimmed white clover   160
S. Dahlin

Within-field variations in, and relations between, grain protein content, grain yield and plant-available soil nitrogen 162
S. Delin

The turnover of organic manure using natural ^{13}C abundance 164
K. Dittert and R. Bol

Nitrification during autumn and winter of ammonium nitrogen in cattle slurry applied to soil at different times during the autumn 166
L. Engström, B. Lindén and L. Ericsson

Effect of rate of cattle slurry at sowing, number of fertigations with separated slurry liquid and rate of mineral N top dressings on yield and N removal by forage maize 168
A. Fernandes, H. Trindade, J. Coutinho and N. Moreira

N balance in fertilizer trials with composted municipal solid wastes in southern Italy 171
D. Ferri, G. Convertini and F. Montemurro

How to enhance crop utilisation of deep subsoil nitrogen supply 173
J. Haberle, P. Svoboda and J. Krejčová

Model SFOM - the first step towards fertiliser and manure use integration 176
T. Jadczyszyn

Straw-rich deep litter manures - can decomposition and N turnover be predicted from quality? 179
L.S. Jensen, A. Jensen and A. Pedersen

Field experiments to determine N accumulation under fertility building crops 182
A. Joynes, D.J. Hatch, A. Stone, S. Cuttle, G. Goodlass and S. Roderick

A comparison of different indices for nitrogen mineralisation 184
A. Kokkonen and M. Esala

A novel approach to regulate nitrogen mineralisation in soil 186
K. Kumar, C.J. Rosen and M.P. Russelle

Variable nitrogen fertilisation by tractor-mounted remote sensing 188
A. Link, J. Jasper and H.-W. Olfs

A dynamic version of the predictive balance sheet method for fertiliser N advice 191
J.M. Machet, S. Recous, M.H. Jeuffroy, B. Mary, B. Nicolardot and V. Parnaudeau

Residual effects of maize nitrogen fertilisation on winter barley crop: N content, yield, yield components and agronomic factors 194
M. Maiorana, F. Montemurro, G. Convertini and F. Fornaro

Turnover of grain legume N rhizodeposits and effect of rhizodeposition on the turnover of crop residues 197
J. Mayer, F. Buegger, E.S. Jensen, M. Schloter and J. Heß

Decomposition of soluble compounds obtained after fractionation of different animal wastes 200
T. Morvan and B. Nicolardot

Is good nutrient management possible by applying more than 170 kg nitrogen per hectare from manure? 203
A. Mulier, G. Hofman and I. Verbruggen

Regulation of N release from polyphenol-protein complexes by fungi in different tropical production systems 205
R. Mutabaruka and G. Cadisch

Modelling nitrogen mineralisation from mature bio-waste compost applied to vineyard soils 207
C. Nendel, S. Reuter, K.C. Kersebaum, R. Nieder and R. Kubiak

Inorganic N dynamics in vineyard soils under different covering strategies 209
B. Nicolardot, P. Thiébeau, C. Herre, A.F. Doledec, A. Perraud and B. Mary

Effect of rate and timing of nitrogen fertiliser on yield and grain protein content in winter wheat. The chlorophyll meter as a nutritional diagnostic tool 212
M.A. Ortuzar, A. Alonso, A. Aizpurua, A. Castellón and J.M Estavillo

Assessing N dynamics of organic wastes in field conditions using a calculation model 214
V. Parnaudeau, P. Robert, C. Herre, F. Millon, B. Mary and B. Nicolardot

^{15}N labelling and use of dairy manure components for N cycling studies 217
J.M. Powell and K.A. Kelling

Predicting soil mineral nitrogen in spring, based on soil mineralisable N in autumn 219
M. Quemada

Effect of seasonal split of N dosage on the fate of ^{15}N fate in citrus trees 222
A. Quiñones, J. Bañuls, E. Primo-Millo and F. Legaz

Methodology of soil incubation studies -comparison between laboratories   224
T. Salo, L.S. Jensen, F. Palmason, T.A. Breland, B. Stenberg, A. Pedersen,
C. Lundström and M. Esala

Attempts to overcome genotypic variability of SPAD readings in winter triticale   226
S. Samborski and J. Rozbicki

Organic nitrogen loads at different scale levels in Spain   229
J. Soler-Rovira and J.M. Arroyo-Sanz

Organic nutrient management indicators at district level in Spain   231
J. Soler-Rovira and J.M. Arroyo-Sanz

**Physical separation of pig slurry has a small effect on the overall utilisation
of nitrogen**   234
P. Sørensen, I.K. Thomsen and B.T. Christensen

**The use of anaerobic incubations to predict net N immobilisation after the
application of organic residues to soils**   236
J.R. Sousa, R. Lagoa, F. Cabral and J. Coutinho

**Nitrogen uptake by ryegrass (*Lolium perenne*) as affected by the decomposition
of apple leaves and pruning wood in soil**   239
M. Tagliavini, G. Tonon, D. Solimando, P. Gioacchini, M. Toselli, P. Boldreghini and
C. Ciavatta

Use of the NDICEA model analysing nitrogen efficiency   242
G.J. van der Burgt

Microbial biomass measurements in soil of the central highlands of Mexico   244
M.S. Vásquez-Murrieta and L. Dendooven

**Natural ^{15}N abundance as an indicator of the effect of management intensity
on nitrogen cycling in montane grasslands**   246
M. Watzka and W. Wanek

Section 3 – Controlling losses to air   249

Controlling losses to air   251
E.A. Davidson and A.R. Mosier

Pitfalls in measuring nitrous oxide production by nitrifiers   260
N. Wrage, G.L. Velthof, H.L. Laanbroek and O. Oenema

The effect of organic and mineral nitrogen fertilisers on emissions of NO, N_2O and CH_4 from cut grassland   268
R. Rees, S. Jones, R.E. Thorman, I. McTaggart, B. Ball and U. Skiba

A new agricultural ammonia emission inventory for Switzerland based on a large scale survey and model calculations   277
B. Reidy, S. Pfefferli and H. Menzi

N_2O emissions from a field trial as influenced by N fertilisation and nitrification inhibitors   285
D. Báez, J. Coutinho, N. Moreira and H. Trindade

Denitrification and interactions between nitrification and methane oxidation under elevated atmospheric CO_2   288
E.M. Baggs, H. Blum, M. Richter, G. Cadisch and U.A. Hartwig

Soil water content as a factor that controls N_2O production by denitrification and autotrophic and heterotrophic nitrification   290
E.J. Bateman, G. Cadisch and E.M. Baggs

Short-term CO_2 and N_2O emission after application of manure and maize residues to three different soil types: a laboratory study   293
D. Beheydt, P. Boeckx, L. Geers, A. Goossens and O. Van Cleemput

Influence of urinary N concentration on N_2O and CO_2 emissions and N transformation in a temperate grassland soil   295
R. Bol, S.O. Petersen, K. Dittert and M.N. Hansen

A model-based evaluation of options for the mitigation of agricultural nitrous oxide emission   298
L. Brown, L. Cardenas, P. Bellamy, S.C. Jarvis, J. Hollis, R.W. Sneath, S. Yamulki and K.W.T. Goulding

Variations in aerobic respiratory and denitrifying activities in the vadose zone: laboratory and field experiments   300
P. Cannavo, F. Lafolie, A. Richaume, B. Nicolardot and P. Renault

Use of a 2-pool model to evaluate the effect of fertiliser application on nitrogen emissions from grassland soils   303
L.M. Cárdenas, J.M.B. Hawkins, D. Chadwick and D. Scholefield

Opportunities for reducing ammonia emissions from pig housing in the UK   306
T.G.M. Demmers, R. Kay and N. Teer

A comparison of nitrous oxide and ammonia fluxes from managed grassland   309
C.Di Marco, M. Anderson, C. Milford, U. Skiba, M.A. Sutton and K. Weston

**Gaseous losses of various nitrogeneous compounds from fertilisation of a
wheat crop** 311
S. Génermont, C. Hénault, P. Laville, M.H. Jeuffroy and D. Flura

A simple denitrification model? Review, sensitivity and application 314
M. Heinen

Nitrous oxide emissions from soil - fertiliser and soil type effects 316
B. Hyde, A. Fanning, M. Ryan, M. Hawkins and O.T. Carton

**The Irish ammonia emission inventory - implications for compliance with the
Gothenburg Protocol** 318
B.P. Hyde, T. Misslebrook and O.T. Carton

Greenhouse gas emissions in meadow mesocosms exposed to elevated O_3 and CO_2 321
T. Kanerva, K. Koivisto,K. Karhu, K. Regina and S. Manninen

**Refining the uncertainty in nitrous oxide emissions from New Zealand
agricultural soils** 323
F.M. Kelliher, A.S. Walcroft, S.F. Ledgard, H. Clark, G. Rys, H. Plume, M. Buchan and
R.R. Sherlock

**Dynamics of N_2O and NO production by Alcaligenes faecalis parafaecalis:
effect of pH, temperature, substrate and oxygen supply** 326
M. Kesik, S. Blagodatsky, H. Papen and K. Butterbach-Bahl

Denitrification and nitrate loss from organic agricultural soil at low temperatures 329
H.T. Koponen, A. Pärnä, H. Silvennoinen and P.J. Martikainen

N_2O emissions from a water-saving rice production system (GCRPS) in North China 331
C. Kreye, K. Dittert, X. Zheng, X. Zhang, S. Lin, H.Tao and B. Sattelmacher

**Controlling nitrous oxide emissions from agriculture: experience from
The Netherlands** 333
P.J. Kuikman, G.L. Velthof and O. Oenema

Fluxes of N_2O from permanent grassland with different levels of nitrogen supply 336
C. Lampe, F. Taube, M. Wachendorf, B. Sattelmacher and K. Dittert

Gaseous nitrogen emissions from effluent irrigated soils 338
Y. Master, R.J. Stevens, R.J. Laughlin, U. Shavit and A. Shaviv

**Reducing losses of nitrous oxide from cattle slurry and mineral fertiliser applied
to grassland by the use of DMPP** 340
P. Merino, A. del Prado, S. Menéndez, L. Careaga, M. Pinto, J.M. Estavillo and
C. González-Murua

Influence of white clover on nitrous oxide fluxes in grassland   342
A. Mori, M. Hojito, H. Kondo, H. Matsunami and D. Scholefield

The effect of agricultural ammonia deposition on nitrous oxide production by soils under coniferous and deciduous woodland cover.   345
T. Morrissey, P. Ineson and D.R. Chadwick

Determination of N source in denitrification studies using stable isotope techniques   347
P.J. Murray, D.J. Hatch, E.R. Dixon, R.J. Stevens, R.J. Laughlin, K. O'Prey and S.C. Jarvis

Using a system of undisturbed, in situ soil lysimeters to determine nitrogen transformations in a sub-surface clay interface   349
P.J. Murray, E.R. Dixon, S.J. Granger, D.J. Hatch, S.C. Jarvis, R.J. Laughlin and R.J. Stevens

Urea concentration affects short-term N turnover and N_2O production in grassland soil   351
S.O. Petersen, S. Stamatiadis, C. Christofides, S. Yamulki and R. Bol

Leached N and the nitrous oxide emission factor   354
D.S. Reay, K.A. Smith and A.C. Edwards

Seasonal subsoil denitrification of leached ^{15}N-labelled nitrate   357
S.M. Thomas, T.J. Clough, G.S. Francis, D.I. Hedderley, R.R. Sherlock and M.H. Beare

N_2O emissions from intensive vegetable production systems   359
S.M. Thomas, H.E. Barlow, G.S. Francis and D.I. Hedderley

Improving New Zealand predictions of N leaching for estimating indirect N_2O emissions   361
S.M. Thomas, S.F. Ledgard and G.S. Francis

Nitrogen losses during storage and following the land spreading of poultry manure   363
R.E. Thorman, B.J. Chambers, R. Harrison, D.R. Chadwick, R. Matthews and R.J. Nicholson

From N_2 fixation to N_2O emission in a grass-clover mixture   365
M. Thyme and P. Ambus

Nitrous oxide emission from an irrigated soil fertilised with pig slurry in Central Spain   367
A. Vallejo, J.A. Díez, L. García-Torres, P. Hernáiz and S. López-Fernández

Can tillage practice affect the contribution of nitrous oxide to the total greenhouse gas production from arable agriculture? 369
C.P. Webster, T.S. Scott and K.W.T. Goulding

Soils as sources of N-trace gases in Germany - results from calculations with biogeochemical models 372
C. Werner, M. Kesik, H. Papen, C. Li and K. Butterbach-Bahl

Denitrification in top soil and sub soil, data and model results 375
K.B. Zwart

Section 4 – Controlling losses to water 379

Controlling losses to water 381
M.A. Shepherd and E.I. Lord

Nitrate leaching from arable crop rotations in organic farming 389
J.E. Olesen, M. Askegaard and J. Berntsen

Nitrogen rate, surplus or residue? Performance of selected indicators for nitrate leaching 397
H.F.M. ten Berge, S.L.G.E. Burgers, M.J.D. Hack-ten Broeke, A. Smit, J.J. de Gruijter, G.L. Velthof, J.J. Schröder, J. Oenema, F.J. de Ruijter, S. Radersma, I.E. Hoving and D. Boels

Safety-nets and filter functions of tropical agroforestry systems 406
G. Cadisch, E. Rowe, D. Suprayogo and M. van Noordwijk

Comparison of the efficiency of different catch crops on potentially leachable nitrate 415
P.Y. Bontemps, R. Lambert, C. Devillers and A. Peeters

Inverse stochastic modelling of water and N drainage from lysimeters 417
M. Decrem, K.C. Abbaspour, J. Nievergelt, F. Herzog and W. Richner

Nitrogen concentrations in an intensively farmed livestock catchment 420
O. del Hierro, M. Pinto, A. Artetxe and A. del Prado

NGAUGE DSS as a tool to assist UK dairy farmers to comply with EU nitrate legislation 423
A. del Prado, L. Brown and D. Scholefield

Comparison between the risk of nitrogen leaching from temporary cut grassland and maize 425
B. Deprez, D. Knoden, H. de Blander, R. Lambert, C. Decamps and A. Peeters

Temporal and spatial denitrification patterns in nitrate retention by three riparian buffer zones 428
K. Dhondt, P. Boeckx, O. Van Cleemput and G. Hofman

Effective reductions in nitrate leaching and nitrous oxide emissions by the use of a nitrification inhibitor, dicyandiamide (DCD), in a grazed and irrigated grassland 431
H.J. Di and K.C. Cameron

Nitrate leaching and N_2-fixation in grasslands of different composition, age and management 434
J. Eriksen and F.P. Vinther

Action programme and practices for reduction of nitrate from agricultural sources in Central Greece 437
T. Georgiou, T. Karyotis, I. Katsilouli, A. Haroulis, M. Toulios and G. Argyropoulos

A field study of nitrate leaching from tillage crops in Ireland 439
K.V. Hooker, K. Richards, C.E. Coxonand R. Hackett

Dissolved organic nitrogen concentration in two grassland soils 441
D.L. Jones, J.F. Farrar and V.B. Willett

Simulation with STICS soil-crop model of catch crop effects on nitrate leaching during the fallow period and on N released for the succeeding main crop 444
E. Justes, F. Dorsainvil, M. Alexandre and P. Thiébeau

Nitrate in soils and water originated from agricultural sources: a case study in Thessaly, Central Greece 447
Ir. Katsilouli, Th. Karyotis, Th. Georgiou, Th. Mitsimponas, A. Panagopoulos, A. Panoras, D. Pateras, A. Haroulis, G. Argyropoulos and M. Toulios

Monitoring and mathematical modelling of nitrate leaching in an experimental field treated with pig slurry 449
P. Mantovi, L. Fumagalli and G.P. Beretta

Nitrate leaching in arable cropland: effects of N-management 452
A. Smit and K.B. Zwart

Reducing nitrate contamination of groundwater from intensive greenhouse-based vegetable production in Almeria, Spain - management considerations 454
R.B. Thompson, M. Gallardo and C. Gimenez

Nitrogen budgets at field, farm and polder scales in a polder used for dairy farming 457
C.L. van Beek, G.L. Velthof and O. Oenema

Nitrate leaching on loess soils as affected by N fertilisation and crop rotation   459
W. van Dijk, P. Dekker and J.R. van der Schoot

Loss of inorganic nitrogen in surface runoff from grazed grassland   461
C.J. Watson, R.V. Smith and E. Chisholm

Nitrogen losses in drainage water following pig slurry applications to an arable clay soil   463
J.R. Williams, B.J. Chambers R.B. Cross and R.A. Hodgkinson

Section 5 – Reconciling productivity with environmental considerations   467

Reconciling productivity with environmental considerations   469
D. Scholefield

Forage maize production as affected by tillage, N source and nitrification inhibitors   434
D. Báez, J. Coutinho, N. Moreira and H. Trindade

Is the N balance a good indicator of nitrogen losses in arable systems?   437
N. Beaudoin, B. Mary, F. Laurent, G. Aubrion and J.K. Saad

A low disturbance technique for applying slurry on forage land   430
S. Bittman, L.J.P. van Vliet, C.G. Kowalenko, S. McGinn, A.K. Lau, N. Patni, T. Forge, N. McLaughlin, D.E. Hunt, F. Bounaix and A. Friesen

Effects of field history on the establishment of white clover in association with perennial ryegrass   433
L.M. Bommelé, D. Reheul, N. Van Eekeren and F. Nevens

Nitrogen losses in relation to rice varieties, growth stages, and nitrogen forms determined with the ^{15}N technique   436
N.C. Chen and S. Inanaga

Effectiveness of alternative managements to reduce N losses from dairy farms   498
S.P. Cuttle and M.M. Turner

Azofert: a new decision support tool for fertiliser N recommendations   500
P. Dubrulle, J.M. Machet and N. Damay

Compost use in vegetable production: impact on gross N fluxes and implications for sustainable management practices   502
T.C. Flavel and D.V. Murphy

Ammonia volatilisation and soil nitrogen dynamics following application of pig deep-litter and pig slurry in different soil tillage systems   504
S.J. Giacomini, C. Aita, E. Guidini, E.B. Amaral and A. Lunkes

Estimation of nitrogen loading in Japanese prefectures and scenario testing of abatement strategies   507
M. Hojito, A. Ikeguchi, K. Kohyama, K. Shimada, A. Ogino, S. Mishima and K. Kaku

Indicators for environmental and economic sustainability on UK dairy farms   510
E.C. Jewkes, D. Scholefield, M.M. Turner and L. Brown

Dairy production using an extended grazing management system - a preliminary assessment of nitrogen flows   513
E.C. Jewkes, D. Scholefield, M.R. Butler, J. Webb, T. Forrester, J. Lapworth, K. Russell, G. Bailey, A. Lathwood and A. Clarke

Decrease in the amount of residual nitrate in cultivated land   516
R. Lambert, V. Van Bol and A. Peeters

Estimating nitrogen losses from animal manures using their phosphorus balance   518
R. Lambert, B. Toussaint and A. Peeters

Encouraging farmers to utilise nitrogen more efficiently   520
K.A. Leach, J.S. Conway, J.P. Morgan, B.F. Pain and D. Munday

Conserving biologically fixed N to increase its utilisation and decrease gaseous losses   523
A.K. Løes, A.K. Bakken, T.A. Breland, R. Eltun and H. Riley

Effect of animal treading intensity on the efficiency of N_2 fixation by clover in mixed grass/clover pasture and potential implications for long-term soil N availability   525
J.C. Menneer, S.F. Ledgard, C.D.A McLay and W.B. Silvester

Effect of type and dose of sewage sludge application in maize+ryegrass rotation in Galicia (NW Spain)   527
M.R. Mosquera-Losada, A. Amador-García and A. Rigueiro-Rodríguez

Cattle slurry and vegetable, fruit and garden (VFG) waste compost in silage maize: fertiliser N no longer needed?   530
F. Nevens and D. Reheul

Applying a ley/arable rotation to reduce N input in forages   532
F. Nevens and D. Reheul

Farm N budgets with estimated nitrogen losses by use of soil N modelling   534
A.H. Nielsen, B.M. Petersen and I.S. Kristensen

Political transformations and nitrogen balances of dairy farms in Poland 536
S. Pietrzak and O. Oenema

Dirty water - a valuable source of nitrogen on dairy farms 539
K. Richards, M. Ryan and C.E. Coxon

**Effect of date and dose of sewage sludge application in grasslands production
and nitrogen soil concentration in Galicia (NW Spain)** 541
A. Rigueiro-Rodríguez, A. Casanova-Vigo and M.R. Mosquera-Losada

Nitrate losses in forestry nurseries using municipal sewage sludge 544
A. Rigueiro-Rodríguez, J. Rasche-Castillo and M.R. Mosquera-Losada

Assessment of measures to reduce nitrogen losses from dairy farms 547
J. Scheringer and J. Isselstein

**Nitrogen use efficiency in a water saving ground cover rice production system
in Beijing, North China** 549
H. Tao, K. Dittert, S. Lin, C. Kreye and B. Sattelmacher

**Effect of herbicide, maize variety precocity and sowing date of three winter
cover crops on forage yield and on herbage N removal from an intensive
double-cropping system** 551
H. Trindade, J. Coutinho and N. Moreira

Effects of grassland renovation on herbage yields and nitrogen losses 554
G.L. Velthof and I.E. Hoving

Nitrogen budgets of Flemish specialised dairy farms during 1990-2000 557
I. Verbruggen, F. Nevens, A. Mulier and G. Hofman

**Substitution of N fertiliser supply for maize with lupin as a winter crop in
rotations under zero and conventional tillage in southern Brazil** 559
L. Zotarelli, B.J.R. Alves, E. Torres, S. Urquiaga and R.M. Boddey

Section 6 – Models and decision support systems 563

**NGAUGE: A decision support system to optimise N fertilisation of UK grassland
for economic and/or environmental goals** 565
L. Brown, D. Scholefield, E.C. Jewkes, A. del Prado and D.R. Lockyer

UK-DNDC: a mechanistic model to estimate N_2O fluxes in the UK 567
L. Brown, B. Syed, S.C. Jarvis, R.W. Sneath, V.R. Phillips, K.W.T. Goulding and C. Li

MAST - a model of ammonia volatilisation with an examination of abatement strategies for a dairy farm 569
E.C. Jewkes, C.A. Ross, D. Scholefield and S.C. Jarvis

Development of the OVERSEER® nutrient budget model to examine implications of pastoral management practices on nitrogen flows and losses 571
S.F. Ledgard, D.M. Wheeler, C.A.M. de Klein, R.M. Monaghan and K. Johns

The MANNER model: predicting the crop available N supply from farm manure applications 573
F.A. Nicholson, B.J. Chambers, E.I. Lord, K.A. Smith, S.A. Anthony and M.M. Gibbons

The nutrient leaching model ANIMO 575
J. Roelsma, P. Groenendijk and O.F. Schoumans

The SUNDIAL Model 577
J.U. Smith, P. Smith, A.G. Dailey, M.J. Glendining, G. Tuck and P.K. Leech

Section 7 – Themed working groups 579

Organic matter: does it matter, or can technology overcome most problems related to soil fertility? 581
Report by S. Recous and G. Cadisch

Optimising N additions: can we integrate fertiliser and manure use? 586
Report by J.J. Schröder and R.J. Stevens

Controlling gaseous N emissions - what is achievable? 594
Report by R. Harrison and E.M. Baggs

Missing N: is the solution in dissolved N? 600
Report by A.J. Macdonald and D.L. Jones

Pollution problems: mitigation or are we swapping one form of pollution for another? 606
Report by B.J. Chambers and O. Oenema

Systems studies; do we need them, or can they be replaced by desktop studies? 609
Report by H.F.M. Aarts and A. Pflimlin

Model answers: can we improve their level of confidence and applicability? 615
Report by S.F. Ledgard and N.J. Hutchings

Author index 619

SECTION 1
DRIVERS TOWARDS SUSTAINABILITY: WHY CHANGE?

Drivers towards sustainability: why change?

J.J. Neeteson[1], J.J. Schröder[1] and C. Jakobsson[2]
[1]*Plant Research International, PO Box 16, 6700 AA Wageningen, The Netherlands*
[2]*Swedish Institute of Agricultural Engineering, PO Box 7033, 750 07 Uppsala, Sweden*

Abstract

'Meeting the needs of the present generation without compromising the ability of future generations to meet their needs' is the most frequently used definition of sustainability. Sustainability contains three elements: society, environment, and economy. Current agriculture is not considered to be sustainable. Environment, nature and landscape are affected, natural resources are depleted, and food security is insufficiently assured. In this paper we discuss reasons why agricultural production systems should be sustainable. Informal and formal types of drivers towards sustainability are distinguished. Producers and consumers generally feel that production should be sustainable. This awareness is an informal driver towards sustainability and is partly reflected in formal drivers. The latter comprise legislation, regulations, treaties, and voluntary schemes. Research at the experimental dairy farm De Marke in the Netherlands and the Baltic Agricultural Run-Off Action Programme (BAAP) are examples of research aiming at sustainable N management. At De Marke, measures were taken to reduce the groundwater NO_3^- concentration from 200 to ca. 50 mg NO_3^- l^{-1}. The objective of BAAP is to improve the water quality of the Baltic Sea and of local surface and groundwater bodies by reducing nutrient run-off from agricultural sources.

Keywords: EU Directives, N flows, NO_3^- leaching, N use efficiency, sustainable N management, triple P

Introduction

There are numerous definitions of sustainability of which the definition of the Brundtland Commission is most frequently used: 'Meeting the needs of the present generation without compromising the ability of future generations to meet their needs' (World Commission on Environment and Development, 1987). A more specific definition which is applicable to agriculture is given by Earth Ethics: 'The ability to provide a healthy, satisfying and just life for all people on earth, now and for generations to come, while enhancing the health of ecosystems and the ability of other species to survive in their natural environments' (www.earthethics.com). It is important that all definitions of sustainability contain three elements: society, environment, and economy. These elements are known under the name triple Ps: People, Planet, Profit. Agricultural systems are sustainable with regard to the People element if they deliver food security, quality and safety, if they contribute to the maintenance of vital rural communities, if they contribute to an enjoyable landscape, and if they provide self-reliance and self-esteem to individuals and communities in the rural area. With regard to the 'Planet' aspects, agricultural systems are sustainable if they preserve the quality of soil, groundwater, surface water, and atmosphere, without depletion of finite non-renewable resources, and without reduction of biodiversity. As far as the 'Profit' element is concerned, agricultural systems are sustainable if farmers receive a fair reimbursement for the goods and services they produce, when consumers pay fair prices for food, and when the result is a proper distribution of income among individuals.

Current agriculture is not considered to be sustainable. It's impacts on the environment, nature and landscape are strongly debated, resources are depleted at a fast pace, animal welfare is challenged, and food security, both qualitatively and quantitatively, is insufficiently assured. Many farmers, but not all of them, know how this could be changed. However, they feel that it is extremely difficult to receive a fair reimbursement for their efforts in a globalising world. Consequently, rural communities feel manipulated and are unable to find a way out of this maze.

In this paper, we discuss reasons why agricultural production systems should be sustainable. Firstly, we identify important drivers towards sustainable agricultural production. Subsequently, we focus on examples of research aiming at sustainable N management, which plays a key role in sustainable food production but also because it affects the quality of the environment (Cartwright et al., 1991; Novotny, 1999; Pretty et al., 2003).

Drivers towards sustainability

Informal and formal types of drivers towards sustainability can be distinguished. Producers and consumers generally feel that production should be sustainable, they are more or less aware of the Triple P concept. This awareness is an informal driver towards sustainability and is partly reflected in formal drivers. The latter constitute legislation, regulations, treaties and voluntary schemes. To be effective, formal drivers usually contain 'carrots' or 'sticks'. Carrots are the positively formulated incentives intended to change things for better in a friendly way: producers are rewarded for producing in a sustainable manner. Subsidies on organic agriculture and on the extension of slurry storage capacity are examples of these carrots. Sticks are less friendly: producers are punished for not producing in a sustainable manner, e.g. fees are imposed when maximum allowable values of N surpluses are exceeded. Targets, indicators and threshold values are needed for the allocation of carrots and sticks to farmers. It should be noted, however, that it is difficult to define appropriate indicators and thresholds (Schröder et al., 2003).

Formal drivers have been established at various political scales. At the global scale, agreements have been made on biodiversity and climate change in the Rio Declaration in 1992 (www.unep.org/documents), on climate change in the Kyoto Protocol in 1997 (unfccc.int/resource/docs) and on sustainable development in the Johannesburg Declaration in 2002 (www.un.org/esa/sustdev/documents). According to the Kyoto Protocol, most European countries have to reduce greenhouse gas emissions, including N compounds, by 8% in 2005 as compared with 1990. At the European scale, the Oslo and Paris Conventions (OSPAR Commission, 1998) and the Helsinki Convention (Helsinki Commission, 1999) have been issued to protect the marine environment of the North-East Atlantic and the Baltic Sea, respectively. Within the European Union the Nitrates Directive (European Council, 1991), the Water Framework Directive (European Parliament and Council, 2000) and the Directive on National Emission Ceilings (European Parliament and Council, 2001) are well-known. One of the objectives of these Directives is a reduction of N emissions to groundwater, surface water and atmosphere. To meet the European directives, Member States have imposed national regulations, e.g. on the period during which it is allowed to apply manures to fields or the obligation to cover storage facilities for manure.

An example of a voluntary agreement is the Baltic 21 agreement of 1998 (www.ee/baltic21) which aims at a sustainable development of the Baltic Sea region (Jakobsson, 2001). It is a good illustration of the Triple P concept. The agenda for sustainable agriculture, which is part of the Baltic 21 agreement, includes the production of high quality food and other agricultural products (People) with consideration for economy (Profit) and social structure (People) in such a way that the resource base of renewable and non-renewable resources is maintained (Planet). On 26th June 2003, the EU ministers of agriculture adopted a fundamental reform of the Common Agricultural Policy (CAP). The new CAP will be geared towards consumers and taxpayers, while giving EU farmers the freedom to produce what the market wants. While current EU subsidies for cereals, sugar, milk and meat are directly related to production volume, future payments, probably as from 2005, will be introduced on the basis of the cross compliance principle: payment will be linked to environmental achievements and standards for food safety, animal and plant health, and animal welfare. In addition to this, all farmland should be kept in good agricultural and environmental condition. Strengthening the link between subsidies and production will make EU farmers more competitive and market- orientated.

Since environmentally friendly production is one of pillars in the new CAP, farmers will only receive payments from the EU when they meet environmental standards. Minimising N emissions from agriculture will become an important issue. Obviously, the EU Nitrates Directive and more recently the EU Water Framework Directive have already during a number of years, encouraged farmers to minimise N emissions with varying success. The foreseen linking of 'single farm payments' to environmentally friendly production will strongly increase farmers' motivation to further reduce emissions. Farmers therefore urgently need cost-effective strategies for sustainable N management.

Strategies for sustainable nitrogen management in agricultural systems

The key issue in sustainable N management is the matching of supply and demand in space and time: i.e. synlocation and synchronisation (Van Noordwijk & Wadman, 1992). This can be achieved by precision farming, for instance by analysing the soil before fertiliser application, by band application of fertilisers, by a deliberate postponement of indicator-based N dressings, or by taking better account of the history of the field (e.g. Schröder et al., 2000). Sustainable N management should also include soil fertility preservation and N recycling. The development of N management strategies should not focus on one scale only, for instance the field scale. The effects at higher levels, such as the farm level or the regional level, should also be taken into account to avoid problem swapping (Jakobsson, 1999; Neeteson et al., 2002; Schröder et al., 2004). Figure 1 shows that the efficient use of N involves more than just the fields on which it is applied. In addition to the field scale, other spatial scales are distinguished such as farm, region and country or maybe even the globe as some N losses have a global impact.

Nitrogen flows to and from the farm include feeds, animal and plant products, fertilisers, deposition, N_2 fixation and losses. Within the farm, N flows from feed to animals, from manure to soil, from soil to crops, and from crops to feed. Each flow is associated with losses. Consequently, each conversion step has its own N use (in)efficiency. It is relatively low for the conversion from feed into milk and meat, and relatively high for the conversion of crops into feed. Variation, however, is wide and hence there is much scope for optimisation on both dairy farms and arable farms (Table 1). The overall N use efficiency of an arable farm is generally higher than that of, for instance, a dairy farm. This has little to do with an inability of livestock farmers to manage N. Dairy farms that produce their own feedstuffs are by nature facing all four conversion steps, including the one pertaining to a proper recycling of manures. Conversely, many arable farmers have to deal with just the relatively easy conversion of mineral fertiliser N to crop uptake.

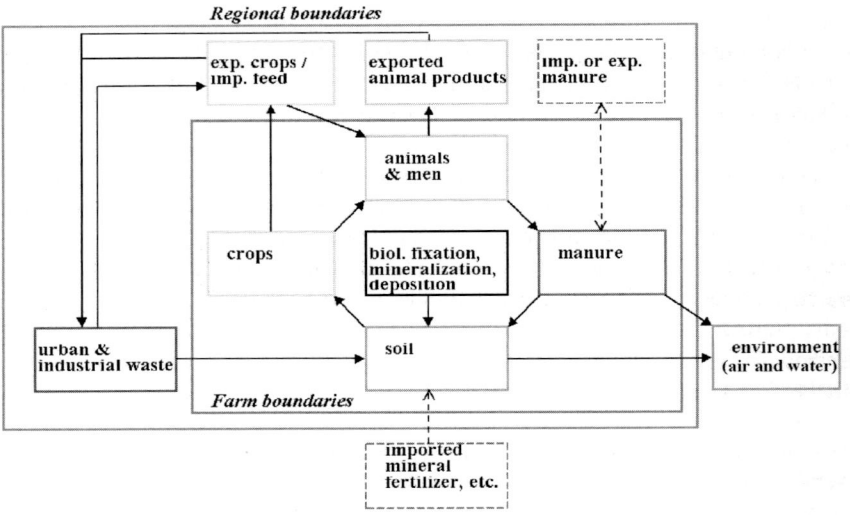

Figure 1. Schematic representation of nutrient flows in mixed farming systems and their relationships with other parts of society and the environment (Neeteson et al., 2002; Schröder et al., 2003).

Table 1. Nitrogen use efficiencies (after Van Der Molen et al., 1990; Bussink & Oenema, 1998; Monteny & Erisman, 1998; Schils et al., 1998; Huijsmans & De Mol, 1999; Aarts et al., 999a, 2000; Neeteson et al., 2002).

Nitrogen flow	Nitrogen use efficiency (%)
From feed to milk and meat	20-40
From manure to soil	50-90
From soil to crop	40-80
From crop to feed	80-90
Whole dairy farm	10-40
Whole arable farm	40-80

How can regulations best be met when they aim at reducing N emissions per ha? The most obvious measure that can be taken is to reduce N application rates. This will generally result in lower N emissions per ha, but yields will be lower so that a larger production area is needed to obtain the same production volume (if desired). Society, however, increasingly demands that agricultural land is used for other purposes, such as water conservation, recreation or nature conservation. Intensification is the most probable option to meet this demand. This means that less land is needed for agricultural production but the environmental load per ha increases. All depends, however, on the definition of recreation and nature. If the focus is, e.g. on the quality of the rural landscape, environmental loads per ha should be reduced and agriculture should, hence, extensify. However, this does not mean that the loads per region or per country are reduced as well. If landscapes with golf courses, heather, wetlands and other 'new wildernesses' are desirable, it may be better to intensify production and reduce the area needed for production. This reasoning is not only valid for N emissions, but also for energy use. Corré et al. (2003) showed that the energy use to obtain a certain volume of production is lower in organic agriculture than in conventional agriculture (Table 2). The area needed to obtain this production volume, however, was much larger in organic farming. When the extra area needed in (extensive) organic farming systems would be used for growing energy crops in (intensive) conventional farming systems, the net energy use in conventional systems would be even lower than that in organic farming. Nitrogen losses were larger in conventional farming when expressed on a per hectare basis, but were lower when expressed on a per capita basis (Table 2). Corré et al. (2003) also showed that the ever expanding preference of humans for animal-derived protein over plant protein, has much more relevance for overall energy use, land use and N losses than the decision to produce food organically or conventionally. Apparently, people prefer to debate how farmers should feed their soils than the way they feed themselves.

The examples presented here illustrate the need to consider various scale levels when sustainable N management strategies have to be designed.

Sustainable agriculture requires drastic changes in farming practices. Farmers need inspiring examples showing them how these transitions can work out. For this reason numerous pilot projects have been set up all over Europe. One of them is the experimental dairy farm De Marke in The Netherlands (Aarts et al., 1999b). One of the aims of the De Marke farm is to meet the requirements of the Nitrates Directive, i.e. not exceeding the concentration limit of 50 mg NO_3^- l^{-1} in groundwater. De Marke was deliberately located on a coarse sandy soil which is

Table 2. Energy use, land use and N loss in an organic farming and a conventional farming system (Corré et al., 2003).

	Farming Type		
	Organic	Conventional	Difference
Annual energy use			
MJ per capita	0.991	1.645	-0.654
Annual land use			
ha per capita	0.109	0.077	+0.032
Total annual N loss			
kg N per capita	14.35	13.99	+0.36
kg N per ha	132	182	-50

extremely susceptible to N leaching. If it is possible to reach a groundwater NO_3^- concentration < 50 mg l^{-1} at De Marke, it would undoubtedly be possible for other dairy farms to obtain similar results. At the onset of De Marke in 1992 the groundwater NO_3^- concentration was 200 mg NO_3^- l^{-1} (Aarts et al., 1999a). This very high value was the result of high annual application rates of slurry to silage maize for a large number of years. Measures taken to reach an acceptable NO_3^- concentration level in the groundwater include a limited milk quota (11,600 l ha^{-1}) a lower N fertiliser input, high quality feed production on the farm itself, rather than somewhere else, a shorter grazing season and a shorter grazing period per day, growing a cover crop after silage maize, alternation of grassland and silage maize to compensate for the loss of organic soil N during maize cultivation, and the rapid incorporation of manure into the soil to reduce the loss of NH_3. Adoption of this system enabled De Marke to reduce the groundwater NO_3^- to ca. 50 mg l^{-1} (Figure 2).

Another European pilot project is the Swedish Baltic Agricultural Run-Off Action Programme (BAAP), running from 1994 to 2002 in several countries, including Estonia (Loigu et al., 2003) and Lithuania (Sileika et al., 2003). The overall objectives of the programme were to improve the water quality of the Baltic Sea and of local surface and groundwater bodies by reducing nutrient run-off from agricultural sources. Various groups of stakeholders in the rural area participated in the programme: farmers, land owners, policy makers, experts from research institutes and universities, the advisory service, agricultural schools, local municipalities, and the general public. BAAP included a project on sustainable crop production. The objectives of the project were to demonstrate the connection between biology, technology and economy towards sustainable crop production. In the project, farming procedures were according to good agricultural practice (GAP), best environmental practice (BEP), and best available techniques (BAT). To develop ways of sustainable agriculture, individual farmers are supported by the programme in drawing up nutrient balance sheets, preparing production plans, and developing

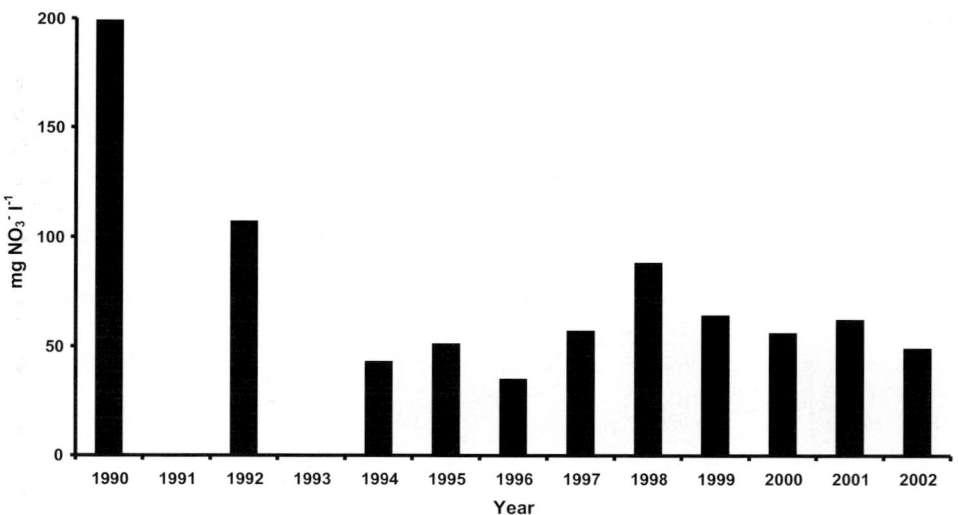

Figure 2. Nitrate concentration in the upper groundwater levels at 'De Marke' (after Aarts et al., 1999a; Neeteson, 2000; personal communication J. Oenema (Plant Research International, Wageningen)).

strategies for the utilisation of manure and urine. Demonstration fields have been set up for knowledge transfer to farmers that are not participating in the programme. Background information and results have been disseminated through lectures, demonstrations, field days, and booklets.

The programme focuses on the scale level of watersheds. All farms within the studied watersheds are involved in the programme (Figure 3).

Many N balance sheets have been drawn up for the entire area in watersheds, e.g. for the watersheds Räpu in Estonia (Loigu et al., 2003) and Graisupis in Lithuania (Sileika et al., 2003). There are 33 private farms in the Räpu watershed with an average size of 21 ha. Two-thirds of the total area of agricultural land in the watershed is used as grassland and one-third as arable land. In the Graisupis watershed, about 70% of the entire area is used for arable cropping, mostly cereals (spring wheat, winter wheat and winter triticale). The agricultural companies grow more winter crops, while the private farmers grow more spring crops.

Annual N balance sheets have been drawn up at all farms in the watersheds, both at field and farm level. Nitrogen balance sheets for the entire watersheds were calculated on the basis of the results of the farm level balance sheets. Table 3 shows examples of N balance sheets at the watershed level.

It is interesting to see in Table 3 that the N surplus was positive in the Räpu watershed whereas it was negative in the Graisupis watershed. Since N losses are included in the output terms, a positive surplus suggests a build-up of N in the soil, whereas a negative surplus results in mining of soil N. It should be noted, however, that all errors associated with overlying terms precipitate in the surplus term, so that conclusions on apparent soil pool changes should be drawn with great caution. The difference between the two watersheds results from the higher N output through agricultural products in Graisupis since almost all agricultural land was used

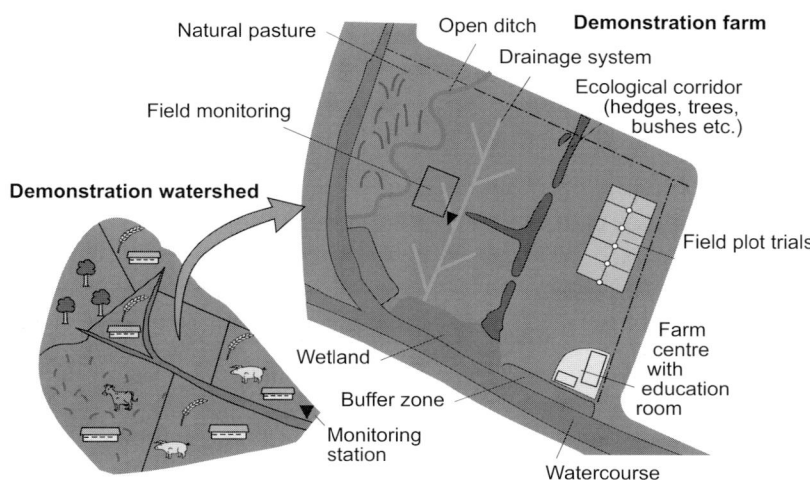

Figure 3. Schematic representation of the BAAP project on sustainable agricultural production at the watershed level.

for arable cropping, whereas in Räpu two-thirds of the area was used as grassland and, hence, livestock production. As discussed above, N use efficiency of arable farms is inherently higher than that of dairy farms. This means that the lower output/input ratio in the Räpu watershed does not point at a lower ability of the farmers involved to use N efficiently (Schröder et al., 2004).

Why change?

Farmers have to change in order to contribute to the sustainability of the production of food and other agricultural products and services because society and consumers demand that they do so. Nowadays, farmers can no longer simply produce what they want in a way that is simply most profitable to themselves. The increasing importance of environment, nature and rural area, and the increasing power of consumers is not only reflected in the new CAP (see the paragraph on drivers towards sustainability), but also in e.g. the change of names of the former Ministries of Agriculture: Department for Environment, Food and Rural Affairs (UK), Ministry for Food, Agriculture and Consumer Protection (Germany) and Ministry for Agriculture, Nature and Food Quality. Moreover, in Sweden there is a Minister for Agriculture and Consumers.

Sustainable agricultural production is with us. All of us, including those involved in the development, transfer and application of N-related knowledge, are nowadays asked to find a proper balance between the various goals resulting from the Triple P concept. The higher the demands for each of the Triple P elements, the smaller the chance that win-win situations occur. As one generally cannot eat the cake and have it too, anyone working on or with N is hence, expected to reveal the exchange rates between Profit, People and Planet, and account for the final choices made with respect to these elements.

Table 3. Nitrogen balance sheets for the Estonian watershed Räpu (in 2002) and the Lithuanian watershed Graisupis (in 2001), kg N ha^{-1}.

		Räpu	Graisupis
Input	Fertilisers + seeds	77	84
	Forage and livestock	9	-
	Precipitation	3	15
	Biological fixation	4	2
	Total input	**93**	**101**
Output	Plant products	11	74
	Animal products	8	-
	Ammonia volatilisation	9	4
	Nitrate leaching	6	20
	Denitrification	15	23
	Total output	**49**	**121**
Surplus (not accounted for)		44	-20
Agricultural output/input		0.20	0.73

References

Aarts, H.F.M., Habekotté, B., Hilhorst, G.J., Koskamp, G., Van Der Schans, F.C. and De Vries, C. (1999a). Efficient resource management in dairy farming on sandy soil. Netherlands Journal of Agricultural Science, 47, 153-167.

Aarts, H.F.M., Habekotté, B. and Van Keulen, H. (1999b). Limits to intensity of milk production in sandy areas in The Netherlands. Netherlands Journal of Agricultural Science, 47, 263-277.

Aarts, H.F.M., Habekotté, B. and Van Keulen, H. (2000). Efficiency of nitrogen (N) management in dairy farming system "De Marke". Nutrient Cycling in Agroecosystems, 56, 231-240.

Bussink, D.W. and Oenema, O. (1998). Ammonia volatilization from dairy farming systems in temperate areas: a review. Nutrient Cycling in Agroecosystems, 51, 19-33.

Cartwright, N., Clark, L. and Bird, P. (1991). The impact of agriculture on water quality. Outlook on Agriculture, 20, 145-152.

Corré, W.J., Schröder, J.J. and Verhagen, A. (2003). Energy use in conventional and organic farming systems. Proceedings No. 511. International Fertiliser Society, London, 23 pp.

European Council (1991). Directive 91/676/EEC concerning the protection of waters against pollution caused by nitrates from agricultural sources. Official Journal L 375, 31/12/1991, 1-8.

European Parliament and Council (2000). Directive 2000/60/EC establishing a framework for Community action in the field of water policy. Official Journal L 327, 22/12/2000, 1-73.

European Parliament and Council (2001). Directive 2001/81/EC on national emission ceilings for certain atmospheric pollutants. Official Journal L 309, 27/11/2001, 22-30.

Helsinki Commission (1999). Helcom Handbook. At: www.helcom.fi/handbook.

Huijsmans, J.F.M. and De Mol, R.M. (1999). A model for ammonia volatilization after surface application and subsequent incorporation of manure on arable land. Journal of Agricultural Engineering Research, 74, 73-82.

Jakobsson, C. (1999). Ammonia emissions - Current legislation affecting the agricultural sector in Sweden. In: Regulation of Animal Production in Europe, KTBL, Wiesbaden.

Jakobsson, C. (2001). Baltic 21 - Towards sustainable development. Sustainable development and water security in the Lake Victoria basin - Buiil;ding bridges between Lake victoria and the Baltic Sea. SIWI Proceedings, Report 15.

Loigu, E., Carlson, G. and Jakobsson, C. (2003). Final report from the BEAROP activities in the Baltic Agricultural Runoff Action Programme BAAP II 1999-2002. Estonia. JTI Report, Swedish Institute of Agricultural Engineering, Uppsala, 36 pp.

Monteny, G.J. and Erisman, J.W. (1998). Ammonia emission from dairy cow buildings: a review of measurement techniques, influencing factors, and possibilities for reduction. Netherlands Journal of Agricultural Science, 46, 225-247.

Neeteson, J.J. (2000). Nitrogen and phosphorus management on Dutch dairy farms: legislation and strategies employed to meet the regulations. Biology and Fertility of Soils, 30, 566-572.

Neeteson, J.J., Schröder, J.J. and Ten Berge, H.F.M. (2002). A multi-scale approach to nutrient management research in the Netherlands. Netherlands Journal of Agricultural Science, 50, 141-151.

Novotny, V. (1999). Diffuse pollution from agriculture: a worldwide outlook. Water Science and Technology, 39, 1-13.

OSPAR Commission (1998). For the protection of the Marine Environment of the north-east Atlantic. At: www.ospar.org.

Pretty, J.N., Mason, C.F. Nedwell, D.B., Hine, R.E., Leaf, S. and Dils, R. (2003). Environmental costs of freshwater eutrophication in England and Wales. Environmental Science and Technology , 37, 201-208.

Schils, R.L.M., Van Der Meer, H.G., Wouters, A.P., Geurink, J.H. and Sikkema, K. (1998). Nitrogen utilization from diluted and undiluted nitric acid treated cattle slurry following surface application to grassland. Nutrient Cycling in Agroecosystems, 53, 269-280.

Schröder, J.J., Neeteson, J.J., Oenema, O. and Struik, P.C. (2000). Does the crop or the soil indicate how to save nitrogen in maize production? Field Crops Research, 66, 151-164.

Schröder, J.J., Scholefield, D., Cabral, F. and Hofman, G. (2004). The effects of nutrient losses from agriculture on ground and surface water quality: the position of science in developing indicators for regulation. Environmental Science and Policy, 7 (in press).

Schröder, J.J., Aarts, H.F.M., Ten Berge, H.F.M., Van Keulen, H. and Neeteson, J.J. (2003). An evaluation of whole-farm nitrogen balances and related indices for efficient nitrogen use. European Journal of Agronomy (in press).

Sileika, A.S., Carlson, G. and Jakobsson, C. (2003). Final report from the BEAROP activities in the Baltic Agricultural Runoff Action Programme BAAP II 1999-2002. Lithuania. JTI Report, Swedish Institute of Agricultural Engineering, Uppsala, 36 pp.

Van Der Molen, J., van Faassen, H.G., Leclerc, M.Y., Vriesema, R. and Chardon, W.J. (1990). Ammonia volatilization from arable land after application of cattle slurry. Netherlands Journal of Agricultural Science, 38, 145-158.

Van Noordwijk, M. and Wadman, W.P. (1992). Effects of spatial variability of nitrogen supply on environmentally acceptable nitrogen fertilizer application rates to arable crops. Netherlands Journal of Agricultural Science, 40, 51-72.

World Commission on Environment and Development (1987). Our common future. Oxford University Press, New York, 400 pp.

Nitrogen balances over seven years on a mixed farm in the Cotswolds

K.A. Leach[1], K.W.T. Goulding[2], D.J. Hatch[3], J.S. Conway[1] and K.D. Allingham[1]
[1]Royal Agricultural College, Soil Science Department, Cirencester, Glos., GL7 6JS, UK
[2]Rothamsted Research, Agriculture and the Environment Division, Harpenden, Herts., AL5 2JQ, UK
[3]Institute of Grassland and Environmental Research, North Wyke Research Station, Okehampton, Devon, EX20 2SB, UK

Abstract

The annual farm gate N balance for a mixed farming system in the Cotswolds was calculated for seven consecutive years using farm records, modelled data on gaseous losses and measurements of leaching losses, including soluble organic N (SON). Years 1-4 (dairy/sheep/arable system - Phase 1) gave an average surplus of total inputs over sold outputs of 141 kg N ha^{-1}, which was reduced to 117 in years 6-7 (arable/sheep only - Phase 2). Total losses averaged 128 kg N ha^{-1} for Phase 1 and 102 kg N ha^{-1} for Phase 2. Nitrogen Use Efficiency (NUE), i.e. N in sold outputs/total N inputs, averaged 45% and 47% for Phase 1 and Phase 2, respectively. Leaching constituted the greatest loss, with a mean of 86 kg N ha^{-1} across all years (range 60-117 kg N ha^{-1}). Accounting for SON added 6% to leaching under arable land and 10% to leaching under grassland. When a full N budget was prepared and all losses were accounted for, there was no significant difference between total inputs and total outputs. Comparing light fraction organic matter N and potentially mineralisable N at the start and end of the project showed that these readily mineralisable soil N fractions were maintained under permanent pasture, but declined over six years under crop rotations.

Keywords: Leaching, mixed farm systems, nitrogen balance, soluble organic nitrogen

Introduction

Farm gate N balances have become an accepted method for calculating N surpluses and the efficiency of N use in farming systems (Brouwer, 1998; Watson & Atkinson, 1999). However, these do not usually apportion N losses between the various loss processes: leaching, denitrification and volatilisation. Since these losses can have detrimental environmental effects (Brouwer & Hellegers, 1997) and represent inefficient use of inputs, quantifying them is an important further step in assessing the potential environmental impact of farming systems and achieving improvements. This paper reports annual N balances for a long term project which aimed to quantify the N losses and N use for a commercial mixed farm system in the Cotswolds. The implications for N balance of annual variation in weather conditions and alterations in the farming system were observed. During the early years of the project, there appeared to be a shortfall between inputs and outputs of N. Attempts were made to identify where this 'missing N' might be found, thus accounting for and understanding more of the N cycle.

Materials and methods

Farming system

The study was carried out at the Royal Agricultural College's 244 ha Coates Farm, 3 km west of Cirencester, Gloucestershire, UK (OS reference ST982008), where Best Management Practice (BMP) was followed. The predominant soils are stony clay loam to clay over limestone of the Sherborne (up to 30 cm deep) and Moreton (30 - 100 cm deep) series (Conway, 1986; Courtney & Findlay, 1978). In years 1- 4 (Phase 1: Feb 1996 - Jan 2000) the system included a dairy herd of 160 cows producing 6200 l per lactation, and arable and sheep production. The dairy herd was relocated during year 5 (a change-over year which must be considered separately), leaving only arable and sheep enterprises in 2001/02 and 2002/03 (Phase 2: Feb 2001 - Jan 2003). The system of Phase 1 has been described in detail by Allingham et al., (2002). The majority of fields were in a 4, or 7 year rotation, with 18 ha permanent pasture. Two years' winter wheat was followed by winter barley. Oilseed rape, short term leys, or other forage crops formed a break from cereals. A single year of a forage crop resulted in a 4 year rotation, whilst a 3 year ley as a break gave a 7 year rotation. The choice of break crops depended upon forage requirements and cereal and oilseed prices. At the end of Phase 1, about 15 ha of long term leys, previously grazed by the dairy herd, were ploughed and entered into cereal production. The sheep enterprise in year 6 consisted of weaned lambs, or ewes visiting from other farms to graze grass or stubble turnips, or for winter housing, but a dedicated Coates flock was established by buying in 500 ewes in February 2002, with additional ewes from other college flocks grazing on stubble turnips at Coates in the winter of 2002/03. Annual stocking rates and forage areas are shown in Table 1. Rainfall data were collected at a Meteorological Office weather station, 1km from the farm.

Table 1. System management details and winter rainfall.

	Phase 1					Phase 2	
	1996-7	1997-8	1998-9	1999-00	00-01	01-02	02-03
Stocking rate (LU ha^{-1})	1.1	0.79	0.69	0.62	0.55	0.23	0.57
c - cows s - sheep	c + s	c + s	c	c + s	c	s	s
% area grass	28	26	35	24	23	19	23
% area forage crops	50	53	48	39	34	35	42
Fertiliser rate arable (kg N ha^{-1})	161	191	193	179	162	169	232
Fertiliser rate grass (kg N ha^{-1})	142	251	229	253	208	253	215
Winter rainfall (mm)	415	462	653	465	788	551	668

Calculation of N balance

Nitrogen balances at the 'farm gate' level were calculated on a 'per ha' basis for the period 1 February - 31 January each year, so that each accounting year began before any spring fertiliser was applied. Total N inputs, outputs as sold produce, and outputs as losses (leaching, volatilisation and denitrification) were either measured or calculated. Total inputs minus saleable outputs were described as N surplus. Saleable outputs, expressed as a percentage of total inputs, were described as N Use Efficiency (NUE). Total inputs minus total outputs (saleable outputs plus losses) were described as missing N. Since leaching measurements did not commence until autumn 1996, the first balance year with a full set of leaching measurements, and therefore loss figures, was 1997-8. Quantities of purchased inputs and sold outputs were taken from farm records. Nitrogen contents for these were taken from official records (fertiliser, compound feeds, milk sold), analysis of total N content by dry combustion (crops and straw sold), or 'book values' (seed and livestock). Inputs by precipitation and dry deposition were measured in years 1-4 using precipitation collectors and diffusion tubes, respectively. After year 5, these labour intensive measurements were discontinued and the inputs were assumed to be equal to those measured in year 5 (close to an average of the preceding years). N fixation by lucerne was estimated from total N offtake. Leaching losses were estimated from chemical analysis of leachate sampled using ceramic cups placed in ten fields ('farmlets') and drainage figures were modelled using a modified Penman-Monteith equation. Leaching from other fields was estimated by selecting values from a 'farmlet' field with the closest corresponding cropping pattern (Allingham et al., 2002). Nitrate and NH_4^+ in leachate were quantified using a Flow Injection Analyser. Gaseous losses from denitrification were measured in the farmlets between 1996-2001 (Allingham et al., 2002) and thereafter denitrification was modelled using a UK version of the DNDC model (Brown et al., 2002). Gaseous losses by volatilisation were modelled annually using the NH_3 inventory for UK emissions (IGER et al., 2001).

Measurement of soluble organic N
During years 2000 and 2003, additional leachate samples were taken with quartz/teflon cups in two fields. The leachate was autoclaved and oxidised using potassium persulphate to oxidise any SON and analysed for NO_3^- and NH_4^+ content to give a value for total soluble N. Subtracting the initial total inorganic N value from this gave SON leached. SON as a proportion of inorganic N was calculated separately for samples from cultivated land and grassland. These values were used to amend Phase 1 leaching data.

Statistical test for significance of difference between inputs and outputs.
Theil's regression (Dhanoa, 1998) was applied to the total input and output data from 1997-2000 (six data points) to test whether the relationship between inputs and outputs differed significantly from 1:1.

Measurements of N in soil
Soil samples (to 20 cm depth) were taken in autumn 1996 and autumn 2002, from ten sites within each of the ten farmlets and analysed for total N by combustion. The values for corresponding sites for 1996 and 2002 were compared using a paired t-test. For a sub-set of four of these fields, the same samples were analysed for potential mineralisable N (PMN) by anaerobic incubation (Lober & Reader, 1993) and N in the light fraction organic matter (LFON). These tests quantify the readily mineralised N: the fraction most likely to alter over the time period studied.

Results

Nitrogen balance for 7 years
The N balance for each year is shown in Table 2. The major input was always in fertiliser (average 75% of inputs for the whole period). Inputs in concentrates were greatly reduced in Phase 2 when the dairy herd was moved. The greatest proportion of saleable outputs was always in

the form of arable crops. Straw outputs varied, depending on the cropping pattern and farm requirements for bedding. In 2002/03, some silage was exported to the new dairy unit. In 2000/01, the largest net export of livestock occurred as the dairy cows were removed from the farm. In 2002/03, livestock feature in the inputs, representing the 500 ewes bought to establish the Coates flock.

NUE ranged from 41% to 55% during the course of the project. The mean value for Phase 1 was 45% and for Phase 2 was 47%. The changeover year (2000-01) showed the highest efficiency

Table 2. Nitrogen balance for Coates Farm over 7 years: kg N ha^{-1}.

	Phase 1					Phase 2	
Year	1996-7	1997-8	1998-9	1999-00	00-01	01-02	02-3
Inputs:							
Fertiliser	142	159	176	178	163	190	199
Precipitation	18	16	17	20	17	17	17
Dry deposition	4	3	3	4	5	5	5
Seed	2	2	2	1	2	2	1
Fixation	14	20	23	26	6	0	0
Concentrates	50	63	40	38	16	0	3
Livestock							4.5
Manure							0.5
Total Inputs	**230**	**263**	**261**	**267**	**209**	**214**	**230**
Outputs:							
Sales							
Grain, pulses & oilseed	68	67	87	73	88	85	85
Straw		13	16	23	7.5	14	9
Milk	22	21	21	23	9.4		
Cows/calves	3	3	2	0.5	10.7		
Sheep	2	3	5			2	3
Wool	1	2					1
Silage							11
Total saleable output	**96**	**109**	**131**	**120**	**116**	**101**	**109**
Losses:							
Leaching Feb-Jan	N/a	64	95	117	60	89	86
Volatilisation	17	25	15	15	8	5	8
Denitrification	13	17	17	18	16	8	8
Total losses		**106**	**127**	**150**	**84**	**102**	**102**
Output total		215	258	269	200	203	211
Balance (missing)		48	3	-2	9	12	20
NUE (%)	41.7	41.4	50.2	44.8	55.5	47.2	47.3
N surplus	134	154	130	147	93	113	121

N/a - not available

value of 55%, with the sale of the cows increasing outputs and reducing the requirement for concentrates. Nitrogen surpluses ranged from 93 kg N ha^{-1} in the changeover year to 154 kg N ha^{-1} in 1997-8. The mean surplus for Phase 1 was 141 kg ha^{-1} and for Phase 2 was 117 kg ha^{-1}.

Soluble organic N
In grass fields, SON in leachate was, on average, 10% of the inorganic N leached, while in arable fields the proportion was 6%. These values were used to adjust leaching losses in proportion to the cropping pattern each year. The average increase in leaching, after accounting for SON, was 7% across all years.

Missing N
In each year except 199-2000, inputs exceeded outputs; i.e. a small proportion of the N inputs was unaccounted for. This 'missing N' ranged from 3 kg N ha^{-1} (year 3) to 48 kg N ha^{-1} (year 2). In year 4, outputs exceeded inputs by 2 kg N ha^{-1}. The mean 'missing N' value was 16 kg N ha^{-1}, representing 6% of inputs.

Statistical significance of difference between inputs and outputs.
Figure 1 shows the relationship between total inputs and total outputs for each complete year (6 years' data). These data were used in Theil's regression analysis.

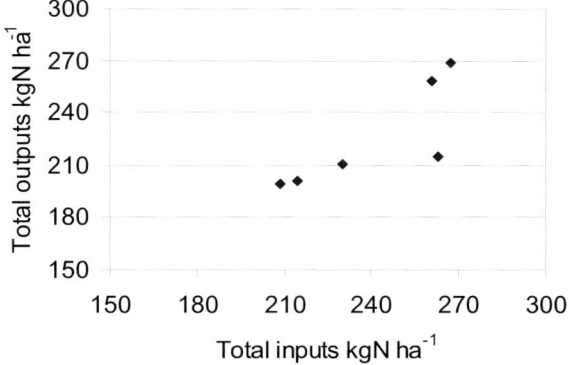

Figure 1. Relationship between N inputs and N outputs.

Theil's regression of total inputs on total outputs gave a fitted relationship of:

y = -35.98 + 1.116x.

The 95% confidence intervals of the fitted slope are 0.9867 and 1.227. Thus, the fitted line is not significantly different from the 1:1 line, where inputs would equal outputs. Within 95% confidence limits, the total inputs are not significantly different from the total outputs for these six years.

Soil N content

Total N analyses showed a slight decrease in total N content in the majority of fields, but this effect was small and only significant in three fields. Detecting, with confidence, a difference in total N soil content on the scale occurring over a six year period would in fact be unlikely. Therefore, measurements were made of the fractions of total N considered to be the most likely to change over this time span, i.e. the N in the light organic fraction, and the potentially mineralisable N. These results indicated that these readily mineralisable fractions were decreasing in arable fields, and stable in permanent grass fields (Table 3).

Table 3. Levels of soil organic matter N at start and end of project (g kg^{-1} soil), mean and (sd).

Field	Light fraction organic matter N		% decrease in N
	Autumn 1996	Autumn 2002	
1 (arable)	0.202 (0.036)	0.061 (0.036)	70*
11(arable)	0.125 (0.018)	0.071 (0.046)	43*
13 (arable)	0.119 (0.073)	0.085 (0.014)	28*
41 (permanent pasture)	0.395 (0.156)	0.377 (0.211)	5
	Potentially mineralisable N		
1 (arable)	0.130 (0.033)	0.064 (0.008)	51*
11 (arable)	0.1445 (0.027)	0.070 (0.028)	52*
13 (arable)	0.115 (0.022)	0.050 (0.013)	56*
41 (permanent pasture)	0.251 (0.017)	0.223 (0.029)	11

* Difference between years significant , $p < 0.05$

Discussion

Fertiliser formed the largest constituent of inputs throughout the study period and represented 63% of total inputs in Phase 1. In Phase 2, fertiliser rates for cereals increased after the adoption of a policy to apply foliar sprays to milling wheat to increase grain N content. Combined with the substantial reduction in concentrate inputs, this meant that fertiliser N increased to 87 % of total inputs. Purchased feed accounted for 15% of inputs in Phase1 but was minimal in Phase 2. Total N inputs were reduced by 13% with the removal of the dairy enterprise. When considering opportunities for improving N efficiency, both fertiliser and purchased feed inputs should be examined to ensure that the requirements of both plants and animals are not being exceeded, and that the expected responses from inputs are being achieved. There may be scope for reducing one or both of these inputs without affecting performance.

Leaching constituted the greatest loss throughout, amounting to an average of 36% of the total inputs (range 24 - 44%), largely dependent upon winter drainage resulting from the large excess rainfall at this time of year. Nitrate concentrations often exceeded the EU limit at the start of the drainage season, regardless of cropping and management (Allingham et al,, 2002), demonstrating that, even with BMP, it is difficult to avoid occasionally breaching the limit. The peak concentration is largely a function of weather . Considering the effects of a change to the farming system, there was little difference between leaching losses, or NUE, for Phase

1 and Phase 2. Both milk and sheep production are inefficient processes in utilising N, with sheep systems being less efficient than dairy systems.

Calculations for enterprises on other local commercial farms showed a range of NUE of 14% to 21% for dairy systems and 10% to 14% for sheep enterprises (Leach, unpublished data). Also, sheep grazed for a larger proportion of the drainage season than cattle, so N returns to grazed fields continued for longer. Thus, replacing the dairy herd with additional sheep grazing days did little to improve the efficiency of the mixed system. The N surplus was somewhat reduced; this could be at least partly explained by a 50% reduction in predicted NH_3 emissions, resulting largely from the elimination of losses from stored slurry and housed cattle. However, when making comparisons between years, it must be remembered that weather conditions also have an important effect, particularly on leaching. Thus, between-year differences may not be attributable solely to changes in farming systems.

Accounting for SON in leachate, by extrapolating from a small number of measurements, improved the reconciliation of N inputs and outputs. When this was included, although there were discrepancies between total inputs and outputs of N, the magnitude of which varied from year to year, on average, only 6% of inputs remained unaccounted for and statistical analysis showed that the difference was not significant. In fact, with the level of precision achievable with farm scale measurements and the need to make certain assumptions (e.g. extrapolating from farmlets to the whole farm area), achieving perfect agreement between inputs and outputs was unlikely. The difficulty in accounting fully for N in farming systems has been experienced in previous modelling exercises (Pionke et al., 1999) and in other systems studies (Watson & Atkinson, 1999). From year to year, it is likely that N levels in the soil will fluctuate, and this will result in an imbalance (viz. 'missing N' in some years and 'excess N' in others). There was a tendency towards an inverse relationship between 'missing N' and winter rainfall, suggesting that N may accumulate in the soil in dry years and be removed in wet years. Total N measurements comparing year 1 with year 7 suggested that soil N levels fell slightly, but given the precision of total N measurements, and the magnitude of the missing N in relation to total soil N, these measurements are not likely to be sensitive enough to provide an explanation for the 'missing N'. Investigation of the most labile soil N components gave the expected result of a general depletion of readily available N in the predominantly arable fields and a relatively steady state under permanent pasture, and gave no evidence for any 'missing N' accumulating in these fractions.

Conclusions

With a combination of careful measurement, detailed farm records and some appropriate modelling, it is possible to account for all the N inputs and outputs on a farm. This exercise showed an average NUE for Coates farm of 47% over six years. It would seem most critical to identify the main loss pathway and effort should be focussed on measuring this as accurately as possible. On Cotswold soils, the greatest N loss occurred through leaching. SON was an important component of leaching losses. Removal of the dairy herd and consequent expansion of arable and sheep enterprises reduced total N inputs by 13%, NH_3 volatilisaton by 50% and N surplus by 17%. Leaching losses and NUE were, however, unaffected. More generally, EU NO_3^- limits for ground water were exceeded in leachate from Coates Farm, despite BMP. While some

NO_3^- may be removed between farm and groundwaters (there is little surface water on, or near the farm) this gives some cause for concern in a Nitrate Vulnerable Zone. Because of the large effect of the weather on N concentrations on individual sampling dates, a measurement based over the whole drainage period may be more appropriate when assessing systems for their likely environmental impact.

Acknowledgements

This study was funded by MAFF and Defra. Co-operation of RAC farm staff and assistance by technical staff at RAC, IGER North Wyke and Rothamsted are gratefully acknowledged. IGER and Rothamsted Research receive funding from BBSRC.

References

Allingham, K.D., Cartwright, R., Donaghy, D., Conway, J.S., Goulding, K.W.T. and Jarvis, S.C. (2002). Nitrate leaching losses and their control in a mixed farm system in the Cotswold Hills, England. Soil Use and Management, 18, 421-427.

Brouwer, F. (1998). Nitrogen balances at farm level as a tool to monitor effects of agri-environmental policy. Nutrient Cycling in Agroecosystems, 52, 303-308.

Brouwer, F. and Hellegers, P. (1997). Nitrogen flows at farm level across EU agriculture. In: E. Romstad, J. Simonsen and A. Vatn (eds). Controlling mineral emissions in European Agriculture. 11 - 26. CAB International, Wallingford.

Brown, L., Syed, B., Jarvis, S.C., Sneath, R.W., Phillips, V.R., Goulding, K.W.T. and Li, C. (2002). Development and application of a mechanistic model to estimate emission of nitrous oxide from UK agriculture. Atmospheric Environment, 36, 917-928.

Conway, J.S. (1986). The Soils of Coates and Eysey Manor Farms. Royal Agricultural College.

Courtney, F.M. and Findlay, D.C. (1978). Soils in Gloucestershire II (Stow on the Wold). Soil Survey Record 52.

Dhanoa, M.S. (1998). A procedure for Theil's regression model. Genstat Newsletter, 34, 21 -26.

IGER, SRI, ADAS and CEH (2001). Inventory of Ammonia Emissions from the UK - 1999. MAFF Project Report AM C108.

Leach, K.A., Morgan, J.P., Pain, B.F. and Munday, D. (2003). Encouraging farmers to utilise nitrogen more efficiently. Book of Abstracts, 12th N Workshop, University of Exeter.

Lober, R.W. and Reader, J.D. (1993). Modified waterlogged incubation method for assessing nitrogen mineralization in soils and soil aggregates. Soil Science Society of America Journal, 57, 400-403.

Pionke, H.B., Rotz, C.A., Sanderson, M.A., Stout, W.L. and Sharpley, A.N. (1999). Nitrogen and phosphorus sources and their importance to pasture-based livestock systems. In: A.J. Corrall (ed). Accounting for Nutrients. British Grassland Society Occasional Symposium No. 33, British Grassland Society, Reading. 13 - 22.

Watson, C. A. and Atkinson, D. (1999). Using nitrogen budgets to indicate nitrogen use efficiency and losses from whole farm systems: a comparison of three methodological approaches. Nutrient Cycling in Agroecosystems, 53, 259-267.

Watershed nitrogen modelling

N.J. Hutchings[1], T. Dalgaard[1], B.M. Rasmussen[1], J.F. Hansen[1], M. Dahl[2], P. Rasmussen[2], L.F. Jørgensen[2], V. Ernstsen[2], F. von Platen-Hallermund[2] and S.S. Pedersen[2]
[1]*Danish Institute of Agricultural Sciences, Dept. of Agroecology, P.O. Box 50, Research Centre Foulum, 8830 Tjele, Denmark*
[2]*Geological Survey of Denmark and Greenland, Øster Voldgade, 1350 Copenhagen K, Denmark*

Abstract

The ARLAS scenario system enables land use scenarios to be constructed and the economic and environmental consequences to be judged at the watershed scale. Using data from national databases, the scenario system classifies each farm within the area under study into one of several types (e.g. pig, dairy cattle). A crop rotation model estimates the crop that is likely to be present on each field in each year. The crop nutrient requirement of each field is calculated in a manure model. If insufficient animal manure is available, then fertiliser is added to cover the deficit. Excess manure is exported to farms with spare capacity. The information on animals, cropping and manuring is passed to the FASSET farm model. The model estimates the flows of N, crop production and farm-scale economics, plus losses of N to the air as NH_3 or in drainage water as NO_3^-. The information is passed to a hydrological model (MODFLOW), which simulates the movement of water and NO_3^- through the watershed, allowing the concentrations in drinking water from boreholes or in streams and rivers to be estimated. Case study results highlight the importance of including interactions within and between farms when predicting the effect of agriculture on N flows.

Keywords: model, Nitrates Directive, nitrogen, watershed

Introduction

For many years, the price of agricultural commodities has been falling in real terms. Farmers wishing to maintain their living standard are responding to this pressure in a number of ways, including expanding and intensifying their enterprises. The EU Nitrates Directive aims to protect drinking water supplies and surface water habitats from the effects of excessive NO_3^- loading. In Denmark, as in most W. European countries, agriculture is now the main source of NO_3^-, and the whole country is classified as a Nitrate Vulnerable Zone. In the future, the Nitrates Directive will be supplemented by the EU Water Framework Directive (WFD), which sets environmental targets for surface water bodies, including marine inshore waters. These targets will vary between watersheds, although the extent of that variation is as yet unclear. To determine whether further regulatory action is necessary within a given watershed, it will be necessary to assess the consequences of current agricultural practice on NO_3^- loading to the target water body. If this assessment suggests that the environmental target will be exceeded, then regulatory options for reducing leaching will need to be investigated. The cost of these options must also be assessed, as the WFD also contains mechanisms to ease or delay implementation of regulatory measures, if they would impose an excessive cost burden on agriculture.

Modelling represents the only realistic method of assessing the consequences of current and future agricultural practices on the quality of surface waters. A number of methods exist for modelling the NO_3^- loading of watersheds. If measurements of NO_3^- loadings already exist, it may be possible to relate them to relevant factors within the watershed. For example, Edwards et al. (1990) could relate NO_3^- concentrations at points along two Scottish rivers to the land use within the watersheds of streams draining into those rivers. Such models could be used to assess the likely impact of major changes in land use (e.g. afforestation) on water quality. However, calibration of the model requires that the desired range of land uses already exist within the catchment, that data exist for drainage from relatively homogenous watersheds containing these land uses and that the land uses are not confounded with such characteristics as soil type or topography. In many situations, satisfactory water quality might be achieved with less radical and costly changes to land use and management. Assessing the consequences of these more subtle changes demands models that are more complex, which invariably also means more demanding in terms of the amount and quality of input data.

Predicting NO_3^- leaching at the field scale is difficult because it represents a relatively small difference between two large numbers; the input of N via fertilisers, fixation and deposition and the output of N in harvested crops or via NH_3 volatilisation and denitrification. Both input and output can be controlled by the farmer, through their choice of crop and crop management. However, the degree of control that can in practice be achieved is constrained by the interaction of management with physical factors such as soil type and climate, and the limited ability to control inputs via N_2 fixation or animal manure. The difficulty with animal manure arises due to variations in the quality of the manure applied and the greater susceptibility to losses via NH_3 volatilisation and denitrification than experienced with artificial fertilisers.

A number of field-scale models exist that have the capacity to predict NO_3^- leaching under different crop, management and physical situations. These range in complexity from relatively simple empirical models such as N-LESS (Simmelsgaard & Djurnuus, 1998), through to dynamic, mechanistic models such as SUNDIAL, DAISY and FASSET (Smith et al., 1996; Hansen et al., 1990; Berntsen et al., 2003). The latter models have the advantage that they can, in theory, describe the interaction between management, soil and climate. Changes in management are generally the cheapest method of reducing NO_3^- leaching and are, therefore, the focus of much attention by policymakers. The Danish ARLAS project developed a system to examine the effect of different land use policy scenarios on farm economics, wildlife and losses of N to the environment. In this paper, that part of the system that considers losses of N to the environment will be described, with particular emphasis on the effect of policy initiatives on NO_3^- loadings to a watershed.

Materials and methods

A schematic representation of the watershed management tool is shown in Figure 1.

At the centre of this tool is a geographic information system (GIS). The task of this GIS is to access a range of databases containing data describing the target watershed and to accept instructions concerning the land use and management options (management scenarios). The data input includes the following;

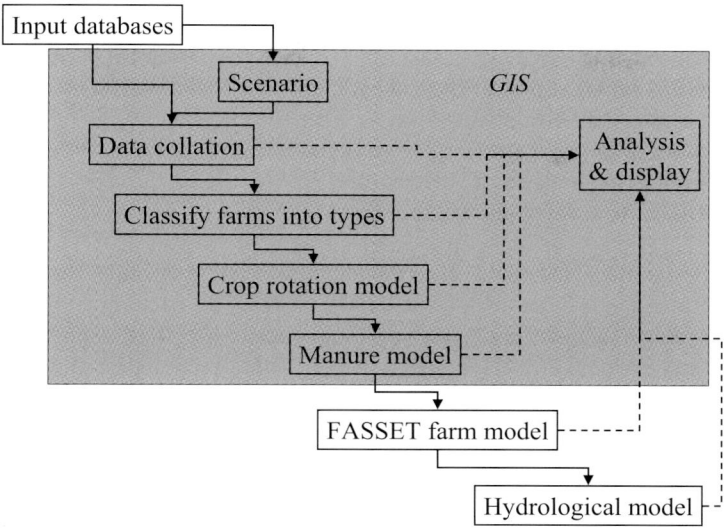

Figure 1. Scenario management tool.

- The number of animals of each animal type (e.g. dairy cattle, sows) on each farm in the target watershed
- The size and location of the fields associated with each farm
- The soil type associated with each field
- A file containing a 30-year record of daily climate data

Based on the animal numbers present, each farm is allocated to one of the following categories; dairy, beef, pig, arable. Using regional agricultural statistics, a typical arable and a typical roughage crop rotation are constructed. The arable rotation is assumed to operate on the pig and arable farms, whilst both rotations are assumed to operate on beef and dairy farms. Embedded within the GIS is a crop rotation model. On the pig and arable farms, this model implements the arable rotation on each farm, allocating a crop to each field for each of 30 years. In doing so, the model attempts to achieve the correct area distribution for each crop in each year. However, fields are never split, so variations from the ideal distribution will occur on most farms in most years. On the dairy and beef farms, the area of the farm allocated to the roughage crop rotation is adjusted to balance the production of roughage with the demand of the cattle for grazing and winter roughage. The fields allocated to the roughage rotation are those closest to the farm buildings, reflecting the need to reduce the distance for grazing animals to walk to and from animal housing. The remaining area is allocated to the arable crop rotation.

Also embedded within the GIS is a manure model. This first calculates the approximate amount of manure produced on each farm with animals, based on the number and type of animals present. In Denmark, limits are imposed on the maximum amount of animal manure and artificial fertiliser that can be applied to different crops ('crop demand'). The manure available on a farm is distributed to the fields on the farm. If the total amount produced on the farm is less than the capacity of the fields to receive manure, the manure is distributed to each field in proportion to its contribution to the total farm capacity. The remaining crop demand is satisfied using artificial fertiliser. If the manure production exceeds the farm's total crop demand, the excess is earmarked for export. Once all the animal farms have been processed, the GIS begins a sweep with increasing radius from those farms with surplus manure, identifying recipient farms (e.g. arable farms or animal farms with spare capacity). Once all farms with surplus manure have been processed, the manure is distributed to the fields on the recipient farms, supplementing with artificial fertiliser where relevant.

The GIS exports the farm, field, animal, crop and manuring information to the relevant input files of the FASSET farm model. The model was run for 29 years for each farm in the watershed. The animal and crop management was the same in all years. The model exports a wide range of environmental and economic data on either a daily and annual basis. The environmental data include NO_3^- leaching, NH_3 volatilisation, denitrification and drainage for each field and farmstead, and production and economic statistics at the whole farm scale. The FASSET output is read by the GIS, which is used for displaying the results and for passing to the hydrological model.

The MODFLOW hydrological model is used to describe the flow of water and NO_3^- within the groundwater. A 100 m x 100 m horizontal grid is laid over the landscape within the watershed and each cell within this grid is divided into five discrete layers, with each layer varying in depth from a few metres to tens of metres, dependent on the structure of the underlying terrain. The upper layer described the surface moraine, the next two layers the oxidised zone and the lower two layers the reduced zone. The NO_3^- leaching and drainage water coming from each field in each year is mapped onto the grid. Non-agricultural areas (e.g. urban, forest and nature areas) are assigned constant values for NO_3^- leaching and drainage. The model was then used to simulate the movement of water and NO_3^- through the watershed, allowing concentrations in drinking water from boreholes or in streams and rivers to be estimated.

Case study

The system was tested in a case study area of 10 x 10 km, located in mid Jutland in Denmark (Figure 2). The soils within the area are a mixture of sand or sandy loam. Depth to groundwater varies between 15 and 20 m and drainage ca. 345 mm per year. There is a mix of dairy, beef, pig and arable farming within the area. A relatively extreme policy scenario was chosen to test the system. In this scenario, the whole of the catchment area of the borehole was extensified. In practice, this meant that the western part of the catchment was planted to forestry while the eastern part was converted to fallow grassland (cut twice per year, with no harvesting of the cut grass). The GIS was used to modify the land use in the appropriate fields and the farms were then reclassified. If up to 25% of the area of the farm was extensified, it was assumed that any animal production on the farm was reduced proportionately. If more than 25%, but less than 75%, of the farm area was extensified then any animal production was assumed to cease and the remaining agricultural area used for arable cropping or fallow grassland. If 75% or more of the farm area was extensified, farming was assumed to cease. This meant that a farm could remain as before, change type (e.g. from livestock to arable) or cease to exist.

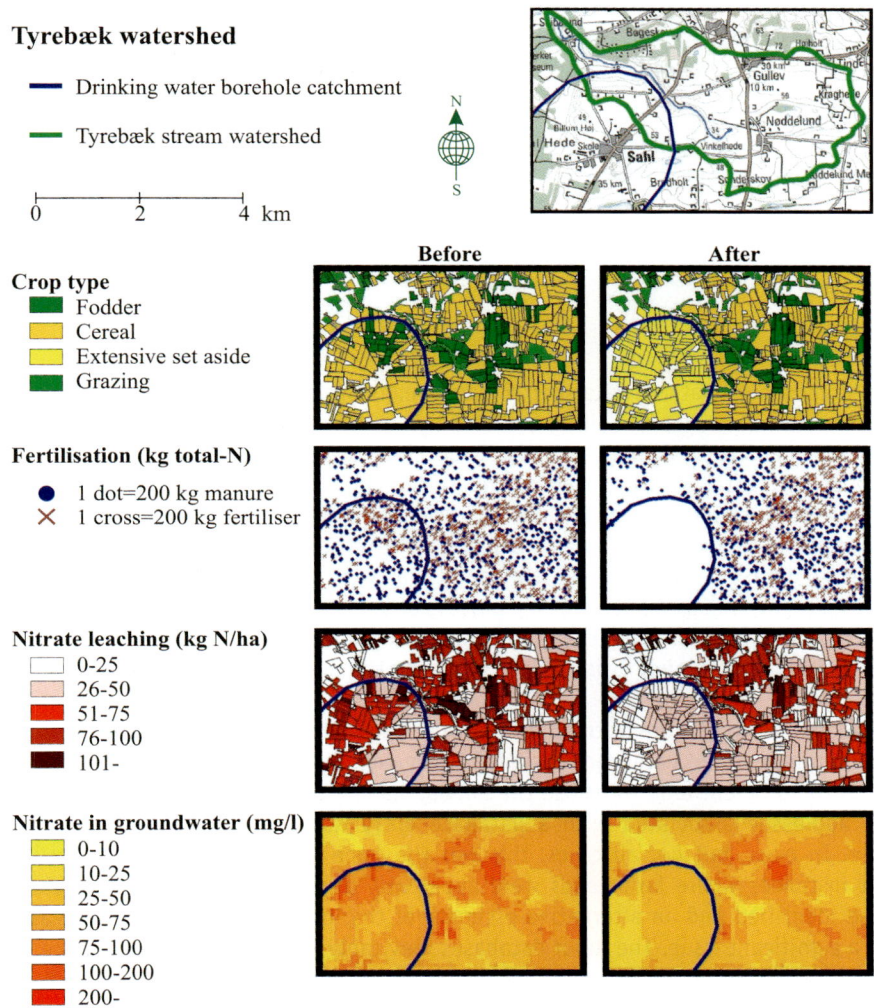

Figure 2. Case study site, with the Tyrebæk stream watershed in green and the catchment area for the drinking water borehole in blue. The 'before' and 'after' maps show the results from A, the crop rotation, B, manure, C, farm and D, hydrological models, before and after extensifying the borehole catchment. The values for NO_3^- leaching relate to water leaving the root zone, whilst the concentrations in groundwater are those for the lowest of the oxidised layers in the hydrological model.

The manure and crop rotation models were then re-run, using the modified land use and farm types. The results were passed to the farm model and then to the hydrological model.

Non-agricultural areas (e.g. urban, forest and nature areas) were assigned a constant rate of leaching of 15 kg N ha^{-1} yr^{-1} and a drainage of 80% of the mean value for agricultural land in the area over the relevant period. The hydrological model was parameterised using groundwater and drainage measurements from the area for the relevant period.

Results

The changes to land use within the catchment area had effects on land management outside the catchment area (Figure 2). Firstly, this was because some farms had land both inside and outside the area and the changes within the catchment were sufficient for the farm to change type. Secondly, the changes in land use and farm type within the catchment changed the pattern of manure production and crop manure demand. The changes in land use also affected both the amount and spatial distribution of NO_3^- leaching from the root zone and NO_3^- concentration in the upper groundwater.

Discussion

The case study shows that the watershed N management system functions, i.e. that it is capable of collating, manipulating, modelling and displaying the consequences of a major change in land use and management. It has not been possible to test the reliability of the resulting estimates of NO_3^- leaching or concentration. Under Danish conditions, such a test would be difficult. The time taken for drainage water to reach rivers or drinking water extraction points can be considerable and a test using historical data would require inputs for periods before good quality data are available electronically. Parts of the model system can be tested independently. The FASSET model is in the process of being tested and data are available within Denmark that would allow both the crop rotation and manure models to be tested. Despite the lack of testing, a number of issues are raised by the use of the system in this case study.

The first relates to the nature of the databases used to generate input to the scenario system. The main databases used here were designed to manage EU subsidies and ensure that farmers had access to sufficient land to utilise the animal manure produced. This meant, for example, that the addressee associated with a database entry was that of the legal owner of the enterprise. In most, but not all cases, this was the same as the farmstead (animal housing, etc.) but instances where farmsteads were apparently divorced from their land holding had also to be identified and an alternative estimate of the location provided. Quality assurance was also a significant issue. The input databases inevitably contained errors and some of these could be trapped by testing for unusual values or inconsistencies. The complexity of the modelling also increased the likelihood of programming errors, so quality control routines such as checking for continuity of N flow had to be included.

The case study highlights the need to consider the farm scale, even when the main interest is in processes that occur at the field scale, because farms operate as coherent entities. For example, the production (and therefore area) allocated to roughage feed production on dairy farms will normally be such that it matches approximately the demand for roughage feed. Within areas directly affected by environmental legislation, such as the Nitrates Directive, land use and management will also be influenced by the need to utilise animal manure. Finally, policy initiatives that impact elsewhere on the farm (e.g. reduction of NH_3 emission from animal housing) will affect the amount of N input to the fields. If a policy advice tool operates solely at the field scale, it will fail to capture the full consequences of policy initiatives related to field management.

If all or a large number of the farms within an area are affected by a change in the economic or legislative climate, it is necessary to consider the interaction between farms. This can be seen in the current example, where the extensification of the catchment area had two effects on farms that were partly or wholly outside its boundary. The opportunities for some livestock farmers to export surplus manure to areas within the catchment were reduced and the supply of manure to some arable farms from livestock farms inside the area was also reduced. Similar inter-farm effects might be expected if scenarios were constructed to examine intensification or expansion of livestock farms.

Finally, it is worthwhile discussing whether it is correct to use complex, dynamic models such as FASSET and MODFLOW in such a scenario system. The arguments for their use are that they can reflect relatively minor changes in agricultural management, that they are more adaptable to different climatic and soil combinations than simpler models and that they can predict the time taken for policy measures to have an effect. The arguments against such models are that they are complex and data-demanding, which increases the risk of errors and makes them difficult to use in data-sparse situations, and that in the longer term, uncertainty concerning the socio-economic factors driving agricultural production will increasingly dominate. There is probably no single system that is appropriate for all circumstances, as the demands made by policymakers and the data available for driving models will vary widely. This argues for the development of a range of models of varying complexity and of a framework for deciding which is most appropriate under a given set of circumstances.

Conclusions

The choice of model or models for predicting the consequences of changing agricultural management on watershed N flows is likely to vary depending on the input data available and on the policy questions asked. However, it is important that the models used can take account of the interactions that occur at the farm level and above.

Acknowledgements

This project was funded by the Danish Ministry of Food, Agriculture and Fisheries and the Danish Ministry of Environment, via the programme 'Land use; the farmer as a land manager'.

References

Berntsen, J., Jacobsen, B.H., Olesen, J.E., Petersen, B.M. and Hutchings, N.J. (2003). Evaluating nitrogen taxation scenarios using the dynamic whole farm simulation model FASSET. Agricultural Systems, 76, 817-839.

Edwards, A.C., Pugh, K., Wright, G., Sinclair, A.H. and Reaves, G.A. (1990). An empirical model for estimating nitrite leaching as affected by crop type and N fertilizer rate. Soil Use and Management, 14, 37-43.

Hansen, S., Jensen, H.E., Neilsen, N.E. and Svendsen, H. (1990). DAISY - Soil Plant Atmosphere System Model. Publ. Royal Veterinary and Agricultural University, Copenhagen, 272 pp.

Simmelsgaard, S.E. and Djurhuus, J. (1998). An empirical model for estimating nitrate leaching as affected by crop type and N fertilizer rate. Soil Use and Management, 14, 37-43.

Smith, J.U., Bradbury, N.J. and Addiscott, T.M. (1996). SUNDIAL: a PC-based system for simulating nitrogen dynamics in arable land. Agronomy Journal, 88, 38-43.

Mitigating non-point source pollution from dairy farms: Economic evaluation on the Don watershed

N. Turpin[1], G. Rotillon[2], P. Bontems[3], T. Bioteau[1] and R. Laplana[4]
[1]Cemagref, UR GERE, 17 avenue de Cucillé, CS 64427, 35044 Rennes cedex, France
[2]University of Paris-X Nanterre, THEMA, 200 avenue de la République, 92001 Nanterre, France
[3]INRA, IDEI, PO Box 27, Chemin de Borde Rouge, 31326 Castanet-Tolosan cedex, France
[4]Cemagref, UR ADBX, 50 avenue de Verdun, Gazinet, 33612 Cestas cedex, France

Abstract

We built a model of regulation for non-point source water pollution through non-linear taxation/subsidisation of agricultural production. Farms are heterogeneous along two dimensions: their ability to transform inputs into final production, and the available area they possess. We propose a framework for describing the ability of farmers, the cost and the emission joint functions from a population of heterogeneous producers on a watershed. The application uses data collected on the Don watershed (France). Several policies are compared *ex ante* on this watershed on their cost and efficiency.

Keywords: cost, dairy, efficiency mitigation, non point source pollution

Introduction

The design of regulating instruments is particularly delicate for agricultural Non Point Source (NPS) pollution from fertilisers, because this problem combines large sets of possible pollutants with heterogeneous de-polluting costs and uncertainty on both production and emissions related to weather conditions and run-off events (Shortle & Horan, 2001). Despite these difficulties, regulators will have to implement the Water Framework Directive (2000/60/EC) and for this purpose they will have to choose from several possible policies on efficiency and cost criteria.

One key issue for the design of mitigating policies is the heterogeneity of farms. It is obvious that farms have a wide range of production factors (e.g. soils, climate, management skills, genetic merit of the herds) and that the farmers' objectives are very different. This results in a wide range of technical choices, such as the degree of production intensity, the amount of inputs used and the techniques implemented. The heterogeneity of the farms has consequences for their behaviour when facing a regulation, and on the amounts of pollutants they emit, i.e. the same technical choice on two different farms may result in different emission rates. When designing a mitigation policy, a regulator will also have to make choices on how the de-polluting effort among the heterogeneous producers will be distributed. For a regulator, designing a policy to mitigate NPS pollution from farms means choosing two different things: first the regulator has to determine which instrument is to be regulated. Many studies conclude that the choice of instrument base can significantly influence the cost-effectiveness of agri-environmental policy (e.g. Weinberg & King, 1996). The basis of this instrument can be an estimation of the individual emissions, or some input related with these emissions, or some specific production technique which is supposed to be polluting, or the production level. Secondly, the regulator has to define the method of applying the policy: is a uniform regulation satisfactory, or an optimally differentiated one required?

Combined modelling of emission decrease and cost of management changes is of great help for a regulator when he/she has to choose among different possible policies on both the instrument and the method of application. The NPS pollution literature focuses on two basic frameworks. The first framework assumes that the soils and climate are of great importance for both farmers' profit and pollutant transfers to water. The models resulting from this approach are generally soil-management based. They are often combined with a hydrological model to describe water and pollutant transfers through a watershed, and are coupled with a simple economic model of some technically defined types of farms. The coupling of the physical and economic models allows a cost-efficiency analysis of different policies. This framework led to several important advancements: there are large differences in emission levels from one farm to another, especially when their outputs (e.g. meat, milk or cereals) differ (Schou et al., 2000); management decisions interact with soil and climate conditions with significant consequences on profit and emissions (Kampas & White, 2003). Thus, a regulating option can be of high value for some types of farms in one given watershed and of no interest elsewhere (Polman & Thijssen, 2002). The second framework goes further and focuses on the description of the heterogeneity of the farms along a watershed or a country. The modelling designs policies that are differentiated among the producers. Usually, regulation strategies (including much information) out-perform uniform regulations and, in most cases, by a large margin (Claassen & Horan, 2001). Because the regulator cannot access all the information, he/she would need to design such a regulation, and has to add some incentive constraints during the design. These constraints are costly but they ensure that each producer chooses the instruments that have been designed for him. The model developed in this paper belongs to this set of literature, according to Laffont (1994) who applied the economical Agency theory to pollution problems. Laffont (1994) showed that as soon as a regulator can correctly characterise the possible polluters, he/she is able to efficiently distribute the de-polluting effort among the producers. Theoretical developments of this concept are numerous (Wu & Babcock, 1996; Goldsmith & Basak, 2001) but empirical applications are still rare.

The objective of this paper is to develop a method to characterise the heterogeneity of dairy farms in a watershed, and to integrate this heterogeneity in the design of policies to mitigate NPS N in waters pollution from these farms.

Materials and methods

Study area
The Don watershed (71,706 ha) is located in the western part of France, in the 'Pays de la Loire' region. Farm production is mainly cattle production (dairy and meat), where cereals are grown for both grain and forage. Indoor rearing is still of low importance, but the number of pig and poultry farms is increasing. Grasslands, associated with dairy production, account for around 50% of total agricultural area. Cereals represent 18% and maize 15% of the total area. The average size of farms in 1999 was 74 ha. The climate is typically maritime, with cool wet winters and warm drier summers. The Don watershed is covered by brown soils resulting from

the alteration of the underlying schist rock. The watershed is flat and soil hydraulic conductivities are low, and these thin soils (60 to 90 cm deep) are frequently hydromorphic. The water coming from the Don watershed is connected to two pumping stations for drinking water, supplying around 150,000 people. The 'Departmental Council of Loire Atlantique' monitors water flows and NO_3^- concentrations at the 'Conquereuil' station (draining 59,306 ha of the whole Don watershed).

In the Don watershed, NO_3^- concentration regularly reached or exceeded the EU guidelines of 50 mg l^{-1} at the 'Conquereuil' pumping station in the mid-nineties when a recovery programme was set up by local extension services. Cropping, fertilising and manuring advice has been given to the farmers who could choose to adopt them or not. For several years, no change in water quality has been observed, so the attention focused on to the rate of adoption of the 'best practices' that had been promoted on this watershed.

Farm survey

The population of farms in this watershed was initially surveyed by an extension service in 1999. This inventory was used to stratify the whole population of 820 farms with production system criteria. The stratification of the population was based on both the production system and on the balance between grazed (mostly pastures) to harvested (mostly corn) areas. A sample of 82 farms was randomly selected from these *strata* (with sample size proportional to each stratum size) and surveyed in more details. The aim of the more detailed survey was to describe the stages of decision making by the farmers faced with environmental questions (Ölmer et al., 1998), i.e. problem detection, problem definition, analysis and choice (observation and searching for options, analysis tools) and eventually the implementation of action. Special attention has been paid to the evolution of these ideas for the last five years. A precise description of the fertilisation practices for each crop in each rotation has been collected for the last ten years. For the farmers who accepted it, gross and net output production costs have been collected for the last three years.

Model principles

Characterisation of the heterogeneity of the farms: According to Laffont (1994), the heterogeneity of the farms is captured by a one-dimensional parameter, i.e. the management system of the farm. New econometric techniques allow the common determination of both the production function parameters and a value for the farm management system (Thomas, 1995). In our case, we aimed to relate the farm system to both production and N emission functions. Thus a technical definition of the type was necessary. The determination of the farm type has been made in a two-step process: first a multiple component analysis has been performed on all the quantitative variables that have been collected during the survey to determine the variable that best represents the variability of the farms. Second, these variables have been combined to mimic the technical qualitative classification of the farms that was performed by the local advisers.

Inclusion of the heterogeneity parameter in the farmer's profit function: The economic model represents the farmers as price-takers (the prices of input and output are not affected by on-farm decisions) and profit maximisers (the farmers wish to maximise their profit, given some constraints on land, capital and labour, on their own farms). The Principal-Agent model represents the farms with continuous cost and yield functions, including heterogeneity among the producers. The model focus on asymmetric information problems between a regulator (the Principal) and the farmers (the Agents). The Agents are supposed to have more precise information on their own farm than the Principal, who has information on the density functions for each parameter only (Laffont, 1994). This representation allows the assessment of both the introduction of technical Best Management Practices (BMPs), defined as a modification of the cost or the yield functions (or both), and the application of various policies: the modelled farms react to a given policy by moving along their profit function (Bontems et al., 2003). The parameters of both the type-dependent cost and profit functions have then been estimated using a maximum likelihood method from the data collected on the Don watershed.

Inclusion of the heterogeneity parameter in the emission function: Individual emissions have been estimated using the SWAT model. This model (Soil and Water Assessment Tool) is a semi-distributed watershed model with a GIS interface (DiLuzio et al., 2002) that outlines the sub-watersheds and stream networks from a digital elevation model and calculates daily water balances from meteorological, soil and land-use data. SWAT simulates each sub-basin separately, according to the soil water budget equation, taking into account daily amounts of precipitation, runoff, riverbed transmission losses, percolation from the soil profile, and evapotranspiration. Once the physical model was calibrated, type-dependent individual emissions were determined, from its estimations, using a quadratic function for the emission function of N into waters and a constrained maximum likelihood method (Bontems et al., 2003). The cost of social damage has been valued as being the cost of water treatment for transforming 'dirty' water into drinking water. Obviously, a complete evaluation of damage should include the consequences of lack of biodiversity, increased eutrophication, decrease in recreation activities and so on. This task is out of the scope of this paper. Thus our valuation will be an underestimation of total social damage.

Designing mitigation regulations

Once the cost, emission and damage functions have been described and their parameters estimated, the regulation design is an optimisation problem: the regulator's objective is to maximise a welfare function, written as the sum of taxpayers surplus, the farmers' total surplus, minus the environmental damage. Feasible allocations are constrained by the information set of the regulator. We also introduce acceptability constraints as part of the constraints that the regulator has to take into account. Basically, the regulator has to satisfy a given proportion of farmers through his/her intervention and farmers are satisfied if they do not loose from regulation compared to the *laissez-faire* situation. Three types of mitigating policies have been tested within this framework:

1. We considered policies which are optimally differentiated and take into consideration the heterogeneity of farms on their cost and emission functions. The regulator proposes a contract with the farmers and designs this contract by maximising their own welfare functions.

2. We compared these differentiated policies with the mandatory application of new, less polluting techniques (BMPs). The associated costs are mostly borne by the farmers. The costs are associated with an increased use of machinery or labour, or related to a risk of yield decrease when lower amounts of polluting inputs are used. Some benefits can be

associated with the lower use of polluting inputs, if they are not replaced by more expensive ones.

3. We compared the differentiated policies with standard economic instruments, such as taxes or quotas applied on inputs or outputs. In the Don watershed, a tax on the amounts of mineral N used by the farms, a quota of mineral N applied, and a linear extensification (decrease of milk yield per hectare) have been tested. These linear standard instruments can be associated with subsidies that increase on-farm profit, such that more farmers benefit from the policy.

For each policy, the level of the instrument (the tax, the subsidy) is determined by the model while maximising the welfare function of the regulator. Thus, the polluting level is not fixed by the regulator as an objective, but it is a result of the maximisation process.

Results and discussion

Determination of the farmer type

The parameter describing the heterogeneity of dairy farms on the Don watershed (denoted θ) has been interpreted as their ability to transform their forage area into milk. This parameter is estimated by combining the technical variables that capture differences across farms and describe 43% of the whole variability among the farms. These variables are: 1) the amount of dry matter produced by grasslands: this variable is a clue for herbage availability and grazing management, which both affect nutrient supply for grazing cows (Kuusela & Khalili, 2002); 2) the amount of milk produced per cow: high yielding dairy cows are more sensitive to feeding management than lower yielding ones (Peyraud & Astigarraga, 1998); 3) the balance in forage feed between proteins and energy: recent whole-animal models of N utilisation in dairy cow show that the availability of energy is crucial to the efficiency of N utilisation by the animals (Kebreab et al., 2002); 4) the amount of required concentrates, estimated as a difference between a theoretical food intake (calculated with high quality forages) and what has been observed on each farm. Increased levels of concentrates supplementation are reported to have a substantial impact on the profitability and nutrient balance for grazing dairy farms.

Soils have their importance in milk production, but we deliberately omitted them in the defining variables (in fact, soil effects are already included in the parameters entering in the definition of our θ-type: fertile and easy-to-manage soils allow high yields for grasslands, provide high quality forages and thus allow high production cows and low levels of concentrates). A farmer who manages to obtain high grass yields, who breeds highly productive cows, gives them good quality forages and needs only low rates of concentrates to produce milk, is considered to possess a good ability to transform his forage crops into milk, and thus the θ-parameter is high. For our sample, the estimation of parameter θ suggests that high θ-type produces more milk with less inorganic N than low ones.

Dairy production functions

In the observed situation, the farmers choose a milk yield when maximising their profit function, given the price of the milk and their own θ-type. Figure 1 represents their dairy production cost (euros ha^{-1}), *versus* the milk yield they have chosen and depicts the production costs after the statistical estimation of the parameters. In our application, the higher the θ-type

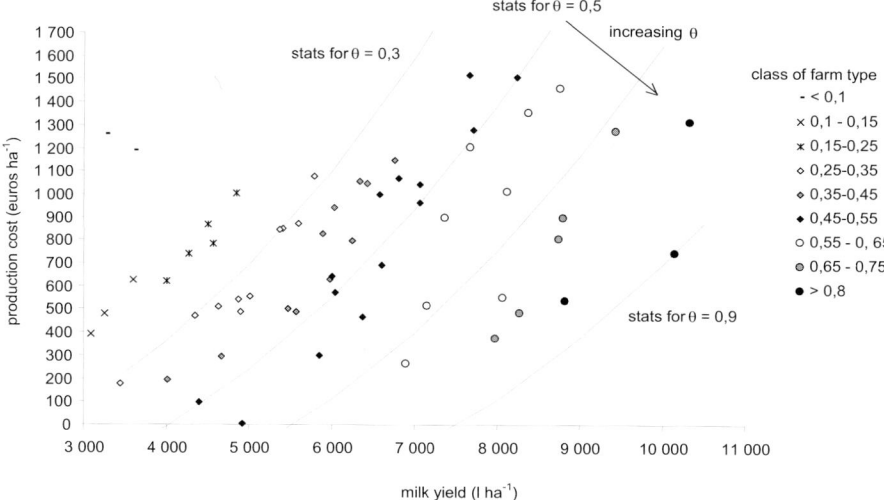

Figure 1. Milk yield and production cost for the surveyed farms, depending on their type and statistics for some values of θ.

parameter, the lower the production cost *cæteris paribus*. On the Don watershed, the modelled on-farm dairy production does not statistically differ from the recorded value, but the aggregated milk production on the watershed is slightly (2%) over-estimated (see Bontems et al., 2003, for details).

Comparing different policies

The different policies tested on the Don watershed have been compared using a two-dimensional graph (see Figure 2): the first axis represents the cost-efficiency ratio (expressed in euros kg^{-1} of avoided pollutant) and the second axis shows the different acceptability levels associated with each BMP.

For 'technical' BMPs, the acceptability level comes from surveys. The acceptability of linear and optimally differentiated policies is assessed by the model as being the percentage of farms who benefit from the regulation. The simulations suggest that some policies, when the level of the instrument is determined such as to maximise the regulator's objective, do not allow the European threshold of 25 mg NO_3^- l^- to be reached. Mandatory technical changes (depicted 'BMPx' in Figure 2) can be performed with relatively low cost, but we did not consider enforcement costs. It is interesting to note that despite a loss of profit, nearly 35% of the farmers consider that they would adopt these technical changes.

A tax on the mineral N used, with its level determined to maximise the regulator's welfare (here this level is 0.23 euros kg^{-1} of bought N, which is relatively high when compared with the price of mineral fertilisers), does not permit the EU threshold to be reached with very high levels of taxation. When the amount of the tax is redistributed to the farmers, so that 50% of them benefit from the regulation, this threshold can be reached. In this case, the regulator's welfare is not optimal at all, and the mechanism has a tremendous cost for Society

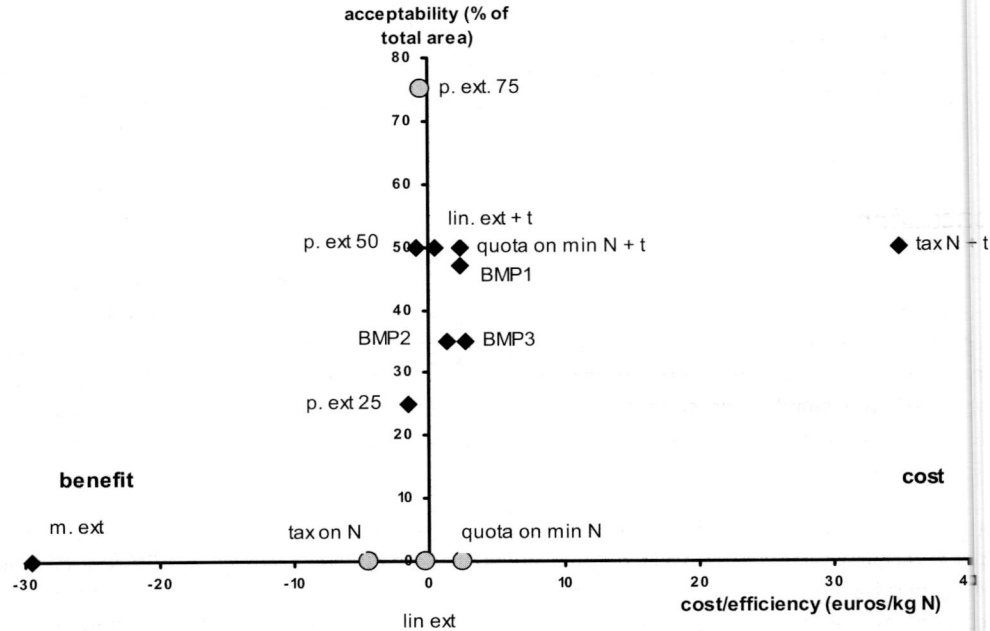

Figure 2. *Cost/efficiency ratio (euros kg^{-1} of avoided N) and acceptability for several regulation policies on the Don watershed.*

Legend :
Mechanisms depicted with grey circles do not result in exceeding the EU threshold of 25 mg NO$_3^-$ l^{-1}. Those depicted by black diamonds do exceed this threshold.

differentiated mechanisms :

m. ext:	*differentiated mandatory extensification (the decrease of the production level depends on the type of the farm)*
p. ext. α:	*differentiated extensification associated to a subsidy built such that α% of the farmers benefit from the regulation*

technical modifications:

BMP1:	*improved fertilisation*
BMP2:	*better use of manures (spread on all the forage area rather than on corn only)*
BMP3:	*BMP1 + BMP2*

uniform regulations:

tax on N:	*tax on mineral N used*
quota on min N	*quota of mineral N used*
lin ext.:	*uniform extensification (the decrease of the production level is the same for all the farms)*
+ t:	*same mechanism, with a level of the instrument built such that the damage is the same as the "p. ext 50" mechanism and associated with a subsidy so that 50% of the farmers benefit from the regulation.*

(here 35 euros kg^{-1} of avoided N). The same results are obtained for a uniform extensification, or a limit to the amount of N used; the limiting mechanism being less costly. We should notice that optimally differentiated policies have a high level of acceptability, perform with an increase of welfare, and allow the EU standard for water quality to be reached until 55% of the farmers benefit from the regulation.

Conclusion

The main results are twofold. Firstly, the differentiated mechanism based on the production level generates the higher welfare increase, when compared with the non-regulated situation. In particular, when all the farmers suffer from this regulation, the benefit is nearly six times the benefit generated by a tax on the amount of N fertilisers used. Second, on the Don watershed, the simulations suggest that there are several mechanisms which decrease the pollution level below the EU threshold of 25 mg NO_3^- l^{-1}. Among all these possibilities, only the differentiated mechanisms lead to a social benefit.

Many other improvements of this research can be foreseen: until now, we have only focussed on the potential effects and costs of particular BMPs. Obviously, the application of a specific BMP generates effects on other practices at the farm scale. Developing a joint approach that incorporates the economical, sociological and physical aspects of the modelling through the building of a Decision Support System is, in our opinion, the key for future research in the area of mitigating nonpoint source pollution from human activities. This would be the best way to help EU Member States to ensure a programme of measures to mitigate water pollution within the Water Framework Directive.

Acknowledgements

We acknowledge financial support from EU through AgriBMPWater project (5th PCRD). We are grateful to David Chadwick for improving a preliminary version of this paper.

References

Bontems, P., Rotillon, G. and Turpin, N. (2003). Acceptability constraints and self-selecting agri-environmental policies, submitted, http://www.bordeaux.cemagref.fr/adbx/ agribmpwater/index.html, pp. 45.

Claassen, R. and Horan, R.D. (2001). Uniform and Non-Uniform Second-Best Input Taxes: The Significance of Market Price Effects on Efficiency and Equity. Environmental and Resource Economics, 19, 1-22.

DiLuzio, M., Srinivasan, R., Arnold, J.G. and Neitsch, S.L. (2002). ArcView Interface for SWAT2000 User's Manual. GSWRL Report 02-03, BRC Report 02-07, Texas Water Resources Institute TR-193y, College Station, TX.

Goldsmith, P.D. and Basak, R. (2001). Incentive Contracts and Environmental Performance Indicators. Environmental and Resource Economics, 20, 259-279.

Kampas, A. and White, B. (2003). Probabilistic programming for nitrate pollution control: comparing different probabilistic constraint approximations. European Journal of Operational Research, 147, 217-228.

Kebreab, E., France, J., Mills, J.A.N., Allison, R. and Dijkstra, J. (2002) A dynamic model of N metabolism in the lactating dairy cow and an assessment of impact of N excretion on the environment. Journal of Animal Science, 80, 248-259.

Kuusela, E. and Khalili, H. (2002). Effect of grazing method and herbage allowance on the grazing efficiency of milk production in organic farming. Animal Feed Science and Technology, 98, 87-101.

Laffont, J.-J. (1994). Regulation of Pollution with Asymmetric Information. In: C. Dosi and T. Thomasi (eds). Nonpoint source pollution regulation: Issues and analysis. Fondazione Eni Enrico Mattei Series on Economics, Energy and Environment, Dordrecht and Boston, 39-66.

Ölmer, B., Olsoin, K. and Brehmer, B. (1998). Understanding farmers' decision making processes and improving managerial assistance. Agricultural Economics, 18, 273-290.

Peyraud, J.L. and Astigarraga, L. (1998). Review of the effect of nitrogen fertilization on the chemical composition, intake, digestion and nutritive value of fresh herbage: consequences on animal nutrition and N balance. Animal Feed Science and Technology, 72, 235-259.

Polman, N.B.P. and Thijssen, G.J. (2002). Combining results of different models: the case of a levy on the Dutch nitrogen surplus. Agricultural Economics, 27, 41-49.

Schou, J.S., Skop, E. and Jensen, J.D. (2000). Integrated agri-environmental modelling: A cost-effectiveness analysis of two nitrogen tax instruments in the Vejle Fjord watershed, Denmark. Journal of Environmental Management, 58, 199-212.

Shortle, J.S. and Horan, R.D. (2001). The economics of nonpoint pollution control. Journal of Economic Surveys 15, 255-289.

Thomas, A. (1995). Regulating pollution under asymmetric information : the case of industrial wastewater treatment. Journal of Environmental Economics and Management, 28, 357-373.

Weinberg, M. and King, C.L. (1996). Uncoordinated agricultural and environmental policy making: an application to irrigated agriculture in the West. American Journal of Agricultural Economics, 78, 65-78.

Wu, J.J. and Babcock, B.A. (1996) Contract design for the purchase of environmental goods from agriculture. American Journal of Agricultural Economics, 78, 935-945.

Relationships between fertiliser nitrogen additions, crop carbon returns and soil quality

A. Bhogal, B.J. Chambers and F.A. Nicholson
ADAS Gleadthorpe Research Centre, Meden Vale, Mansfield, Notts. NG20 9PF, UK

Introduction

Arable cropping has been widely implicated in the deterioration of soil quality through the depletion of soil organic matter (SOM) reserves because of oxidation following cultivation. Reductions in the use of fertiliser N may also result in decreased SOM levels through lower crop residue returns, whilst conversely, applications of fertiliser N at, or in excess of, the optimum economic level, can result in a gradual increase in SOM status, which could potentially improve soil quality and fertility (Bhogal et al., 1997; Nicholson et al., 1997). Soil organic matter is fundamental to the maintenance of soil fertility and function, and has been identified as a key indicator of soil quality, with influences on a wide range of physical, chemical and biological properties (DoE, 1996; Carter et al., 2001). However, many of the claimed improvements in soil physical and biological properties. As a result of organic C inputs, are largely based on anecdotal evidence. The aims of this study were to determine relationships between N fertiliser rate (and associated crop organic C returns) and SOM status, and to evaluate the effects of fertiliser N additions on 'key' indicators of soil quality.

Materials and methods

The study was undertaken at three existing experimental sites, on contrasting soil types, where inorganic fertiliser N had been applied at differential rates for 17-23 years. At ADAS Gleadthorpe (6% clay) and Morley Research Centre (13% clay), 6 rates of fertiliser N (0-250 kg N ha^{-1}) had been applied annually since 1984, with both sites growing combinable crops for the duration of the experiment. At Ropsley (27% clay), 8 rates of fertiliser N (0-245 kg N ha^{-1}) had been applied annually from 1978-1990, with continuous winter wheat grown since 1990. The effects of these fertiliser N additions and associated crop organic C inputs on key indicators of topsoil physical, chemical and biological quality were measured in spring 2000 (Gleadthorpe and Morley) and 2001 (Ropsley). Carbon and N balances were constructed each year (1998-2001) in order to compare the C and N inputs in crop residues, with measured changes in soil organic C and total N content.

Results

Positive relationships (*P*<0.001) between inorganic fertiliser N addition rate and crop organic C returns (in straw, stubble and roots), and changes to topsoil total organic C (TOC) and light fraction organic C (LFOC) were measured. On average, ca. 0.5 t ha^{-1} of organic C was returned in the form of straw, stubble and roots for every 100 kg ha^{-1} fertiliser N applied. This, in turn, was related to increases in topsoil organic C levels, with topsoil TOC and LFOC increasing by ca. 0.4 and 1.1%, respectively, for every 1 t ha^{-1} C incorporated, or *c.*200 kg ha^{-1} N applied (Figure 1). The fertiliser N additions and associated crop organic matter returns also had important and measurable beneficial effects on other key indicators of soil quality. The size and activity of

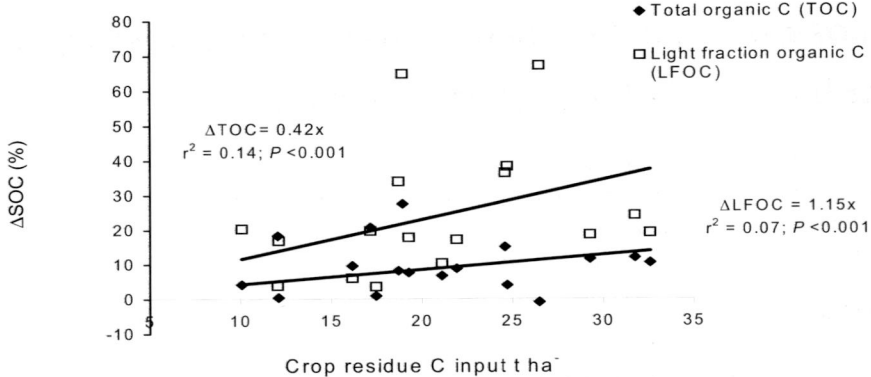

Figure 1. Relationship between crop residue C inputs and topsoil organic C content.

the soil microbial biomass increased with increasing organic C inputs ($P<0.001$) i.e. by 1.0% and 1.8%, respectively, for every 1 t ha^{-1} C incorporated. Potentially mineralisable N (measured by anaerobic incubation) increased by ca. 3% with every 1 t ha^{-1} N returned to the soil in crop residues ($P<0.001$), indicating that long-term soil N supply was increased by the fertiliser additions. Also, topsoil shear strength decreased with increasing C inputs ($P<0.05$) indicating that soils which have received repeated fertiliser N applications (and associated increased crop C returns) would be easier to cultivate, but would potentially be less trafficable, than soils that had not received fertiliser N.

Conclusions

Medium-term applications of inorganic fertiliser N and associated increased organic C returns had important and measurable beneficial effects on a range of soil chemical, biological and physical properties, which are important indicators of soil quality. These results provide valuable information on how fertiliser N use and crop organic C returns can replenish SOM reserves and maintain the inherent fertility of arable soils.

Acknowledgements

This project was funded by the Department for Environment, Food and Rural Affairs.

References

Bhogal, A., Young, S.D., Sylvester-Bradley, R., O'Donnell, F.M. and Ralph, R.L. (1997). Cumulative effects of nitrogen applications to winter wheat at Ropsley, UK from 1978-1990. Journal of Agricultural Science, Cambridge, 129, 1-12.
Carter, M.R. (2001). Organic matter and sustainability. In: R.M. Rees, B.C. Ball, C.D. Campbell and C.A. Watson (eds.) Sustainable Management of Organic Matter. CAB International, Wallingford, Oxford, p. 9-22.
DoE (1996). Indicators of Sustainable Development in the United Kingdom. Chapter 4, Soil, p. 142-146. Department of the Environment.
Nicholson, F.A., Chambers, B.J., Mills, A.R. and Strachan, P.J. (1997). Effects of repeated straw incorporation on crop fertiliser nitrogen requirements, soil mineral nitrogen and nitrate leaching losses. Soil Use and Management, 13, 136-142.

Fluxes of nitrogen following clearing of Brazilian Amazonian tropical forest for pasture

M.C. Piccolo[1], C. Neill [2], C.C. Cerri [1] and J.M. Melillo[2]

[1]Centro de Energia Nuclear na Agricultura, Universidade de São Paulo, mail box 96, zipcode 13400970, Piracicaba, SP, Brazil
[2]The Ecosystems Center, Marine Biological Laboratory, zipcode 02543, Woods Hole, MA, USA

Introduction

Soil inorganic N concentrations and rates of net N mineralisation and net nitrification can be indicators of both soil fertility and the potential for losses of N in soil solution and in N oxide gaseous emissions that disturbances such as forest clearing causes.

Materials and methods

We examined how forest clearing for pasture affected soil N cycling and fluxes of N in soil solution and gaseous N emissions at Nova Vida (10° 30 S, 62° 30 W), in the Amazon state of Rondônia. We followed soil and soil solution inorganic N concentrations, net N mineralisation and net nitrification rates in a forest, in an area cleared and converted to pasture (cut forest) and in established pasture (*Brachiaria brizantha*) of 5 and 22 years-old on Kandiudults soils. Soil solution samples were collected by tension lysimeters (Soil Moisture Equipment, Santa Barbara, California, USA) at 100 cm depth. Soil samples for inorganic N concentrations and net mineralisation and nitrification rates were sampled from 0-5 cm and 5-10 cm using a 5 cm diameter corer (Piccolo et al., 1994). Soil and soil solution NH_4^+-N and NO_3^--N were measured using an automated flow injection system. Ammonium was measured colorimetrically and NO_3^- was measured colorimetrically as NO_2^-, following Cd reduction. Solution inorganic N fluxes were calculated for each site as the result of total water flux mean monthly NH_4^+ and NO_3^- concentrations at 100 cm depth.

Results and discussion

Pools of soil inorganic N and rate of net nitrification increased during the first 8 months following forest cutting and burning. Rates of net N mineralisation and net nitrification were similar in forest, cut forest and after pasture installation. Established pastures had higher NH_4^+ pools, lower NO_3^- pools and lower net nitrification rates than the forest. Annual solution fluxes of NH_4^+ from the forest and cut forest were zero (Table 1). Annual NO_3^- flux increased 10-fold in cut forest compared with forest (Table 1), but the difference between the cut forest and forest declined with time after the start of the rainy season. In established pastures, solution NH_4^+ fluxes ranged from 0.1 to 0.8 kg N ha^{-1} yr^{-1} and NO_3^- fluxes ranged from 0.7 to 1.3 kg N ha^{-1} yr^{-1}. Total annual dissolved inorganic N (DIN) fluxes to below 1 m increased 10-fold in cut forest compared with forest and 22-year-old pasture. In pastures, total annual DIN decreased with installation time of pasture.

We compared soil solution DIN fluxes with N_2O losses to atmosphere from forest to pasture sequence as measured by Melillo et al. (2001), and fluxes of NO from forest and pastures

Table 1. Comparison of solution inorganic N fluxes and N oxides fluxes (kg N ha^{-1} yr^{-1}) from study areas. The solution inorganic N fluxes were calculated from water flux and concentrations in tension lysimeters installed at 100 cm.

Study Areas	kg N ha^{-1} yr^{-1}							%	
	NH$_4^+$	NO$_3^-$	Total DIN	NO*	N$_2$O*	Total gaseous flux	Total DIN and N oxides	N flux as solution	N flux as gaseous
Forest	0	2.5	2.5	1.4	3.2	4.6	7.1	35	65
Cut forest	0	24.4	24.4	2.3	5.3	7.6	32.0	76	24
5-year-old pasture	0.8	1.3	2.1	0.2	3.1	3.3	5.4	39	61
22-year-old pasture	0.1	0.7	0.8	0.2	1.2	1.4	2.2	36	64

*N$_2$O losses were from Garcia-Montiel et al. (2001) and Melillo et al. (2002). NO fluxes from the cut forest were estimated from the ratios of N$_2$O/NO measured in the intact forest.

measured by Garcia-Montiel et al. (2001). Total gaseous emissions were 4.6 kg ha^{-1} yr^{-1} in forest and approximately double that in cut forest (7.6 kg N ha^{-1} yr^{-1}), while solution fluxes increased nearly10-fold in the first year after clearing (Table 1). The established pastures total gas fluxes were 3.3 and 1.4 kg N ha^{-1} yr^{-1} in 5 and 22-year-old pasture, respectively. Forest cutting and burning had a greater influence on solution than on gaseous N fluxes. The gas fluxes comprised 65% of total N fluxes in forest and 61-64% of fluxes in established pastures.

Our results on N pools indicate that pasture establishment is associated with increases in extractable NH$_4^+$, decreases in extractable NO$_3^-$, and decreases in rates of net N mineralisation and net nitrification. The distribution of gaseous and solution fluxes were similar among the forest and pastures of different ages, but total gas and solution fluxes from pastures were lower than from forest.

Conclusions

Understanding the N availability in tropical forest and pasture will be important for understanding N losses to the atmosphere and in soil solution in response to disturbances. Both gaseous and solution losses of N may become more important if fertilisation of pastures increases in the future.

Acknowledgements

We gratefully acknowledge the support provided by the FAPESP (Brazil) and NASA (EUA).

References

Garcia-Montiel, D.C., Steudler, P.A., Piccolo, M.C., Melillo, J.M., Neill, C. and Cerri, C.C. (2001). Controls of soil nitrogen oxide emissions from forest and pastures in the Brazilian Amazon. Global Biogeochemical Cycles, 15, 1021-1030.

Melillo, J.M., Steudler, P.A., Feigl, B.J., Neill, C., Garcia-Montiel, D., Piccolo, M.C., Cerri C.C. and Tian, H. (2001). Nitrous oxide emissions from forests and pastures of various ages in the Brazilian Amazon. Journal of Geophysical Research, 106 (D24), 34, 179-34, 188.

Piccolo, M.C., Neill, C. and Cerri, C.C. (1994). Net nitrogen mineralisation and net nitrification along a tropical forest-to-pasture chronosequence. Plant and Soil, 162, 61-70.

Nitrogen losses from forage legumes and effects on succeeding cereal crops in organic farming under pannonic climate conditions in Eastern Austria

R. Farthofer[1], J.K. Friedel[1], G. Pietsch[1], W. Loiskandl[2] and B. Freyer[1]
[1]*Institute of Organic Farming, University of Natural Resources and Applied Life Sciences, Vienna, Austria*
[2]*Institute of Hydraulics and Rural Water Management, University of Natural Resources and Applied Life Sciences, Vienna, Austria*

Introduction

In organic farming, the replenishment of soil N mainly depends on biological N_2 fixation (BNF) via forage legumes. In Austria where stockless organic farming is common, ensuring sufficient N supply to cereal crops, such as winter wheat and winter rye, is very important. Because of the dry climate with high annual average temperatures (9.8°C) and low precipitation (550 mm yr^{-1}), alfalfa is the most commonly grown forage legume. In stockless farming systems, the legume shoots are mainly mulched and used as green manure. During the wet winter season, the high amounts of organic matter added to the soil might create a high potential for N leaching. Growing legumes in mixtures with grass and/or removing the shoots after cutting may help to reduce the content of inorganic N (N_{in}) in soil. Therefore, these methods may be suitable to minimise the NO_3^- leaching potential while still supplying sufficient N to the following cereal crops to ensure appropriate yield and quality. The aim of this project was to optimise the management of forage legumes in stockless organic farming with respect to N losses and N availability to succeeding crops. The specific objectives were to test the potential of i) alfalfa grass mixtures vs. pure alfalfa stands and ii) a cutting regime instead of mulching to reduce the risk of NO_3^- leaching and to ensure N supply to succeeding cereal crops.

Materials and methods

This project was conducted on an organically managed area near Vienna in the Marchfeld region and started in autumn 2000 and ended in autumn 2003. Alfalfa was used either in pure stands an alfalfa-grass mixture (80% alfalfa, 20% grass - coverage at sowing). Winter rye and a pure grass stand served as reference crops. The variants were tested in a randomised block design with 4 replicates. After ploughing the legume stands, winter wheat and winter rye were cultivated in two successive years. Mineral N in soil (0-120cm soil depth) was investigated by soil sampling, 5 times a year. Soil water content was monitored by FDR-probes and tensiometers (Hauer et al., 2003). The potential for NO_3^- leaching was monitored by a soil water balance (Allen, 1998) and NO_3^- in the soil solution, which was obtained with suction cups. Grain yields and protein contents were recorded. Means were tested by 2-way ANOVA (Tukey Test, $P<0.05$).

Results and discussion

The results (after 2 years) showed that fodder legumes, especially under mulching, led to high mineral N contents in soil after ploughing (Table 1). High N contents in soil were the result of mineralisation of incorporated plant residues and soil organic matter during the autumn and winter season and created a high risk for NO_3^- leaching into ground water. Cultivation of legumes in mixtures with grass caused the same amounts of soil mineral N as pure legume stands at a lower level. A low grass content in legume mixtures, therefore, was not suitable for decreasing the risk of NO_3^- leaching. No soil water samples could be collected by suction cups during the winter seasons in 2000/01 and 2001/02. This was due mainly to low precipitation during the first year of investigation. Additionally, the water resources in deeper soil layers were heavily exploited by the preceding alfalfa crop. Considering the high water holding capacity of loess soils, and the low precipitation-level, NO_3^- leaching into the ground water is unlikely in the first two years of investigation. The soil water balance for the year 2002 (Table. 2) was positive in March, April and June. According to the measured water tensions (matric potential) no downward gradient for soil water could be observed and thus no leaching of NO_3^- occurred in 2002 and 2001. One reason for the different outcome of water balance and the monitoring of the soil water potential seems to be based on an underestimation of the actual evapotranspiration by the *Penman-Monteith* equation for pannonic climate conditions. During the winter season 2002/03, soil water samples containing NO_3^- concentrations ranging from 10 up to 50 mg NO_3^--N could be collected in 120cm soil depth. Results of the water balance for 2003 are not yet available. An influence of mulching on soil water content has not been observed so far. The high N content after preceding legume crops did not affect grain yield, but resulted in increased protein contents in the winter wheat crop in 2002. In

Table 1. Soil inorganic N (N_{in}) contents after legumes and reference crops. Values in one column with same letter are not significantly different (P<0,05).

Variants	October 2001 kg N ha⁻¹ (0-120cm)	March 2002 kg N ha⁻¹ (0-120cm)	July 2002 kg N ha⁻¹ (0-120cm)
Alfalfa (mulch)	114 b	134 b	33 ab
Alfalfa (cutting)	83 b	97 ab	23 ab
Alfalfa-grass (mulch)	81 b	77 ab	37 b
Alfalfa-grass (cutting)	10 b	75 ab	20 ab
Reference crop 1 (grass)	17 a	54 a	7 a
Reference crop 2 (rye)	57 a	79 ab	12 ab

Table 2. Soil water balance during 2002.

Variants	Jan (mm)	Feb (mm)	Mar (mm)	Apr (mm)	May (mm)	Jun (mm)
Alfalfa (mulch)	-12	-13	16	11	-3	15
Alfalfa (cutting)	-6	-4	18	13	-3	13
Alfalfa-grass (mulch)	5	-9	25	9	-1	14
Alfalfa-grass (cutting)	-1	-11	16	22	8	15
Reference crop 2 (rye)	3	-15	25	17	10	27

the dry year 2001, no increased protein contents were observed. It is likely that drought periods during the vegetation period reduced the effects of legume N on the succeeding cereal crop. It can be concluded, that under such conditions the cultivation of forage legumes and particularly mulching of legume crops are an appropriate way to raise grain quality.

References

Allen, R. (1998). Crop evapotranspiration - Guidelines for computing crop water requirements. FAO Irrigation and drainage paper 56. FAO - Food and Agriculture Organization of the United Nations. Rome, 1998.

Hauer, G., Kammerer, G., Sokol, W., Loiskandl, W., and Kastanek, F. (2003). Ermittlung der Bodenwasserbilanz mit einem virtuellen Lysimeter. Wien: Österreichische Wasser- und Abfallwirtschaft, 55, 104-112.

Your farm and NVZs: A decision support system for farmers and consultants

C.P. Fawcett[1], M.M. Gibbons[1], F.A. Brown[1], P.M.R. Dampney[2] and S.J. Richardson[2]
[1]ADAS, Woodthorne, Wergs Road, Wolverhampton, WV6 8TQ, UK
[2]ADAS, Boxworth, Cambridge, CB3 8NN, UK

Introduction

Agricultural land is the main source of NO_3^- in most rivers and groundwaters. The EC Nitrates Directive (EU, 1991), an environmental measure adopted in December 1991, requires Member States to reduce NO_3^- pollution by introducing controls on agriculture in water catchments where the NO_3^- concentration in the water exceeds a limit of 50 mg l^{-1} or is at risk of doing so. To comply with the Directive the UK government designated an additional 47% of agricultural land area in England as Nitrate Vulnerable Zones; 8% was originally designated in 1996. An NVZ is an area of land in which surface waters or ground waters are identified as NO_3^--polluted or could become NO_3^--polluted if preventative action is not taken. Individual farms within NVZs are legally obliged to comply with the NVZ Action Programme rules which came into force in the new NVZs on 19 December 2002 (DEFRA, 2002). These rules introduce limits to the loading of organic manure N that may be applied each year across all of the farmed land, and controls on the use of artificial N fertiliser and organic manures to individual fields. In addition, records must be kept for each field for cropping and for the application of N fertilisers and organic manures. A computer based decision support system was developed to help farmers assess the NVZ compliance of their own businesses and for keeping the necessary farm and field records.

'Your farm and NVZs'

'Your Farm and NVZs' is a Windows®-based Decision Support System that provides a straightforward means for farm businesses and their advisors to assess whether they comply with the Action Programme rules. In addition to assessing the current farm circumstances, the system may be used to check the effect of planned changes in livestock numbers on the farm or the area of farmed land. To help maximise its use by farmers, the DSS has been developed with ease of use in mind and it is suitable even for users with only a basic level of computer literacy. It can be used in either metric or imperial units. The software is divided into five modules, each addressing a different aspect of the rules.

Farm details - this is the entry module of the system where the user enters simple farm based information which can be used to identify which other modules need to be completed for the farm. Field details may also be entered if the user wishes to use the record keeping facility.

Organic Manures: N Loading - this module assesses whether the farm is within its permitted annual manure total N capacity. The system asks for details on the area of farmed land and the number of livestock on the farm. Where relevant, details of imported or exported manures can also be entered. Using this information, the system calculates the farm's manure N capacity (based on the area of land) and its manure N loading. A compliance statement is produced which identifies if the farm meets the requirements of the NVZ Action Programme rules. The use of manufactured fertilisers containing N is not considered in these N loading calculations.

Field-Based Nitrogen - the NVZ Action Programme rules for the use of fertiliser nitrogen and organic manures are summarised in this module and the user is asked if they comply with these rules. Although the rules require an assessment of the amount of fertiliser nitrogen required on each field each year, the system does not provide specific recommendations.

Farm Manures: Slurry Storage - The NVZ Action Programme rules restrict the spreading of any animal slurries, poultry manures or liquid digested sludge during the autumn closed spreading periods, but only on sandy or shallow soils. There is no closed spreading period for farmyard manures on any soil type. This module calculates the production of slurry and dirty water and assesses if the existing slurry storage capacity is sufficient for annual closed spreading periods for manures.

Records - this module provides a record keeping facility for the cropping, fertiliser N and organic manure applications to individual fields. Records of any imported and exported manures may also be kept. A report can be generated that is acceptable to the Environment Agency, the official body responsible for enforcing the NVZ Action Programme rules in England.

Compliance Assessments
Where the compliance assessments made in the software show that a farm does not comply with the NVZ Action Programme rules, or is borderline, then further advice will be needed to find the best way to achieve compliance. The software does not provide such advice, as this will be specific to individual farms, but some guidance is given and sources of further information can be found from the Help menu.

Documentation Library and Further Help
The software contains a library of relevant technical documentation providing background and further information to NVZs and manure management. The documents can be viewed in an integral Adobe® PDF (Portable Document Format) viewer such that they can be referred to alongside the software when it is in use. In addition to this library of technical documentation, on-screen footnotes within each module provide guidance for the user and the software is also accompanied by an HTML (Hyper Text Markup Language) based Help system.

Conclusion

'Your Farm and NVZs' provides a simple way for farmers to check compliance with the NVZ Action Programme rules. It is available free of charge either as a CD-ROM (telephone 08456 023864) or by downloading from the DEFRA nitrate webpage (www.defra.gov.uk/environment/water/quality/nitrate/). The software was released in September 2002 and over 4000 copies were distributed to the agricultural industry in the first four months from release.

Acknowledgements

Funding for this work from DEFRA, London is gratefully acknowledged.

References

E.U. (1991) - Council Directive 91/676/EEC(**c**) of 12 December 1991 concerning the protection of waters against pollution caused by nitrates from agricultural sources.

DEFRA (2002). Guidelines for Farmers in NVZs - England. DEFRA Publications, PB5505

Fate of applied nitrogen in a heathland ecosystem

E.R. Green[1], S.A. Power[1] and E.M. Baggs[2]

[1]*Department of Environment Science & Technology, Imperial College London, Silwood Park, Ascot, Berks., SL5 7PY, UK*
[2]*Department of Agricultural Sciences, Imperial College London, Wye Campus, Wye, Ashford, Kent, TN25 5AH, UK*

Introduction

Nitrogen cycling is a fundamental ecological process and is the focus of much scientific attention because of concerns about the impact of excess N on ecosystems. Deposition inputs of N to ecosystems are elevated in northern Europe. The impact of this excess N on ecosystems is of great concern because there is strong evidence to show that previously nutrient poor, semi-natural ecosystems are becoming less limited by N. This increased availability can have compounding effects on sensitive communities such as lowland heaths, as many of the species adapted to live in these systems are unable to compete in high nutrient environments. Lowland Heathland is one of the most threatened habitats in Europe, and an involvement of N deposition is suspected. At present there is much uncertainty concerning how this additional N is processed within the plant-soil system. However, the fate of added N can be determined using stable isotope tracers, such as ^{15}N. Previous studies employing these tracers have provided detailed measurements of short-term N flux through environmental systems; this is readily achievable through simple quantification of the tracer as it passes into distinct partitions of the ecosystem.

The aims of this pilot study were to determine the fate of deposited atmospheric N using stable isotope techniques, quantifying how much of the added N is immobilised in the microbial pool, taken up by plants, leached or lost as N_2O. This is being achieved by applying ^{15}N-labelled ammonium sulphate to a long-term N manipulation study at Thursley Common, a lowland heathland in Surrey, UK.

Materials and methods

The experiment was carried out over a two week period in October 2002. The enriched $(^{15}NH_4)_2SO_4$ (60 atom % excess ^{15}N) was applied at 30 kg ha^{-1} yr^{-1} Gas samples for N_2O analysis were taken from closed flux chambers with gas tight syringes, and were stored in evacuated gas vials (Baggs et al., 2003). Soil, leachate and vegetation samples were also taken during this study. Gas samples were analysed for $^{14+15}$N-N_2O, CO_2 and CH_4 in an Agilent 6890 gas chromatograph. The ^{15}N enrichment of N_2O was determined using a Europa 20/20 isotope ratio mass spectrometer following condensation/cryofocusing of the samples in an ANCA TGII gas module. Soil was analysed for inorganic N, microbial biomass N and gravimetric water content. Leachate samples were collected using porous cup tension lysimeters, stored at 5 °C and then NH_4^+-N and NO_3^--N concentrations were determined colorimetrically by continuous flow analysis on a Burkard SFA2 autoanalyser. ^{15}N enrichment of heather (*Calluna vulgaris*. L) was also analysed using a Europa 20/20 isotope ratio mass spectrometer.

Results

Emissions of N_2O were generally low but increased slightly on the first 2 days following addition of [15]N. Total N_2O emissions over 14 days were greater from the control plots. ^{15}N-N_2O emissions were low, with >1% of ^{15}N applied lost as ^{15}N-N_2O. Fluxes of ^{15}N-N_2O did not significantly increase until 5 days after N addition (Figure 1). There was no significant effect of N application on soil NO_3^- concentrations; although the presence of NO_3^- does suggest that nitrifying bacteria were active. Negligible [15]NO_3^- or [15]NH_4^+ was measured in leachate samples. Microbial biomass ^{14}N increased after N addition and then declined rapidly (Figure 2). The decrease in microbial biomass ^{14}N corresponded to a small increase in soil ^{14}N-NH_4^+ concentration.

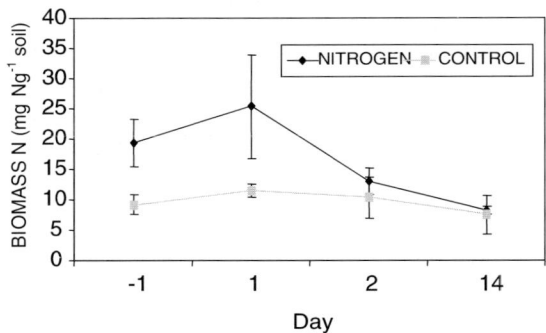

Figure 1. [15]N_2O fluxes from + N treatment. Error Bars represent ± 1 SEM.

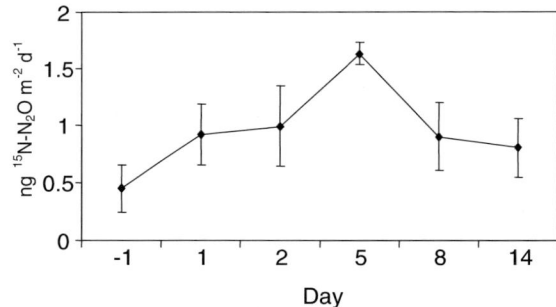

Figure 2. Microbial biomass N g^{-1} soil. Error Bars represent ± 1 SEM.

Conclusions

The following preliminary conclusions can be made:
- Low emissions may have resulted from low soil water contents in the N limiting system.
- Microbial biomass ^{14}N data suggest that microbial populations were quick to react to N additions.
- Nitrification occurred in this acid soil, but at a slow rate.

Results from this study will contribute to future policy decisions on critical loads for N deposition. Future work, which is currently underway, will compare varying management techniques and study N dynamics over a longer (60 day) time scale.

References

Baggs, E. M., Richter, M., Cadisch, G. and Hartwig, U.A. (2003). Denitrification in grass swards is increased under elevated atmospheric CO_2. Soil Biology and Biochemistry, 35, 729-732.

Effects of intensification of dairy farming in New Zealand on nitrogen efficiency, energy use and environmental emissions

S.F. Ledgard[1], J.D. Finlayson[1], M.S. Sprosen[1], D.M. Wheeler[1] and N.A. Jollands[2]
[1]*AgResearch Ruakura Research Centre, Private Bag 3123, Hamilton, New Zealand*
[2]*AgResearch Grasslands, Private Bag 11008, Palmerston North, New Zealand*

Introduction

In the 1970s, New Zealand (NZ) dairy farms were predominantly low cost systems based on clover N_2 fixation, no N fertiliser use, and year-round grazing of permanent ryegrass/white clover pastures. Since then, NZ dairy farms have intensified by increasing their inputs of fertiliser N and integrating maize silage into their farm systems. This trend has become more rapid in the last five years, with farm systems moving closer to EU dairy systems. Sheep and beef farms have also become more closely integrated with dairy farms by providing grazing and growing maize silage under contract or for sale. This paper considers the impacts of these intensification practices on resource use efficiency and environmental emissions.

Methods

To perform this study, data were obtained from a variety of sources. Farmlets (small farms of 20 cows) were used to consider detailed effects of intensification on production, economics, resource use, N inputs and losses. More extensive data were obtained for average and intensive dairy farms and maize cropping using dairy industry statistics, farmer and crop-contractor interviews, and N losses were estimated using the OVERSEER® nutrient budget model (Ledgard et al., 2003). Emissions of N_2O and total greenhouse gas emissions were calculated using IPCC methodology modified for NZ conditions. This information was used to evaluate changes to sheep and beef farms as well as dairy farms. It was also used to assess the effect of increasing farming intensity on the land and resources used to grow maize silage and the contribution of N fertiliser to total energy use.

Evaluation of the total resource use and emissions for the dairy industry also requires inclusion of transport, milk processing and indirect contributors. Life Cycle Assessment (LCA) provides a systematic approach to do this (e.g. Vigon et al., 1994). An LCA study of the Waikato dairy industry, from farms through to processed dairy products, was used to identify potential inefficiencies and define where the major contributors to resource use and environmental emissions occur within the industry.

Results and discussion

Data from the farmlet study were used to calculate efficiency indices. For example, the $NZ profit per kg N leached for farmlets receiving urea at 0, 200 or 400 kg N ha^{-1} yr^{-1} was 65, 35 and 18, respectively.

Urea manufacturing also requires a high energy input and at the current average use of ca. 100 kg N ha^{-1} yr^{-1}, the energy component of urea on farms now equates to almost twice the total energy component of fuel and electricity use on farms. At this level of use, N fertiliser represents over 10% of the total embodied energy use by the dairy industry (including factory processing and indirect contributors).

Calculations based on IPCC methodology indicate that N_2O emissions from dairy farms represent approximately 32% of total greenhouse gas emissions from farms and factories. The N_2O emissions from farms were twice those of energy-based emissions from the dairy factories and two-thirds of farm animal CH_4 emissions.

Two intensification options, using either N fertiliser (+200 kg N ha^{-1} yr^{-1}) or forage (+2 t DM ha^{-1} yr^{-1} equivalent on the dairy farm, as maize+oats silage), were compared for a 20% increase in on-farm milk production. Farm N leaching per m^3 milk was estimated to increase by about 70% for the +200N system. In contrast, the forage system was estimated to decrease N leaching per m^3 milk by 10% on the dairy farm, but this efficiency gain was reduced when total land use was considered (i.e. including land used to produce the forage crop). Greenhouse gas emissions per m^3 milk were similar for the base farm and +forage system, but increased by about 15% for the +200N system. The latter was mainly due to increased N_2O emissions. Thus, the choice of intensification method influences the potential for gain in dairy farm system efficiency.

A comparison was made of an average NZ dairy farm and a Swedish dairy farm (Cederberg, 1998) for resource use efficiency and environmental emissions. Both farms were estimated to give similar GHG emissions per m^3 milk. While the energy-related CO_2 emissions were greater for the Swedish farm, this was countered by lower CH_4 emissions per unit of milk from high-producing Swedish cows. Total energy use per unit of milk production on the Swedish farm was over 5-fold higher than that of the Waikato farm on a whole-system basis. This was mainly due to high fuel use in the Swedish farm system for crop production, feeding and heating the farm dairy. The NZ farm system with all-year-round grazing of long-term permanent legume-based pastures is a low energy requiring system, but this advantage may diminish with intensification. Leaching of N from grazed pasture was calculated to be 35% lower on the Swedish farm because of fewer cows per ha, a lower dietary N concentration and removal of cows from pasture during winter when leaching is greatest. However, grazed pasture constituted only 5% of the total Swedish farm area and when the whole system was accounted for, including N leaching from the land used for growing forage crops and concentrates, total N leaching per m^3 milk was approximately 25% higher for the Swedish farming system. This highlights the importance of evaluating the whole farm system and not just the dairy farm only.

References

Cederberg, C. (1998). Life cycle assessment of milk production - a comparison of conventional and organic farming. The Swedish Institute for Food and Biotechnology. SIK, Gothenburg, Sweden.

Ledgard, S.F., Wheeler, D.M., de Klein, C.A.M., Monaghan, R.M. and Johns, K. (2003). Development of the OVERSEER® nutrient budget model to examine implications of pastoral management practices on nitrogen flows and losses (this volume).

Vigon, B.W., Tolle, D.A., Cornaby, B.W., Harrison, C.L., Boguski, T.L. and Hunt, R.G. (1994). Lifecycle assessment: Inventory guidelines and principles. Lewis Publishers. CRC Press, Florida.

Effect of groundnut stover and weed management methods in the dry season on N cycling and maize yield

S. Promsakha na sakonnakhon[1], B. Toomsan[1], V. Limpinuntana[1], P. Vityakon[1], E.M. Baggs[2] and G. Cadisch[2]
[1]*Faculty of Agriculture, Khon Kaen University, Khon Kaen, Thailand*
[2]*Department of Agricultural Sciences, Imperial College London, Wye Campus, Wye, Ashford, Kent, UK*

Introduction

It is generally thought that grain legume residues make a substantial net N contribution to soil fertility in temperate as well as tropical crop rotation systems. Groundnut (*Arachis hypogaea*) is particularly interesting since it produces large quantities of high quality stover. To maximise the use of N_2 fixed by groundnut, we must understand how efficiently N is recycled under different managements. The main objective of this study was to improve the efficiency of groundnut stover recycling in a groundnut-maize rotation by testing the impact of different stover management methods and altering weed composition during the dry season.

Materials and methods

A groundnut crop was planted in the 2001 rainy season (June-October) on a sandy soil in Northeast Thailand. After groundnut harvesting, different groundnut stover and weed managements were established during the dry season (November-May). Assessments were made by evaluating the effect of: i) dry season stover management (surface application, incorporation or storage until beginning of the next rainy cropping season), combined with ii) weed management (no weeds, only grass weeds, only legume/broadleaf weeds, mixed weeds). Soil mineral N was monitored during both dry and wet seasons. In the 2002 rainy season, maize was planted and yield determined.

Results

Recycling groundnut stover (5 t DM ha^{-1}; 100 kg N ha^{-1}) improved the yield of the subsequent maize crop compared with the treatment without stover. Stover removed and applied before maize planting resulted in higher yields than the other stover management treatments (Table 1). Weeds grown during the dry season and incorporated during maize seedbed preparation, suppressed yield, particularly where they were mainly grasses. However, leguminous and broadleaf weeds improved maize yield, compared with the no weed treatment.

Application of groundnut stover (particularly where removed and applied before maize planting) increased mineral N availability, compared with the stover surface application treatment (Figure 1). During the dry season, mineral N in weed treatments was lower than in the no weed treatments, as weeds took up 50-70 kg N during this period.

Table 1. *Maize dry weight (kg ha⁻¹) and harvest index (HI) at final harvest (first crop) under different stover (S.A = surface application) and weed management.*

Treatments	Dry matter (kg ha^{-1})				
	Stover	Seed	Cob	Total DW	HI
S.A + No weeds	5353	2323	492	8168	0.27
S.A + Grass weeds	3750	1705	360	5814	0.30
S.A + Legume/broadleaf weeds	6630	3731	529	10889	0.34
Removed and applied before crop	7369	3106	618	11093	0.28
SED	957*	482**	99^{ns}	1398**	0.02^{ns}
C.V (%)	23.20	24.05	26.62	21.34	11.64

SED = Standard error of the differences between means. *,**,^{ns} = F-test significance.

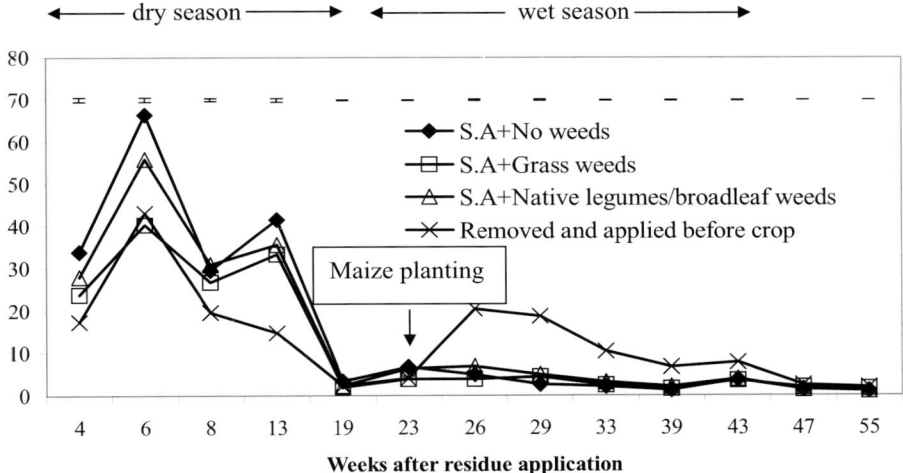

Figure 1. *Effect of different groundnut stover (S.A = surface application) and weed management option on mineral N at 0-15 cm soil depth.*

Discussion

Weeds grown during the dry season effectively captured excess mineral N and reduced N leaching. However, maize yield was often suppressed where weeds were incorporated. This was because of microbial immobilisation of soil mineral N in the presence of the low quality (high C:N) grass weeds. However, weeds predominately consisting of native legumes and broadleaf species were beneficial for the subsequent maize yield crop. Our study shows that both stover and weed management during the dry season are crucial to optimise benefits of recycled groundnut stover and minimise N losses. This necessitates a better understanding of weed ecology, i.e. what factors determine weed composition. Investigations are underway to establish long-term effects (i.e. with the second maize crop) and total N losses (^{15}N balances).

Conclusions

Groundnut stover removed and applied before maize planting resulted in higher maize yields than the other stover management treatments. However, weed management during the dry season is crucial if the benefits of recycling groundnut stover are to be obtained by farmers.

Acknowledgements

This project was funded by the Royal Golden Jubilee (RGJ) PhD Programme under the Thailand Research Fund (TRF).

Development of a decision support tool for global diffuse nitrogen pollution assessment

C. Macleod[1], J. Dela-Cruz[2], P. Haygarth[1], G. Glegg[2], D. Scholefield[1] and L. Mee[2]
[1]*Soil Science and Environmental Quality Team, Institute of Grassland and Environmental Research, North Wyke Research Station, Okehampton, Devon, EX20 2SB, UK*
[2]*Institute of Marine Studies, University of Plymouth, Drake Circus, Plymouth, Devon, PL4 8AA, UK*

Introduction

This paper sets out our approach to the development of a loading model as part of a decision support tool (DST) to aid the assessment and management of diffuse N losses from land-based activities to surface waters at the global scale. Human activities have significantly changed the natural flux of N from land to coastal waters around the world. For example, the release of reactive N from land via rivers to oceans has doubled from 35 to 76 Tg N yr^{-1} since preindustrial times (Galloway et al., 1995). A growing number of global estimates of N loss from land to surface waters have been produced recently because of the pivotal role of N in climate change and eutrophication of coastal waters (Meybeck, 1982; Caraco & Cole, 1999; Smith et al., 2003).

Modelling approach

Step 1: Review previous systems approaches and establish model criteria. Our approach to the development of the DST has been determined by the requirement for the model to function at the chosen scale (global) (Addiscott, 1993), experiences gained from previous regional and global estimates of N sources, N movement through the landscape and the availability of world-wide spatial loading, transport and monitoring data sets.

Step 2: Dataset collection. In order to produce a model, there was a need to search for and collect the data on which the model is to be based and the data that it will be validated against. We have focussed our efforts on collecting spatial datasets with a global coverage at a 0.5° latitude and longitude resolution for the conversion of our qualitative conceptual model into a parameterised grid-based model. Table 1 contains examples of loading, transport and monitoring datasets chosen for the global scale approach.

Step 3: Establishment of monitoring point catchments at 0.5° resolution. The latitude and longitude of the LOICZ (dissolved inorganic N (DIN)) and GLORI (total suspended solid (TSS)) monitoring sites were used to generate map layers. These data layers were compared against the 0.5° catchment boundary data to establish catchments for each monitoring point. Reference

Table 1. Examples of loading, transport and monitoring data sets.

Dataset type	Variable / dataset name	Temporal period	Reference
Loading	N fertiliser use	1995	(Drecht et al., 2001)
Transport	Global simulated topological network at 30-minute spatial resolution	Early 1990s	(Vorosmarty et al., 2000)
Monitor	LOICZ river DIN	1980-2000	(Smith et al., 2003)

against catchment statistics along with ground truthing with an independent data source (John Bartholomew, 1980) were used to reduce errors.

Step 4: Global bio-climatic zones. To enable the production of accurate estimates of global fluxes of N, there was a need to delineate and classify ecologically homogenous areas (Meybeck, 1982; Ludwig et al., 1996). We have reviewed the available ecological and climatic global classification systems and have used the Holdridge Life Zone system to subdivide the monitoring point catchments. This system is an empirical and objective classification scheme for ecosystem functioning delineated by biotemperature, precipitation, potential evapotranspiration ratio and elevation (Holdridge, 1947).

Step 5: Establishment of empirical relationships for each bio-climatic zone. The catchment boundary data (Step 3) were used to extract data from the loading and transport data layers (Table 1). These data were collected in a database for model development and testing. To establish empirical models for DIN and particulate organic N, data on the main source and transport variables were regressed against the global monitoring data sets for dissolved N (Smith et al., 2003) and TSS datasets, respectively. These relationships will then be used to establish global estimates of N loss. This type of approach has been used to generate global estimates of organic C (Ludwig et al., 1996).

Future steps

The N loading model will be tested against available datasets. Sensitivity analysis will be combined with finer resolution modelling on selected catchments to gain greater confidence in the model. The N loading model will be combined with loading models for a range of diffuse pollutants in a DST designed for use around the world.

References

Addiscott, T.M. (1993) Simulation Modeling and Soil Behaviour. Geoderma, 60, 15-40.

Caraco, N.F. and Cole, J.J. (1999) Human impact on nitrate export: An analysis using major world rivers. Ambio, 28, 167-170.

Drecht, G.V., Bouwman, A.F., Knoop, J.M., Meinardi, C., and Beusen, A. (2001). Global pollution of surface waters from point and nonpoint sources of nitrogen. In (eds) J.M. Galloway, E. Cowling, J.W. Erisman, J. Wisniewski and C. Jordan. Optimizing nitrogen management in food and energy production and environmental protection. A A Balkema Publishers, Lisse.

Holdridge, L. (1947) Determination of world plant formations from simple climatic data. Science, 105, 367-368.

John Bartholomew, S.L. (1980) 'The Times' atlas of the world. Times Books Limited, Edinburgh.

Ludwig, W., Probst, J.L., and Kempe, S. (1996) Predicting the oceanic input of organic carbon by continental erosion. Global Biogeochemical Cycles, 10, 23-41.

Meybeck, M. (1982) Carbon, Nitrogen, and Phosphorus Transport by World Rivers. American Journal of Science, 282, 401-450.

Smith, S.V., Swaney, D.P., Talaue-McManus, L., Bartley, J.D., Sandhei, P.T., McLaughlin, C.J., Dupra, V.C., Crossland, C.J., Buddemeier, R.W., Maxwell, B.A., and Wulff, F. (2003) Humans, hydrology, and the distribution of inorganic nutrient loading to the ocean. Bioscience, 53, 235-245.

Vorosmarty, C.J., Fekete, B.M., Meybeck, M., and Lammers, R.B. (2000) Global system of rivers: Its role in organizing continental land mass and defining land-to-ocean linkages. Global Biogeochemical Cycles, 14, 599-621.

Organic fertilisation in silvopastoralism and nitrate losses: sustainable systems

M.R. Mosquera-Losada, S. Rodríguez-Barreira and A. Rigueiro-Rodríguez
Crop Production Department, Escuela Politécnica Superior, Universidad de Santiago de Compostela, Lugo, Spain

Introduction

The use of organic fertilisers, such as sewage sludge, has been encouraged by EU policy in order to provide sustainable management of this residue. The main problem is the heavy metal content and the variability of the N fraction content. Heavy metals are even more important in acid soils because their mobility is increased. Moreover, soil N incorporation, as well as organic matter mineralisation, is limited by low pH and it causes accumulation on the soil surface. Application of mineral N will reduce the C:N relationship and aid the incorporation of this residue.

Materials and methods

The experiment started in 1998 in an acid soil in Galicia with the establishment of sward (25 kg ha^{-1} *Dactylis glomerata* and 4 kg ha^{-1} *Trifolium repens*) under eight year old *Pinus radiata* plantation with a randomised block of 10 treatments and three replicates. Five treatments were limed before spreading fertilisers and five were not. Fertiliser treatments consisted of three sewage sludge doses to give 160, 320 and 480 total N ha^{-1}, with and without the mineral fertiliser regime usually employed in the area. Fertilisers were spread during three consecutive years, and in the fourth year (2002) all plots received 25 kg N ha^{-1} as a 8:24:16 compound. Soil samples were taken in September 2002 at 4.5 and 25 cm depth, tree height and diameter were measured on December 2002 and pasture production (0.3 m side) was sampled in July 2002 (three harvests were made each year except the first year when only one was made). Organic matter, pH (water) and NO$_3^-$ (KCl) was analysed.

Results and discussion

Annual pasture production ranged between 2.9 and 5.6 t ha^{-1}, i.e. the usual range for forest land in Galicia (Rigueiro et al., 2001). There was a positive effect of mineral N fertiliser, previous liming and previous sewage sludge applications on pasture production (Figure 1), being higher with medium sewage sludge applications, liming and mineral N applied in 2001. Tree height and diameter also benefited from sewage sludge application; as found in other studies (Rigueiro et al., 2000 and 2001) (Figure 2). pH was between 4 and 5.5, the lower values associated with inorganic N fertilisation made in the previous year. Sewage sludge added 384, 768 and 1152 kg Ca^{2+} ha^{-1} in the previous years, for the low, medium and high ratio of sewage sludge. Organic matter content was between 11 and 18%, resulting from the low pH (limiting the microorganism growth) and from the important pinewood fall. Mineral fertilisation reduced organic matter content of the soil significantly and reduced C:N ratio, during the experiment. Nitrate content was between 4 and 9 mg NO$_3^-$-N kg^{-1} dry soil, i.e in accordance with those found for long-term grass (Whitehead, 1995).

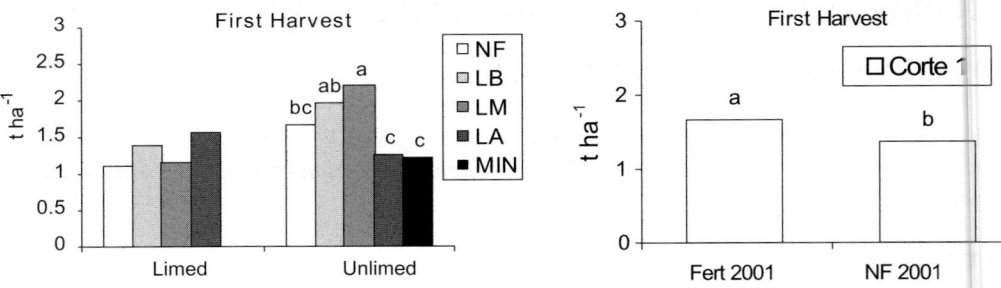

Figure 1. *Pasture production per treatment in the first harvest. Different letters indicate significant differences between sludge treatments. LA: high dose; LM:medium dose; LB:low dose; Min: Mineral treatment and NF: no fertilisation.*

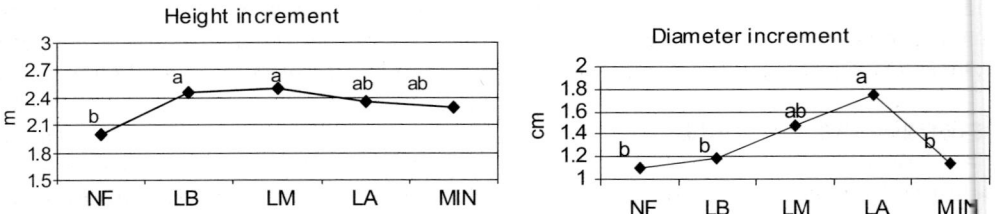

Figure 2. *Tree diameter and height increment per treatment in the first harvest. Different letters indicate significant differences between sludge treatments. LA: high dose; LM:medium dose; LB:low dose; Min: Mineral treatment and NF: no fertilisation.*

Pasture production and tree height and diameter were strongly related to pH, organic matter and both surface and to a lesser extent, depth NO_3^--N. *Pinus radiata* growth was related more to NO_3^--N at depth because of deeper roots. This supports the idea of sustainability of silvopastoral systems, in comparison with agronomic systems, as root trees can develop a layer that reduces NO_3^- leaching and allows tree growth and therefore a more profitable system.

Acknowledgments

We would like to thank to Gestagua, S.A., Agroambprodalt and J.Javier Santiago Freijanes, M.T. Piñeiro-López , Divina Vázquez-Varela and José Alberto Lamas-Díaz for their aid in laboratory and the field.

References

Rigueiro, A., Mosquera, M.R. and Gatica, E. (2000). Pasture production and tree growth in a young pine plantation fertilized with inorganic fertilisers and milk sewage in northwestern Spain. Agroforestry Systems, 48, 245-254.

Rigueiro, A., Mosquera, M.R. and López, M.L. (2001). Crecimiento del arbolado y producción de pasto en un sistema silvopastoral fertilizado con lodos de depuradora urbana en una zona agrícola abandonada en Galicia. Actas del III Congreso Forestal Español, 5, 703-709.

Whitehead, D. C. (1995). Grassland nitrogen. CAB International. Guildford. 397 pp.

How nitrate vulnerable zones can reduce the adverse environmental impact of productive arable agriculture

J.U. Smith[1], P. Smith[1], A.G. Dailey[2] and M.J. Glendining[2]
[1]School of Biological Science (Plant & Soil Science), Aberdeen University, Cruickshank Building, St. Machar Drive, Aberdeen, AB24 3UU, UK
[2]Rothamsted Research, Harpenden, Herts, AL5 2JQ, UK

Introduction

The Nitrates Directive (91/676/EEC) requires that Nitrate Vulnerable Zones (NVZs) be established in catchments where NO_3^- from agricultural land is causing pollution of water sources. Action Programme measures are required to be implemented in these zones to reduce NO_3^- pollution. The establishment of NVZs has great potential to reduce the adverse environmental impacts of agriculture, potentially reducing other forms of N pollution, such as N_2O emissions, as well as NO_3^- leaching. In December 2002, 55% of the land area in England was designated as NVZ: in Scotland it was 14%. This represents a large proportion of the arable land in the UK, where the Action Programme restricts application of inorganic fertilisers and storage and spreading of manures and slurries. NVZs also extend over a large proportion of the arable land in Europe. Rules relating to inorganic fertilisers, amount to little more than Good Agricultural Practice: rules relating to manures and slurries are more stringent. Therefore, it is widely accepted that both the benefits to the environment and costs to the farmer will be in the livestock sector. In many areas, arable agriculture covers the major proportion of the catchment, so it is important to make the Action Programme work in the arable as well as the livestock sector. Here we use a catchment version of the SUNDIAL model (Smith et al., 1996) to examine the potential of NVZs to reduce the adverse environmental impact of productive arable agriculture in Europe.

Methods

In predicting the impact of NVZs on productivity and environmental pollution from arable agriculture, a number of issues arise: the complexity of the model, the need to predict, the availability of data, and the uncertainty of the calculations. The SUNDIAL model has been used because it provides results of appropriate complexity, uses data at the catchment scale, and includes a procedure for calculating the uncertainty of simulations. Sources of uncertainty include inherent model error, data shortage due to gaps in the data and incomplete knowledge of spatial distribution of data at the catchment scale. Inherent model error can be quantified using standard model evaluation procedures (Smith et al., 1997). Gaps in the data can be filled by entering the data as a distribution. The spatial distribution of data can be simulated by entering all possible combinations within the catchment. These sources of uncertainty are then combined to determine the standard error associated with the simulated results. The impact of moving from currently observed agricultural practice to practices within Action Programme rules are simulated for catchments with a range of climates and soil types.

Results and discussion

The results indicate that even after NVZ restrictions on fertiliser use, spring barley and potatoes are associated with high risk of NO_3^- leaching (Figure 1a) and denitrification (Figure 1b). The average peak concentration from the whole catchment is greater than the EU limit unless cropping of potatoes and spring barley is restricted to less than 20% of the total cropping in warm/dry regions or 80% in other regions (Figure 2).

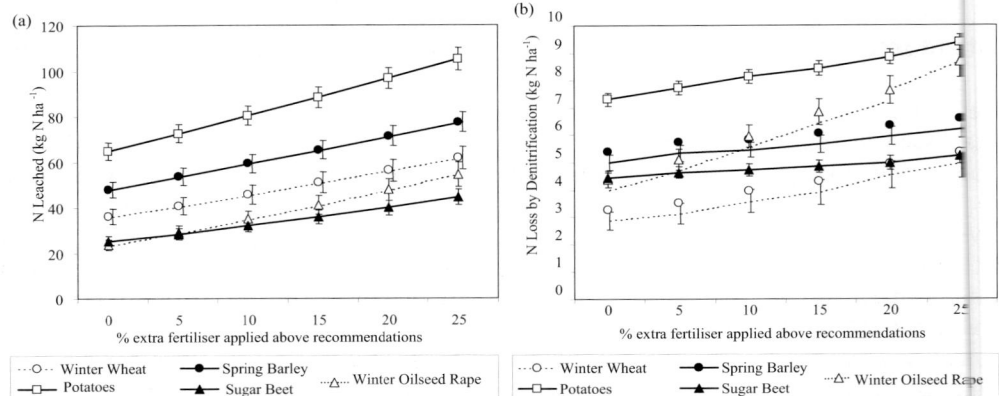

Figure 1. N loss with increasing rates of fertiliser application (a) Leaching (b) Denitrification.

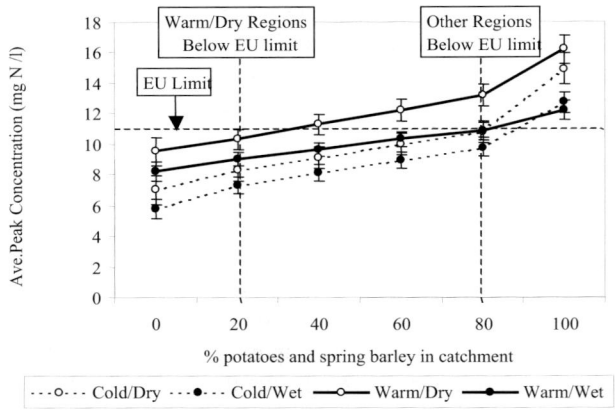

Figure 2. Average nitrate peak concentration with increasing potatoes and spring barley.

Conclusions

The results quantify potential environmental benefits from the introduction of the NVZ Action Programme in arable agriculture. If cropping of potatoes and spring barley is restricted to less than 20% of the total cropping in warm/dry regions or 80% in other regions, these benefits can be achieved while maintaining highly productive agriculture.

Acknowledgements

This work was funded by NERC, UK. Rothamsted Research receives grant-aided support from the Biotechnology and Biological Sciences Research Council of the United Kingdom

References

Smith, J.U., Bradbury, N.J. and Addiscott, T.M. (1996). SUNDIAL: A PC-based system for simulating nitrogen dynamics in arable land. Agronomy Journal, 88, 38-43.

Smith, P., Smith, J.U., Powlson, D.S., et al. (1997). A comparison of the performance of nine soil organic matter models using seven long-term experimental datasets. Geoderma, 81, 153-225.

Decline in species richness in acid grasslands along a gradient of nitrogen deposition

C.J. Stevens[1,2], N.B. Dise[1], D.J. Gowing[1] and J.O. Mountford[2]
[1]*Open University, Walton Hall, Milton Keynes, UK*
[2]*Centre for Ecology and Hydrology, Monks Wood, Huntingdon, Cambridgeshire, UK*

Introduction

Excess N deposition is a major pollutant in the UK and N now constitutes almost two thirds of the total acid deposition that the UK receives. Total annual deposition of inorganic N is 38×10^7 kg N, averaging approximately 17 kg N ha^{-1} yr^{-1} over the country as a whole (NEGTAP, 2001).

This study investigates the effects of N deposition on acid grasslands. Acid grasslands are an extremely important semi-natural habitat in the UK for agriculture, recreation and conservation. They cover approximately 1,166,000 ha, although in recent years there have been considerable losses because of agricultural intensification (Haines-Young et al., 2000). Field experiments clearly show that N addition has the potential to reduce species richness (e.g. Wilson & Tilman, 2002) and change soil processes (Morecroft et al., 1994) in semi-natural grasslands, but little evidence has been gathered to show that these changes are actually occurring in the UK as a result of atmospheric deposition.

Methods

Sixty randomly selected sites belonging to the acid grassland community U4 *Festuca ovina-Agrostis capillaris-Galium saxatile* (Rodwell, 1992) were surveyed along a gradient of anthropogenic atmospheric N deposition in the UK. Vegetation data were collected using five 2×2 m quadrats and topsoil and subsoil samples were collected. Total modelled N deposition (Smith et al., 2000) at the sites ranged from 6 to 37 kg N ha^{-1} yr^{-1}.

Results

Vegetation quadrat data showed a linear decline in species richness with increasing N deposition, indicating a loss of one species for every additional 3 kg N ha^{-1} yr^{-1}. Of 22 potentially important variables tested in a stepwise multiple regression, total N deposition explained most of the variability in species richness. Soil pH also showed a positive correlation. The loss in species richness is predominantly due to a reduction in the number of forb species with increasing levels of pollution. The cover of forbs also decreases.

Species that were adversely affected by N deposition were identified using Canonical Correspondence Analysis (CCA) (ter Braak, 2002). These include a moss (*Hylocomium splendens*), Ribwort Plantain (*Plantago lanceolata*) and Harebell (*Campanula rotundifolia*) (Figure 1). Mean monthly temperature was also important for species composition in the CCA.

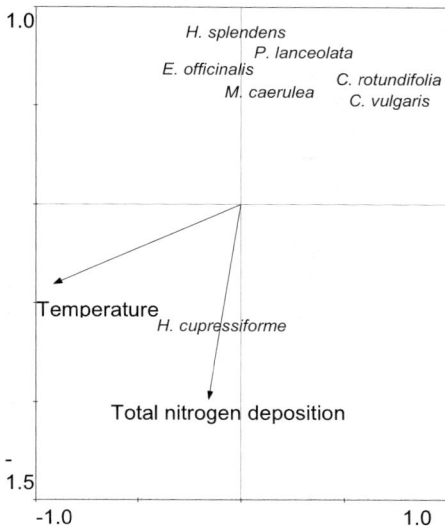

Figure 1. Canonical correspondence analysis ordination diagram showing species strongly negatively correlated with nitrogen deposition.

Conclusions

According to our study, adhering to the current critical load for N would result in the loss of 40% of the species richness of unpolluted acid grassland. We suggest 'target loads' as an alternative to critical loads. Target loads accept a degree of ecosystem change in line with the strong linear trend, however, an 'acceptable' level of damage must be decided upon.

Acknowledgements

This work has been funded by the Open University and is associated with the NERC thematic programme GANE. We are grateful to English Nature, Scottish Natural Heritage, Countryside Council for Wales and landowners who permitted access to sites.

References

Haines-Young, R. H. et al. (2000). Accounting for Nature: assessing habitats in the UK countryside. DETR, London.

Morecroft, M. D., Sellers, E. K. and Lee, J. A. (1994). An experimental investigation into the effects of atmospheric deposition on two semi-natural grasslands. Journal of Ecology, 82, 475-483.

NEGTAP (2001). Transboundary air pollution: Acidification, eutrophication and ground-level ozone in the UK. CEH, Edinburgh.

Rodwell, J. S. (1992) Grasslands and montane communities. University Press, Cambridge.

Smith, R.I., Fowler, D., Sutton, M.A., Flechard, C. and Coyle, M. (2000). Regional estimation of pollutant gas dry deposition in the UK: model description, sensitivity analyses and outputs. Atmospheric Environment, 34, 3757-3777.

ter Braak, C. F. J., and Smilauer, P. (2002). CANOCO 4.5. Biometris, Wageningen.

Wilson, S. and Tilman, D. (2002). Quadratic variation in old-field species along gradients of disturbance and nitrogen. Ecology, 83, 492-504.

Development of a modelling system for prediction and regulation of livestock waste pollution in the humid tropics

T.P. Tee, I.J. Lean, E.M. Baggs and G. Cadisch
Imperial College London, Wye Campus, Department of Agricultural Sciences, Wye, Ashford, Kent, TN25 5AH, UK

Introduction

In Malaysia, excessive nutrients from livestock waste management systems are currently released to the environment. Particularly, large amounts of manure from intensive pig production areas are being excreted daily and sedimented in ponds whilst their overflow contaminates rivers. Alternatively, the excess manure could be applied as an organic fertiliser source in neighbouring cropping systems (either on the small landholdings of the pig farms or in plantations nearby) to improve soil fertility, so that the nutrients will be available for crop uptake instead of being discharged into water streams. Thus, there is a need for better tools to analyse the present situation, to evaluate and monitor alternative livestock production systems and manure management scenarios, and to support farmers in the proper management of manure and fertiliser application. Although models of animal waste management systems exist, it is felt that more development is needed to adapt such models to the humid tropics and conditions of the pond system of Malaysia and other developing countries in the region.

Modelling approaches

The aim is to develop a novel model to evaluate nutrient emission scenarios and the impact of livestock waste at the landscape or regional level in the humid tropics. The study will link and improve existing models to evaluate emissions of N to the atmosphere and leaching of nutrients to groundwater and surface water. The simulation outputs of the model will be integrated with a GIS spatial analysis to assess the distribution of nutrient emissions, leaching and appropriate manure application on neighbouring crop lands, and as an information and decision support tool for the relevant users. The model concept and initial simulations are presented.

Model input

The Pig Production/Manure Model was divided into three-sub modules according to the different types of manure management systems, namely: fresh manure, storage and composting. Housing system, feed intake and quality, and classes of animals determine the manure N outputs. The different manure handling methods and storage types lead to different N loss processes and N loss determined in the sub-modules. In the slurry based housing system, a 25% NH_3-N loss from inorganic N was adopted according to Chambers et al., (2002). In the pond storage system (Figure 1) it was assumed that 20% of the organic N would be mineralised to NH_4^+ N within a year and that there was no N loss by denitrification because of prevailing aerobic conditions.

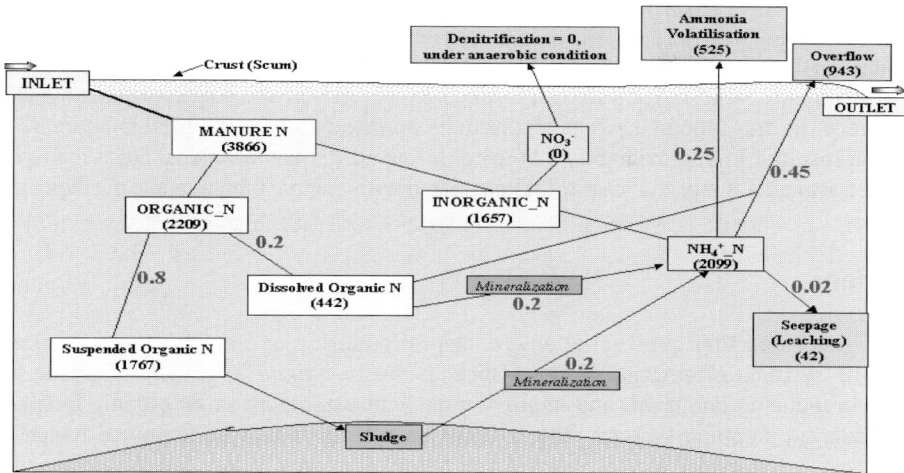

Figure 1. N pathways (kg N yr⁻¹) and fluxes (proportional) in an anaerobic lagoon system supplied with pig slurry from 30-sow unit.

Average losses of mineral N of 25%, 45% and 2% for NH_3-N, overflow and seepage from the pond system were estimated according to data from Dewes et al., (1990) and Ham & Desutter (1992). In the composting system, corresponding losses of 20%, 49% and 2% for leaching, NH_3-N and denitrification losses were adopted according to the results from Martins & Dewes (1992).

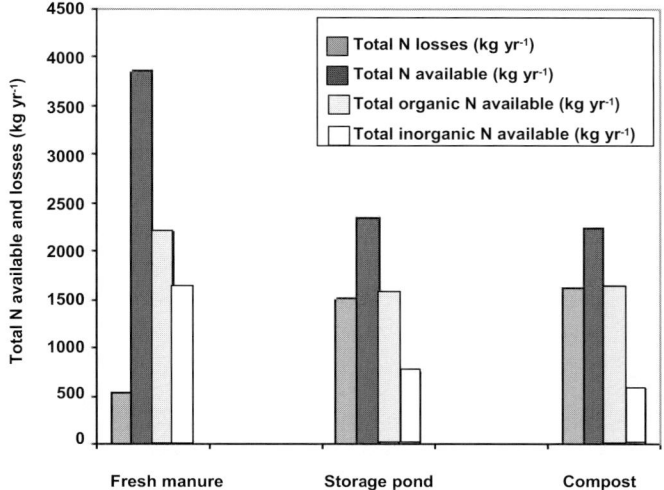

Figure 2. Simulation results of total N available and total N losses in each SubModule.

Model simulation output

On average, a 30-sow pig farm produces approximately 4t of N per year in pig manure of which 1t of N is lost by NH_3 volatilisation in the housing system. Simulations suggest that conventional waste storage in the lagoon system resulted in approximately 40% total manure N loss. Composting resulted in approximately 42% total N loss to the environment. Fresh manure had the greatest available N and the least N loss compared with compost manure and storage manure (Figure 2).

Conclusions

The model suggested that substantial environmental pollution occurs with the current waste management system. Alternative options such as fresh manure application appear to be promising in reducing pollution, and improving nutrient utilisation as an organic fertiliser in the neighbouring cropping systems. The composting system should be improved to reduce N losses. However, the model has to be validated before firm conclusions can be reached

References

Chambers, B., Williams, J. and Philips, R. (2002). Ammonia emissions from pig farming. In: Ammonia in the UK. DEFRA Publication, UK.

Dewes, T., Schmitt, L., Valentin, U. and Ahrens, E. (1990). Nitrogen losses during the storage of liquid livestock manures. Biological Wastes, 31, 241-250.

Ham, J.M. and DeSutter, T. M. (1999). Seepage losses and nitrogen export from swine waste lagoons: A water balance study. Journal of Environmental Quality, 28, 1090-1099.

Martins, O. and Dewes, T. (1992). Loss of nitrogenous compounds during composting of animal wastes. Bioresource Technology, 42, 103 - 111.

The agricultural area survey as a tool for implementing the European Nitrates Directive in the Walloon Region of Belgium

C. Vandenberghe, A.C. Mohimont, T. Garot and J.M. Marcoen
GRENeRA, Gembloux Agricultural University (FUSAGx),5030 Gembloux, Belgium

Introduction

In the Walloon Region of Belgium, the EEC 91/676 European Directive is implemented through the Programme for Nitrogen Sustainable Management in Agriculture ('Programme de Gestion Durable de l'Azote en agriculture', P.G.D.A.). This P.G.D.A. limits the application of organic N to 210 kg ha^{-1} on grasslands and to 120 kg ha^{-1} on crop lands (80 kg ha^{-1} within vulnerable zones). A derogation of these limits can be granted if the farmers undertake to follow a 'Quality Approach' (Démarche Qualité - DQ) with very close surveillance and the implementation of specific N-control techniques. Such evidence is provided on the basis of the NO_3^- - N concentration profiles established in autumn. In order to determine the annual standards for these concentration profiles, a network of 24 pilot farms was monitored by GRENeRA - FUSAGx and its partner, the Laboratoire d'Ecologie des Prairies - UCL. These 24 pilot farms constitute the reference standards for the agricultural area survey (Survey Surfaces Agricoles - SSA). This network, which is distributed throughout the Walloon Region, takes into account the annual meteorological conditions, geographical location (soil types) and type of culture.

Materials and methods

The annual standards of NO_3^- - N concentration profiles were measured in 186 selected fields over the network of pilot farms (with an average of 7 fields per farm). These fields are selected to fit:
1. soil types: the soils of the pilot farms' fields are representative of soils of the areas where they belong,
2. type of management: only managements accounting for more than 5% of the total utilised agricultural area of the Region are taken into consideration.

To provide sustainable agriculture, N fertilising of the main crops on these fields is controlled to minimise post harvest residues. The development of the N residues is monitored by taking three samples to 90 cm depth (three 30-cm layers) taken at the beginning of the drainage period. The samples are taken and analysed according to the strict technical specifications defined by GRENeRA.

Results and discussion

The medians of NO_3^-- N concentration profiles are calculated for each management. These median residue levels are an indicator of good N management. The results confirm the impact of management type on the concentration profiles (Table 1). For each type of management, a decrease is observed throughout the leaching period. Catch crops show their effectiveness after

Table 1. Nitrate-N concentration profiles (median of n measures) per crop type (kg NO_3^- - N ha^{-1}).

	October		November		December	
	kg N-NO_3^-	n	kg N-NO_3^-	n	kg N-NO_3^-	n
Cereals + catch crop	7288	2540	3062	2641	2247	2339
Cereals + organic matter	90	7	88	7	100	8
Potatoes	1104	7	78	7	63	7
Sugar beets	31-	1-	2934	2021	3034	1918
Maize	78-	5-	6878	720	6562	720
Mixed gGrasslands	1755	2938	1140	3442	929	3339

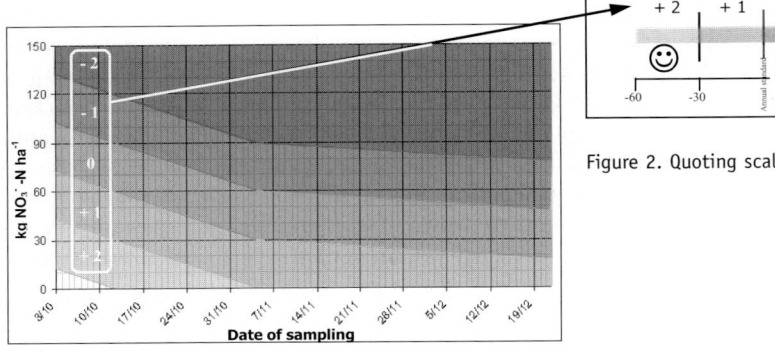

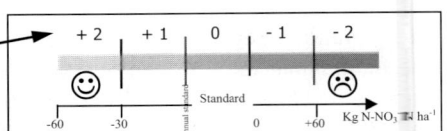

Figure 2. Quoting scale for farms in D.Q.

Figure 1. Evolution of NO_3^- - N concentration profiles after cereals and catch crop.

the cereal harvest and manure application. The measures taken three times from beginning of October to end of December allows us to compare the values in the D.Q. farms with the annual standards at any time.

Based on the measures realised in the 24 pilot farms, the graph shown in Figure1 allows the evaluation of the NO_3^- - N concentration profiles measured in the D.Q. farms. The D.Q. farms can be assessed according to the quoting scale (Figure 2) on the basis of the gap between the sample measured and the annual standard. By the end of a 4-year period, an evaluation will be drawn. As a result of this evaluation, farmers are, or are no longer, allowed to apply the 'Quality Approach' (130 kg organic N ha^{-1} on crop lands and 250 kg organic N ha^{-1} on grasslands).

Conclusion

The network of pilot farms allows us to verify that the N managements within the D.Q. farms comply with a sustainable agriculture.

Acknowledgements

This research is granted by the Walloon Government in the framework of the P.G.D.A.

References

Lambert, R., Van Bol, V., Maljean, J.F. and Peeters, A. (2002). Prop'Eau Sable - Rapport final d'activités. MRW- UCL 107p.

Vandenberghe, C., Mohimont A.C. and Marcoen, J.M. (2002). MRW-FUSAGx. Rapport d'activités annuel intermédiaire 2002, 84 pp.

Anonyme. (2002). Arrêté du Gouvernement wallon du 10 octobre 2002 relatif à la gestion durable de l'azote en agriculture.

Environmental economics of soil organic matter management

N. Wrage and O. Oenema
Wageningen University and Research Centre, Soil Quality, PO Box 8005, 6700 EC Wageningen. The Netherlands

Introduction

Economics deals with the use of scarce resources by individuals or groups (Gilpin, 2000). Environmental economics deals with the consequences of resource use that do not directly cost money or bring profit and that are therefore, not considered by consumers or producers. These consequences are external to the market. With regards to soil organic matter management, external consequences are, for example, increased N losses to groundwater and surface waters and increased emission of N_2O from soils following large additions of organic matter. Often, external consequences are related to the *Tragedy of the Common Lands*. In 1968, Hardin described how each 'rational herdsman' seeks to increase the profit from his cattle grazing on a pasture open to all: he adds more and more animals to the pasture. At some stage, the carrying capacity is surpassed and the system collapses (Hardin, 1968). So far, few farmers would refrain from trying to increase their gains, even if this might threaten the common good. 'Open access property' that has no owner and, like air and water, is especially vulnerable to overuse and pollution.

Obviously, overuse of resources and pollution of 'open access property' is not sustainable. Sustainable resource use 'meets the needs of the present, without compromising the ability of future generations to meet their own needs' (WCED, 1987). To identify sustainable use we differentiate between renewable and non-renewable resources. If the resources are renewable, they should be used at a rate that does not exceed the capacity for renewal. If the resources are non-renewable, depletion should be minimised by using only sparingly, if possible supported by recycling or by switching to renewable substitutes (Gilpin, 2000). The latter is stressed by the *Hartwick rule*, which states that: 'net profit derived from the use of non-renewable resources should be invested in reproducible capital, i.e. a renewable substitute for the resource' (Hartwick, 1977).

Dilemma of soil organic matter management

Soil organic matter can be seen as a natural resource that needs to be maintained for biosphere functioning and for future generations, if agriculture is to be sustainable. To what extent is the soil organic matter resource renewable? The answer to this question depends on the scale, time frame, environmental conditions and the functions of soil organic matter we are investigating. For example, the functions of soil organic matter related to chemical soil fertility are, in most cases, more easily renewable than those related to soil structure and erosion control.

The objective of soil organic matter management should be to sustain an adequate level and quality of soil organic matter and to synchronise the release of N, through mineralisation to the demand of the crop. Currently, most farmers are interested in maximising soil fertility and soil organic matter, to maximise productivity and to minimise the risk of yield losses. Further, policy makers are now interested in maximising soil organic matter to maximise soil organic C sequestration (IPCC, 2000). However, elevated levels of soil organic matter can cause pollution of neighbouring systems (e.g. Gerke et al., 1999). Thus, while an increase in the amount of soil organic matter seems favourable, a careful balance has to be established to minimise losses to the environment. This is the dilemma of soil organic matter management.

Impacts

To stimulate environmental stewardship and careful management and to prevent problems related to the *Tragedy of the Common Lands*, it could be beneficial to put a price on soil organic matter. However, is this possible? For C sequestration, prices of about $30 per ton of C are discussed (Johnson & Keith, 2004). Carbon sequestration is only one aspect of soil organic matter. Others include: soil fertility, binding of pollutants, role in geochemical cycles, beauty of the landscape, soil biodiversity and soil conservation. An optimal level of soil organic matter depends on the amount of clay and the underlying geology and hydrology, but to be able to create a good soil structure and to maintain a steady-state situation. Uncertainty about the optimal amount of soil organic matter as well as the various roles it plays seem to make the determination of a price for it very complex. Furthermore, a market would be difficult to establish, since soil organic matter itself cannot be traded.

Conclusions

Economic use of soil organic matter implies internalisation of all environmental costs (e.g. eutrophication, greenhouse gas emissions). Thus, the full cost of maintaining desirable levels of soil organic matter and of controlling pollution should be included in the production process. This would then be reflected in the market price of the product, so that the public could take an informed decision (Gilpin, 2000). However, both the environmental costs and the various services of soil organic matter are difficult to determine. Therefore, determining the price would be difficult. Perhaps the promotion of the economic value of compounds that help to build soil organic matter, such as green manures, could help to at least raise awareness of the value of soil organic matter. This would not, however, help to internalise the environmental costs.

References

Gerke, H. H., Arning, M. and Stöppler-Zimmer, H. (1999). Modelling long-term compost application effects on nitrate leaching. Plant and Soil, 213, 75-92.

Gilpin, A. (2000). Environmental economics: A critical overview. John Wiley & Sons Ltd, Chichester, 334 pp.

Hardin, G. (1968). The tragedy of the commons. Science, 162, 1243-1248.

Hartwick, J.M. (1977). Intergenerational equity and the investing of revenues from exhaustible resources. American Economic Review, 66, 972-974.

IPCC (2000). Land use, land-use change, and forestry. Cambridge University Press, Cambridge, 377 pp.

Johnson, T.L. and Keith, D. W. (2004). Fossil electricity and CO_2 sequestration: how natural gas prices, initial conditions and retrofits determine the cost of controlling CO_2 emissions. Energy Policy, 32, 367-382.

WCED (1987). Our common future. Oxford University Press, Oxford.

SECTION 2
MATCHING SUPPLY WITH DEMAND

Matching supply with demand

D.V. Murphy[1], E.A. Stockdale[2], F.C. Hoyle[1], J.U. Smith[3], I.R.P. Fillery[4], N. Milton[1], W.R. Cookson[1], L. Brussaard[5] and D.L. Jones[6]

[1]*Centre for Land Rehabilitation, Faculty of Natural and Agricultural Sciences, The University of Western Australia, Crawley WA 6009, Australia*
[2]*Agriculture and Environment Division, Rothamsted Research, Harpenden, Hertfordshire, AL5 2JQ, England, UK*
[3]*Department of Plant and Soil Science, University of Aberdeen, Cruikshank Building, Aberdeen AB24 3UU, Scotland, UK*
[4]*CSIRO, Plant Industry, Private Bag PO Wembley, WA, Australia*
[5]*Wageningen University, Environmental Sciences, Soil Quality Section, PO Box 8005, 6700 EC, Wageningen, The Netherlands*
[6]*School of Agricultural and Forest Sciences, University of Wales, Bangor, Gwynedd, LL57 2UW, Wales, UK*

Abstract

Nitrogen is the primary nutrient limiting crop production in farming systems throughout the world. Matching the supply of N (via fertiliser applications and biologically mediated processes) to the amount, timing and location of crop demand is critical to enable the development of environmentally acceptable farming systems that increase productivity through greater efficiency of N use and minimise losses of N. Supply of fertiliser N is defined by the amount, timing and type of N applied. Although fertiliser application is under the control of the farmer, lack of knowledge of future climatic conditions introduces a large degree of uncertainty when deciding fertiliser application rates. Also, availability of biologically derived inorganic-N to plants is difficult to predict and manage since it is regulated by a number of variables including (i) the quantity/quality of the residue material, (ii) rate of fragmentation and soil incorporation of residues by fauna, (iii) amount and timing of microbial N release, (iv) microbial requirements for N and (v) N loss from the rooting zone. These processes are influenced by both environmental factors (e.g. temperature, wet-dry cycles), which can be difficult to predict, and agricultural management practices (e.g. tillage, green manures, residue management), which can be manipulated to enhance supply and demand. Although we have a good mechanistic understanding of the individual factors regulating soil N supply, our capacity to design, and more importantly have grower adoption of farming systems that harness this N to its fullest, is still questionable.

Keywords: leaching, mid infra red, mineralization, models, synchrony

Introduction

Despite our current understanding of N supply dynamics, completely matched N supply and demand is rarely achieved, particularly in arable and horticultural systems, and consequently the losses of N associated with farming systems can be substantial (Hatch et al., 2003). The challenge to the research community is to address this issue through the development of improved farming practices that will gain farmer adoption. This is likely to be achieved, where appropriate, through (i) the adoption of reduced tillage practices that reduce the rate of soil organic matter mineralisation, (ii) crop rotations/mixed cropping that cycle N more efficiently (iii) a flexible fertiliser strategy (timing, amount) based on estimated yield for the predicted season rainfall, (iv) development of cost effective fertilisers/products that slow nitrification or increase nutrient retention in the rooting zone, (v) increasing plant N demand through removal of soil constraints to root development, (vi) new plant cultivars that increase capture of water and nutrients, (vii) site specific decision support systems (DSS) based on soil and crop information that can be acquired rapidly and cheaply and (viii) precision agriculture that enables variable rate technology for seed and fertilisers. We cannot here consider in detail all the factors that contribute to the synchrony/asynchrony of N supply and demand, nor describe all the possible approaches that have been taken in practice to reduce losses and increase plant N uptake within farming systems. However, we aim to highlight, using examples from the Mediterranean-type and temperate climates of Australia and Europe, some of the challenges that remain for scientists and farmers in understanding the principles, practice and prediction of matching N supply and demand in farming systems.

Principle: What factors govern the supply of N from soil to plant?

Mineral N can be supplied to the soil as fertilisers, or returned in animal excreta, as well as through atmospheric deposition and from decomposition of soil organic matter and crop residues (SOM). The cycling of N in soil, although influenced by chemical and physical factors, occurs primarily as a result of a series of connected and interacting biological processes. The gross N mineralisation rate determines the supply of mineral N from organic matter, while both immobilisation and nitrification rates represent microbial demands for N within the system. Some processes such as N fixation and nitrification are performed by very specific groups of organisms, whilst other processes, such as decomposition, are carried out by very diverse and distinct groups including bacteria, fungi and a wide range of invertebrate phyla. The direct and indirect effects of trophic (food webs) and non-trophic interactions (manipulation of habitat) between soil organisms and growing plants at a range of spatial scales within the soil result, at an ecosystem level, in the processes we measure as the soil N cycle. However, at each level the contest for each individual is one of survival and maximisation of resources over evolutionary time (Brussaard, 1998). The role of each group or species in determining the amount of N cycling in soil (and its fate) has been studied in a number of agricultural systems (see Moore, 1994) and there is some indication that more intensive agricultural systems impact to a greater extent on the diversity of soil organisms compared with conservation techniques such as minimum tillage and integrated farming practices (Stockdale & Cookson, 2003). However, there is little unequivocal evidence that directly links soil biodiversity and nutrient cycling in soils (Swift et al., 1998). The critical biological feedbacks that influence the abiotic factors at an ecosystem level must be considered (Brussaard, 1998) as it is important to determine keystone species amongst the bewildering diversity of organisms contributing to N cycling processes in soils. The inclusion of these species in models may be critical, particularly if global change affects them differentially (Andrén et al., 2001) and models are increasingly used at a wide range of temporal and spatial scales.

Few processes directly lead to the loss of NH_4^+ produced in soil (rather than applied at the surface); in contrast, NO_3^- is readily lost from soils in drainage (by leaching) and/or by denitrification at anaerobic microsites. Nitrification is therefore a key process in regulating the potential for loss of plant available N from soil (Stockdale et al., 2002). The amount of N leached as dissolved organic N (DON) can also be considerable, yet the factors regulating the size, composition and turnover of this pool remain poorly understood, particularly in agricultural soils (Murphy et al., 2000). Although Owen & Jones (2001) have shown that crops such as wheat, beans and maize are able to take up low molecular weight DON directly from the soil in the form of amino acids, in most high production agricultural systems this is thought to constitute only a small proportion of the total N taken up by the plant. Further, in contrast to NO_3^-, intense competition for labile DON exists between plant roots and soil microorganisms. Consequently, most of the labile DON is consumed by rhizosphere microorganisms with the subsequent release of NH_4^+ back into the soil which, in many cases, is rapidly nitrified and becomes plant available. While some researchers have indicated that DON uptake by plants may be significant (Chapin, 1995) they have often only considered DON fluxes in the rhizosphere as a unidirectional influx. However, the efflux of DON from plant roots is also well documented (i.e. root exudation; Jones & Darrah, 1994). The actual net flux of DON at the soil-root interface remains unknown and certainly warrants further research.

In many farming systems the amounts of N cycling through soils during a year are more than enough to satisfy crop N demand, even where no fertiliser is applied (Fillery, 2001; Table 1). However, asynchrony is a feature of many agricultural systems and the use of N fertilisers is commonly employed to overcome the difficulties of matching biological supply with plant N demand. On a global scale, the application of N fertilisers is now a greater source of N supply

Table 1. Nitrogen balance sheets (kg N ha^{-1} for plough layer) for a wheat crop following rotations in experimental trials from Australia (East Beverley, 0-10 cm) and the UK (Woburn, 0-23 cm).

	Australia[a]		United Kingdom[b]	
	No legume-in system Wheat:Wheat	Legume based system Lupin:Wheat	Legume-based system Beans:Wheat	Grass-ley rotation Grass:Wheat
N pools in soil				
Soil total N	1008	1002	3027	3376
Microbial biomass-N	64	68	84	163
Inorganic N in soil profile (0-90 cm at sowing)	28	32	48	13
Nitrogen supply (per year)				
Gross mineralisation	100	120	440	689
Gross immobilisation	57	61	357	511
Net mineralisation	43	59	83	178
Fertiliser applied	46	0	0	0
Crop N Demand				
Plant N uptake (total above ground at harvest)	13	44	36	54

[a]data from Murphy et al. (1998); [b]unpublished data from D.V. Murphy and E.A. Stockdale

to agricultural crops than that via biological processes (Jenkinson, 2001). In the UK, about half of crop N uptake is supplied directly by fertilisers applied in that season (Jenkinson, 2001), whilst in southern Australia, because of the lower fertiliser N applications and lower fertiliser efficiency, on average 80% of crop uptake is supplied via biological processes (Angus, 2001). It is clear, however, that in both systems fertiliser N supply remains an important component of crop N uptake.

Asynchrony is characterised by (i) the accumulation of plant available N in soil in excess of current crop and microbial demand, at a time when those N pools are subject to N loss processes or (ii) characterised by insufficient availability when crop demand is high. Abiotic factors, such as climate and soil parent material, are widely recognised as the major determinants of ecosystem composition and function; their effects on N fluxes often overshadow even large differences between management treatments (Andrén et al., 1999). Whilst these abiotic factors affecting asynchrony cannot be controlled, it is important that if we are to better understand the timing and location of both N supply and demand, their effects on N fluxes should be understood. The development of ^{15}N isotopic dilution techniques has provided an opportunity to obtain a much clearer understanding of soil N cycling by enabling the process rates that form specific pathways within the soil N cycle to be quantified (see review, Murphy et al., 2003). Studies of the processes controlling soil N supply confirm that there is a strong seasonality in gross N fluxes in soil which have been related to the complex interacting seasonal patterns of soil temperature, moisture content and residue quality (Murphy et al., 2003). Climatic conditions also strongly influence crop growth and development (and hence N demand) in any season, which in turn determines the quality of residue inputs. Given the inherent variability in temperature and rainfall patterns, it is therefore clear that, even where the same treatments and management practices are applied, synchrony between soil N supply and crop demand may be achieved more effectively in some cropping seasons than others (Ma et al., 1999).

Practice: Can systems be managed to match N supply with crop N demand?

Management of organic inputs to improve synchrony of soil N supply

In agro-ecosystems, management of cropping patterns and inputs of fertilisers (as inorganic and organic materials) and crop residues have a significant role, alongside the abiotic factors described briefly above, in determining the timing and spatial location of both N supply and demand. The influence of these management practices may have a range of both short-term (this season) and longer-term effects that are more difficult to quantify. For example, Magdoff et al. (1997) reported data showing that although crop recovery of fertiliser N (40%) was higher than legume residue N (17%), much more legume N (47%) was retained in the soil compared with fertiliser N (17%). The size of the microbial biomass in the soil, but not necessarily its activity, was also increased. Manipulation of N release in soil through the application of combinations of residues and organic fertilisers has been more often studied in tropical than in temperate climates (e.g. Tian et al., 1995). However, using typical temperate crop residues, Gunnarsson & Marstorp (2002) showed that it was possible to change both the total amount of N mineralised and the time-course of mineralisation by combining different quality residues during decomposition. In the future, genetic manipulation of crop plants may lead to increased opportunities to manipulate the quality and structure of crop residues and consequently their decomposition in soil (Hopkins et al., 2001).

Historically, spatial and temporal planning of crop rotations has not normally considered the impact of crop residues on N cycling. However, evidence of significant nutrient flushes under windrowed crops and the export of nutrients from hay paddocks has resulted in a heightened awareness of the role of crop residues in nutrient cycling. Primary manipulation of the soil N cycle is largely associated with retention or export of crop residues. Fine manipulation of the soil N cycle would be difficult to achieve under field conditions, since the chemical composition of plant materials is not constant but strongly dependent on climate, soil and management factors (Gunnarsson & Marstorp, 2002). While a range of timing and incorporation options have been considered for organic materials and leguminous residues in arable systems (Korsaeth et al., 2002), there is clearly less flexibility to alter the synchrony of N supply and demand in legume/grass pastures themselves. However, the management and timing of cutting and grazing, particularly in respect to the significant spatial impact of excretal returns on N cycling, can be an important tool in manipulating the balance between N_2 fixation, soil N supply and plant N uptake (Ledgard, 2001).

Management options to increase plant N demand
There are considerable differences between the N requirements of different crops. Both the timing of a crops maximal demand for N and its characteristic rooting structure (rate of growth, depth, surface area, distribution and architecture) affect the efficiency with which it can use soil N. This information can be used to assist in the selection of crops during rotation planning, where patterns of soil N supply can be predicted. Within any growing season, crop N requirements are also influenced by climate (length of season, available water, temperature) as well as chemical (e.g. nutrients, pH, EC), physical (e.g. soil compaction, leaching potential) and biological (e.g. disease) constraints to plant production. Consequently crop N demand can vary significantly from season to season despite almost identical crop management practices. Cultivar differences in N efficiency from fertiliser and soil sources have also been seen (Anderson & Hoyle, 1999). These have been linked to differences in rooting patterns. Until recently, little has been known of the optimal root pattern for NO_3^- capture and few breeding programmes have even considered rooting structure. Simulations using the model ROOTMAP, with inclusion of soil water and nutrient dynamics (Dunbabin et al., 2002) in the low density cereal stands in Mediterranean climates, have highlighted that in sandy soils prone to leaching, NO_3^- in mobile water can evade capture when root density is low (Dunbabin et al., 2003). To improve N efficiency in Mediterranean climates, it is important to use combinations of crops which (i) develop high root densities in the surface layer of soil early in the growing season and so are able to capture more early season NO_3^- and (ii) have a vigorous tap root which is able to access deep stored water and NO_3^- later in the season (Dunbabin et al., 2003). For these reasons, current plant breeding efforts in Australia have been directed at increasing early root vigour through selective breeding (Richards & Lukacs, 2002).

The ability of plant roots to force their way into the soil is often an important factor limiting plant growth; roots grow in soil partly through existing large pores and root channels, and partly by moving aside soil particles. Poor soil structure can therefore inhibit root growth. Consequently tillage practices that remove soil physical constraints, such as compaction layers resulting from vehicle traffic and/or grazing, can have a major impact on NO_3^- and water use efficiency (Figure 1). Deep ripping is a common practice on coarse-textured soils in southern Australia to remove vehicle compaction layers that generally form at depths between 10 and

40 cm. Predictions using the N and root model 'Select Your Nitrogen' (SYN, Western Australian Department of Agriculture) indicate that this can result in an additional 20-40% of the soil water and NO_3^- being accessed by the growing crop (Figure 1). In Mediterranean climates improving water use efficiency and N use efficiency are often closely allied.

Management of fertilisers to improve utilisation

Best practice in N fertiliser management depends in part on an ability to predict the likely timing and location of soil N supply throughout the cropping season, so that fertiliser N is only applied when and where it is necessary. Splitting the total amount of fertiliser N between a number of applications linked to the achievement of particular plant growth stages is becoming common; this maximises the opportunity for crop uptake at the right time and minimises the risk of direct losses of fertiliser. A development of this approach, which has been applied successfully in grassland management, is to regularly measure the mineral N pool in the soil and to add only sufficient fertiliser to meet predicted crop demand at that time (Brown et al., 1997).

The timing of N release from fertilisers in relation to crop use efficiency can also be manipulated by careful choice of both appropriate fertiliser products and application methods.

While nitrification inhibitors have been shown to be effective, they are not and seem unlikely to become cost effective for use at a field scale. Various other approaches have been taken to alter the release characteristics of N fertiliser including the production of slow release fertiliser granules (Shaviv, 2003). Adoption will be based on cost and ease of use. It is also possible to delay the release of available N from NH_4^+ fertilisers through deep placement, or banding of fertilisers below the seed. This means that early root growth is unlikely to be affected by the high concentration of NH_4^+, which would lessen the benefit of early rainfall events, and also restricts nitrifier activity, so that NO_3^- release is delayed (Angus, 2001). Here adoption is dependent on the availability/modification of farm equipment and the avoidance of damage to the germinating seedling, which requires accuracy.

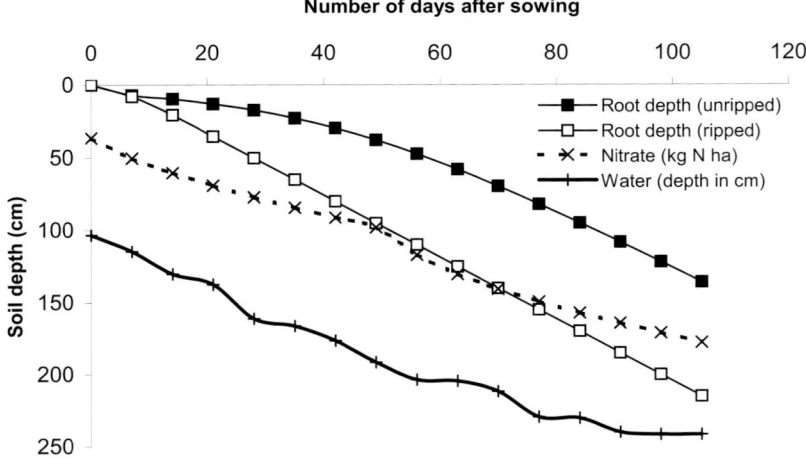

Figure 1. Model simulation of the impact of soil ripping on the rooting depth of wheat in a sandy loam soil at Avondale, Western Australian using rainfall data for 2002 (i.e. in a non-leaching year). (F.C. Hoyle, unpublished data).

Prediction: Do recommendation systems accurately encompass soil N supply?

Rapid measurement of soil properties for inclusion in DSS and Precision Agriculture

A range of in-field plant analysis techniques are used to support N management decisions, including petiole NO_3^- analysis, spectral reflectance and chlorophyll meters. Such plant measurements are related to sufficiency indices, or compared with well-fertilised plots under the same growing conditions to provide the necessary basis for variable N application to supplement soil N supply (Bronson et al., 2003; Liu et al., 2003). Detailed knowledge of the spatial and temporal dynamics of N-cycling processes (mineralisation, immobilisation and nitrification; e.g. Corre et al., 2002) would also aid with improving fertiliser recommendations for a system. However, in practice it is cost prohibitive and time consuming to measure these process rates on a spatial and/or temporal scale. Instead, under non-leaching environments, the measurement of deep soil pre-plant mineral N is an effective way of adjusting fertiliser rates based on soil available N. This is particularly relevant in rotations where increased active SOM pools, augmented through incorporation of leys or use of animal manure, are expected to have released large but variable amounts of N through mineralisation. Alternatively, where leaching continues throughout the growing season, this measurement is not considered as useful. Here measurements of a labile pool of SOM could be used to identify the soil N supply capacity in a field for the purpose of site specific parameterisation of fertiliser decision support systems (DSS). For example, potentially mineralisable N (PMN) provides a simple index of soil N supply which has been used recently in the UK to determine the spatial characteristics of soil N supply within a field with the intended purpose of adjusting fertiliser rates (Baxter et al., 2003). Few attempts have been made to make fertiliser DSS field or site specific through initial parameterisation with soil measurements. However, this approach certainly has an application in the strategic and variable management of fertilisers if a suitable soil initialising characteristic can be developed which is cost effective and rapid (Smith, 2001). Whilst both pre-plant mineral N and PMN can be assessed simply in the laboratory, delays in turn-around times at the beginning of the field season (when fertiliser strategies are being developed) have reduced the popularity of these assays with some growers.

Recently, near infrared (NIR) spectroscopy has been applied to the prediction of soil N supply in Australian rice soils, through calibration against measures of PMN (Russell et al., 2002) and mid infrared (MIR) calibration curves, have been developed for a range of soil biological, chemical and physical soil attributes (Janick et al., 1998). The advantage of this technology is that, once calibrated, soil samples can be scanned rapidly (minutes) to provide predictions for a range of soil properties. Although not as accurate as measuring each soil property using standard techniques, the technology does have a place in the development of soil spatial maps for the purpose of zoning fields and defining variable management strategies. For example, in Figure 2, soil was collected using a 25m x 25m sampling grid (180 sampling points over 10 ha). An initial MIR calibration curve was developed for the prediction of PMN from an independent data set using a Perkin Elmer FT-IR spectrometer. On the 180 grid samples, PMN was measured by 7-day anaerobic incubation at 40°C (Figure 2b, range 3-27 mg N kg^{-1}; Keeney & Bremner, 1966). Mid infrared predictions of PMN were from ground sub-samples of the same 180 soils (Figure 2c). Spatial maps were then generated using an inverse distance weighting. An assumption of this deterministic method for spatial interpolation is that sampling points close together are more alike than those farther apart. Map surfaces were thus created based on the extent of similarity between grid sample points. There was good agreement between predicted and measured PMN ($r^2 = 0.70$). PMN was shown to be spatially related to soil mineral N content (range 3-187 mg N kg^{-1}; Figure 2a) illustrating the relevance of this measurement to soil N supply and crop demand.

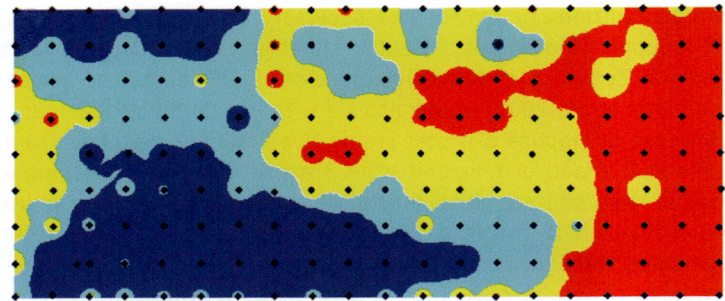

(a). Soil Mineral N ($NH_4^+ + NO_3^-$) and 25m x 25m grid sampling points.

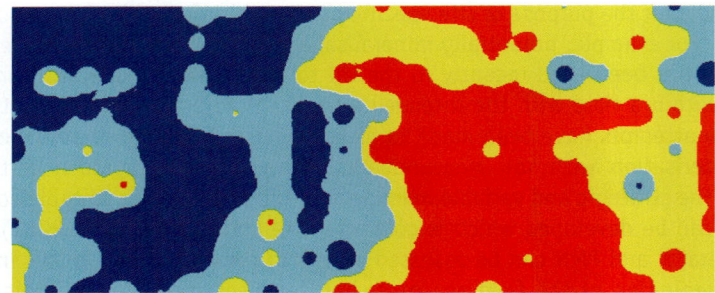

(b). Measured potentially mineralisable N.

(c). Mid infrared prediction of potentially mineralisable N.

Figure 2. Spatial maps of (a) mineral N, (b) potentially mineralisable N and (c) mid infrared predicted potentially mineralisable N for the 0-10 cm layer of a Western Australian agricultural soil. Colours represent data catergorised into quartile ranges where red=0-25%, yellow=25-50%, light blue=50-75% and dark blue = 75-100%. (D.V. Murphy & N. Milton, unpublished data).

Models as fertiliser recommendation systems

Currently, most prediction systems used to support farmers in matching N supply and demand have been developed as N fertiliser recommendation systems (FRS). Once these models have been evaluated against independent data, they then provide a tool for making practical use of the huge body of research information (Stockdale, 1999). Many recommendation systems have used a static calculation, not allowing for deviations from the expected weather conditions or plant growth. Static models usually require less input data to run, and less computing power, and the calculations are more transparent to the user, but provide only minimal site and season specificity (e.g. N-CYCLE; Scholefield et al., 1991).

However, we now have sufficient confidence in our models of N dynamics to incorporate them into DSS for practical use. Dynamic simulation models can be used, not only to optimise the amount and timing of fertilisers applied, but also to redesign crop rotations, so that the growing crop is able to take up N as it becomes available (Smith & Glendining, 1996). Dynamic models (e.g. SUNDIAL-FRS; Smith et al., 2001 and N-Able; HRI, 1994) are more site and season specific, but require more input data to run, and are dependent on the predicted values of a notoriously difficult factor to predict: the weather. However, their potential value depends on the accuracy of the model, short-term weather forecasts, and the wide uptake of the models by farmers and/or their advisors.

There are a number of FRS available to farmers in the UK; each has been developed, at least initially, with particular farming systems (arable, horticulture, mixed farming) in mind. Field trials on 24 working farms across the UK were used to compare the current accuracy of fertiliser recommendations provided by SUNDIAL-FRS, N-Able, RB209 and expert farmer judgement (Smith et al., 2001). The overall results showed no clear statistical difference between the 3 DSS and the farmer's own judgement (Figure 3). However, when the most accurate model for each crop and soil type was selected, the three DSS together provided significantly more accurate recommendations than the farmer, suggesting a combined DSS would be more successful. While each shows limited success in predicting the N optima for a wide range of crops, greater accuracy can be achieved by combining a number of models within a DSS which allows the most appropriate model for the soil/crop combination to be identified (Smith et al., 2001; Figure 3). The adoption of DSS and FRS is a major problem. In the UK, uptake of systems is limited by failure of the developers to establish suitable delivery agreements. Where this has been overcome, there is a high level of uptake of such systems by farmers. For example, the EMA environmental auditing package (Lewis et al., 1997) has been distributed to over 10,000 farmers, and the MANNER manure management system (ADAS, 1997) to over 7,500 farmers.

Challenges for the future

In general, only a small proportion of growers readily adopt new technology and best practice; the majority still rely on their understanding of the principles of soil and fertiliser management gained through education, or passed down through the generations. Associated cost and ease of adoption of management changes into current farming practice (particularly where significant capital expenditure is required) are also major limitations to adoption of new management strategies by growers. For example, Australian farmers readily adopt new crop

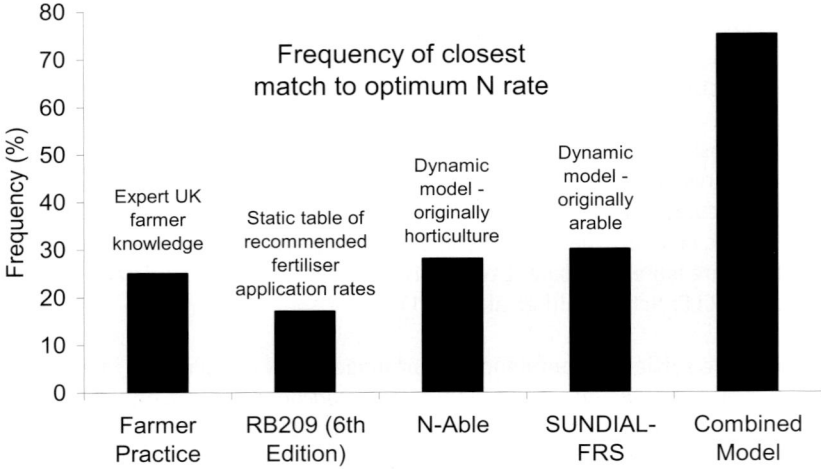

Figure 3. Comparison of farmer 'best practice', UK static table of fertiliser recommendations and models (individual and combined performance) in determining the optimal rate for N application on 24 UK farms. (J.U. Smith & E.A. Stockdale, unpublished data).

varieties, whilst the purchase of machinery such as GPS navigation systems to enable controlled traffic strategies are still not commonplace, although the benefits to the soil are well documented. Technology transfer is therefore a major challenge to the research and extension community. Increasing the adoption rate for DSS that enable farmers to better match N supply and demand in practice, will only be achieved when the constraints on farm practice and limitations to farmer adoption are well understood and their implications integrated into the design of research programmes.

Acknowledgements

Australian research was funded through the Australian Grains Research and Development Corporation. UK research was funded in part by the Department of Environment Food and Rural Affairs. Rothamsted Research also acknowledges the grant aided funding of the UK Biological and Biotechnological Sciences Research Council.

References

ADAS (1997). Welcome to MANNER. MANNER (Version 2.1) USER GUIDE. ADAS Gleadthorpe Research Centre, Mansfield, UK.

Anderson, W.K. and Hoyle, F.C. (1999). Nitrogen efficiency of wheat cultivars in a Mediterranean environment. Australian Journal of Experimental Agriculture, 39, 957-965.

Andrén, O., Brussaard, L. and Clarholm, M. (1999). Soil organism influence on ecosystem-level processes - bypassing the ecological hierarchy? Applied Soil Ecology, 11, 177-188.

Andrén, O., Kätterer, T. and Hyvönen, R. (2001). Projecting soil fauna influence on long-term soil carbon balances from soil faunal exclusion experiments. Applied Soil Ecology, 18, 177-186.

Angus, J.F. (2001). Nitrogen supply and demand in Australian agriculture. Australian Journal of Experimental Agriculture, 41, 277-288.

Baxter, S.J., Oliver, M.A. and Gaunt, J. (2003). A geostatistical analysis of the spatial variation of soil mineral nitrogen and potentially available nitrogen within an arable field. Precision Agriculture, 4, 213-226.

Bronson K.F., Chau, T.T., Booker, J.D., Keeling, J.W. and Lascano, R.J. (2003). In-season nitrogen status sensing in irrigated cotton: II. Leaf nitrogen and biomass. Soil Science Society of America Journal, 67, 1439-1448.

Brown, L, Scholefield, D., Jewkes, E.C. and Preedy, N. (1997). Integrated modelling and soil testing for improved fertiliser recommendations for grassland. In: O. van Cleemput, S. Haneklaus, G. Hofman, E. Schnug and A. Vermoesen (eds). Fertilisation for Sustainable Plant Production and Soil Fertility. 11th International World Fertilizer Congress. Proceeding Volume III. CIEC Editorial Board with FAL. Braunschweig-Volkenrode, Germany, 248-255.

Brussaard, L. (1998). Soil fauna, guilds, functional groups and ecosystem processes. Applied Soil Ecology, 9, 123-135.

Chapin, F.S. (1995). New cog in the nitrogen-cycle. Nature, 377, 199-200.

Corre, M.D., Schnabel, R.R. and Stout, W.L. (2002). Spatial and seasonal variation of gross nitrogen transformations and microbial biomass in a Northeastern US grassland. Soil Biology and Biochemistry, 34, 445-457.

Dunbabin, V., Diggle, A. and Rengel, Z. (2003). Is there an optimal root architecture for nitrate capture in leaching environments? Plant, Cell and Environment, 26, 835-844.

Dunbabin, V.M., Diggle, A.J., Rengel, Z. and van Hugten, R. (2002). Modelling the interactions between water and nutrient uptake and root growth. Plant and Soil, 239, 19-38.

Gunnarsson, S. and Marstorp, H. (2002). Carbohydrate composition of plant materials determines N mineralisation. Nutrient Cycling in Agroecosystems, 62, 175-183.

Fillery, I.R.P. (2001). The fate of biologically fixed nitrogen in legume-based dryland farming systems: A review. Australian Journal of Experimental Agriculture, 41, 361-381.

Hatch, D., Goulding, K. and Murphy, D. (2002). Nitrogen as a pollutant from agriculture in surface and ground water. In: P. Haygarth and S. Jarvis (eds). Agriculture, Hydrology and Water Quality. CABI. Section 1 - Agriculture, 7-27.

Hopkins, D.W., Webster, E.A., Chudek, J.A. and Halpin, C. (2001). Decomposition in soil of tobacco plants with genetic modifications to lignin biosynthesis. Soil Biology and Biochemistry, 33, 1455-1462.

HRI (1994). WELL-N Nitrogen Advisory Model User Guide. Horticultural Research Industry, Wellesbourne, Warwick, U.K.

Janik, L.J., Merry, R.H. and Skjemstad, J.O. (1998). Can mid infrared diffuse reflectance analysis replace soil extractions? Australian Journal of Experimental Agriculture, 38, 681-696.

Jenkinson, D.S. (2001). Nitrogen in a global perspective, with focus on temperate areas - state of the art and global perspectives. Plant and Soil, 228, 3-15.

Jones, D.L. and Darrah, P.R. (1994). Amino-acid influx at the soil-root interface of Zea mays L. and its implications in the rhizosphere. Plant and Soil, 163, 1-12.

Keeney, D.R. and Bremner, J.M. (1966). Comparison and evaluation of laboratory method of obtaining an index of soil nitrogen availability. Agronomy Journal, 58, 498-503.

Korsaeth, A., Henriksen, T.M. and Bakken, L.R. (2002). Temporal changes in mineralisation and immobilisation of N during degradation of plant material: implications for the plant N supply and losses. Soil Biology and Biochemistry, 34, 789-799

Ledgard, S.F. (2001). Nitrogen cycling in low input legume-based agriculture, with emphasis on legume/grass pastures. Plant and Soil, 228, 43-59.

Lewis, K.A., Tzilivakis, J., and Bardon, K.S. (1997). Environmental best practice advisory system for agriculture. In: First European Conference for Information Technology in Agriculture, Copenhagen, 15-18 June 1997, EFITA, Copenhagen, 153-158.

Liu, X.J., Ju, X.T., Zhang, F.S. and Chen, X.P. (2003) Nitrogen recommendation for winter wheat using N-min test and rapid plant tests in North China Plain. Communications in Soil Science and Plant Analysis, 34, 2539-2551.

Ma, B.L., Dwyer, L.M. and Gregorich, E.G. (1999). Soil nitrogen amendment: Effects on seasonal nitrogen mineralisation and nitrogen cycling in maize production. Agronomy Journal, 91, 1003-1009.

Magdoff, F., Lanyon, L. and Liebhardt, B. (1997). Nutrient cycling, transformations, and flows: Implications for a more sustainable agriculture. Advances in Agronomy, 60, 1-73.

Moore, J.C. (1994). Impact of agricultural practices on soil food web structure: Theory and application. Agricultural Ecosystems and Environment, 51, 239-248.

Murphy, D.V., Fillery I.R.P. and Sparling G.P. (1998). Seasonal fluctuations in gross N mineralisation, ammonium consumption and microbial biomass in a Western Australian soil under different land use. Australian Journal of Agricultural Research, 49, 523-535.

Murphy, D.V., Macdonald, A.J. Stockdale, E.A. Goulding, K.W.T. Fortune, S. Gaunt, J.L. Poulton, P.R., Wakefield, J.A., Webster, C.P. and Wilmer, W.S. (2000). Soluble organic nitrogen in agricultural soils. Biology and Fertility of Soils, 30, 374-387.

Murphy, D.V., Recous, S., Stockdale, E.A., Fillery, I.R.P., Jensen, L.S., Hatch, D.J. and Goulding, K.W.T. (2003). Gross nitrogen fluxes in soil: Theory, measurement and application of ^{15}N pool dilution techniques. Advances in Agronomy, 79, 69-118.

Owen, A.G. and Jones, D.L. (2001). Competition for amino acids between wheat roots and rhizosphere microorganisms and the role of amino acids in plant N acquisition. Soil Biology and Biochemistry, 33, 651-657.

Richards, R.A. and Lukacs, Z. (2002). Seedling vigour in wheat - sources of variation for genetic and agronomic improvement. Australian Journal of Agricultural Research, 53, 41-50.

Russell, C.A., Angus, J.F., Batten, G.D., Dunn, B.W. and Williams, R.L. (2002). The potential of NIR spectroscopy to predict nitrogen mineralisation in rice soils. Plant and Soil, 247, 243-252.

Scholefield, D., Lockyer, D.R., Whitehead, D.C. and Tyson, K.C. (1991). A model to predict transformations and losses of nitrogen in UK pastures grazed by beef cattle. Plant and Soil, 132, 165-177.

Shaviv, A. (200). Impact of environmentally friendly N- fertilisation techniques: evaluation and modelling. Book of Abstracts, 12th N Workshop, Exeter, 2003.

Smith, J.U. (2001). Prediction of soil nitrogen supply by combined modelling and measurement. In: Proceedings of the SCI meeting on assessing soil fertility, modelling, measurement or both? London, 20/03/2001.

Smith, J.U. and Glendining, M.J. (1996). A decision support system for optimising the use of nitrogen in crop rotations. Aspects of Applied Biology No. 47, Rotations and Cropping Systems, 103-110.

Smith, J.U., Burns, I., Draycott, A., Glendining, M.J., Jaggard, K., Rahn, C., Stockdale, E.A., Stone, D. and Willmott, M. (2001). Decision support system to design whole farm rotations that optimise the use of available nitrogen in mixed arable and horticultural systems: On-farm testing. Final Report. MAFF Link, 164 pages.

Stockdale, E.A. and Cookson, W.R. (2003). Sustainable farming systems and their impact on soil biological fertility - Some case studies. In: L.K. Abbott and D.V. Murphy (eds), Soil biological fertility: A key to sustainable land use in agriculture. Kluwer Academic Publishers, in press.

Stockdale, E.A. (1999). Predicting nitrate losses from agricultural system: Measurements and models. In: W.S. Wilson, A.S. Ball and R.H. Hinton (eds), Managing risks of nitrates to humans and the environment. Royal Society of Chemistry, Cambridge, 21-41.

Stockdale, E.A., Hatch, D.J., Murphy, D.V., Ledgard, S.F. and Watson, C.J. (2002). Verifying the nitrification to immobilisation ratio (N/I) as a key determinant of potential nitrate loss in grassland and arable soils. Agronomie, 22, 831-838.

Swift, M.J., Andrén, O., Brussaard, L., Briones, M., Couteaux, M-M., Ekschmitt, K., Kjoller, A., Loiseau, P. and Smith, P. (1998). Global change, soil biodiversity and nitrogen cycling in terrestrial ecosystems: Three case studies. Global Change Biology, 4, 729-743.

Tian, G., Brussaard, L. and Kang, B.T. (1995). An index for assessing the quality of plant residues and evaluating their effects on soil and crop in the (sub-) humid tropics. Applied Soil Ecology, 2, 25-32.

Can N mineralisation be predicted from soil organic matter? Carbon and gross N mineralisation rates as affected by long-term additions of different organic amendments

A. Herrmann, E. Witter and T. Kätterer
Department of Soil Sciences, Section for Soil Fertility and Plant Nutrition, Swedish University of Agricultural Sciences, Box 7014, SE-750 07 Uppsala, Sweden

Abstract

Mechanistic models are often used to predict N mineralisation from soil organic matter. Such models usually estimate N mineralisation from C mineralisation, often necessitating assumptions about the C:N ratio of the decomposing organic matter. We tested, in a laboratory experiment, whether the latter was reflected in the C:N ratio of bulk soil organic matter. Soils that, since 1956, had received organic amendments of widely different C:N ratios were sampled in May 2001, approximately 18 months after the last applications of the amendments. Carbon and gross N mineralisation were determined consecutively over a 17-week incubation period at 20°C. Gross N mineralisation rates were estimated using equations based on first-order kinetics for NH_4^+ consumption. Despite differences in C:N of the bulk soil organic matter, across the soils gross N mineralisation was approximately proportional to C mineralisation. The proportionality factor between C and gross N mineralisation was approximately 5 (gross N mineralisation in relation to C mineralisation), suggesting that C:N of the decomposing organic matter of ca. 9 in all soils.

Keywords: C:N ratio, mechanistic models, mineralisation, soil organic matter.

Introduction

N mineralisation from soil organic matter is an important source for crop uptake (Shepherd et al., 1996). For example, soil organic matter has been found to contribute up to 50% of N uptake by spring barley grown on soils receiving 120 kg fertiliser N ha^{-1} for maximum yield (McTaggart & Smith, 1993). Predicting N mineralisation from soil organic matter is difficult because it is affected by several abiotic factors, while being the net outcome of concurrent processes that produce and consume mineral N (Jansson & Persson, 1982). A wide range of mechanistic models has been suggested for predicting N mineralisation (see e.g. Smith et al., 1997). The principle in these models is similar, in so far that they divide soil organic matter into several organic C pools and assume a certain C:N value for each pool (e.g. Parton et al., 1987; Hansen et al., 1991; Kätterer & Andrén, 2001). Each organic C pool is treated as a homogenous substrate with a specific turnover rate based on first-order kinetics. These rates are then modified by factors related to temperature, soil moisture and soil texture using empirical relations. Gross N mineralisation is usually estimated from C mineralisation, while the C:N value of organic matter in the source and sink pools determines whether net mineralisation or net immobilisation occurs. C:N values of these pools are, however, usually not known, necessitating these to be estimated or assumed, for example on the basis of the value for bulk soil organic matter.

The objective of this paper is to quantify the relation between gross N and C mineralisation as affected by the amount and quality of organic matter in a clay soil. Both C and gross N mineralisation were therefore, simultaneously determined in soils that for 45 years had received either no C input, input from only crop residues, or input from both crop residues and organic amendments with differing C:N.

Materials and methods

Soils
Soil samples were taken from the Ultuna Long-Term Soil Organic Matter Experiment (Uppsala, Sweden; 60 °N, 17 °E) (Persson & Kirchmann, 1994). The experiment was started in 1956 on a post-glacial clay loam classified as a Typic Eutrochrept (Soil Survey Staff, 1987) or an Eutric Cambisol (FAO, 1988). In this experiment, soils have been treated with different N fertilisers or organic amendments. All treatments were replicated in four blocks in a semi-randomised design. Six treatments were selected: (a) bare fallow (Fallow); (b) $Ca(NO_3)_2$ (N-fertilised); (c) Straw + $Ca(NO_3)_2$ (Straw + N); (d) Farmyard manure; semi-solid manure from dairy cattle (FYM); (e) Sawdust (coniferous) (SD); (f) Sawdust + $Ca(NO_3)_2$ (SD + N). In the selected treatments, the N fertiliser has been applied every year at the time of sowing as calcium nitrate at a rate of 80 kg N ha^{-1} yr^{-1}, and the organic amendments every other year (in the autumn) at a rate of 8 t ash-free organic matter per ha. Soil data are given in Table 1. Six soil sub-samples to a depth of 0-7 cm were taken from each block in May 2001 (approximately 18 months after the last application of organic material), sieved through a 4-mm sieve, bulked together to give one representative sample per treatment and block, thoroughly mixed, and stored at 4°C for three weeks.

Incubation of soil samples
Soil samples were wetted to 47% of their water holding capacity (WHC) and pre-incubated at 20°C for 2, 7 or 17 weeks. This WHC was chosen, so that the addition of a ^{15}N-labelled solution

Table 1. Soil characteristics. Mean values of four replicates. Means suffixed by a different letter are significantly different at $P < 0.05$ (Duncan's multiple range test).

Soil [a]	Soil C (%)		Soil N (%)		C:N ratio		Organic amendment N (%) [b]	WHC (%)		pH (H_2O)	
Fallow	0.9	a	0.09	a	10	A	-	40	a	6.2	a
N-fertilized	1.4	b	0.15	b	9	B	-	39	a	6.8	b
Straw + N	2.0	c	0.19	c	11	A	0.39	45	b	6.7	b
FYM	2.2	d	0.22	d	10	A	2.48	45	b	6.6	b
SD	2.1	ce	0.15	b	14	C	0.02	46	b	6.6	b
SD + N	2.2	de	0.17	e	13	D	0.02	48	c	6.7	b

[a] The mineral fractions consist of 36.5% clay, 41% silt and 22.5% sand.

[b] Organic amendments were applied in autumn 1999 and all amendments had a total C content of ca. 45%. Single determination of total N contents of organic amendments.

(see below) did not cause any smearing or risk of denitrification. Carbon, net and gross N mineralisation rates were determined over a 120 h or 72 h incubation period after each pre-incubation period. To determine gross N mineralisation two sets of soil samples (equivalent to 150 g dw) from each treatment were weighed into 0.5 l glass honey jars. The soils were amended with 3 ml per 100 g dw soil of a solution containing ^{15}N-labelled $(NH_4)_2SO_4$ (2.0 atom% [15]N) giving 5 μg N g^{-1} dry soil and bringing the soil moisture content up to 52.5% WHC. The solution was added drop-wise onto the soil surface, after which the soil samples were thoroughly mixed. Carbon mineralisation was measured in a different set of soil samples (equivalent to 50 g dw) weighed into 100 ml beakers, amended with $(^{15}NH_4)_2SO_4$, as described above, and placed in airtight 0.5 l jars together with a beaker containing a NaOH solution.

Analytical methods

The evolved CO_2 trapped in the NaOH solution was determined over a 120 h (after 2- and 7-week pre-incubation) or 72 h incubation (after 17-week pre-incubation) by titration with 0.l M HCl after addition of excess $BaCl_2$, with airtight jars without soil as control (Zibilske, 1994). Carbon mineralisation rates (μg C g^{-1} soil d^{-1}) were calculated assuming zero-order kinetics. Mineral N in soil samples (equivalent to 50 g dw) was extracted after an incubation period of 2 and 120 h (after 2- and 7-week pre-incubation) or 2 and 72 h (after 17-week pre-incubation), by shaking with 200 ml 2 M KCl for 2 h, followed by centrifugation and filtration using Munktell M00 filter paper. All extracts were stored at -18°C prior to colorimetric analysis for NH_4^+-N and NO_3^--N on a TRAACS 800 auto-analyser (Bran and Luebbe, Germany).

The atom% [15]N of mineral N in KCl soil extracts was determined using a diffusion technique (Goerges & Dittert, 1998). Briefly, 15 ml KCl soil extract was placed into a 20 ml scintillation vial (Packard, The Netherlands) and ca. 200 mg MgO, or ca. 200 mg MgO and ca. 400 mg Devarda's alloy, was added to generate NH_3 for the determination of the atom% [15]N of the NH_4^+-N and the NH_4^+-N + NO_3^--N pool, respectively. The atom% [15]N of the NO_3^--N pool was determined by differences between these two N pools. The evolved NH_3 was trapped by 10 μL of 2.5 M $KHSO_4$, which was placed on an acid (HCl) washed paper disk (Whatman No. 42) of 5-mm diameter. The paper disk was placed between a double layer of PTFE tape (Garco, Art. Nr PE 70, Taiwan) stretched over the top of the scintillation vials which were then capped. All samples were shaken at room temperature on a vertical rotary shaker at 50 revs. min^{-1}. After 72 h, the paper disk was removed, dried overnight at 40°C over silica gel and ca. 50 ml conc. H_2SO_4 was added, to prevent cross-contamination during the drying process, and then placed in tin capsules for ^{15}N analysis. In some soil extracts the amounts of NH_4^+-N were too small for accurate measurement of the atom% [15]N. Tin capsules for these samples were therefore, first spiked with 10 μg N $(NH_4)_2SO_4$ solution (0.3668 atom% [15]N) and dried overnight in a desiccator over ca. 50 ml conc. H_2SO_4 at 40°C, before inserting the dried paper disk. Blank samples (15 ml of 2 M KCl) were used to adjust for any background ^{14}N and [15]N. In all samples, the atom% [15]N of the dried paper disks was determined after conversion of mineral N to molecular N_2 using an Automatic Elemental Analyser (EA 1110) coupled to an isotope ratio mass spectrometer (Finnigan MAT Delta[plus], Thermo Finnigan, USA). The adequacy of the diffusion and spiking techniques was checked on KCl extracts with known ^{14}N and ^{15}N amounts of NH_4^+-N and NO_3^--N. Preliminary tests showed that isotopic discrimination between ^{14}N and ^{15}N when diffusing KCl extracts was not of importance with an atom% [15]N recovery between 97-99 % (data not shown).

Total recovery of ^{15}N added to the soil was carried out at the end of the incubation period after 2-week pre-incubation. Soil samples (not used for extraction) were oven-dried at 105°C, milled and the atom% [15]N of the entire soil samples was determined after converting organic N to molecular N_2, as described above.

Calculation of gross N mineralisation rates

Gross N mineralisation rates in soils were calculated from the dilution of the ^{15}N-labelled soil NH_4^+-N pool using equations (1) and (2) (Herrmann, 2003), which are based on first-order kinetics for consumption rates:

$$^{15}A(t) = {}^{15}A_0 \exp(-c\,t) \tag{1}$$

$$^{Total}A(t) = \frac{m}{c} + \left({}^{Total}A_0 - \frac{m}{c} \right) \exp(-ct) \tag{2}$$

where A is the amount of N in the NH_4^+ pool (µg N g^{-1} soil), m is the gross N mineralisation rate (µg N g^{-1} soil h^{-1}), c is the first-order rate constant for consumption of NH_4^+ (h^{-1}) and t is time after addition (h). The superscripts *total* and *15* denote the amount of total N and the amount of ^{15}N in excess of natural abundance in the NH_4^+ pool (µg N g^{-1} soil), respectively. It was assumed that the natural abundance of the NH_4^+ pool was the same as that of total soil N. The suffix *0* denotes the start of the incubation period (time of addition). The first-order rate constant c was estimated by non-linear regression (equation (1)) from the disappearance of ^{15}N in excess of natural abundance during the incubation period using the amount of added ^{15}N in excess of natural abundance as a starting value for $^{15}A_0$ (µg N g^{-1} soil).

Statistical analysis

All experiments were carried out on soil samples from each of the four replicate blocks of the field experiment. The mean values reported are for the four blocks with the significance of treatment effects tested by analysis of variance for a completely randomised design. Equations (1) and (2) were fitted to the data for each replicate separately, using procedure NLIN in the SAS software (1999-2000). The resulting set of parameter values was subjected to analysis of variance.

Results and discussion

The ^{15}N isotope dilution technique for estimating gross N mineralisation rates is based on a number of assumptions (Powlson & Barraclough, 1992). Violation of these assumptions may, however, have a large impact on the estimates of gross N rates and testing these assumptions is therefore an important part of the evaluation of such estimates.

Loss of added N

Recovery of added N in the soil at the end of the incubation period after 2-week pre-incubation was slightly, but statistically significant ($P < 0.05$), less than 100% in the Fallow, FYM and SD + N treatment (Table 2). Nevertheless, we assumed no gaseous losses of added N in the calculation of gross N mineralisation rates. Gaseous losses from denitrification would not have

Table 2: Recovery of added $^{15}NH_4^+$-N as a percentage of total soil (after 2-week pre-incubation) and mineral N pool (NH_4^+-N and NO_3^--N pool, respectively) after 2 and 120 h (after 2- and 7-week pre-incubation) or 72 h (after 17-week pre-incubation) incubation period. Mean values and standard errors of the mean of four replicates.

Recovery ^{15}N (%)

| | 2-week pre-incubation | | | | | 7-week pre-incubation | | | | 17-week pre-incubation | | | |
| | Total soil | NH_4^+-N | | NO_3^--N | | NH_4^+-N | | NO_3^--N | | NH_4^+-N | | NO_3^--N | |
	120 h	2 h	120 h	2 h	120 h	2 h	120 h	2 h	120 h	2 h	72 h	2 h	72 h
Fallow	83±5	84±2	4±1	1±2	74±5	89±3	7±2	7±3	89±3	113±9	24±6	8±3	83±7
N-fertilised	105±2	61±2	0.05±0.01	10±2	78±2	67±4	0.21±0.02	14±2	90±1	94±7	0.23±0.01	17±1	102±1
Straw + N	104±5	53±2	0.09±0.01	26±1	77±1	64±3	0.11±0.03	22±2	86±0.4	68±1	0.05±0.01	16±2	85±2
FYM	96±1	54±2	0.05±0.01	24±2	74±2	66±2	0.07±0.01	24±3	81±2	79±2	0.04±0.03	16±2	86±4
SD	98±8	70±2	0.14±0.01	11±2	50±8	77±2	0.24±0.02	9±1	71±0.3	77±6	0.08±0.01	15±2	82±6
SD + N	88±2	49±3	0.10±0.01	26±4	64±4	63±2	0.21±0.02	12±1	79±1	70±4	0.05±0.01	22±1	78±5

affected the calculations, whereas NH_3 volatilisation was deemed unlikely as the pH of the soil samples was below 7 while the added acidic solution was mixed into the soil. It is possible that incomplete reduction of NO_3^- to N_2 gas during dry combustion may have led to the incomplete recovery of the added N from the soil. In the calculation of gross N mineralisation, we assumed that added NH_4^+-N is completely extractable after addition (0 h). The latter assumption was based on the fact that we obtained 100% recovery of added NH_4^+-N in sterilised gamma irradiated soil samples (12 kGy) (data not shown).

Discrimination

Even though there was very little extractable NH_4^+ at the end of the incubation period (Table 2), it is unlikely that isotopic discrimination introduced a bias. The $\delta^{15}N$ in the NH_4^+ pool at the end of each incubation period was still at least 3- to 5-fold greater than any discrimination that may have occurred during N transformations in the soil, which is thought to be at most 20-30‰ $\delta^{15}N$ (Högberg, 1997).

Process kinetics

The assumption of first-order kinetics for NH_4^+ consumption rates in the present study is underpinned by the fact that already 2 h after addition substantial amounts (up to approximately 25% of added NH_4^+-N) were recovered in the NO_3^--N pool (Table 2), indicating that the added N was rapidly nitrified. Moreover, virtually all added N was consumed after 1-3 d with < 0.3% of added ^{15}N recovered in the NH_4^+ pool at 120 h or 72 h (excepting the Fallow treatment; Table 2). These amounts were significantly greater than zero in all treatments. The goodness of fit of the first-order equation compared with the measured data points is illustrated for the Fallow and Straw + N treatment in soils pre-incubated for 2 weeks (Figure 1). In contrast, assuming zero-order kinetics would have greatly underestimated NH_4^+ consumption. Clearly, the incubation period was too long in order to meet the assumption of constant zero-order rates during the incubation period.

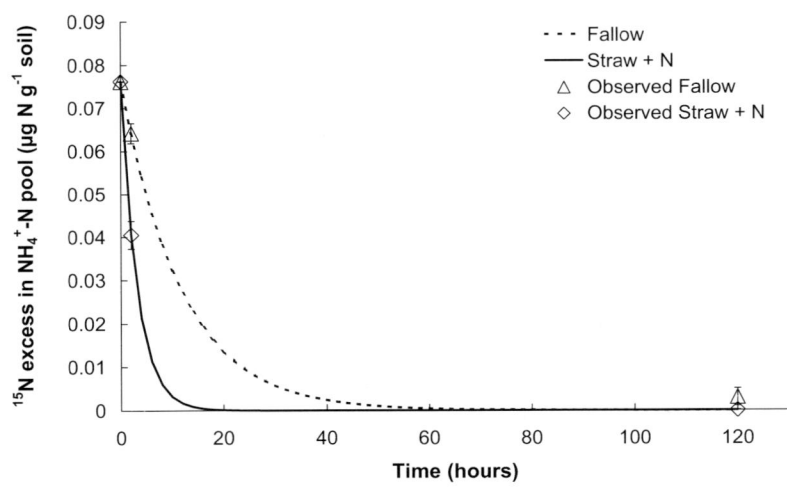

Figure 1. *Amounts of predicted and observed ^{15}N in excess of natural abundance in the NH_4^+ pool (µg N g^{-1} soil) for the Fallow and Straw + N treatment. Mean value and standard deviation of four replicates.*

Preferential use of added N

Preferential use of added N may be a more common occurrence in ^{15}N isotope dilution studies than hitherto thought and can lead to a 2- to 3-fold overestimation of gross N mineralisation rates (Herrmann, 2003). Nevertheless, gross N mineralisation rates in our study were similar to those found in other studies of arable soils (Murphy et al., 1999; Recous et al., 1999; Andersen & Jensen, 2001; Sørensen, 2001) and, in the absence of evidence for preferential use, were therefore, assumed to reflect true mineralisation rates.

Relation between C and gross N mineralisation rates

Plotting of C vs. gross N mineralisation across all treatments and samplings suggested proportionality between C and gross N mineralisation rates, irrespective of incubation time and treatment (Figure 2). The SD treatment showed, however, lower gross N mineralisation rates per amount of C mineralised in the first half of the incubation (i.e. after 2- and 3-week pre-incubation) and these points were considered as outliers (Figure 2).

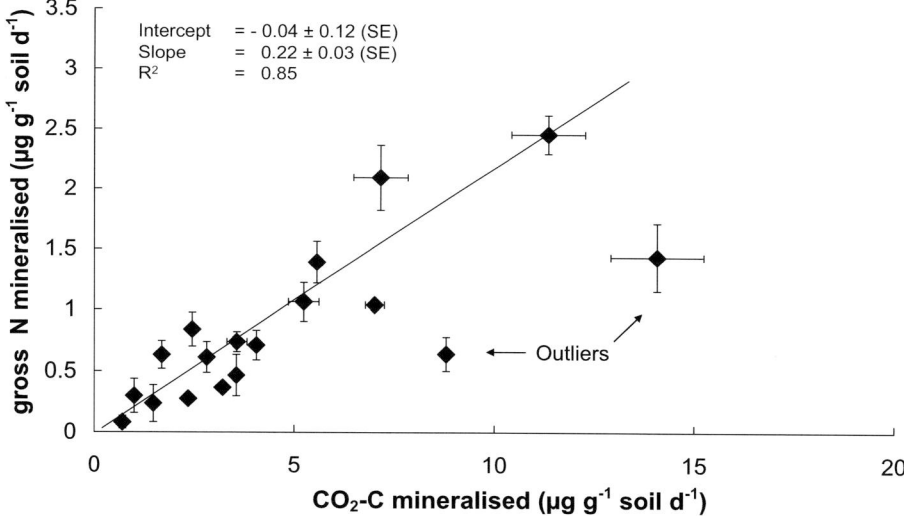

Figure 2. Gross N mineralisation rates in relation to C mineralisation rates among all soils. Mean value and standard deviation of four replicates.

The proportionality between C and gross N mineralisation rates underpins the general assumption in mechanistic models (e.g. Smith et al., 1997) that gross N mineralisation can be estimated from C mineralisation. For example, the SUNDIAL model (Smith et al., 1997) estimates gross N mineralisation from C mineralisation with knowledge of the C:N of the decomposing material, the microbial C use efficiency (α) and the fraction of soil organic matter stabilized (β). This model has been found to give adequate prediction of gross N mineralisation from C mineralisation in the field (Murphy et al., 2003). Using our gross N and C mineralisation data would suggest an average C:N value of 9 for the decomposing material in all soils (with the exception of the SD treatment for which the average C:N value over the entire incubation period would be 18). The C:N of the material undergoing decomposition was largely unaffected by the wide differences in the N contents of the organic amendments and the resulting differences in the values for the soil organic matter (Table 1). This suggests that the decomposing organic matter was of a similar nature in all soils, while low C:N suggests that the material may be of microbial origin. There are few studies where both C and gross N mineralisation have been measured simultaneously. Recous et al. (1999) also found a linear relation between C and gross N mineralisation in a loamy soil in the field over one year, but the proportionality factor was smaller compared with that obtained in the present study (0.1 versus 0.2, cf. Recous et al. (1999) and Figure 2). It is possible that under the conditions of our experiment (a relatively high constant temperature of 20° C and no C inputs) microbial products of a low C:N are a more dominant source for C and N mineralisation than under field conditions where there are inputs from crop residues.

Conclusions

Despite large differences in the C:N of the organic amendments and the resulting differences in the C:N of bulk soil organic matter, organic material undergoing decomposition was of a similar nature in all soils. Gross N mineralisation was proportional to C mineralisation with a proportionality factor of about 5 (gross N mineralisation in relation to C mineralisation).

Acknowledgments

This project was funded by the Swedish Research Council for Environment, Agricultural Sciences and Spatial Planning (FORMAS) and the Foundation of Swedish Research in Plant Nutrition (Stiftelsen Svensk Växtnäringsforskning).

References

Andersen, M.K. and Jensen, L.S. (2001). Low soil temperature effects on short-term gross N mineralisation-immobilisation turnover after incorporation of a green manure. Soil Biology and Biochemistry, 33, 511-521.

FAO (1988). Soil Map of the World. Revised legend. World Soil Resources Report 60, FAO Rome. 117 pp.

Goerges, T. and Dittert, K. (1998). Improved diffusion technique for N-15 : N-14 analysis of ammonium and nitrate from aqueous samples by stable isotope spectrometry. Communications in Soil Science and Plant Analysis, 29, 361-368.

Hansen, S., Jensen, H.E., Nielsen, N.E. and Svendsen, H. (1991). Simulation of nitrogen dynamics and biomass production in winter wheat using the Danish simulation model DAISY. Fertiliser Research, 27, 245-259.

Herrmann, A. (2003). Predicting Nitrogen Mineralisation from Soil Organic Matter - a Chimera? Doctoral Thesis, Swedish University of Agricultural Sciences, Department of Soil Sciences, SE-750 07 Uppsala, Sweden. Agraria 429. ISSN 1401-6249; ISBN 91-576-6468-4.

Högberg, P. (1997). Tansley review No 95 - N-15 natural abundance in soil-plant systems. New Phytologist, 137, 179-203.

Jansson, S.L. and Persson, J. (1982). Mineralisation and Immobilisation of Soil Nitrogen. In: F.J. Stevenson (ed). Nitrogen in Agricultural Soils. American Society of Agronomy Monograph no. 22, South Segoe Road, Madison, WI 53711, USA. pp. 229-252.

Kätterer, T. and Andrén, O. (2001). The ICBM family of analytically solved models of soil carbon, nitrogen and microbial biomass dynamics descriptions and application examples. Ecological Modelling, 136, 191-207.

McTaggart, I.P. and Smith, K.A. (1993). Estimation of potentially mineralizable nitrogen in soils by KCl extraction. II. Comparison with plant uptake in the field. Plant and Soil, 157, 175-184.

Murphy, D.V., Bhogal, A., Shepherd, M., Goulding, K.W.T., Jarvis, S.C., Barraclough, D. and Gaunt, J.L. (1999). Comparison of ^{15}N labelling methods to measure gross nitrogen mineralisation. Soil Biology and Biochemistry, 31, 2015-2024.

Murphy, D.V., Recous, S., Stockdale, E.A., Fillery, I.R.P., Jensen, L.S., Hatch, D.J. and Goulding, K.W.T. (2003). Gross nitrogen fluxes in soil: Theory, measurement and application of ^{15}N pool dilution techniques. Advances in Agronomy, 79, 69-118.

Parton W.J., Schimel, D.S., Cole, C.V. and Ojima, D.S. (1987). Analysis of factors controlling soil organic matter levels in great plains grasslands. Soil Science Society American Journal, 51, 1173-1179.

Persson, J. and Kirchmann, H. (1994). Carbon and nitrogen in arable soils as affected by supply of N fertilisers and organic manures. Agriculture, Ecosystems and Environment, 51, 249-255.

Powlson, D.S. and Barraclough, D. (1992). Mineralisation and Assimilation in Soil-Plant Systems. In: R. Knowles and T.H. Blackburn (eds). Nitrogen Isotope Techniques. Academic Press, Inc.; Harcourt Brace Jovanovich, San Diego, New York, Boston, London, Sydney, Tokyo, Toronto. pp. 209-242.

Recous, S., Aita, C.and Mary, B. (1999). In situ changes in gross N transformations in bare soil after addition of straw. Soil Biology and Biochemistry, 31, 119-133.

SAS software (1999-2000). SAS Institute Inc., Cary, NC, USA.

Shepherd, M.A., Stockdale, E.A., Powlson, D.S. and Jarvis, S.C. (1996). The influence of organic nitrogen mineralisation on the management of agricultural systems in the UK. Soil Use and Management, 12, 76-85.

Smith, P., Smith, J.U., Powlson, D.S., McGill, W.B., Arah, J.R.M., Chertov, O.G., Coleman, K., Franko, U., Frolking, S., Jenkinson, D.S., Jensen, L.S., Kelly, R.H., Klein-Gunnewiek, H., Komarov, A.S., Li, C., Molina, J.A.E., Mueller, T., Parton, W.J., Thornley, J.H.M. and Whitmore, A.P. (1997). A comparison of the performance of nine soil organic matter models using datasets from seven long-term experiments. Geoderma, 81, 153-225.

Soil Survey Staff (1987). Keys to Soil Taxonomy (3rd printing). SMSS Technical Monograph no. 6, Ithaca, New York.

Sørensen, P. (2001). Short-term nitrogen transformations in soil amended with animal manure. Soil Biology and Biochemistry, 33, 1211-1216.

Zibilske, L.M. (1994). Carbon Mineralisation. In: R.W. Weaver (ed). Methods of Soil Analysis, Part 2 - Microbiological and Biochemical Properties. SSSA Book Series, no.5, Madison. pp. 835-863.

N mineralisation from decomposition of catch crop residues under field conditions: measurement and simulation using the STICS soil-crop model

E. Justes[1] and B. Mary[2]
[1]*INRA, UMR ARCHE, Auzeville, BP27, 31326 Castanet-Tolosan, France*
[2]*INRA, Unité d'Agronomie Laon-Reims-Mons, 02007 Laon Cedex, France*

Abstract

The objective of this work was to evaluate the ability of the STICS soil-crop model to predict N dynamics in soil with catch crop residues incorporated in field plots. STICS was evaluated in bare soils, with or without catch crop residues, using parameters from incubation studies. Because of discrepancies between observed and simulated data, the decomposition submodel was re-calibrated using some components of available field experiments and was finally validated successfully with the remaining treatments. However, further studies on the effect of low temperatures on catch crop residue decomposition are required.

Keyword: catch crop residues, field conditions, mineralisation, modelling, N, STICS model

Introduction

As well as good N fertiliser management, the control of N in soil in the fallow period between two main crops is required to limit NO_3^- concentration in drainage water in temperate climates. The efficiency of catch crops (CC) to prevent NO_3^- pollution has been widely demonstrated. However, additional work is needed to optimise their management in order to simultaneously minimise water transpiration, maximise N removal from soil and N release for the next crop (Dorsainvil, 2002). An efficient solution to achieve this goal is to use a crop model able to simulate N dynamics during CC growth and during decomposition of CC residues. In fact, the CC can induce a pre-emptive competition for N and thereby, reduce N uptake of the succeeding crop: because N mineralised from CC residues during their decomposition may not compensate for the depletion in soil mineral N through CC growth (Thorup-Kristensen, 1994). Thorup-Kristensen et al. (2003) pointed out the importance of using a carefully tested and calibrated model in order to trust the simulation outputs. The objective of this work was to evaluate the ability of the STICS model (Brisson et al., 1998; 2003) to predict N mineralisation and the evolution of mineral N in soil with CC residues incorporated in actual field conditions. The evaluation was firstly made in fallow soil, without CC residues, in order to evaluate the processes of organic matter mineralisation and NO_3^- transfer in soil. Finally, model evaluation was carried out after CC incorporation for a wide range of C:N values of CC residues.

Materials and methods

Overview of STICS model

The soil-crop model STICS simulates water and NO_3^- transport using a capacity-type approach. It simulates water drainage and evaporation, N mineralisation and leaching. Nitrate transport is simulated using the mixing cells approach which accounts for NO_3^- dispersion by varying the thickness of soil layers (Mary et al., 1999). Nitrogen mineralisation comes from two main sources: soil organic matter (SOM) and crop residues. Both processes are affected by soil temperature and moisture conditions. N mineralisation from SOM depends on the amount of organic-N in the tilled soil layer and on its clay and limestone contents. N release from crop residues is a function of the amount and their C:N ratio. It also depends on the location of the residues in the soil profile, since mineral N can be a limiting factor of the decomposition. The model can simulate net mineralisation or immobilisation.

The decomposition of organic residues in soil considers three organic pools: residues, (zymogenous) microbial biomass and humified SOM (Figure 1). The decomposed C is either mineralised as CO_2 or assimilated by the soil microflora; microbial decay producing both C humification and secondary C mineralisation. N dynamics is governed by C rates and the C:N ratio of the compartments which remain constant in the absence of N limitation. The model was calibrated using a data set produced under laboratory conditions for mature residues. Three parameters of the model were correlated with the C:N of residues: i) the decomposition rate constant of residues (k), ii) the C:N of biomass ($C:N_{MB}$) and iii) the humification rate (h). The C:N of organic residues ($C:N_R$) was then used as the single criterion of 'quality' (Nicolardot et al., 2001). The model has been further tested and recalibrated for immature residues derived from catch crops (Justes et al., 2001).

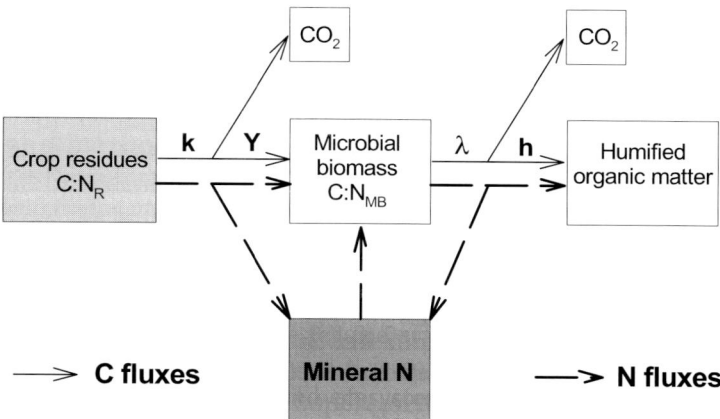

Figure 1. Conceptual diagram of the organic residue decomposition module of the STICS model (Nicolardot et al., 2001). k: residue decomposition rate; y: assimilation yield; λ: biomass decomposition rate; h: humification rate; $C:N_{MB}$: biomass C:N; $C:N_r$: residues C:N.

Field experiments

Five field experiments were carried out during 1998-1999 and 1999-2000 in soils mainly located in northern France (Table 1). Three species were used as CC: white mustard (*Sinapis alba* L.), Italian ryegrass (*Lolium multiflorum* L.) and radish (*Raphanus sativus* L.). Treatments with or without N fertiliser or irrigation were used to obtain a large range of crop biomass and C:N ratios. All experiments included a treatment without CC residues as a control for N mineralisation from SOM. We also used a data set from another experiment with incorporated rapeseed volunteers (Justes et al., 1999).

In order to parameterise the decomposition module of STICS for these immature residues, the 14 CC residues (C:N ranging from 11 to 32) coming from the two experiments located in Warmeriville (Table 2) were incubated with the corresponding soil in the laboratory at 15°C, at soil moisture close to field capacity and for non-limiting N conditions (Justes et al., 2001). In field experiments, the incorporated biomass of CC (shoots + roots) varied widely, from 0.9 to 9.9 t ha^{-1}, as did its C:N ratio, i.e. from 10 to 37 (Table 2). Each experiment started at the date of CC destruction which varied from November to early April. CC were all incorporated firstly by crushing or disk ploughing and finally, conventional ploughing immediately afterwards to 20-25 cm depth. The soil was then maintained as a bare fallow using herbicide applications.

Table 1. Characteristics of cover crop field experiments.

Site	Year	Soil type	Species	Treatment No.	Date of incorporation
Warmeriville 1	1998-1999	Chalk	Mustard	4	5 Nov & 3 Dec 1998
		(hypercalcareous	Ryegrass	2	03 Dec 1998
Warmeriville 2	1999-2000	rendosol)	Mustard	6	29 Nov 1999
			Radish	2	29 Nov 1999
Boigneville	1999-2000	Silty clay loam	Mustard	2	06 Dec 1999
			Ryegrass	1	11 Apr 2000
Lusignan	1999-2000	Silty loam	Mustard	1	17 Dec 1999
			Ryegrass	2	04 Apr 2000
Somme-Tourbe	1999-2000	Chalk	Mustard	2	06 Dec 1999

Table 2. Description of CC residues data set and ranges in their C and N composition.

Experiment	Species	Treatment No.	Total N (% DM)	NO$_3$-N (% DM)	Total C (% DM)	organic C:N ratio	Field DM (t ha^{-1})
Lab+Field	Mustard	12	1.4 to 4.7	0.0 to 1.1	39 to 43	11 to 32	0.9 to 9.9
Lab+Field	Ryegrass	2	1.4 & 2.8	0.0 & 0.2	42	30 & 16	1.9 & 3.3
Lab+Field	Radish	2	2.1 & 3.2	0.0 & 0.3	40 & 39	19 & 13	3.6 & 8.7
Field	Mustard	5	1.8 to 4.6	0.0 to 0.5	40 to 42	10 to 23	1.5 to 3.4
Field	Ryegrass	3	1.1 to 1.5	0.0	41 to 42	28 to 37	3.2 to 8.6
Field	Rapeseed	1	2.9	0.1	40	14	1.4

DM = dry matter; Field DM = amount of CC residues (shoots+roots) incorporated in field.

Soil sampling and analysis
Eight soil samples were collected to depths of 90 or 120 cm in each block, using a hydraulic coring device, every 2-4 weeks from the date of incorporation until next June or September (for Boigneville and Lusignan sites). Each core was divided into three or four layers of 30 cm. Bulk densities were measured once, in each soil layer. The gravimetric water content (oven-dry basis, 105°C, 48h) was measured in each layer. Fresh samples (100 g) were mixed with 200 ml KCl 1M, shaken for 30 min. and filtered. Nitrate-N and NH_4^+-N concentrations in the extract were determined by colorimetry (Skalar Analytical) using a modified Griess method and the Berthelot method, respectively.

Nitrogen mineralisation rates (humus + residues) were calculated on the basis of measured soil water and mineral N contents using the calculation model LIXIM (Mary et al., 1999). This model calculates the most probable values of soil evaporation and N mineralisation by fitting the water and mineral N profiles measured at each sampling date. Net N mineralisation (or immobilisation) due to decomposition of CC residues was obtained by subtracting the mineralisation calculated in the control soil.

Approach used to evaluate STICS decomposition submodel for CC
The measured soil water and NO_3^- contents were compared with STICS simulations. Four steps were carried out:
- Step 1: STICS was evaluated in bare soils without CC residue; the input soil values for each site were either measured or fitted
- Step 2: STICS was evaluated in soils with CC residues; this was done for the 14 CC that were also incubated; the simulated net N mineralisation from CC residues was also compared with calculations based on field data
- Step 3: if needed, the decomposition submodel was re-calibrated using part of available treatments carried out in field experiments
- Step 4: STICS was independently evaluated using the remaining treatments

Three statistical criteria were used: model efficiency (*EF*), the relative root mean square error (*RRMSE*) and the relative coefficient of residual mass (*CRM*), as an indicator of bias.

Results

Net N mineralisation in field experiments
Surprisingly, soil mineral N rapidly increased after CC incorporation even in winter for December incorporation (results not shown). This indicates that a net N release from CC occurred within 2 or 3 weeks after soil tillage, particularly for the CC with a low biomass and a low C:N ratio. The net N release continued for 3 or 4 months but the rate of mineralisation decreased rapidly. No significant N release was measured after 6 months.

Step 1. Evaluation of STICS model in bare soil without CC residues
Both water and NO_3^- contents in soil profile were satisfactorily predicted by STICS without significant bias in the 6 experiments without CC residues (Figure 2). No difference in prediction was obtained between chalky and loamy soils. These results suggest that water and NO_3^- transport and N mineralisation from humus were correctly simulated.

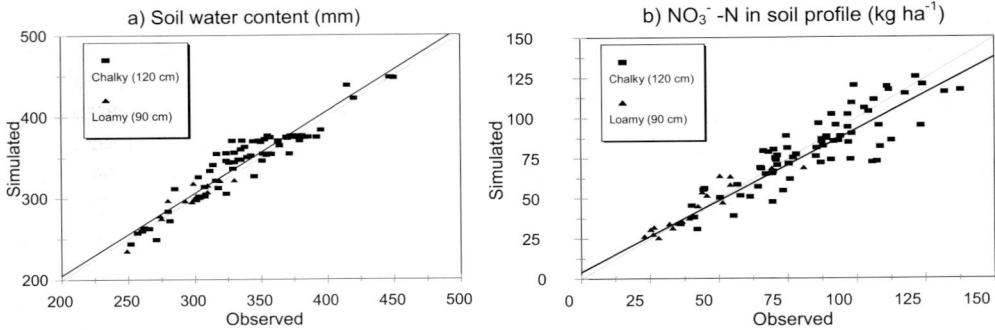

Figure 2. First step of STICS evaluation: simulated vs. observed data in bare soil without CC residues. a) soil water content; b) soil NO_3^- content (full line = linear regression; dotted line = first bisector).

Step 2. Evaluation of STICS in field experiments with CC residues

The soil water content (*SWC*) was simulated correctly as before without CC residues (results not shown), whereas the amounts of NO_3^- in soil (*SMN*) were underestimated by the model, both in the whole profile and in the upper 0-30 cm layer (Figure 3b; Table 3). This is attributed to an underestimation by the model of the N mineralisation derived from CC residues, for the majority of the treatments (Figure 3c).

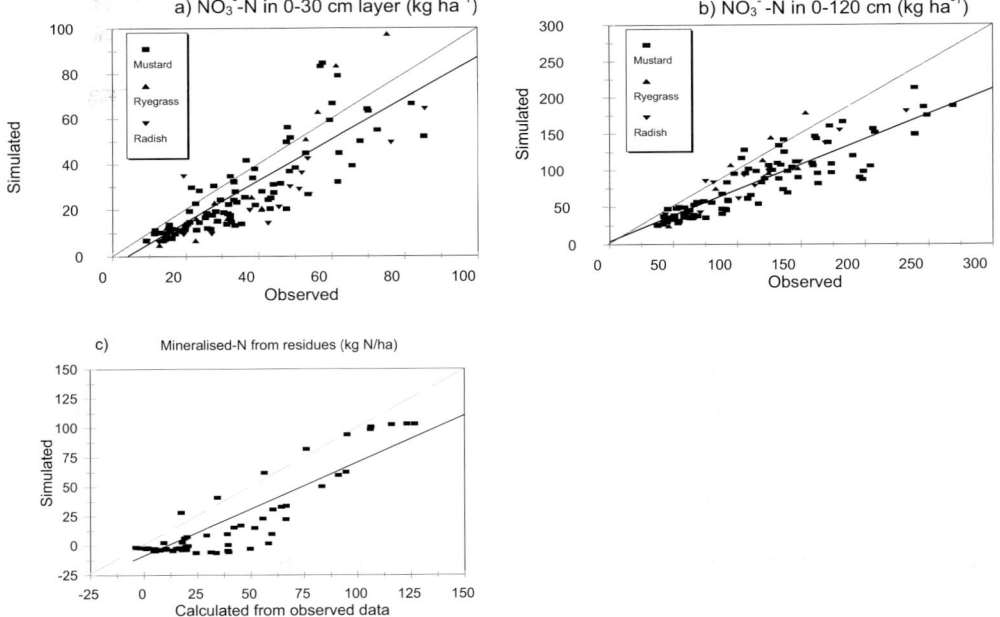

Figure 3. Second step of STICS evaluation: simulated vs. observed data in bare soil with CC residues. a) NO_3^- content in the 0-30 cm layer; b) NO_3^- content in the 0-120 cm layer; c) N mineralised during the whole experiment (about 6 months).

Step 3. Re-calibration of STICS with field experiments

The residue decomposition submodel of STICS was recalibrated by optimising the five parameters governing residue decomposition using a part of the field database: we used the data from 10 out of 23 treatments. The optimisation was carried out with the STICS optimiser. The N mineralised from CC residues could be much better predicted than before, without bias (Figure 4c). The amounts of NO_3^- in soil were also better simulated both in the upper layer and in the whole profile (Figure 4). The optimisation procedure led to changes in three parameters: the decomposition rates from residues (k) and microbial biomass (<f"symbol">I>λ<f"symbol">I>) (see Figure 1) which were greatly increased. For example, for a residue with C:N of 20, the decomposition rates were multiplied by a 2.5 fold factor. The third parameter affected was the humification yield coefficient (h) which was decreased by 20%. An important point is that the mean soil temperature was ca. 5°C in the field during the first 4 months after autumn incorporation, whereas it was 15°C in laboratory conditions.

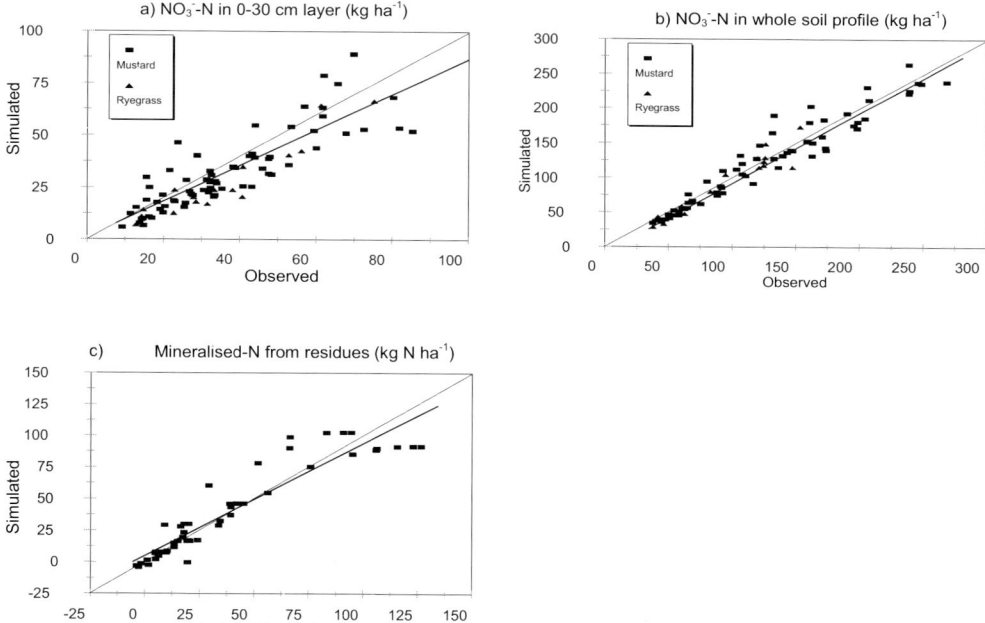

Figure 4. Third step of STICS evaluation: simulated vs. observed data in bare soil with CC residues in 10 treatments out of 23. a) NO_3^- content in the 0-30 cm layer; b) NO_3^- content in the 0-120 cm layer; c) N mineralised.

Step 4. Evaluation of the newly calibrated model

The newly-calibrated model was evaluated independently on the 13 remaining treatments. The amounts of NO_3^- in the soil profile were well predicted both in the upper layer and in the whole soil profile (Figure 5; Table 3). Moreover, no significant effect of CC species was found amongst the four tested here.

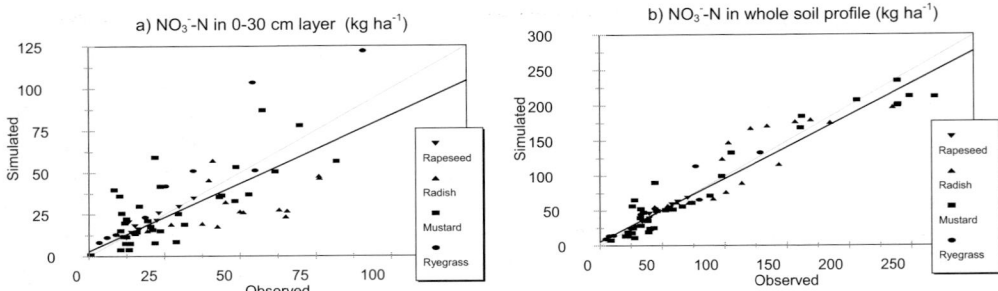

Figure 5. Fourth step of STICS evaluation: simulated vs. observed data in bare soil with CC residues in the 13 remaining treatments. a) NO_3^{-1} content in the 0-30 cm layer; b) NO_3^- content in the 0-120 cm layer.

Table 3. Statistical criteria obtained for STICS model evaluation for steps 2 and 4.

	Step 2			Step 4		
	EF	RRMSE %	CRM %	EF	RRMSE %	CRM %
SWC 0-120 cm	0.61	6	-3	0.92	3	-1
SMN 0-120 cm	0.47	35	+26	0.95	24	0
SMN 0-30cm	0.42	38	+22	0.61	42	-4
Mineralised N*	*0.53*	*92*	*+52*	*0.79*	*59*	*-2*

* N mineralised from CC residues. *EF:* model efficiency; *RRMSE:* relative root mean square error; *CRM:* relative coefficient of residual mass.

Discussion

The amount of N mineralised after the addition of CC residues varied in our field experiments from -5 to +125 kg N ha^{-1} in about 6 months. The N release depended mainly on the amount of CC residues incorporated and their C:N ratios. The STICS model was previously calibrated using laboratory incubation data performed with some of the CC residues and could adequately reproduce the C and N mineralisation kinetics measured under laboratory conditions (Justes et al., 2001). However, N release measured in field conditions was greater than that predicted by the model thus calibrated. Thorup-Kristensen et al. (2003) also pointed out the fact that large differences between field results and simulated values extrapolated from laboratory incubation can be obtained, and mentioned that the extent of error which may occur is not always clear. The N release in our experiments was underestimated probably because the decomposition rates were underestimated. Four possible explanations can be put forward:

i. the fact that N rhizodeposits and some undecomposed dead leaves are not accounted for in incubation studies;

ii. the smaller contact between soil and residues *in situ* compared with laboratory conditions may have reduced the net N immobilisation;

iii. the decomposition rates could be faster at low temperatures for catch crops than for mature crops;

iv. gross N immobilisation may be more sensitive to low temperatures than gross N mineralisation (Henriksen & Breland, 1999; Magid et al., 2001).

The last two hypotheses seem the most plausible. In fact, Breland (1994) showed that decomposition of clover green manure was rapid even at temperatures below 5°C. Van Schöll et al. (1997) observed that 20% of organic N of rye CC had been mineralised after only ten weeks of incubation at 1°C; their model also underestimated the mineralisation during the first weeks of incubation. Magid et al. (2001) also found that the C mineralisation rates of three green manures were not much reduced when the temperature dropped from 25 to 9 and 3°C. They did not observe a concomitant retardation of N mineralisation, and nitrification was not deterred at 3°C. Breland (1994) concluded that the predictions of N mineralisation were sensitive to the type of function applied for correction of decay rates at temperatures below 0°C. It seems necessary to establish precise temperature functions at low temperature for unmature crop residues, since the present temperature functions used in STICS were established for mature residues (maize) in the range 1 to 25°C.

Conclusion

This work has enabled us to evaluate the formalism of the decomposition submodel of STICS and the approach used to parameterise the model. It confirms both the interest and the difficulty in parameterising models using incubation data (Van Schöll et al, 1997; Thorup-Kristensen & Nielsen, 1998). Since low temperatures are frequently encountered in northern Europe during the decomposition of CC residues, our results suggest that further studies on the effect of low temperatures on decomposition of immature plant residues are required for a better understanding and prediction of N mineralisation in cropping systems, including catch crops. This work also indicated that a satisfactory prediction of N dynamics was obtained by using the newly calibrated model. It would be preferable to modify the temperature function at low temperatures in order to obtain more robust results whatever the date of CC incorporation. However, considering these results and other evaluations not shown here, it seems possible to simulate correctly the short term effect of CC on soil mineral N, N leaching and N release for the next crop. The STICS model can be used to optimise catch crop management, at least in the short-term (Dorsainvil, 2002).

References

Breland, T. (1994). Measured and predicted mineralisation of clover green manure at low temperature at different depths in two soils. Plant and Soil, 166, 13-20.

Brisson N., Mary, B., Ripoche, D., Jeuffroy, M-H., Ruget, F., Nicoullaud, B., Gate, P., Devienne-Barret, F., Antonioletti, R., Dürr, C., Richard, G., Beaudoin, N., Recous, S., Tayot, X., Plenet, D., Cellier, P., Machet, J.M., Meynard, J.M. and Delécolle, R. (1998). STICS: a generic model for the simulation of crops and their water and nitrogen balances. Agronomie, 18, 311-346.

Brisson, N., Gary, C., Justes, E., Roche, R., Mary, B., Ripoche, D., Zimmer, D,. Sierra, G., Bertuzzi, P., Burger, P., Bussière, F., Cabidoche, Y.M, Cellier, P., Debaeke, P., Gaudillère, J.P., Hénault, C., Maraux, F., Seguin, B. and Sinoquet, H. (2003). An overview of the crop model STICS. European Journal of Agronomy, 18, 309-332.

Dorsainvil, F. (2002). Evaluation, par modélisation, de l'impact environnemental des cultures intermédiaires sur les bilans d'eau et d'azote dans les systèmes de culture. PhD Thesis, INA P-G, Paris, 124pp.

Henriksen, T.M. and Breland, T.A. (1999). Decomposition of crop residues in the field: evaluation of a simulation model developed from microcosm studies. Soil Biology and Biochemistry, 31, 1423-1434.

Justes, E., Mary, B. and Nicolardot, B. (1999). Comparing the effectiveness of radish cover crop, oilseed rape volunteers and oilseed rape residues incorporation for reducing nitrate leaching. Nutrient Cycling in Agroecosystems, 55, 207-220.

Justes, E., Nicolardot, B. and Mary, B. (2001). C and N mineralisation of catch crop residues: measurements and evaluation of STICS model. Proceedings of the 11th Nitrogen Workshop, 9-12 September 2001, Reims, France, INRA, 113-114.

Magid, J., Henriksen, O., Thorup-Kristensen, K. and Mueller, T. (2001). Disproportionately high N-mineralisation rates from green manures at low temperatures - implications for modelling and management in cool temperate agro-ecosystems. Plant and Soil, 228, 73-82.

Mary, B., Beaudoin, N., Justes, E. and Machet, J.M. (1999). Calculation of nitrogen mineralisation and leaching in fallow soil using a simple dynamic model. European Journal of Soil Science, 50, 549-566.

Nicolardot, B., Recous, S. and Mary, B. (2001). Simulation of C and N mineralisation during crop residue decomposition: a simple dynamic model based on the C:N ratio of residues. Plant and Soil, 228: 83-103.

Thorup-Kristensen, K. (1994). The effect of nitrogen catch crop species on the nitrogen nutrition of succeeding crops. Fertilizer Research, 37, 227-234.

Thorup-Kristensen, K., Magid, J. and Jensen, L.S. (2003). Catch crops and green manures as biological tools in nitrogen management in temperate zones. Advances in Agronomy, 79, 228-302.

Thorup-Kristensen, K. and Nielsen, N.E. (1998). Modelling and measuring the effect of nitrogen catch crops on the nitrogen supply for succeeding crops. Plant and Soil, 203, 89-89.

Van Schöll, L., Van Dam, A.M. and Leffelaar, P.A. (1997). Mineralisation of nitrogen from incorporated catch crops at low temperature: experiment and simulation. Plant and Soil, 188, 211-219.

Gross N transformation rates and soil organic matter contents in grassland soils of different age

F. Accoe[1,2], P. Boeckx[1], O. Van Cleemput[1] and G. Hofman[2]
[1]*Laboratory of Applied Analytical Chemistry - ISOFYS, Ghent University, Belgium*
[2]*Department of Soil Management and Soil Care, Ghent University, Belgium*

Introduction

In this study we investigated the soil organic N contents in different size and density fractions of soil organic matter (SOM) and the gross N transformation rates in two soil layers (0-10 and 10-20 cm depth) from three grassland soils of different age with a sandy loam texture. These soils had been converted from cultivation to permanent grassland: 6, 14 and approximately 50 years ago, respectively.

Materials and methods

The SOM was separated into the following size and density fractions, by the method of Meijboom et al. (1995): the light density (d<1.37 g cm^{-3}) macro-organic matter fraction (LF 150-2000 µm), the heavy density (d>1.37 g cm^{-3}) macro-organic matter fraction (HF 150-2000 µm) and the size fractions 50-150 µm and <50 µm. The gross N mineralisation, nitrification and immobilisation rates were determined by means of a fully mirrored ^{15}N isotope dilution experiment in the laboratory. Freshly collected soil samples were sieved, homogenized and pre-incubated for 7 days at 15°C before the start of the experiment. Differentially ^{15}N-labelled (10 at%) NH_4NO_3 was applied at a rate of 30 mg N kg^{-1} soil. The incubations were performed at 50% water-filled pore space, at the bulk density in the field and 15°C. After 3 and 7 days of incubation, soil samples were extracted with 2 M KCl for determination of the concentration and ^{15}N-enrichment of the NH_4^+ and NO_3^- pools. The gross rates of N mineralisation, nitrification and immobilisation were calculated by means of the equations of Kirkham and Bartholomew (1954).

Results and discussion

In the 0-10 cm layer, the total N contents increased with increasing age of the investigated grasslands (ranging from 1.8, 2.1 to 3.7 g N kg^{-1} soil). With increasing age of the grasslands, the largest relative increase in soil organic N content was observed in the HF 150-2000 µm fraction, followed by the 50-150 µm and <50 µm size fractions (Figure 1).

In the three grasslands, both gross N mineralisation and nitrification rates tended to be larger in the 0-10 cm layer than in the 10-20 cm layer (Table 1). Gross rates of mineralisation, NH_4^+immobilisation and nitrification showed an increase with increasing age of the investigated grasslands in both layers. The gross N mineralisation rates were strongly correlated with the soil organic N contents in the size fraction <50 µm (y = 3.3x - 1.9, R^2=0.95, p<0.001) and the HF 150-2000 µm fraction (y = 4.3x + 0.4, R^2=0.95, p<0.005). In the three grasslands, the potential NH_4^+ immobilisation/gross mineralisation ratio was considerably smaller in the 0-10

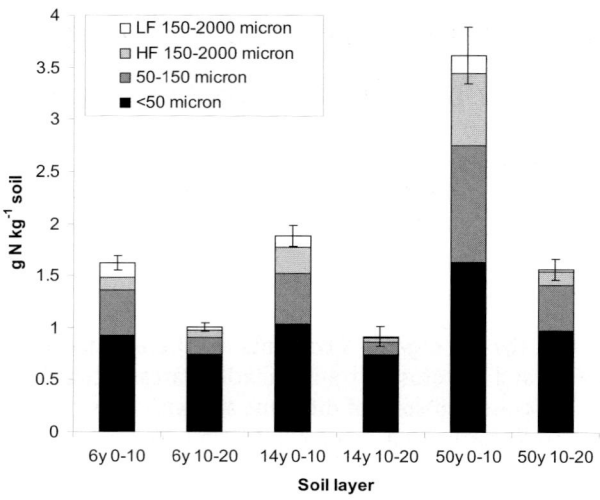

Figure 1. Soil organic N contents in the light (LF) and heavy density (HF) macro-organic matter fraction 150-2000 µm, the size fraction 50-150 µm and the size fraction <50 µm in three grassland soils (0-10 and 10-20 cm depth) of different ages.

Table 1. Gross rates of N mineralisation, NH_4-immobilisation and nitrification in the three grassland soils (0-10 and 10-20 cm depth) between day 3 and day 7 of a ^{15}N isotope dilution experiment (averages of 3 replicates, standard deviations in brackets).

Age of grassland (yr)	Depth (cm)	Mineralisation	NH_4^+ immobilisation (mg N kg^{-1} soil d^{-1})	Nitrification
6	0-10	0.62 (0.28)	0.36 (0.28)	1.09 (0.17)
	10-20	0.46 (0.14)	0.68 (0.09)	0.68 (0.14)
14	0-10	1.49 (0.16)	1.15 (0.33)	1.68 (0.15)
	10-20	0.71 (0.10)	0.90 (0.30)	1.11 (0.07)
50	0-10	3.39 (0.31)	1.39 (0.60)	3.73 (0.08)
	10-20	1.37 (0.16)	4.08 (1.34)	3.02 (0.86)

cm layer (0.6, 0.8 and 0.4, respectively, with increasing age of the grasslands) than in the 10-20 cm layer (1.5, 1.3 and 3.0, respectively), indicating a larger N immobilisation potential in the 10-20 cm layer in relation to the 0-10 cm layer.

Acknowledgements

This study was supported by the IWT of the Ministry of the Flemish Community, Belgium.

References

Kirkham, D., and Bartholomew, W.V. (1954). Equations for following nutrient transformations in soil utilizing tracer data. Soil Science Society of America Proceedings, 18, 33-34.

Meijboom, F.W., Hassink, J. and Van Noordwijk, M. (1995). Density fractionation of soil macro-organic matter using silica suspensions. Soil Biology and Biochemistry, 27, 1109-1111.

The relative efficiency of fertiliser N in NK blends and concentrated complex fertilisers

N.A. Akhonzada and J.S. Bailey
The Queen's University of Belfast, Department of Agricultural and Environmental Science, Newforge Lane, Belfast, BT9 5PX, United Kingdom

Introduction

The relative efficiency of N in NK blends and concentrated complex fertilisers (CCF's) has received little attention, despite increased awareness that fertiliser N efficiency, and hence the potential for N loss to the environment, is dependent on interactions at microsite or granule level. Given the fundamental role of K^+ in NO_3^- uptake by plant roots (Minotti et al., 1968) it is hypothesised that this process should be most efficient if both ions were supplied in the same fertiliser granule. A pot experiment was conducted to see if supplying N and K in the same pellet (NK-CCF), or in separate pellets (NK-blend) with different spacing between N and K fertiliser entities, influenced the recovery of $^{15}NH_4^+$ and $^{15}NO_3^-$ by perennial ryegrass.

Materials and methods

The experiment was carried out in a controlled environment cabinet, which mimicked the average conditions of day length and temperature in April (day length 13 hours, day temperature 15°C and night temperature 10°C). It had a factorial design with: 3 fertiliser forms, F_1 (N and K in the same pellet, i.e. 12 pellets per pot), F_2 (N and K in separate pellets, i.e. 12 N pellets and 6 K pellets per pot) and F_3 (N and K in separate pellets, i.e. 12 N pellets and 1 K pellet per pot); 2 locations of the ^{15}N label, L_1 ($^{15}NH_4^+$) and L_2 ($^{15}NO_3^-$); and 4 harvests, i.e. days 6, 14, 22 and 37 after treatment application. In each F treatment, N was supplied at 100 mg per pot as NH_4NO_3, and K was supplied at 50 mg per pot as KCl. Each treatment had 4 replicates, giving a total of 96 pots. Six perennial ryegrass seedlings were transplanted into plastic pots, 125 mm in diameter, containing 1 kg of fresh soil, which was maintained at 80% of field water capacity. Plants were grown to maturity over two preliminary growth periods, totalling 12 weeks, before the experimental treatments were applied. At each of the 4 experimental harvest days, plants were destructively harvested, and shoots and roots were separated, dried, weighed, milled and analysed for ^{15}N and other nutrients. Chloride was not determined in plant roots because of interferences from residual soil mineral matter.

Results

At day 37, shoot dry matter (DM) yield in the 6-K pellet treatment (F_2) was 12.8% (P<0.05) greater than that in the one-K pellet treatment (F_3) and 9.8% greater than that in the NK-CCF treatment (F_1) (Table 1). At day 22, ^{15}N utilisation by shoots was significantly greater in the 6-spot blend treatment (F_2) than in either of the other two treatments. At day 37, ^{15}N utilisation by shoots and whole plants (data not shown) was also significantly greater in F_2 than in F_3 (Table 1). At day 14 of regrowth, shoot and whole plant utilisation of ^{15}N-labelled NO_3^--N was 16% greater in F_2 than in F_1 (data not shown since the $F \times L \times$ harvest interaction was not significant), and the concentration of Cl in shoots was 19% less in F_2 than in F_1 (Table 2). The concentration of K in plant shoots was significantly greater in F_1 and F_2 than in F_3, whereas, the concentration in plant roots was unaffected by F treatment (data not shown). The concentration of Cl in root tissue was not determined because of technical difficulties with the method.

Table 1. The effect of fertiliser form (F) on (DM) yield and ^{15}N utilisation by plant shoots at four points in time during a 37-day regrowth period (averaged over both L treatments).

Sampling day	DM yield (g pot^{-1})			LSD 95%	Signif. t-test	^{15}N utilisation (%)			LSD 95%	Signif. t-test
	F_1	F_2	F_3			F_1	F_2	F_3		
6	1.41	1.36	1.21	0.406	NS	12.88	12.45	12.31	7.170	NS
14	1.46	1.47	1.49	0.406	NS	27.26	28.83	27.79	7.170	NS
22	2.00	2.20	2.25	0.406	NS	50.90	59.01	51.42	7.170	*
37	3.99	4.38	3.88	0.406	*	65.02	71.86	63.75	7.170	*

Table 2. The effect of fertiliser form (F) on the concentrations of K and Cl in plant shoots at four points in time during a 37-day regrowth period (averaged over both L treatments).

Sampling day	Shoot K (g kg^{-1})			LSD 95%	Signif. t-test	Shoot Cl (g kg^{-1})			LSD 95%	Signif. t-test
	F_1	F_2	F_3			F_1	F_2	F_3		
6	22.43	23.98	24.37	2.804	NS	10.80	10.68	8.49	2.877	NS
14	31.34	31.11	27.56	2.877	*	16.90	13.64	12.50	2.877	*
22	33.29	31.81	32.29	2.877	NS	14.26	16.02	14.72	2.877	NS
37	19.12	20.30	20.79	2.877	NS	12.14	13.00	12.92	2.877	NS

Discussion

The results run contrary to the original hypothesis. Plants in F_1, the NK-CCF, produced less shoot DM and recovered less labelled N than those in F_2, i.e. the treatment with a similar ratio of N to K-fertiliser entities as a commercial blend. There is a strong possibility that Cl$^-$ in the NK-CCF pellets had competed with NO$_3^-$ for plant uptake during the first two weeks of regrowth (Deane-Drummond & Glass, 1982) as evidenced by the lower recovery of ^{15}NO$_3^-$ and the higher recovery of Cl$^-$ in plants shoots in the NK-CCF treatment at day 14, compared with the recoveries of these nutrients in shoots in F_2. Since Cl$^-$ has an inhibitory effect on nitrate reductase (NR) activity (Richharia et al., 1997), it is possible that the high concentration of this ion in shoots in F_1 at day 14 had suppressed NR activity and NO$_3^-$ assimilation and thereby hampered subsequent N (NH$_4^+$ plus NO$_3^-$) uptake by plants.

Conclusions

Antagonisms between Cl$^-$ and NO$_3^-$ in relation to NO$_3^-$ uptake and assimilation appear to outweigh any advantages of supplying K$^+$, Cl$^-$ and NO$_3^-$ in the same fertiliser pellet (NK-CCF) as opposed to separate fertiliser pellets (NK-blend).

References

Deane-Drummond, C.E. and Glass, A.D.M. (1982). Studies of nitrate influx into barley roots by the use of $^{36}ClO_3^-$ as a tracer for nitrate. 1. Interaction with chloride and other ions. Canadian Journal of Botany, 60, 2147-2153.

Minotti, P.L., Craig, W.D. and Jackson, W.A. (1968). Nitrate uptake and reduction as affected by calcium and potassium. Soil Science Society of America Proceedings, 32, 692-698.

Richharia, A., Shah, K. and Dubey, R.S. (1997). Nitrate reductase from rice seedlings: Partial purification, characterization and the effects of in situ and in vitro NaCl salinity. Journal of Plant Physiology, 151, 316-322.

Nitrogen mineralisation from fish sludge as affected by plant uptake

M.A. Alfaro, F.J. Salazar and A. Valdebenito
National Institute for Agricultural Research (INIA-Remehue). PO Box 24-0, Osorno, Chile

Introduction

The fish farming industry, with cages located in the sea and lakes, produces large quantities of sludge. Volcanic soils of southern Chile have low nutrient availability and high P fixation capacity. The recycling of fish sludge as fertiliser represents an opportunity to reduce the risk of water pollution. Nutrients in fish sludge can either be immediately available to plants or require mineralisation of the organic fraction before becoming available. In southern Chile, N is a strategic nutrient for grassland production. An understanding of N mineralisation from fish sludge is required to predict both short and long term release of plant available N, and to avoid high levels of soil N accumulation which may be subject to losses. The main aim of this study was to determine N mineralisation of the organic N fraction of fish sludge on two different volcanic soils as affected by plant uptake.

Materials and methods

A pot experiment was carried out under greenhouse conditions. The treatments tested were the origin of the fish sludge (sea or lake) and the soil type (andisol and ultisol). Both materials were collected in the Lake Region of Chile in January 2003. The treatments were randomised in a 2 (sludge type) x 2 (soil type) factorial design, plus controls, with four replicates for each treatment. Treatments provided 0.35 mg total N g^{-1} dry soil. The fish sludge was dried at 37° C for one week, to eliminate NH_4^+-N (Table 1). Pots were filled with air-dried soil and the fish sludge (2 mm sieve), which were mixed before seeding with ryegrass (*Lolium multiflorum cv.* Sabalan). Soils were maintained at 80% of field capacity throughout the experiment. Plants were harvested three times over an evaluation period of 90 days, leaving a stubble of 25 mm. The harvested material was weighed and dried to obtain the dry matter yield and then ground and analysed for N (Kjeldahl). The percentage of mineralised N from sludge was calculated as the difference between the total N plant uptake for the period on sludge treated pots and that from the control treatments, divided by the total N applied. Analysis of variance (ANOVA) was used to determine differences between treatments.

Results

Total N offtake ranged between 102 ± 1.4 and 186 ± 0.7 mg kg^{-1} of dry soil in the sludge treatments. The andisol mineralised 47% more than the ultisol ($p \leq 0.01$, Figure 1). The sea sludge mineralised both more (27%) and faster than the lake sludge ($p \leq 0.01$), so that after 30, 60 and 90 days, the sea sludge had mineralised 64%, 32% and 4% of the total sludge N, whereas the lake sludge had mineralised only 51%, 30% and 19%, respectively. Of the total N applied, the sea sludge treatment mineralised 52% over the 90 days period, while the lake sludge mineralised only 39%.

Table 1. Chemical characteristics of the fish sludge used in the experiment (dry matter basis).

Fish sludge	DM (%)	Total N (kg t^{-1})	NH$_4^+$-N (kg t^{-1})	C (kg t^{-1})	C: total N
Sea	94.7 ± 0.01	8.8 ± 0.004	0.05 ± 0.001	71.3 ± 0.53	8.1 ± 0.11
Lake	95.8 ± 0.09	14.3 ± 0.003	0.60 ± 0.001	124.0 ± 0.27	8.7 ± 0.04

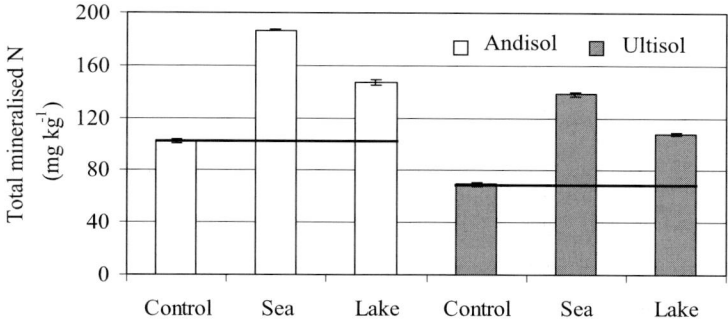

Figure 1. Total mineralised N measured as plant uptake for the different treatments.

Discussion

Both sludges had a low C:N ratio, which explains why no immobilisation was observed. Mineralisation values measured during this experiment were higher than those reported widely in the literature for cattle slurry (e.g. Kyvsgaard et al., 2000). This agrees with the low C:N ratio of fish sludge in relation to that of cattle slurry, which would allow a greater N mineralisation. The greater mineralisation in the andisol can be related to the greater organic matter concentration of this soil (ca.16%), compared with that of the ultisol (ca. 10%), which could contribute a greater biomass and microbial activity (Aguilera et al., 1997) and, in turn, also a greater N availability to plants. Results of this study showed that less than 50% of the organic N of the fish sludge was potentially available for plant uptake during the first crop. The remaining N could be available for the following crop.

Conclusions

Sea sludge had a greater potential for N mineralisation than lake sludge. This material also mineralised more and faster N on the andisol than on the ultisol.

Acknowledgements

This research was funded by FONDEF D01I11-13.

References

Aguilera, S., Borie, F., Peirano, P. and Galindo, G. (1997). Organic matter in volcanic soils in Chile: chemical and biochemical characterization. Communications in Soil Science and Plant Analysis, 28, 899-912.

Kyvsgaard, P., Sørensen, P., Møller, E. and Magid, J. (2000). Nitrogen mineralization from sheep faeces can be predicted from the apparent digestibility of the feed. Nutrient Cycling in Agroecosystems, 57, 207-214.

Modelling of N-cycle for processes in soil using soil core experiments

Á. Bálint[1], Cs. Mészáros[1], Gy. Heltai[1], E. Nótás[1] and K. Jung[2]
[1]*Szent István University, H-2103 Gödöllö, Hungary*
[2]*Department of Chemical Ecotoxicology, UFZ, 04318-Leipzig-Halle, Germany*

Introduction

Crop-model calculations spread widely from the nineties, were developed to forecast long-term (seasonal) production of agro-ecosystems. The kinetic description of N-transformation processes is a significant sub-system of these models, and is a complex component. This paper reports the development of a simplified model based on the pool principle and experimental verification in a soil core incubation experiment.

Materials and methods

A rust brown forest soil from Gödöllö, Hungary was used. It is an acidic and low humus content cambic soil (pH(H_2O) 5.4 and pH(KCl) 4.7); humus content: 1.2%; N content: 108 mg 100g^{-1} soil; C content: 1.4 g 100g^{-1} soil; C/N ratio: 12.7; NH_4^+-N content: 0.32 mg 100g^{-1} soil; NO_3^--N content: 0.45 mg 100g^{-1} soil). The air-dried soil was moistened to 25, 50, 75% of the water holding capacity, and placed in PVC tubes (40 cm high and 3.6 cm in diameter). $^{15}NH_4^{15}NO_3$ fertiliser solutions of 10 atom% ^{15}N-content corresponding to a 120 kg N ha^{-1} fertiliser dose (24 mg N column^{-1}) were injected into the upper 3-4 cm layer. Half of the soil cores were treated with fertiliser and the others were used as controls. Three parallel series were prepared. Soil cores were incubated for 30 days at 27°C, at 25, 50 and 75% WHC. Samples were taken on the 10th, 20th and 30th days of incubation and cores were divided into 10 cm segments. The concentrations of different N-forms were determined in each segment of the control and treated soil samples. Ammonium-N and NO_3^--N were analysed after extraction with 1M KCl solution. Active organic N content of soil was calculated from mineralisable N, determined by treatment with 0.25 M sulphuric acid followed by reduction and digestion by Parnass-Wagner distillation and titration techniques, minus mineral-N content.

Data gained from the incubation experiment were used for verification of the mathematical model, which makes the dynamic description of transformation processes possible for any time. The three-pool model from the literature (Myrold and Tiedje, 1986) was used and first-order chemical kinetics were taken into account. For effective solution of the partial differential equation (PDE) system describing diffusing and chemically reacting components in soil cores, one of the most general, direct variational methods of irreversible thermodynamics (Gyarmati, 1970) was applied together with its numerical variant (Stark, 1974). For the determination of reaction rate constants, however, linear and non-linear regression methods were not suitable, because the sampling frequency in our experiment was too low. However, this variational method developed in non-equilibrium thermodynamics is applicable for systems already at stationary states. During the first 10 days when quick changes occur, no sufficiently robust information was available about the transport processes being examined.

Results

The model being set up can be described by following the **coupled** PDE system:

$$\frac{\partial c_1}{\partial t} = D_1 \nabla^2 c_1 - (k_{31} + k_{21})c_1 + K_{12}c_2 = 0$$

$$\frac{\partial c_2}{\partial t} = D_2 \nabla^2 c_2 + k_{21}c_1 - (k_{12} + k_{32})c_2 = 0$$

$$\frac{\partial c_3}{\partial t} = D_3 \nabla^2 c_3 + k_{31}c_1 + k_{32}c_2 - k_4 c_3 = 0$$

For the solution of this system, the usual operational-calculation procedures and the direct variational-calculus algorithm developed for stationary transport processes were combined. For a part of the soil core, diffusion coefficients and chemical reaction rate constants of the component were estimated with the aid of approximate polynomials derived from variational analysis on the basis of the experimental data assuming a near stationary state. These simplified estimated values were used with the equations which were then solved. The calculated functions showed good agreement with the trends in the data in most cases.

Conclusions

The model developed can be used for predicting the time-dependent behaviour of all components of the complete three-N-pool. Although we performed our modelling calculations in relatively low-order approximations, the results show a remarkable accuracy and good agreement with experimental results. Moreover, since it is based on the most general variational principles of the irreversible thermodynamics and numerical mathematics, it can be applied to much more complex systems of coupled transport processes, too.

Acknowledgements

The Hungarian Scientific Research Foundation (38450 and 24146 projects) and Hungarian Science & Technology Foundation (D-9/00) supported this work. Cs. Mészáros gratefully acknowledges support of the Eötvös Fellowship (417/2002.) and Békésy György Postdoctoral Fellowship (148/2002.). Á. Bálint expresses her thanks to the Hungarian Scientific Research Foundation's Travel fellowship, the Hungarian Ministry of Education ("Mecenatura" fellowship, Széchenyi István Fellowship (245/2003)) and the "MAG" Foundation, SZIU, Faculty of Agricultural and Environmental Sciences. We would like to thank Mrs. Kiss, C. David and G. Halász for their hard work.

References

Gyarmati, I. (1970). Non-equilibrium thermodynamics (Field Theory and Variational Principles). Berlin-Heidelberg-New York: Springer-Verlag.

Myrold, D.D. and Tiedje, J.M. (1986). Simultaneous estimation of several nitrogen cycle rates using ^{15}N: theory and application. Soil Biology and Biochemistry, 18, 559-568.

Stark A. (1974). Approximation methods for the solution of heat conduction problems using Gyarmati's principle. Annalen der Physik (Leipzig), 31, 53-75.

Mineralisation of nitrogen in relation to climatic variation and soil

H. Björnsson
Agricultural Research Institute, Keldnaholti, 112 Reykjavík, Iceland

Introduction

Soils in Iceland are of volcanic origin and are rich in organic matter. Mineralisation of N supports appreciable plant growth when other nutrients are not limiting. The N available for plant growth is though, highly variable. The annual variation of N supply in grass fields is related to temperature and, in arable soils, N mineralisation is related to soil C content.

Materials and methods

The relationship of grass yield to temperature was estimated in long-term experiments at four experimental stations in Iceland from 1951-1983 (Table 1, derived from Björnsson & Helgadóttir, 1988). Temperature is most variable outside the summer season and grass yield was most strongly related to mean temperature from September to June. The harvest period was from late June to late August (or mid September when cut twice). The experiments were fertilised in spring and they were usually cut twice. The regression model included adjustments for cutting dates in the harvest year and the preceding year.

Results and discussion

Uptake of mineralised N in long-term experiments and the effects of temperature

A number of N analyses are available from 1955-1976 and 1983. N content was negatively correlated with yield and this may, at least partly, explain the observed temperature effect on N response (Table 1). Ignoring this relationship and assuming 2.3% N in dry matter, the yield response to temperature can be transformed into N response. The result was 14.3 kg N ha^{-1} °C^{-1} in the ON-treatment. This corresponds to a difference of 51 kg N ha^{-1} between the warmest and the coldest year at the station where temperature was most variable. The most likely explanation is that mineralisation is active for a longer period as temperature, especially during winter, increases. The effect of temperature on the length of the growing season has been roughly evaluated as 27 days °C^{-1} (Björnsson & Helgadóttir, 1988) and the increase in N uptake per day appears to be ca. 0.5 kg N d^{-1} ha^{-1}. A lower estimate would be obtained if the relationship between yield and N content is duly taken into account. Assuming 2.3% N in dry matter as before, the N uptake on plots without N fertiliser was 53 kg ha^{-1}. This includes a small contribution from white clover. This uptake can be compared with the length of the growing season. In Reykjavík, the mean temperature 1961-1990 was $\geq$ 5.0°C for 155 days and $\geq$ 7.0°C for 124 days (Icelandic Meteorological Office, unpublished data). The growing season is similar to Reykjavík at one of the experimental stations and shorter at the others. The conclusion was that 0.3-0.4 kg N ha^{-1} d^{-1} may be a realistic average of N uptake for the whole season. Another estimate of the rate of N uptake was obtained by dividing the N uptake in the 2nd harvest by the number of days between cuts, assuming that the fertiliser applied in

Table 1. Regression of grass yield on mean temperature from September to June in Iceland.

	Mean fertiliser (N kg ha^{-1})	Mean yield (DM t ha^{-1})	DM kg ha^{-1} °C^{-1} Regression coefficient	Standard error
All experiments	95	5.1	735	92
Nitrogen treatments	100	5.2	842	111
Without nitrogen	0	2.3	617	99
N response (100N - 0N)			171	55

spring was taken up prior to 1st harvest. An average over stations is 0.61 kg N ha^{-1} d^{-1}, ranging from 0.48 to 0.69. In many cases, the 2nd harvest was cut late, when the N uptake had probably come to an end. A considerably higher rate is therefore expected in mid summer. In some cases, the experiments were cut only once or the regrowth was cut well before the end of the growing season. Nitrogen taken up by the forage, following harvest, appears to be stored and is available for growth the following season (Björnsson, 1998). A detailed study of soil profiles in one of the long-term experiments indicated accumulation of N in soil organic matter, although some 90 kg N ha^{-1} were removed by herbage annually in excess of N uptake from fertilisers. This was interpreted as evidence of rhizospheric fixation of N (Gudmundsson et al., 2003). Although this experiment was the highest yielding in the whole series, the result indicates that the N release in soils may be a sustainable process.

Nitrogen mineralisation in arable soils
Barley is often grown on light soils where mineralisation is limited. These soils require ca. 90-120 kg fertiliser N ha^{-1} for optimal yield when barley follows barley. On soils with a high content of organic matter, 30 kg N ha^{-1} may be sufficient for barley, following grass or green fodder crops. Uptake of N from soil alone (in nine experiments 2000-2001) ranged from 16-124 kg N ha^{-1} and was, on average, 0.8% of organic N in the top 30 cm of the profile. Nitrogen ranged from 0.14-1.8% in the soils and the bulk density, in inverse order, from 0.91-0.25 kg dm^{-3}. Measurements of mineralisation in incubation experiments at 15°C gave comparable results. In a soil with 0.9% N, the release corresponded to 0.8 kg N d^{-1} ha^{-1} (Pálmason et al., 2003). With 50% N uptake efficiency, the rate of uptake would be 0.4 kg N d^{-1} ha^{-1}.

Conclusions

N mineralisation is a significant factor in soil fertility. It is related to soil organic matter and to the length of the growing season which, in turn, is a function of temperature. In Icelandic soils, mineralisation commonly supports N uptake at the rate of 0.3-0.4 kg N d^{-1} ha^{-1} over the season.

References

Björnsson, H. (1998). Application of nitrogen fertilizers in autumn. In: G. Nagy and K. Petö (eds) Ecological Aspects of Grassland Management. Proceedings of the 17th General Meeting of the European Grassland Federation. Debrecen, Hungary, Agricultural University Debrecen, May 18-21, 1998: 639-642.

Björnsson, H. and Helgadóttir, A. (1988). The effect of temperature variation on grass yield in Iceland, and its implications for dairy farming. In: M. L. Parry, T. R. Carter and N. T. Konijn (eds) The Impact of Climatic Variations on Agriculture. Vol. 1 Assessments in Cool Temperature and Cold Regions. Kluwer Academic Publishers Group, Dordrecht, 445-474.

Gudmundsson, Th., Björnsson, H. and Thorvaldsson, G. (2003). Organic carbon accumulation and pH changes in an andic gleysol under a long-term fertilizer experiment in Iceland. Catena, (in press).

Pálmason, F., Björnsson, H. and Hermannsson, J. (2003). Nýting niturs í kornökrum (Nitrogen utilisation in barley fields). Ráðunautafundur, 2003, 173-177 (in Icelandic).

Resolving differences in N cycling between more polluted and pristine forests using ^{15}N isotope dilution

P. Boeckx[1], R. Godoy[2], C. Oyarzún[2], J. Bot[1] and O. Van Cleemput[1]
[1]Ghent University, Laboratory of Applied Physical Chemistry - ISOFYS, 9000 Gent, Belgium
[2]Universidad Austral de Chile, Valdivia, Chile

Introduction

Indications are emerging that N cycling differs between pristine and more disturbed or polluted forest ecosystems. In 100 pristine old-growth forest catchments of South Chile, having a low N deposition, N losses between 0.2 and 3.5 kg N ha^{-1} yr^{-1} were observed. The DON and DIN drained from these catchments was 20 and 80%, respectively (Perakis & Hedin, 2002). Rainfall was the main factor affecting the DON:DIN loss ratio. The contribution of DON to N losses from forests of more polluted regions of the world, on the other hand, is negligible. MacDonald et al., (2002) showed that in forests of Europe (N deposition 1-60 kg N ha^{-1} yr^{-1}) the predominant N loss mechanism was NO_3^- leaching (1-40 kg N ha^{-1} yr^{-1}). Nitrate formation in N polluted forests seemed to be largely controlled by N input and N enrichment of the forest floor. Organic layer C:N ratios of ca. 25 seem to be crucial to initiate more intense nitrification and subsequent NO_3^- losses. It has been suggested that forest soils may possess a substantial biotic or abiotic NO_3^- or NH_4^+ retention mechanism controlling DON and DIN losses. The exact mechanisms, the capacity and time frame of these alternative N retention processes in forest soils are still unclear. Here we present an initial study comparing N turnover in a pristine and a more polluted forest.

Materials and methods

A fully mirrored ^{15}N isotope dilution experiment has been used to resolve differences in gross and net N mineralisation and nitrification in a forest receiving elevated N deposition in Belgium (*Aelmoeseneie*, A) and a pristine forest in southern Chile (*Puyehue National Park*, P). Forest A showed relatively high NO_3^- losses (up to 18.5 kg N ha^{-1} yr^{-1}; De Schrijver et al., 2000). In forest A, the F and H layer and the 0-5 cm layer of the mineral soil were sampled. Forest P showed significant DON losses, but negligible NO_3^- losses (DON = 5.2 kg N ha^{-1} yr^{-1}, NO_3^- = 0.6 kg N ha^{-1} yr^{-1}; Oyarzún & Godoy, 2002). In forest P, the mineral soil was sampled from 0-10 cm (the fermentation (F) or humic (H) layers did not exist). In a first laboratory experiment, ^{15}N-NH$_4^+$ was added to study net and gross N mineralisation rates. In a second experiment, ^{15}N-NO$_3^-$ was applied to study net and gross nitrification rates. ^{15}N-NH$_4^+$ or ^{15}N-NO$_3^-$ was applied so that the initial NH_4^+ and NO_3^- concentrations were doubled. When ^{15}N-NH$_4^+$ and ^{15}N-NO$_3^-$ were applied, ^{14}N-NO$_3^-$ and ^{14}N-NH$_4^+$ were also added, respectively, to assure similar mineral N concentrations in both experiments. Incubations were carried out at 15°C and 50% water holding capacity for 7 days.

Results and discussion

The pristine and more polluted forest showed clear differences in N turnover. In forest A, the net N mineralisation rates were much smaller than the gross N mineralisation rates, especially for the F and H layer (Table 1). In this experiment, a higher gross N mineralisation indicates a substantial (biotic or abiotic) NH_4^+ immobilisation capacity of the soil. The net and gross nitrification rates were not significantly different (Table 1). A similar net and gross nitrification rate indicates that all the NO_3^- that was produced was potentially available for plant uptake or N leaching, i.e. no NO_3^- immobilisation is likely to occur. In forest A, net mineralisation rates were similar to net nitrification rates. Thus, NH_4^+ and NO_3^- was equally available for uptake but, in the field, most of the NO_3^- which is not taken up by the trees, will not be retained in the ecosystem. Nitrate leaching has been observed in field monitoring campaigns (see above).

With forest P, a different pattern was observed. Both the net N mineralisation rate and the net nitrification rate were lower than their respective gross rates (Table 1). Because the net nitrification rates are much lower than the net N mineralisation rates, trees will preferentially take up NH_4^+ and leave NO_3^- in the ecosystem. However, since gross nitrification is larger than net nitrification, microbial or abiotic NO_3^- immobilisation is likely to occur. Thus, NO_3^- can be retained in the pristine ecosystem, preventing leaching. A field monitoring campaign confirmed that the amount of NO_3^- drained from forest A is negligible (see above). However, NO_3^- immobilisation might be a source of DON.

Table 1. Gross and net N mineralisation and nitrification rates for a more polluted forest in Belgium (Aelmoeseneie, A) and a pristine forest in southern Chile (Puyehue National Park, P).

| | Puyehue (P) | | Aelmoeseneie (A) | | | | | |
| | Mineral layer (0-10 cm) | | F layer | | H layer | | Mineral layer (0-5 cm) | |
	Gross	Net	Gross	Net	Gross	Net	Gross	Net
Mineralisation (mg N kg^{-1} day^{-1})	51.7	30.6	29.0	5.2	11.2	2.7	1.9	1.6
Nitrification (mg N kg^{-1} day^{-1})	1.8	0.9	6.5	7.0	3.7	4.2	0.3	0.2

Acknowledgements

We thank the Ministry of the Flemish Community for funding for a bilateral cooperation between Flanders and Chile (BIL99/4) entitled "comparison of ecosystem functioning and biogeochemical cycles in temperate forests of southern Chile and Flanders" and the Flemish Interuniversity Council (VLIR) for funding for travel for J. Bot.

References

De Schrijver, A., Van Hoydonk, G., Nachtegale, L., De Keersmaecker, L., Mussche, L. and Lust, N. (2002). Water, Air and Soil Pollution, 122, 77-91.

MacDonald, J.A., Dise, N.B., Matzner, E., Armbruster, M., Gundersen, P., Forsius, M. (2002). Global Change Biology, 8, 1028-1033.

Oyarzún, C.E. and Godoy, R. (2002). In: A. De Schrijver, V. Kint and N. Lust (eds) Comparison of ecosystem functioning and biogeochemical cycles in temperate forests in southern Chile and Flanders. Academia Press, Gent, p. 13-18.

Perakis, S.S. and Hedin, L.O. (2002). Nitrogen loss from unpolluted South American forests mainly via dissolved organic compounds. Nature ; 415: 416-419.

Effects of tillage and crop rotation on the microbial population and dynamics of soil organic matter

C. Carranca[1], A. Oliveira[2], A. de Varennes[2], M. Pampulha[2], M. Costa[2], A. Prazeres[1], J. Baeta[1], C. Neto[1], M.P. Andrada[1] and M.O. Torres[2]
[1]*Quinta do Marquês, Av República, Nova Oeiras, 2784-505 Oeiras, Portugal*
[2]*Instituto Superior de Agronomia, Tapada da Ajuda, 1349-017 Lisboa, Portugal*

Introduction

Crop rotations that include legumes are a traditional practice in Mediterranean agriculture. The benefits derived from legumes depend largely on total plant biomass produced, amount of N_2 fixed, and amount of N returned to the soil by plant residues. Interest in conservation tillage has increased as a result of state-funded programmes to reduce soil erosion. Other potential benefits are significant reductions in production costs and labour requirements, and increase in soil organic matter (SOM). Equilibrium levels of SOM are determined by biological, chemical, and physical properties of soils that control biological activity. Total organic matter content varies more slowly than its active fraction (Wani et al., 1995). Several long-term studies have documented the decline in SOM with tillage, but very few have reported on the combined effects of crop rotation and tillage on SOM. The objectives of this study were to quantify the short-term effects of tillage and crop rotation on soil organic N and C and biological activity in a Haplic Podzol of Portugal. Net N mineralisation *in situ,* and soil biomass and activity were evaluated under no till and chisel plough soil management with two crop rotations: lupin/oat and oat/oat.

Materials and methods

The experiment started in 2001 in Pegões (Portugal), on a Haplic Podzol (Table 1). Lupin (*Lupinus albus* L. cv. Estoril) and oat (*Avena sativa* L. cv. Sta. Eulália) were grown in a simple rotation and compared with an oat/oat system, both under no till (NT) or conventional (chisel plough) till (CT). Chopped crop residues were left on the surface in the NT treatment and incorporated in the soil for CT in a split-plot design with three replicates.

At harvest, soil samples were collected from each plot, at a depth of 0-20 cm, and analysed for C and N contents. Residues were analysed for N and C. PVC cores were inserted into the soil without disturbance. At the bottom of each cylinder, a nylon bag filled with an anion-exchange resin was placed to intercept any NO_3^- leaching through the soil core. One soil core was removed monthly (July to December 2002) from each plot to evaluate N mineralisation *in situ*. Soil samples were taken during the vegetative cycles and period of soil incubation in the inter-row spaces of each plot at a depth of 0-15 cm. These samples were tested for dehydrogenase activity, and bacteria and fungi numbers.

Results

In the first growth cycle, oat residues were significantly greater under CT (8594 kg dry weight (DW) ha^{-1}) than under NT (6813 kg DW ha^{-1}), and greater than lupin residues (1200 kg DW ha^{-1} in both soil practices). Total C in lupin and oat residues was, respectively, 480 and 470 g C kg^{-1} DW, but no significant differences were found for total lignin content (data not shown). Total N content was greater in lupin (10.5 g N kg^{-1} in the stems and 37.5 g N kg^{-1} in the leaves) under CT than in oat residues (8.4 g N kg^{-1} for both NT and CT). Organic C and N contents in the soil after harvest were greater under CT than NT, but only organic N differed between crops (Table 2). The soil C:N ratios were greater for CT than NT for both crops. Under both treatments (tillage and crop rotation), no significant differences were found for dehydrogenase activity. The greatest activity was observed in December 2001 (59.5 µg Triphenyl Formazan (TPF) g^{-1} dry soil) and the smallest in December 2002 (15.2 µg TPF g^{-1} dry soil). At this time, bacterial numbers were smaller (4.7 x 10^6 Colony Forming Units (CFU) g^{-1} dry soil) than in April 2002 (10.5 x 10^6 CFU g^{-1} dry soil). The number of soil fungi in December 2002 was ca. 50% that of April 2002 (2.5 x 10^5 CFU g^{-1} dry soil). No significant effects of treatments were found on soil biomass.

There were no significant differences in mineralisation between NT and CT, but the first crop in the rotation (lupin and oat) affected net N mineralisation. Oat residues led to an overall immobilisation of 590 g NO$_3$-N ha^{-1} month^{-1} while lupin residues favoured mineralisation at an average rate over the whole period (June-November 2002) of 490 g NO$_3$-N ha^{-1} month^{-1}. The seasonal pattern of mineralisation was similar for both crops, with greater net mineralisation taking place in August and November 2002.

Table 1. Main characteristics of the Haplic Podzol of Pegões (Portugal).

Depth (cm)	pH$_{(H20)}$	Total C (g kg^{-1})	Total N (g kg^{-1})	Available P (mg kg^{-1})	Available K (mg kg^{-1})	CEC [cmol$_{(+)}$kg^{-1}]
0-10	5.01	6.18	0.48	73	70	0.88
10-20	4.87	6.06	0.44	75	57	0.82

CEC= Cation Exchange Capacity.

Table 2. Contents of soil organic C and N (g kg^{-1}) in the soil at a depth of 0-20 cm, after harvest (May 2002) under no till (NT and chisel plough (CT) treatments.

Crop	Organic C (g kg^{-1})		Organic N (g kg^{-1})		C/N	
	NT	CT	NT	CT	NT	CT
Lupinus albus 'Estoril'	6.4 a	7.0 a	0.54 b	0.57 b	11.9	12.3
Avena sativa 'Sta. Eulália'	6.8 a	8.5 a	0.60 a	0.65 a	1.3	13.1

NT= No till; CT= Conventional till; means in the same column with the same letter did not differ at $p<0.05$.

Discussion

After the first crop season, organic C and N increased in the surface layer, particularly under CT. Dehydrogenase activity seemed to be affected by rainfall (130 mm, March/02 and 51 mm, April/02) and temperature (10 °C, December/01 and 14 °C, April/02) but not by the crops and tillage. Net N mineralisation *in situ* was influenced by the crop residues: fallen leaves of lupin during crop growth plus the greatest crop N content should have improved mineralisation rates.

Acknowledgements

This study was funded by PIDDAC 166/01 and POCTI 42616/01 grants.

References

Wani, S. P., Rego, T. J., Rajeswari, S. and Lee, K. K. (1995). Effect of legume-based cropping systems on nitrogen mineralisation potential of Vertisol. Plant and Soil, 175, 265-274.

The changes in microbial biomass nitrogen when different rates and forms of N were applied in a long-term experiment with maize

J. Černý and J. Balík
Czech University of Agriculture in Prague, Department of Agrochemistry and Plant Nutrition, Kamycka 957, 165 21 Prague, Czech Republic

Introduction

Soil microbial biomass is a small and labile component of soil organic matter. It is thought to exert a key controlling influence on the rate at which N, C and other nutrients cycle through agricultural and other ecosystems (Jenkinson, 1988). The interest in estimating soil microbial biomass is related to its function as a pool for subsequent delivery of nutrients, and its role in structure formation and stabilisation of soil and as an ecological marker (Smith & Paul, 1990). Soil microbial biomass can be affected by different N management, particularly in the long term (Lovell & Jarvis, 1998).

Materials and methods

The experiment with continuous cultivation of silage maize was established in 1990 and the experiment consisted of seven treatments: 1) control, 2) ammonium sulphate - AS, 3) urea ammonium nitrate solution - UAN, 4) UAN + straw, 5) manure, 6) slurry, and 7) unfertilised fallow. Each treatment was repeated four times. All plots (2 - 6) were treated with 120 kg N ha^{-1}. Soil samples were collected from the soil profile (0 - 30 cm) in spring (April) every year from 1999 - 2002. The soil is a Luvisol, containing 1.66 % total C, 0.23 % total N, pH 6.6. Microbial biomas N was estimated by the fumigation-extraction method after pre-extraction (Brookes et al., 1985; Mueller et al., 1992). Microbial biomass N was calculated as the difference in N content in fumigated and nonfumigated sample (E_N) using a coefficient, k_{EN} (microbial biomass N = $E_N : k_{EN}$). The value k_{EN} = 0.54 was used to calculate microbial biomass N (Brookes et al., 1985; Jenkinson, 1988).

Results

The content of microbial biomass N fluctuated between 5.2 and 22.7 mg N kg^{-1}. The highest content of microbial biomass N was estimated in the variant treated with manure (17.2 to 22.7 mg N kg^{-1}). The average content of microbial biomass N in this treatment was 38% higher compared with the control. A positive influence on the contents of the microbial biomass N came from application of slurry. The content of microbial biomass N was estimated to be 29% higher than in the control. With UAN, the average content of microbial biomass N reached 83% of that in the control and with amonium sulphate it was 61%. There were no differences between control and UAN + straw treatments. However, the addition of straw caused an increase in the content of microbial biomass N by 36% compared with UAN treatment. In unfertilised

Table 1. The average contents of microbial biomass nitrogen (1999 - 2002).

Treatments	Control	AS	UAN	UAN+Straw	Manure	Slurry	Fallow
mg N kg^{-1}	13.9	8.5	12.1	14.1	19.1	17.9	14.2

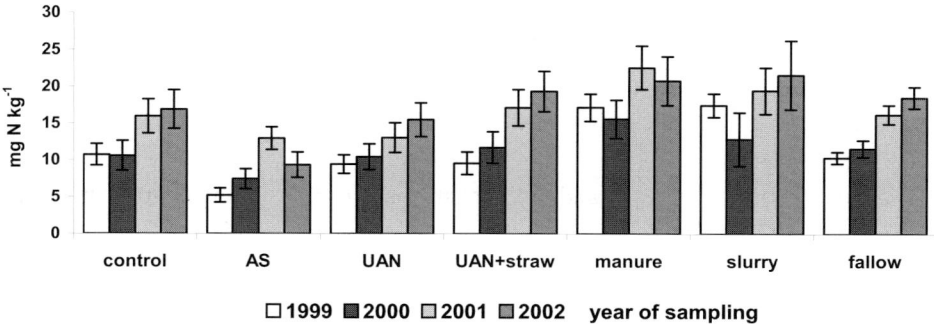

Figure 1. The contents of microbial biomass nitrogen (1999 - 2002).

fallow soil, the content of microbial biomass N was higher than in the planted control, but changes were not significant.

The year of sampling markedly influenced the content of microbial biomass N in topsoil. Yearly variations in microbial biomass are a result of the complex of physical, chemical, biological and anthropogenic factors, and a result of variability in these factors, which along with the short generation time of microorganisms, is reflected in the large fluctuations of microbial biomass content.

Conclusions

The contents of microbial biomass depended on the treatment and year of sampling. Higher contents of microbial biomass were found in treatments with organic fertilisers compared with the control. In treatments with mineral N fertilisers, there was a tendency towards a lower content of microbial biomass, compared with the control.

Acknowledgements

This research was supported by Project No. 521/02/D128 of Grant Agency of CR and by Project No. 21140/1312/3139 of Grant Agency of FA CUA.

References

Brookes, P.C., Landman, A., Pruden, G. and Jenkinson, D.S. (1985). Chloroform fumigation and the release of soil nitrogen: a rapid direct extraction method to measure microbial biomass nitrogen in soil. Soil Biology and Biochemistry, 17, 837-842.

Jenkinson, D.S. (1988). The determination of microbial biomass carbon and nitrogen in soil. In: J.R. Wilson (ed) Advances in nitrogen cycling in agricultural ecosystems. CAB International, Wallingford, 368-386.

Lovell. R.D. and Jarvis, S.C. (1998). Soil microbial biomass and activity in soil from different grassland management treatments stored under controlled conditions. Soil Biology and Biochemistry, 30, 2077-2085.

Mueller, T., Joergensen, R.G. and Meyer, B. (1992). Estimation of soil microbial biomass C in the presence of fresh roots by fumigation - extraction. Soil Biology and Biochemistry, 17, 179-181.

Smith, J.L. and Paul, E.A. (1990). The significance of soil microbial biomass estimations. In: B. Metting (ed) Soil Microbial Ecology. Marcel Dekker, New York, 357-396.

Screening of organic biological waste products for their potential to manipulate the N release from crop residues

B. Chaves[1], S. De Neve[1], G. Hofman[1], P. Boeckx[2] and O. Van Cleemput[2]
[1]*Ghent University, Department of Soil Management and Soil Care, Coupure Links 653, 9000 Ghent, Belgium*
[2]*Ghent University, Department of Applied Analytical and Physical Sciences , Coupure Links 653, 9000 Ghent, Belgium*

Introduction

Vegetable crop residues may release large amounts of mineral N in the soil, resulting in high NO_3^- leaching risks. There are indications that certain organic waste products can be used to manipulate the N mineralisation of crop residues. Rahn et al., (1999) found that the N release from sugar beet and Brussel sprouts leaves doubled when mixed with molasses, but found a strong decrease with straw or green waste compost. The objective of this research was a systematic screening of several organic waste products for their potential either to slow down the N release from crop residues, or to stimulate remineralisation at a time when crop demand starts increasing again.

Materials and methods

The incubation period consisted of two stages. In the first stage (0-99 days), celery residues were mixed with organic wastes aimed at immobilising celery-N (=1st category wastes). At the start of the second stage (100-190 days) half of the tubes received an organic waste aimed at stimulating remineralisation (= 2nd category wastes), and the others received no additional waste (=control treatments). The incubation tubes were sampled destructively during both stages at regular time intervals, and the mineral N contents were measured. The chemical composition of the organic materials was determined by standard chemical analyses (Table 1).

Results

First category wastes
The immobilising capacity of 1st category wastes was calculated as the mineral N content in the '1st category waste treatment' minus the mineral N content in the 'control celery treatment' (Figure 1, A). The addition of 1st organic wastes caused significant ($P < 0.01$) N immobilisation at all sampling dates only upon addition of straw. Nevertheless, the net N release patterns of

Table 1. Composition of the organic materials.

	Celery leaves	Straw	GCP1*	GCP2*	Saw dust	Tannic acid	Paper sludge	Molasses	Vinasses	Dairy sludge	Malting sludge
DM	13.2	91.4	52.5	55.5	95.0	94.0	29.0	72.3	65.0	8.8	9.1
N	4.68	0.47	1.45	1.3	0.06	0.06	0.61	2.26	5.89	5.14	2.99
C:N	7.7	99.4	19.0	18.7	863.3	876.6	50.7	19.6	6.7	8.3	8.4

*GCP1 was one month older than GCP2. (GCP: green waste compost, DM: dry matter (%), N: total N content (% on DM))

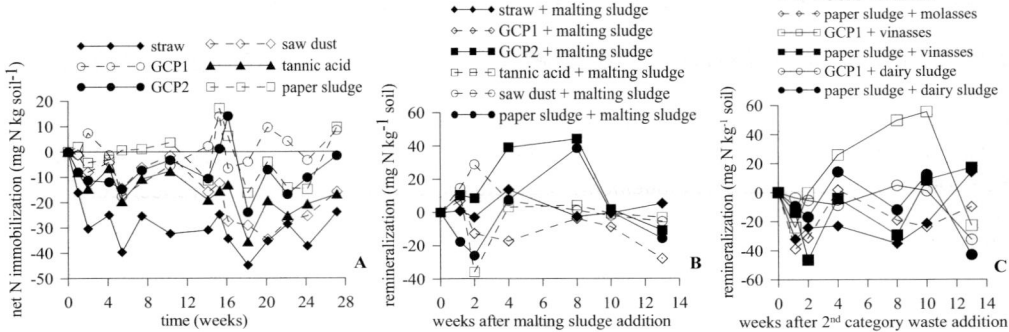

Figure 1. N immobilisation of 1st category wastes (A) and remineralisation primed by the 2nd category wastes: malting sludge (B), molasses, dairy sludge and vinasses (C).

GCP1, GCP2, tannic acid and sawdust also showed a tendency towards N immobilisation. Quite unexpectedly, paper sludge did not immobilise N, probably because of its low C:N ratio compared with paper sludge in other studies (C:N 86, Aitken et al., 1998).

Second category wastes

The remineralisation primed by the 2nd category wastes was calculated as the mineral N content in the '1st + 2nd category waste treatment' minus the mineral N content in the 'control treatment' minus the net N release of 2nd category waste (Figure 1, B and C). The greatest remineralisation occurred for vinasses + GCP1 and malting sludge + GCP2. Malting sludge + sawdust and malting sludge + paper sludge gave a shorter remineralisation period. Although remineralisation seemed temporary, a new crop can profit from this extra N release if it is sown closely after the addition of the 2nd category waste.

Conclusion

There seems to be scope for manipulating the N release from crop residues by the addition of organic wastes, in order to reduce nitrate leaching risks. Both intensity and time of immobilisation and remineralisation seem manageable with the right choice of organic waste.

Acknowledgements

The research was funded by the Belgian Ministry of Small Enterprises and Traders and Agriculture, Division Research and Development (project S-6059).

References

Aitken, M.N., Evans, B. and Lewis, J.G. (1998). Effect of applying paper mill sludge to arable land on soil fertility and crop yields. Soil Use and Management, 14, 215-222.

Rahn, C., Bending, G.D., Turner, M.K. and Lillywhite, R. (1999). Use of carbonaceous amendments of contrasting lability for the management of nitrogen losses from high N content crop residues. BSSS Symposium, poster abstract (http://www.sac.ac.uk/envsci/external/bsss/poster2.htm).

Nitrogen dynamics in soil with mulch and incorporated crop residues

F. Coppens[1,2], S. Recous[1], P. Garnier[1] and R. Merckx[2]
[1]*INRA, Unité d'Agronomie LRM, rue Fernand Christ, 02007 Laon Cedex, France*
[2]*K.U.Leuven, Dept. of Land Management, Kasteelpark Arenberg 20, 3001 Heverlee, Belgium*

Introduction

Crop residue management in agricultural soils has received much attention to control soil erosion and C sequestration. One option for a more sustainable agriculture is a change towards reduced or no tillage. This implies a change in the management of fresh organic matter that influences hydrodynamic fluxes and the dynamics of C and N in soil. Tillage practice determines the initial placement of added organic matter in soil that, by turn, acts directly on soil physical properties (water flux, solute transport and soil temperature) (Guerif et al., 2001) and on the activity and composition of the soil microbial biomass. These soil physical and biological changes interact and modify the soil structure by aggregation and influence the decomposition of soil organic matter and the associated biotransformations of C and N. As part of a larger programme that aims at understanding and modelling physical and biological interactions with various qualities of crop residues and various managements, the present work examines the influence of the initial placement of crop residues on N mineralisation and water fluxes in soil, and their combined effect on transport and distribution of N in a soil profile.

Materials and methods

Fresh, loamy soil (13.5% clay, 0.7% organic C, 0.09% N) was sieved to 2 mm. Soil was compacted in PVC columns (25 cm, $\varnothing$ 15 cm, 1.3 g cm^{-3}) and received labelled rape residue (^{13}C and ^{15}N) either left as a mulch at the soil surface or incorporated homogeneously in the upper 10 cm. The added residue consisted of a mixture of leaves, stalks, seed heads and pods (particle size 1 cm obtained by hand cutting), equivalent to the return of 8t dry weight ha^{-1}. Control soils without added residues were included. The soil columns were incubated over 9 weeks at 20°C. They went through 3 dry-wet cycles, induced by the application of artificial rain (2.5h at 12 mm h^{-1}), followed by a 3-week evaporation period.

Several variables were measured: evaporation rate, soil and residue moisture content, soil matric potential and composition of the soil solution at different depths in the soil columns. This allows changes in water retention, hydraulic conductivity and transport of soluble C and N through the soil profile to be followed. Soil respiration and residue decomposition were determined by measuring the CO$_2$-flux and the concentration of ^{13}C-CO$_2$ at the soil surface. At different depths, soil was analysed for total C and N (with their isotopic excess), soluble C and mineral N.

Results and discussion

Leaving crop residues at the soil surface reduced the soil water evaporation rate by ca. 50%. This resulted in higher soil humidity under the mulch and promoted the development of a soil water gradient when residues were incorporated. The decrease in water content of the residue particles themselves was faster and more pronounced when they were applied as a mulch. These observed differences in soil and residue humidity between the two treatments are key factors determining the decomposition rate of the residues.

The initial placement of the applied residues determined the distribution of residue-^{15}N in the soil during decomposition. After 9 weeks of incubation, 77.7% of the initial residue-^{15}N was recovered in the soil fraction (< 2 mm) while only 17.9% was left in the residue fraction (> 2 mm) when the residues were incorporated. With the mulch, 48.6% of the initial residue-N was found in the soil fraction and 45.0% remained in the residues at the surface. The ^{15}N balance suggests that 4.4% of the initial residue-N was lost by gaseous emissions with residue incorporation and 6.4% when left as a mulch. Residue incorporation resulted in a net immobilisation of N during the whole incubation period and mainly took place in the upper 5 cm of the soil profile; the decomposition of residues left as a mulch led to net N mineralisation only (Figure 1).

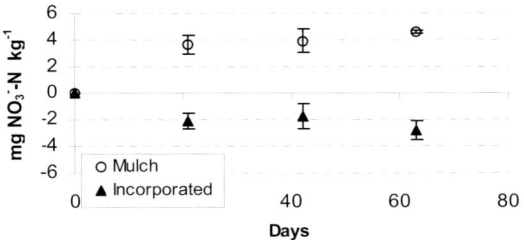

Figure 1. Release of NO_3^--N in the soil for the mulch (0) and the incorporated treatment ($\bigcirc$) during the 9-week incubation period. Error bars give the standard deviation of 3 replicates. NO_3^--N was calculated by difference at each sampling date between control and residue treatments, and by difference from the initial NO_3^--N content.

Conclusion

The distribution of mineral N through the soil profile varied greatly between mulch and incorporated residue treatments. It was the result of the combined effects of soil water content on microbial activities and NO_3^- transport, and of C availability on microbial growth and N immobilisation. Modelling will help to identify and quantify the contribution of the various processes to the resulting N dynamics (Garnier et al., 2003).

Acknowledgements

The collaboration between K.U.Leuven and INRA was supported by the bilateral French- Flemish Tournesol Project (T2001.013). The PhD scholarship of F. Coppens was funded by INRA/Région Picardie.

References

Garnier, P., Néel, C., Aita, C., Recous, S., Lafolie, F. and Mary, B. (2003). Modelling carbon and nitrogen dynamics in a bare soil with and without straw incorporation. European Journal of Soil Science, 54, 555-568.

Guerif, J., Richard, G., Durr, C., Machet, J.M.., Recous, S. and Roger-Estrade, J. (2001). A review of tillage effects on crop residue management, seedbed conditions and seedling establishment. Soil and Tillage Research, 61, 13-32.

Prediction of nitrogen mineralisation from organic residues and supply to ryegrass

C.M.d.S. Cordovil[1], J. Coutinho[2] and F. Cabral[1]
[1]*Inst Sup Agronomia, Dep Química Agrícola e Ambiental, 1349-017 Lisboa, Portugal*
[2]*Univ Tras-os-Montes, Dep Soil Science - CCEA, ap. 1013, 5000-911 Vila Real, Portugal*

Introduction

The production of on- and off-farm residues has increased at a very high rate. These organic residues can be recycled to agricultural land as a source of organic matter and plant nutrients, namely N, as well as to enhance the future crop production by improving soil fertility. This study investigates further the mineralisation of N from two types of residues with different origins (industrial and animal), evaluates their performance as N suppliers when compared with mineral fertiliser used for the growth of ryegrass, and tests the reliability of N fate, predicted by laboratory incubation experiments.

Materials and methods

Aerobic incubations were carried out using a cambic Arenosol (FAO) mixed with different rates of dried and ground pulp mill sludge (PMS) and horn meal (HM) equivalent to 0, 40, 80, 120, 160 and 200 kg N ha^{-1}, based on residue Kjeldhal N (TKN). Mixtures were placed in black plastic bags, wetted at 70% soil field capacity, kept at 24±1°C, and aerated weekly. Nine samplings at 0, 7, 14, 21, 35, 67, 102, 175 and 244 days, were performed to determine mineral N (N-NH_4^++N-NO_3^-), by extraction with 2 MKCl solution (Mulvaney, 1996). To estimate organic N mineralisation, the Stanford & Smith (1972) model ($N_m = N_0*(1-\exp^{-k*t})$ was fitted to the incubation data. N_m represents the mineral N accumulated with time t, N_0 the potentially mineralisable N (PMN) and k the mineralisation rate constant. A pot experiment with ryegrass (*Lolium perenne* L.) was conducted, with the same soil and residues, which were added in 2 doses (80 or 160 kg TKN ha^{-1}, respectively, treatments A and B). Control and half of the pots received NH_4NO_3 (30 kg ha^{-1} basal dressing + 30 kg ha^{-1} top dressing after each of the 3 cuts). The remaining pots received only residue as N supply. All pots received a mineral basal dressing without N. Soil was kept at 70% field capacity. Results from the pot experiment were correlated to equations arising from incubation data fitting.

Results and discussion

Good fits of the incubation data (g N kg^{-1} residue) to the Stanford & Smith (1972) model were obtained and resulted in the equations: $N_{m(PMS)} = 1.44 + 10.27*(1-\exp^{-0.071*t})$ and $N_{m(HM)} = 23.84 + 37.96*(1-\exp^{-0.052*t})$, corresponding, respectively, to the following $r^2 = 0.94$ and $r^2 = 0.83$ (Figure 1). The model showed a better fit for PMS, since this residue has a higher proportion of N recalcitrant compounds, and this model was originally developed for SOM mineralisation prediction. In fact, PMS has a lower PMN than HM, as can be seen from the equations.

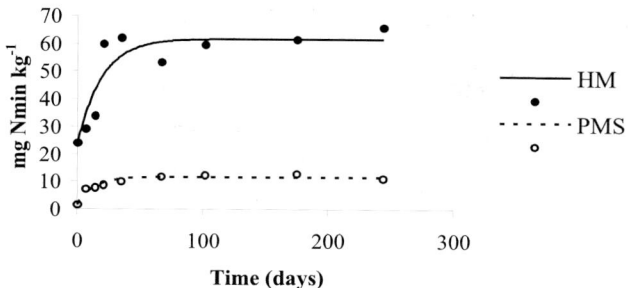

Figure 1. Kinetics of net N mineralisation of both residues added to soil.

Treatments without mineral fertilisation promoted lower plant growth than the control. However, N uptake from HM was similar (A) or greater (B) than the control at the first cut of ryegrass, which reflects a flush in N mineralisation from this residue. The same was observed by Cordovil et al. (2001) on potato and ryegrass crops. Plant growth was well correlated with N uptake by ryegrass, both from PMS (r = 1.00) and from HM (r = 0.99), reflecting growth dependence from N supply to the crop. Correlations between calculated N_m values along the growth period and N uptake by the plants at each cut, are presented in Table 1.

Table 1. Correlation coefficients (r) for the relationships: yield × N uptake and N uptake × Nm.

Treatments	Yield × uptake	N uptake × N_m	
		80 kg N ha^{-1}	160 kg N ha^{-1}
PMS	1.00	0.79	0.88
HM	0.99	0.86	0.90

Trends in N mineralisation observed in the aerobic incubations with both residues were consistent with the results obtained in the pot experiment with ryegrass. In fact, both data sets were well correlated. The joint application of mineral fertiliser and organic residues seems to be the best solution for plant nutrition because, while the former enhances soil fertility, the later allows a better synchronisation between N supply and plant needs.

Acknowledgements

This work was partially funded by FCT and partially funded by the project POCTI/AGG/46559.

References

Cordovil, C. M. d. S., Cabral, F. and Dachler, M. (2001). Fertilising value and mineralisation of nitrogen from organic fertilisers (pot and incubation experiments) Acta Horticulturae, 563, 139-145.

Mulvaney, R. L. (1996). Chemical Methods. In: Methods of Soil Analysis. 3ª ed. Part 3. SSSA, pp 1123-1184.

Stanford, G. and Smith, S. J. (1972). Nitrogen mineralisation potential of soils. Soil Science Society of America Proceedings, 109, 190-196.

Reliability of a chemical method to assess nitrogen uptake by winter wheat

C.M.d.S. Cordovil[1], J. Coutinho[2] and F. Cabral[1]
[1]*Inst. Sup Agronomia, Dept Química Agrícola e Ambiental, 1349-017 Lisboa, Portugal*
[2]*Univ. Tras-os-Montes, Dept Soil Science - CCEA, ap. 1013, 5000-911 Vila Real, Portugal*

Introduction

Over the last few decades there has been considerable research on the factors and conditions determining N mineralisation from soil organic matter. On the other hand, there has been recently a strong increase in on- and off-farm residue production. These organic residues can be recycled to agricultural land as a source of organic matter and plant nutrients, namely N, as well as to enhance the future crop production by improving soil fertility. The main practical objective of the present work has been to develop a suitable method to predict the soil and organic residue N supply to crops, in order to improve fertiliser recommendations.

Materials and methods

A pot experiment with wheat (*Triticum aestivum* L.) was performed, with a cambic Arenosol soil and six different residues (municipal solid waste MSW, pulp mill sludge-PMS, horn meal - HM, poultry manure-PM, solid phase from pig slurry-SPPS and composted pig manure-CPM), that were added in 2 doses (80 or 160 kg Kjeldahl N ha^{-1}). Control treatments and half of the pots received NH_4NO_3 (90 kg ha^{-1} basal dressing + 30 kg ha^{-1} top dressing at tillering). The remaining pots received only residue as a N supply. All pots received a mineral basal dressing without N. Soil was kept at 70% field capacity. At setting time, soil samples from the pots without mineral N fertilisation were collected, in order to perform a 4h N- extraction in cold (20°C) and hot (100°C) 2M KCl (Gianello and Bremner, 1986). Mineral N in hot KCl extract was correlated with N uptake by the plants. $N-NH_4^+$ derived from organic N hydrolysis was determined by difference from that extracted with cold KCl N-.

Results and discussion

The application of residues to the soil increased wheat biomass production, as well as N uptake by the plants, as a consequence of the supply of higher amounts of N resulting from organic residue mineralisation (data not shown), as reported before by Cordovil et al., (2002). The most effective residue as N supplier to plant growth and uptake was poultry manure (Table 1). In fact, a higher amount of potentially available N (PAN) arose from the organic residues produced from animal production (Cordovil et al., 2001).

Table 1. Percentage of N uptake from total N applied as organic residue.

	MSW	PMS	HM	PM	SPPS	CPM
% N uptake	7.7	11.0	26.4	44.5	15.9	29.3

The correlation coefficient between wheat N uptake data from all the six organic residues tested (g N pot[-1]) and the amount of PMN (potentially mineralisable N) extracted by hot 2M KCl was 0.14. When the total amount of PAN extracted by hot KCl solution was considered, the correlation coefficient was 0.73 (Figure 1). In this case, the correlation was performed between the PAN determined by the chemical method and the wheat N uptake from the organic residues (mineral + mineralised). When the correlation was obtained from N extracted by hot solution minus the mineral N determined by cold 2M KCl extraction, **r** = -0.04.

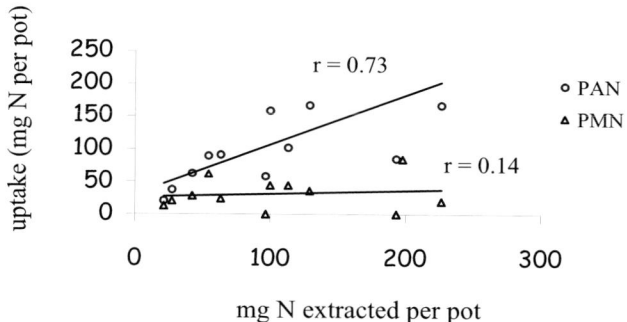

Figure 1. Correlation between hot KCl N extraction and PAN and PMN for wheat uptake.

This chemical method presented a better correlation when mineral N was not subtracted, as observed before by Rodrigues (2000). Hot KCl has so far, been recommended by several authors as providing the best index for N mineralisation prediction, both from soil alone (Gianello & Bremner, 1986; Gouveia & Coutinho, 2001), as well as from mixtures of soil and organic residues (Rodrigues, 2000). This method seems to be well suited for laboratory routine, as it is simple and relatively quick. Because this method only provides an index of PAN, further study, namely field experiments, is required for a more accurate calibration.

Acknowledgements

This work was partially funded by FCT and partially funded by the project POCTI/AGG/46559.

References

Cordovil, C. M. d. S., Cabral, F. and Dachler, M. (2001). Fertilising value and mineralisation of nitrogen from organic fertilisers (pot and incubation experiments). Acta Horticulturae, 563, 139-145.

Cordovil, C. M. d. S., Cabral, F. Coutinho, J. and Vasconcelos, E. (2002). Effect of organic residues application to soil on mineral nutrition of winter wheat. VII ESA Congress pp. 347-348.

Gianello, C. and Bremner, J. M. (1986). Comparison of chemical methods of assessing potentially available organic nitrogen in soil. Communications in Soil and Science Plant Analysis, 17, 215-236.

Gouveia, J. P. and Coutinho, J. (2001). Comparação de métodos laboratoriais para estimar a mineralização de azoto em solo vitícolas. (Comparison of laboratorial methods to predict nitrogen mineralisation on vineyard soils). I Congresso Nacional das Ciências do Solo. Livro de resumos. pp. 41.

Rodrigues, M. A. (2000). Gestão do azoto na cultura da batata. (Management of nitrogen on potato crop- PhD thesis). Tese de Doutoramento. UTAD, Vila Real. 277 p.

Rhizodeposition and symbiotic N_2 fixation in trimmed and untrimmed white clover

S. Dahlin
Swedish University of Agricultural Sciences, Department of Soil Sciences, Box 7014, S-750 07 Uppsala, Sweden

Introduction

Legumes may contribute large amounts of N to pastures, leys and green manures, and are particularly promoted in organic farming. Our knowledge is limited on how management practices (such as trimming) affect total amounts of symbiotically fixed N, including below-ground, plant-derived N (BGN, i.e. N in live roots and rhizodeposition). In this study, the effects of trimming of white clover on BGN and N_2 fixation were examined.

Materials and methods

White clover (*Trifolium repens* L. cv Ramona) was grown in two pot experiments in growth chambers. At the late vegetative stage, plants were either trimmed at 4 cm and the trimmings removed, or left untreated. Samplings were made at this time (named Before trimming) and after four weeks of regrowth. All plant and soil fractions were dried, ground and analysed on a Finnigan Delta Plus mass spectrometer. Rhizodeposition and BGN were determined through *in situ* leaf labelling (McNeill et al., 1997) with ^{15}N-urea (0.5% w/w, 98 atom%) on three occasions during the growth period. After each labelling, the leaves used were isolated on the outside of the pots to avoid direct transfer of label from leaves to soil, and removed one week later. At sampling, the soil was sectioned into 0-10 cm and 10-20 cm depth, and sorted into clean root, soiled root, root-free soil (i.e. no visible root fragments), and a residual fraction. N_2 fixation was determined through ^{15}N natural abundance with perennial ryegrass (*Lolium perenne* L. cv Leia) as a reference plant. Roots were washed free of soil in deionised water. B-values (i.e. no soil) were derived from plants grown under similar conditions, where the soil was substituted by washed quartz sand and N-free nutrient solution. Separate B-values, used for calculating % Ndfa, were derived for all plant fractions and both sampling dates.

Results and discussion

Below-ground, plant-derived N was approximately twice as high as recovered root N for all times and treatments, although absolute amounts of BGN were significantly higher in the untrimmed treatment than in the trimmed treatment and 'Before trimming' (Figure 1). BGN was up to 30-35% of total plant-derived N present at the time of sampling, irrespective of treatment and time of sampling (Table 1). In contrast, BGN was a significantly lower proportion (19%) of total plant-derived N when the plant material removed by trimming was included in the calculations.

The $\delta^{15}N$ vs air of the B-value plants was significantly different between the treatments and times, as well as between roots and shoots, confirming that in this type of study, an individual

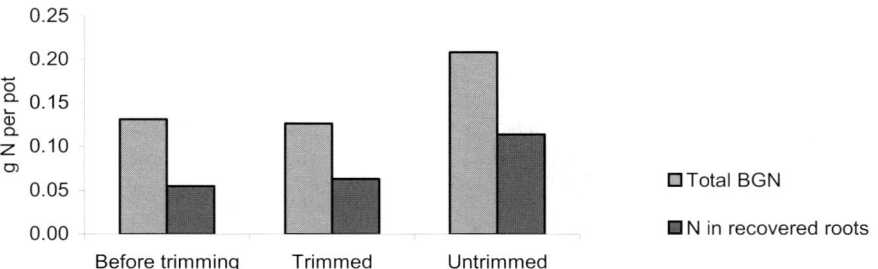

Figure 1. Total below-ground N (BGN) and N in recovered roots.

Table 1. Below-ground plant-derived N (BGN), BGN in % of total plant-derived N (PDN), percentage of N derived from air (Ndfa), mg Ndfa, and ΔNdfa (Ndfa including rhizodeposition minus Ndfa including recovered roots only) (significantly different values in italics).

Time/Treatment	BGN mg per pot	BGN % of PDN	% Ndfa	Ndfa mg per pot	ΔNdfa mg per pot
Before trimming	131	30.4	78.3	288	39
Trimmed (excl./incl. trimmings)	126	34.9/ *19.2*	77.9	425	43
Untrimmed	*209*	31.3	74.6	*549*	47

B-value must be used for each sample fraction (data not shown). The percentage of N derived from air (% Ndfa) was [H1]similar in all plants (Table 1). However, the untrimmed treatment had higher amounts of Ndfa than the trimmed treatment, as it had a larger biomass and total N (734 mg N per pot in the untrimmed treatment compared with a total of 545 mg N per pot in the trimmed treatment, including trimmings). The difference between the treatments in total biomass accumulation and Ndfa may be expected to decrease in the longer term though, the trimmed plants were still actively growing at the end of the experiment, whereas the untrimmed plants had started to mature.

Conclusions

Inclusion of rhizodeposition in the estimates of Ndfa produced significantly higher values than inclusion of recovered roots only, with an increase of approximately 9-14% above the root only values. Although the absolute levels of fixed N_2 may not be directly relevant to the field situation, the results indicate the importance of including rhizodeposition in N budgets.

Acknowledgements

This project was funded by The Swedish Research Council for Environment, Agricultural Sciences and Spatial Planning (FORMAS).

References

McNeill, A.M., Zhu, C. and Fillery, I.R.P. (1997). Use of *in situ* ^{15}N-labelling to estimate the total below-ground nitrogen of pasture legumes in intact soil-plant systems. Australia Journal of Agricultural Research, 48, 295-304.

Within-field variations in, and relations between, grain protein content, grain yield and plant-available soil nitrogen

S. Delin

Swedish University of Agricultural Sciences, Department of Agricultural Research Skara, P.O. Box 234, SE-532 23 Skara, Sweden

Introduction

Variations in grain crude protein concentration (CP) can be considerable within a field (Reyns et al., 1999). In uniformly fertilised cereal fields, high yielding areas and areas with smaller amounts of plant-available soil N (Np) may have lower CP because of smaller amounts of N available per kg grain yield. Site specific fertilisation demand, in relation to these parameters, should depend on how, and if, CP, yield and Np are related to each other. Therefore, the size of variation, the linear relationships between CP, yield and Np and the consistency in maps between years within an arable field were investigated.

Materials and methods

CP, grain yields and Np were measured at 34 sites within a 15 ha field in southwest Sweden during 1998-2000. Half of the field was dominated by sandy loams in both topsoil and subsoil and the rest, by loams in the topsoil and silty clays in the subsoil. Winter wheat was grown in 1998 and 2000 and spring barley in 1999. CP and grain yields were measured in areas (10 m^2) harvested separately at each site. Np was estimated by measuring N content in the crop cut at ground level in six squares (0.25 m^2) in adjacent large areas (60 m^2), not receiving any fertiliser N. The correlations between variables and years were calculated.

Results

Np correlated positively with CP in 1998, with both CP and yield in 1999 and with neither yield nor CP in 2000 (Table 1). CP had a negative correlation with yield in 1998, but the variations were small. In 1999, CP and grain yield were positively correlated. In 2000, CP was lower, where yield was higher at most sites (Figure 1), but CP did not correlate significantly with either yield or Np.

The maps of grain yield and CP were not consistent between years (Figure 1) and only Np showed some consistency. Np correlated significantly between all years, CP only between 1998 and 1999 and yield did not correlate significantly between any of the years.

Conclusions

Within-field variations in CP relate differently to yield depending on year. When N is limiting both yield and CP (as in 1999 in this field), then both are likely to be positively correlated.

Table 1. Summary statistics for the different variables in different years and their correlations (r) to other variables in the same year.

Variable	Mean	Standard deviation	r (Np)	r (Yield)
CP 1998 (%)	10.4	0.4	0.32*	0.45**
CP 1999 (%)	9.4	1.1	0.71**	0.77***
CP 2000 (%)	13.5	0.9	-0.13	-0.16
Yield 1998 (kg ha^{-1})	7030	800	-0.08	
Yield 1999 (kg ha^{-1})	4080	740	0.69***	
Yield 2000 (kg ha^{-1})	6360	810	0.26	
Np 1998 (kg N ha^{-1})	83	26		
Np 1999 (kg N ha^{-1})	45	16		
Np 2000 (kg N ha^{-1})	92	21		

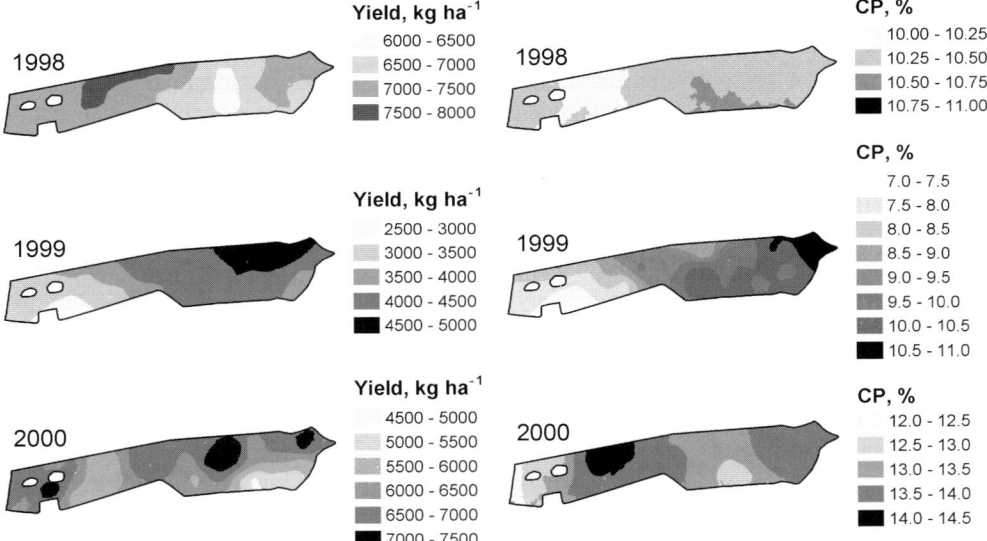

Figure 1. Maps of grain yield and grain protein concentration (CP) in the different years.

When N only limits CP (as in 1998), CP and yield are instead, likely to be negatively correlated. In either case, fertiliser N demand is higher in areas with low CP. However, CP and yield may not be correlated (as in 2000). This can happen when both CP and yield are limited simultaneously at some sites, because of limited utilisation of supplied N caused by, for example, fungal diseases or competition from weeds. Then fertiliser N demand is probably not related to CP. Since CP and yield are not very well correlated between years, historical maps are not of significant use to estimate future fertiliser demand.

References

Reyns, P., De Baerdemaeker, J. and Ramon, H. (1999). In: J.V. Stafford (ed). Precision Agriculture '99: Proceedings of the 2nd European Conference on Precision Agriculture, (Sheffield Academic Press, Sheffield, UK), 665-676.

The turnover of organic manure using natural ^{13}C abundance

K. Dittert[1] and R. Bol [2]
[1]*Institute for Plant Nutrition and Soil Science, Christian Albrecht University, Kiel, Germany*
[2]*Institute of Grassland and Environmental Research, North Wyke Research Station, Okehampton, Devon, UK*

Introduction

Efficient use of N from organic manures is often limited by difficulties in predicting the time course of mineralisation of organic N compounds and their plant availability. Initially, after manure application, easily decomposable manure compounds play an important role, as they drive organic matter turnover in soil. Mineralisation of nitrogenous compounds is therefore closely related to their breakdown, and soil respiration provides insight into these processes. The objective of the present work was to evaluate the use of the ^{13}C natural abundance tracer technique to dynamically quantify which proportion of the respired CO_2 was derived from applied slurry or from the 'native' soil organic matter.

Experimental

From soil (sandy loam), which developed under C3 vegetation, and which was then turned into grassland, 12 intact soil cores were taken for a pot experiment. After equilibration, pots were amended with water (control), or one of two cattle slurries that were collected from cows feeding on either a normal dairy cow diet (concentrate and hay; C3 plants), or exclusively on maize silage (C4 plant). Therefore, these slurries differed in their δ^{13}C. For a period of 23 days, respired CO_2 from the soil cores was collected over 45 minutes (at 15 min intervals) using the closed-chamber method. CO_2 concentrations and δ^{13}C were analysed on a trace gas mass spectrometer coupling, TGII Europa 2020. The percentage of respired CO_2 derived from slurry ($CO_{2\ dfs}$) was calculated as:

$$CO_{2\,dfs} [\%] = 100 * (\delta^{13}C\text{-}CO_{2\ Control} - \delta^{13}C\text{-}CO_{2\ sample})/(\delta^{13}C\text{-}CO_{2\ Control} - \delta^{13}C\text{-}CO_{2\ from\ pure\ slurry}) \quad (1)$$

This is similar to the calculation proposed by Balesdent & Mariotti (1996) but entirely based on δ^{13}C signatures of respired CO_2. Because of the variable contribution of air and respiration derived δ^{13}C in background CO_2 at T0 and throughout the experimental period, the contribution of background δ^{13}C was not corrected for, but instead δ^{13}C was measured at T0 and used as the baseline.

Results and discussion

After application of slurry, total soil respiration showed a rapid increase and a maximum 2 days after slurry addition. Respiration of C4 slurry-derived C was slower, showing its maximum also at day 2, but remaining at an elevated level over the next 6-7 days. A small priming effect was observed on day 2, when slurry addition led to soil C-derived respiration that was higher than control respiration (Figure 1). Then until day 5, total respiration fell back and almost returned to the

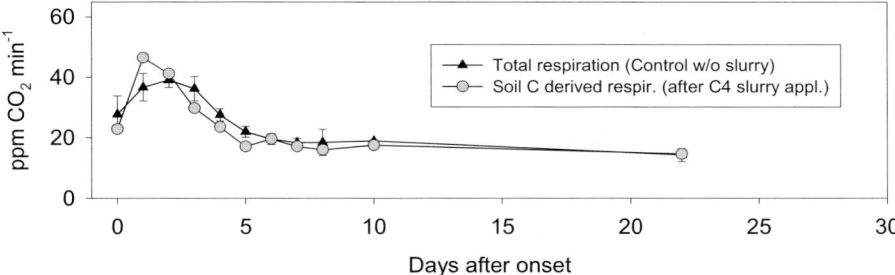

Figure 1. Comparison of native soil C derived respiration in control and slurry amended pots. Calculated rates based on $\delta^{13}C$ signatures of evolved CO_2.

initial range in all treatments. At this time, the maximum share of C4 slurry-derived CO_2 of 33% was found in total respired CO_2. Slurry-derived C remained in the range of 25-30% of total respiration for the next 10 days, then decreased to a final level of 20% at the end of the experiment.

In the present study of liquid manure turnover in soil, the natural $\delta^{13}C$ differences in slurry and soil organic matter were used to differentiate C sources from respired CO_2. A similar approach has been used for a study of maize crop residue decomposition under field conditions (Rochette et al., 1999). Here the absolute $\delta^{13}C$ ‰ values of respired CO_2 were in a higher range (-14 to -24 $\delta^{13}C$ ‰) than those measured in our experiment, where most data ranged between -14 to -27 $\delta^{13}C$ ‰. We attribute this difference to higher turnover rates in our short-term manuring experiment, leading to a low contribution of atmospheric CO_2 (-7.8 $\delta^{13}C$ ‰). The difference (Δ) in $\delta^{13}C$-CO_2 of C3-derived and C4-derived CO_2 ($\Delta = 16.8$ ^{13}C ‰) matched the Δ in $\delta^{13}C$ of the respective substrates, as measured according to the Balesdent & Mariotti (1996) approach used in the work of Flessa et al. (2000). Initially after slurry addition to soil, higher respiration activity was found, as compared with the control treatment, and a large increase in soil organic matter turnover through slurry addition might have been expected. The natural ^{13}C tracer technique revealed that the observed additional respiration almost entirely resulted from slurry C turnover. Only a small initial priming effect was found, but in the final balance, it was fully compensated for by a slightly lower SOM degradation later-on.

Acknowledgements

We thank Christian Pflieger for his skilful help during the experiment and the European Science Foundation for a travel grant to K. Dittert.

References

Balesdent, J. and Mariotti, A. (1996). Measurement of soil organic matter turnover using ^{13}C natural abundance. In: T.W. Boutton and S.I. Yamasaki (eds). Mass spectrometry of soils. Marcel Dekker, New York. 83-111.

Flessa, H., Ludwig, B., Heil, B. and Merbach, W. (2000). The origin of soil organic C, dissolved organic C and respiration in a long-term maize experiment in Halle, Germany, determined by ^{13}C natural abundance. Journal of Plant Nutrition and Soil Science, 163, 157-163.

Rochette, P., Angers, D.A. and Flanagan, L.B. (1999). Maize residue decomposition measurement using soil surface carbon dioxide fluxes and natural abundance of carbon-13. Soil Science Society of America Journal, 63, 1385-1396.

Nitrification during autumn and winter of ammonium nitrogen in cattle slurry applied to soil at different times during the autumn

L. Engström[1], B. Lindén[1] and L. Ericsson[2]
[1]*Swedish University of Agricultural Sciences (SLU), Skara, Sweden*
[2]*Swedish University of Agricultural Sciences (SLU), Umeå, Sweden*

Introduction

Application of slurry or other types of manure in the autumn usually increases N leaching during the winter period in maritime, temperate regions, such as south Sweden. The reason is mainly that NH_4^+ in the manure is transformed to NO_3^- which is easily lost from the soil. As nitrification of NH_4^+ declines with decreasing temperatures, one theoretical way (in northern Europe) to reduce NO_3^- leaching after autumn application of manure would be to supply the manure as late as possible in the autumn, at soil temperatures near the freezing point or shortly before the soil freezes. To investigate the possibilities in this respect, a field incubation study was started in October 2001 with determination of the rate of nitrification of applied NH_4^+-N in soil at natural temperature conditions, following incorporation of dairy slurry into soil in the autumn. The aim was to find, if possible, environmentally acceptable periods for autumn application of slurry in different parts of Sweden.

Materials and methods

The study involved incubations of dairy slurry in soil in watertight, ventilated plastic bottles placed in the 15-20 cm soil layer in open fields during autumn and winter. Slurry was mixed into a loamy sand soil corresponding to about 70 kg NH_4^+-N ha^{-1} in a 10-cm soil layer. The moisture content was adjusted to 65% WHC (Jansson, 1958). Bottles were placed in the soil on three occasions in the autumn: early October, late October and in mid to late November. The nitrification process was studied by taking out bottles frequently for analysis of NH_4^+ - and NO_3^--N. In order to consider soil N mineralisation, bottles with soil (but without slurry) were put into the topsoil with the bottles with soil-slurry mixtures. The investigation was performed at four sites: Röbäcksdalen in Västerbotten (63°50[1]N), Hamre in Dalarna (60°17[1]N), Lanna in Västergötland (58°22[1]N) and Lilla Böslid in Halland (56°37[1]N) in north, central, south-west and south Sweden, respectively.

Results and discussion

At all sites the nitrification rate was slower at the later application dates. The results indicate, however, that application of slurry, even as late as in mid to late November led to complete nitrification of the added NH_4^+-N within 2-3 months at all sites under the weather conditions of autumn of 2001. A corresponding laboratory incubation at 5, 2 and 0°C showed that NH_4^+ was completely nitrified after 50, 70 and 90 days, respectively, and 80 % was left after 90 days at -2 °C. The results indicated a rate slightly slower than nitrification at fluctuating

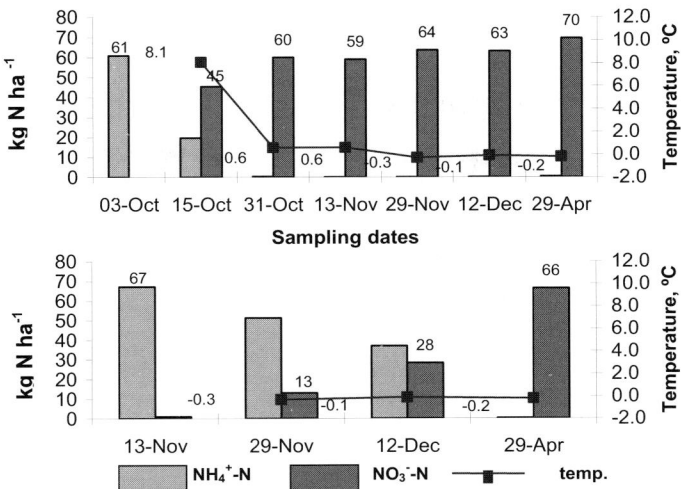

Figure 1. Nitrification rate in slurry following application at 3 October and 13 November 2001. Incubation studies at the northern site, Röbäcksdalen, in relation to mean soil temperature between the sampling dates.

temperatures in the field, but also that temperatures as low as about -2 °C were necessary to stop nitrification. In the field incubation study, the lowest average temperature in the soil was -0.6 °C between two occasions of analysis. This obviously allowed the nitrification process to continue. At the two most northern sites (Hamre and Röbäcksdalen), snow covered the ground during most of the winter period (December - April) which explains the relatively high temperatures in the soils.

Conclusions

The results demonstrate increased risks of NO_3^- leaching following autumn application, even in northern Sweden, when slurry has been incorporated into the soil as late as the end of November. Despite this, the risk for N leaching is less in the northern areas (Hamre and Röbäcksdalen) than in the south. In the north, the soil is more likely to be constantly frozen until spring, than in the south (here represented by L. Böslid and Lanna) where very mild periods during the winter can be frequent, with increased N leaching as a result. As winter wheat sown in late September in south and central Sweden only takes up about 5-10 kg N ha[-1] before the winter (Lindén, 2000), NO_3^- leaching cannot be reduced by sowing winter cereals after application of slurry in these regions. Consequently, slurry, as well as other types of manure, should preferably be distributed in spring or during the growing season when the risk of NO_3^- leaching is at a minimum, or before sowing a winter oilseed rape crop which has a large capacity to take up N in the autumn.

References

Lindén, B. (2000). Nitrogen uptake of winter cereals during autumn. Swedish University of Agricultural Sciences, Department of Agricultural Research Skara. Report 5, (In Swedish, with English summary), 23 pp.

Jansson, S.L. (1958). Tracer studies on nitrogen transformations in soil with special attention to mineralisation-immobilisation relationships. Annals of the Royal Agricultural College of Sweden, 24, 101-361.

Effect of rate of cattle slurry at sowing, number of fertigations with separated slurry liquid and rate of mineral N top dressings on yield and N removal by forage maize

A. Fernandes[1], H. Trindade[2], J. Coutinho[2] and N. Moreira[2]
[1]DRAEDM, S: Pedro de Merelim, 4700-859 Braga, Portugal
[2]Dept. of Plant Sci. and Agric. Engineering, UTAD, Ap. 202, 5001-911 Vila Real, Portugal

Introduction

Solid-liquid slurry separation techniques expand the possibilities to improve slurry use efficiency and to reduce its negative environmental impacts. The aim of this work was to evaluate the feasibility of supporting N fertilisation of maize grown for silage by the application of slurry at sowing and fertigations with the slurry-liquid fraction during the crop growth period.

Materials and methods

The experiment was carried out at Braga in the NW region of Portugal between May and September 2002. The soil was a deep well-drained sandy loam derived from granite and classified as dystric cambisol. The trial was laid out as a three factor factorial design with three replications and a split-split plot layout. Three rates of cattle slurry applied just before sowing (main plots), corresponding to 0, 70 and 140 kg total N ha^{-1}, were factorially combined with four fertigation treatments (F0, F1, F2 and F3) of cattle slurry liquid fraction separated by sieving and with the application of 0 and 100 kg N ha^{-1} as ammonium nitrate, top-dressed at the maize 6-7th leaf stage. Fertigation treatments consisted of: F0- nil; F1- one fertigation (at the maize 6-7th leaf stage); F2- two fertigations, (the first as in F1 and the second 2 weeks later); and F3- three fertigations, (the first two as in F2 and the third, 2 weeks later). The amount of slurry-liquid fraction for each fertigation supplied ca. 50 kg total N ha^{-1} with the corresponding volume spread over the soil, and the following crop sprinkler-irrigated to avoid any misleading effects associated with the small amount of water added by treatments. Maize cv. Estoria (400 FAO) was sown on May 20 and harvested on September 19.

Results and discussion

Maize production was significantly affected by the interaction of the three factors under study (Table 1). Dry matter yield ranged from 12.8 to 23.0 t DM ha^{-1}. Top dressing of 100 kg mineral N ha^{-1} markedly increased DM yield and reduced the response to rate of slurry applied at sowing. The effect of the number of fertigations applied, showed an unexpected pattern. Treatments where only a 2nd fertigation was applied (F2), showed a decrease in DM yield, whilst when a 3rd fertigation was applied (F3), yields returned to values similar to those obtained at F1. Mineral N top dressing was the only factor that significantly affected forage N content. Application of 100 kg mineral N ha^{-1} at 6-7th leaf stage increased maize N content from 9.3

Table 1. Maize dry-matter yield (t DM ha^{-1}) as affected by treatments.

Mineral N top dressed	Slurry at sowing	Fertigation treatments			
(kg N ha^{-1})	(kg total N ha^{-1})	F0	F1	F2	F3
0	0	12.8 k	15.8 ij	14.9 ijk	19.1 defg
	70	15.5 ij	18.8 efgh	15.2 ijk	18.5 fgh
	140	13.3 jk	19.7 cdef	16.5 hi	16.9 ghi
100	0	20.0 bcdef	22.3 ab	21.7 abcd	23.0 a
	70	20.0 bcdef	22.0 abc	21.1 abcdef	22.9 a
	140	21.1 abcde	21.7 abc	23.0 a	16.3 hi

Data followed by the same letters do not differ at p<0.05 level, Tukey test.

to 11.0 g N kg^{-1} DM. Interaction effects between rate of slurry at sowing and fertigation treatment (Figure 1) and between rate of slurry at sowing and rate of mineral N top dressed (Figure 2) were observed for maize N removal. Highest plant N removal occurred when no slurry, or a medium rate of slurry, was used at sowing, together with the application of three fertigations, and when no slurry was used at sowing and 100 kg mineral N ha^{-1} was applied as a top dressing. The results indicate that, when high rates of slurry were applied at sowing, the amount of N removed by the crop decreases if significant quantities of N are applied at maize 6-7th leaf stage (either as mineral or organic N).

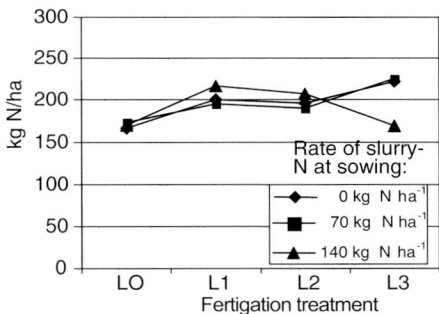

Figure 1. Effect of rate of slurry-N at sowing and fertigation treatment on maize N removal.

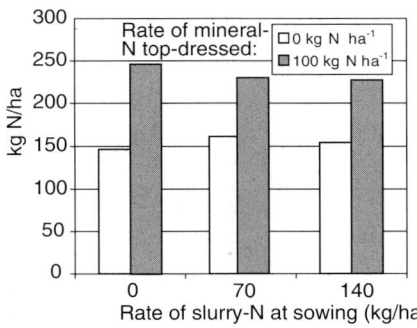

Figure 2. Effect of rate of slurry-N at sowing and rate of ineral N top dressed on maize N removal.

Controlling nitrogen flows and losses

Conclusions

When no slurry, or a moderate rate of slurry was applied at crop sowing, part of the maize mineral-N top dressing may be replaced by fertigations with the slurry-liquid fraction separated by mechanical sieving. However, no explanation was found for the depressive effect of the 2nd fertigation.

Acknowledgements

This work was supported by INIA - Project AGRO no. 177.

N balance in fertilizer trials with composted municipal solid wastes in southern Italy

D. Ferri, G. Convertini and F. Montemurro
Istituto Sperimentale Agronomico (MiPAF), Via C. Ulpiani, 5, 70125 Bari, Italy

Introduction

The development of composting Municipal Solid Waste (MSW) with an appropriate C/N ratio is providing an innovative management to supply N in organic form to crops in southern Italy's experimental conditions. Two cropping systems (sugar beet - durum wheat and tomato - durum wheat) were investigated in Southern Italy using two types of MSW compost obtained from general and selected sources of Municipal Solid Wastes. The effects of MSW compost effects on yield response, N use efficiency and N balance were evaluated

Materials and methods

Three field experiments were carried out in Mediterranean conditions on a Chromic Vertisol, typical of the flat land of the Apulia region (Foggia - South Italy). The soil is an alluvial vertisol, silty-clay textured (Typic Chromoxerert). The first trial was carried out during 1994-1999 (6 years) on sugar beet and durum wheat, cropped in a 2-year rotation, treated with differing doses of MSW compost corresponding to three N doses (40, 80, 120 kg N ha^{-1}). The MSW treatments were compared with other types of fertiliser (mineral fertiliser; organic + mineral fertiliser) in a completely randomised block experimental design with three replications. The second experiment, was carried out during the period 1997-2002 (6 years) on the same crops, but comparing different treatments using general sources of MSW (N120= MSW compost corresponding to 120 kg N ha^{-1}; N240= MSW compost corresponding to 240 kg N ha^{-1}; N0=control). The third experiment began in 2001 and is still in progress. It is being carried out on two typical rotations (industrial tomato-durum wheat and sugar beet-durum wheat) and compares the following N treatments using a selected source of MSW: MSW-compost application (N100) with 100 kg ha^{-1} of organic N for sugar beet; compost and mineral application

Table 1. Effects of treatments on yield and quality of sugar beet.

	Treatments	Yield	Sugar	α - N	Alkalinity
		t ha^{-1}	%	meq 100 g^{-1} pulp	index
First Trial	N 0	51.84	15.43	5.24	0.72
	N 80	60.90	14.52	5.65	1.00
	N 120	58.53	16.02	5.84	0.96
Second Trial	N 0	41.70	11.82	2.06	4.02
	N 120	43.60	11.68	2.24	3.67
	N 240	41.16	12.25	2.21	4.05
Third Trial	N 0	30.00	9.35	8.60	0.40
	N 50	33.36	9.06	8.51	0.48
	N 100	30.90	9.84	8.56	0.42

(N50) with 50 kg ha^{-1} of organic N and 50 kg ha^{-1} of mineral N for sugar beet; unfertilised plots (N0). MSW compost was applied about 1 month before sowing (sugar beet), whereas mineral N was applied during growth.

Results

The results are presented in Tables 1 and 2. Sugar beet yield, quality and N balance were affected by soil treatments. Moreover, the research showed that the N balance in the 'sugar beet - soil' system stayed more or less constant if 120 kg N ha^{-1} as organic compost was supplied to the soil. However, the application of N doses from 0 to 120 kg N ha^{-1}, as compost, increased the organic pool of the soil without pollution risks.

Table 2. Sugar beet N balance affected by soil treatments.

First Trial	N 0	N 80	N 120
N before sowing	28.33	27.03	35.33
N as a top dressing	0	80	120
N uptake	130.81	159.69	148.80
N after harvest	20.73	11.55	15.88
Δ N	-81.75	-41.11	22.41
Second Trial	N 0	N 120	N 240
N before sowing	24.09	23.94	22.14
N as a top dressing	0	120	240
N uptake	117.37	131.87	122.88
N after harvest	19.87	16.73	21.19
Δ N	-73.41	28.8	160.45
Third Trial	N 0	N 50	N 100
N before sowing	14.92	7.38	7.51
N as a top dressing	0	50	100
N uptake	176.25	120.76	163.22
N after harvest	8.82	20.31	13.05
Δ N	-152.51	-43.07	-42.66

Conclusions

Even though the results of the three trials refer to different periods of time, they show that the N balance in the 'sugar beet - soil' system is almost similar when it is supplied to the soil at about 120 kg N ha^{-1} as organic N. Lower N quantities could impoverish the soil, while more elevated N doses increase the organic pool of the system and do not seem to increase the risk of pollution. In an experiment of 6 years, this rate of MSW compost has increased the yields (first trial) in comparison with the control and it has not raised the α-amino N. Finally, the use of a new type of MSW compost from selected sources did not negatively influence the yield and quality of the sugar beet.

How to enhance crop utilisation of deep subsoil nitrogen supply

J. Haberle, P. Svoboda and J. Krejčová
The Research Institute of Crop Production, Plant Nutrition, 161 06 Prague, Czech Republic

Introduction

Distribution of mineral N (N_{min} = NO_3^- + NH_4^+ N) in a soil profile affects its utilisation during crop growth. There is rarely optimal synchronisation and localisation of available N supply, root system distribution and crop demand. Maximum depletion of available N (mostly NO_3^-) from the subsoil is desirable and thanks to its mobility, it is possible even at low root densities. Generally, depletion of N from deep soil layers may be enhanced by increasing uptake rates from, or by stimulating root growth (depth, density) in, the soil layers. Some authors observed stimulation of root growth and/or rooting depth under suboptimal N nutrition. Locally increased NO_3^- concentration causes root proliferation and increases the rate of N absorption. In contrast, there is a popular notion that high rates of N fertilisation (or irrigation water) reduce the size and depth of a root system. Functional equilibrium, (i.e. shift in allocation of growth to root system when nutrients or water are limited), has been the subject of a long-term discussion. These indices suggest that rooting depth and the depletion of available N may be manipulated to achieve maximum utilisation of NO_3^- leached from top soil during the inter-growth period. The environmental benefits of depletion of a deep subsoil N are evident, however, fertilisation schemes, or other measures to achieve the goal, need to be specified. Further, the procedures may result in reduction of yield, or quality, that may not be compensated for by savings of relatively cheap N fertiliser.

Materials and methods

Apparent depletion of mineral N from top and subsoil layers by winter wheat was observed in a field experiment on typical hapludoll soil carried out from 1995. Treatments were: unfertilised with N (N0), fertilised in spring with 100 kg N ha^{-1} (N1) or 200 kg N ha^{-1} (N2). From 2001/2002, the content of N in subsoil in N1 and N2 treatments was increased by autumn application of 100 kg N ha^{-1}, thus no N and 100 kg N ha^{-1} were applied in spring in N1 and N2 treatments, respectively. N_{min} distribution down to 120-150 cm was determined several times during growth; rooting depth and root length distribution were determined at anthesis or grain filling. The progress of the rooting front was estimated from observation of soil samples taken for the determination of N_{min} and soil moisture. Accumulation of N in above ground parts and other traits were observed.

Results

We observed mostly insignificant and inconsistent changes in root density in subsoil and rooting depth after fertilisation in the eight year experiment. As expected, the supply of mineral N in subsoil down to at least 90 cm was fully (apparently) utilised by the unfertilised wheat crop and with some delay, also in treatment N1. We observed a good agreement of distribution of apparent N uptake with the progress of the rooting front, N accumulation in crop biomass and water depletion. The application of 200 kg N ha^{-1} resulted in a higher content of residual N in both subsoil and topsoil and a reduction of rooting depth and root length density in subsoil in some years.

Discussion

A low N_{min} content in both N0 and N1 treatments, about the time when root growth to a deep subsoil is to be stimulated (heading, anthesis), may be the reason for a weak effect of N fertiliser on root growth. In N2, a higher N_{min} content in enriched topsoil could not be fully utilised by plants due to a dried up topsoil. The notion is supported by the fact that the critical level of the N dilution curve for wheat (Justes et al., 1994) was not reached either in N0 and N1 plants, or, convincingly, in N2 (Svoboda et al., 2000). Water content in top layers decreased often near to wilting point about, and after heading in experimental years. Enhanced shoot growth in treatments N1 and N2 increased water depletion of subsoil layers in comparison with the unfertilised treatment that might stimulate root growth to depth and thus disguise the possible effect of different N_{min} contents. Observed apparent depletion may be also affected by other factors (denitrification, mineralisation and immobilisation of N, NO_3^- leaching). However, these factors were probably of little importance in the subsoil (Kuhlman et al., 1989) under these soil-climate conditions. Mass flow and diffusion of N from under-rooted layers was not important because of a low N_{min} in the layers.

Conclusions

N fertilisation had a weak and inconsistent effect on root depth and density in a deep subsoil. Thus, maximum depletion of subsoil N was ensured by crop demand rather than modification of the root system by different N_{min} content. Slightly suboptimal N supply (N1 as opposed to the N2 'insurance' dose) guaranteed a low residual N in both top and subsoil. The inevitable, small risk of a yield/quality penalty with such a strategy is compensated by environmental benefits. Fine tuning of N rates in relation to available soil N (+ mineralisation), estimated N demand and accounting for water supply is needed. As extensive sampling of deep subsoil mineral N is not feasible, mathematical models, if reliable, offer an alternative for estimating N_{min} supply and N uptake distribution (Haberle & Svaboda, 2000). Further, accounting for spatial variability of N_{min} and/or root growth in the field (Haberle et al., 2003) may improve management of soil N depletion.

Acknowledgements

The project was supported by MZe ČR research plan M01-01-01.

References

Haberle, J., Kroulík, M., Svoboda, P., Krejčová, J. and Cerhanová, D. (2003). Spatial variability of mineral nitrogen content in topsoil and subsoil. In: Programme book of the joint conference of ECPA-ECPLF, Berlin 2003, 415-416.
Haberle, J. and Svoboda, P. (2000). The simulation of nitrogen depletion in winter wheat crop. In: 3rd International Crop Science Congress, Hamburg 2000, p.152.

Justes, E., Mary, B., Meynard, J.-M., Machet, J.-M. and Thelier-Huche, L. (1994). Determination of a critical nitrogen dilution curve for winter wheat crops. Annals of Botany, 74, 397-407.

Kuhlman, H., Barraclough, P.B. and Weir, A.H. (1989). Utilization of mineral nitrogen in the subsoil by winter wheat. Zeitschrift Für Pflanzenernährung und Bodenkunde, 152, 291-295.

Svoboda, P., Haberle, J. and Krejčová, J. (2000). The relation between nitrogen supply in rooted soil volume and plant nitrogen status. In: 3rd International Crop Science Congress, Hamburg 2000, p.166.

Model SFOM - the first step towards fertiliser and manure use integration

T. Jadczyszyn

Institute of Soil Science and Plant Cultivation, Czartoryskich 8, 24-100 Pulawy, Poland

Introduction

Manure is a valuable source of nutrients for plants. In sustainable agricultural systems, the recommended doses of fertiliser should supplement the gap between plant demand and the amount of nutrients applied in manure. However, it is extremely difficult to obtain the necessary information on manure (especially solids) production, as well as its chemical composition from farms. This model enables the calculation of the quantity and quality of manures available at the farm, based on live-stock numbers.

Model description

The input data required by the model are livestock numbers and type of barn or pig sty. Calculations are based on the standard quantity of mineral elements excreted by different categories of animals in dung and urine (Table 1). Information on the amount of straw used for bedding in individual farms is also required, otherwise standard amounts of straw are taken into account (Jadczyszyn, 2001). The model also considers the standard quantity of water used for washing dairies etc. and in manure. The quantity of nutrients and matter collected in manures are reduced if the animals are periodically kept on pasture.

The type of manure produced on the farm depends on the animal housing system. In the case of a deep litter system, the only product is the solid manure which consists of faeces, urine, straw and water. All components are collected together during a period of up to 6 months. Usually manure from deep litter houses is removed directly onto the field. In shallow litter barns, the straw is changed regularly (often every day). The small portion of straw does not

Table 1. The average amount of nutrients excreted from livestock per year.

Animals	Faeces kg				Urine kg			
	Fresh matter	N	P	K	Fresh matter	N	P	K
Calf 3 - 6 months	1680	5.04	2.46	2.34	1780	12.2	0.02	9.40
Heifer, 6 -12 months	2630	7.30	2.23	3.51	2260	14.9	0.10	20.7
Heifer, 12-24 months	4790	14.9	5.38	7.04	3327	27.5	0.19	40.2
Heifer above 24 months	5125	15.9	5.68	7.54	3560	29.5	0.20	43.2
Bull above 24 months	4620	16.0	5.07	8.66	3430	29.7	0.16	34.6
Cow	12717	37.8	11.2	21.7	5260	48.9	0.27	69.9
Sows and boars	1116	4.60	2.09	1.86	1850	15.3	0.85	3.70
Sow with 18 piglets to 20 kg	1170	11.4	5.75	4.40	4410	25.8	2.46	8.80
Piglets 20-30 kg	240	1.98	1.12	1.12	500	4.71	0.49	1.60
Porker 30 - 70 kg	680	3.33	1.57	1.48	1350	7.74	0.67	2.30
Porker 70 - 110 kg	820	4.44	2.40	2.22	1640	10.3	1.02	3.40

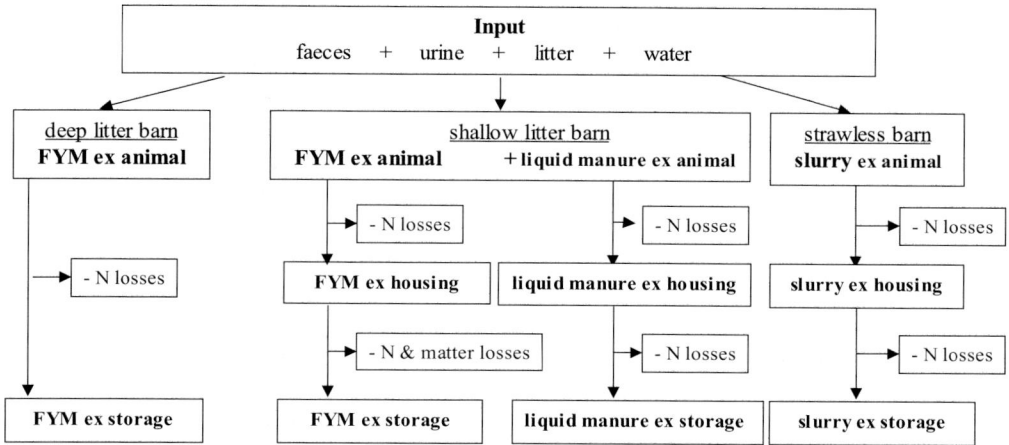

Figure 1. The scheme of nutrient and matter flow during manure production.

absorb the whole amount of excreta and water, so these partly percolate and flow outside to a container. Therefore, two kinds of manure are produced: solid and liquid. Slurry is produced in strawless systems, and this is a mixture of faeces, urine and water. It is obvious that the chemical composition of manure will change during manure storage, mainly because of N losses. The SFOM model simulates mass and nutrient flow step by step, i.e. ex-animal, ex-housing and ex-storage, considering standard N losses at each stage (Figure1). The results of simulations are for fresh matter and NPK content in manure produced at the farm.

Model validation

There are no available standards for the chemical composition of manures produced by a particular category of animals. So the average nutrient content for cattle and swine manures was calculated and compared with the standards (Table 2). The model substantially overestimated the NPK content in cattle liquid manure. The accuracy of the nutrient distribution between solid and liquid manures in shallow litter barns needs to be verified.

Table 2. Simulated and standard nutrient contents in manures.

Manure	%N		%P		%K	
	simulated	standard	simulated	standard	simulated	standard
Cattle FYM	0.45	0.45	0.07	0.12	0.50	0.50
Swine FYM	0.43	0.45	0.11	0.13	0.45	0.50
Cattle slurry	0.35	0.30	0.05	0.05	0.26	0.25
Swine slurry	0.35	0.35	0.10	0.09	0.16	0.21
Cattle LYM*	0.56	0.30	0.07	0.00	0.83	0.42
Swine LYM*	0.27	0.25	0.04	0.00	0.12	0.17

* liquid manure

Acknowledgements

The work was done as a part of the EU Inco-Copernicus Programme "MAINTAINE" (Contract IC15 - CT98 - 0108).

References

Jadczyszyn, T. (2001). Model for calculation of the amounts of nutrients in manure "SFOM"Fertilizers and Fertilization, 1, 40-50.

Fotyma, M., Jadczyszyn, T. and Pietruch, Cz. (2001). A decision support system for sustainable nutrient management on farm level: MACROBIL. Fertilizers and Fertilization, 2, 7-26.

Straw-rich deep litter manures - can decomposition and N turnover be predicted from quality?

L.S. Jensen, A. Jensen and A. Pedersen
The Royal Vet. and Agricultural University, Plant Nutrition and Soil Fertility Laboratory, Thorvaldsensvej 40, 1871 Frederiksberg C, Denmark

Introduction

There is a demand for optimised utilisation of animal manures on farms with intensive animal husbandry systems. For animal welfare reasons, there is an increasing interest in open housing systems built with solid floors strewn with straw and these are expected to increase the amount of deep litter produced. As deep litter consists of a mixture of urine, faeces and bedding material, this is a highly heterogeneous material and increases the variability of the chemical composition. One major difficulty, in this context, is predicting the plant availability of deep litter N. The objectives of the present study were to investigate if i) the variability in deep litter chemical composition could be explained by easily accessible properties, such as estimated age and dry matter content and ii) the C and N mineralisation patterns of the deep litters could be predicted from simple characteristics (total-N, NH_4^+-fraction, C/N ratio) and whether this could be further improved by data on gross biochemical composition (solubles, cellulose, hemi-cellulose and lignin).

Materials and methods

Twenty five deep litter samples (pig and cattle, fresh and composted) were collected from different farms in Denmark. The bedding material for all deep litters was straw. The average age of the sampled deep litter was estimated from the information given by the farmers. All samples were analysed for inorganic N, total C and N, and a subset also for cellulose, hemi-cellulose and lignin. Six of the litters were selected for an incubation experiment in a sandy soil at 15°C and field capacity. C and N mineralisation was determined periodically over 80 days. Empirical kinetic parameters of the cumulative C and N mineralisation of the deep litter materials were determined and correlated to the measured quality variables.

Results and discussion

There was a large variation in dry matter content for both cattle and pig deep litters (18-70%) and also large variation in the content of total N, NH_4^+-N and for C/N ratio (Figure 1). Only small (and insignificant) differences between cattle and pig deep litter characteristics were found. However, a significant ($p<0.001$) difference between fresh or composted deep litter was found for the NH_4^+ content (absolute and in percent of total N, Figure 1C). Dry matter content could not be used to predict the $C:N_{org}$ ratio, or the NH_4^+ to total N ratio. However, NH_4^+ to total N and C/N_{org} ratios seemed to be narrower, the higher the estimated age of the deep litters (Figure 1D and E). The deep litters differed markedly in gross biochemical composition, with most of the C in deep litters in the neutral detergent soluble fraction (NDS) and with lignin accounting for the least part of total C for most of the samples. However, no

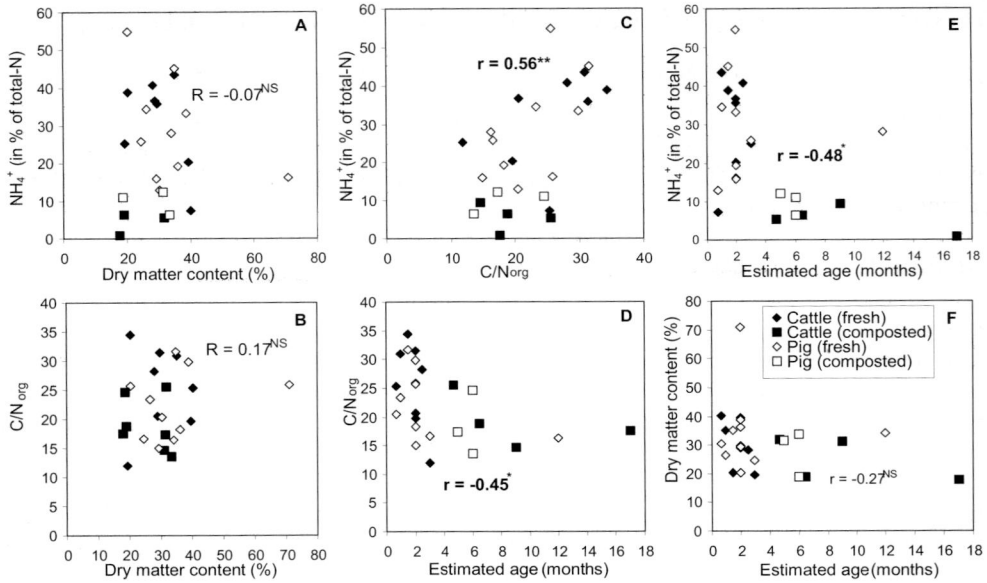

Figure 1. Influence of estimated age and dry matter on C/N-ratio and NH_4^+ content.

Table 1. Pearson correlation coefficients between deep litters characteristics and C- and N-mineralisation.

	Carbon mineralisation		Nitrogen mineralisation			
	C_{max}	$t_{50\%}$	$N_{max.imm.}$	$t_{N-max.imm.}$	m_{re-min}	Net $N_{min\ -80d}$
Dry matter	0.35	-0.32	0.47	0.01	0.35	-0.32
NH_4^+:Total N	0.59	**-0.83***	0.51	0.48	-0.09	-0.78
$C:N_{tot}$	-0.06	0.42	0.05	0.44	0.50	0.05
$C:N_{org}$	0.57	-0.33	0.59	**0.81***	0.51	-0.68
ND Solubles	0.03	0.29	-0.59	-0.45	-0.15	0.45
Hemi-cellulose	0.40	-0.27	(0.85*)	0.06	0.60	-0.68
Cellulose	0.67	**-0.86***	0.46	0.68	-0.02	-0.73
Lignin	**-0.95****	0.73	-0.75	-0.71	-0.41	**0.85***

**, * Indicates significance at 1% and 5% level, respectively, correlations without asterisk are non significant. The significant correlation shown in parentheses is due to a single high value.

significant difference between cattle and pig deep litters was found. On average, 86% of the total N was in the NDS fraction (r^2=0.93, p<0.0001), and although some of this was NH_4^+, a significant proportion of the organic N in the deep litters was found to be soluble.

As seen in Table 1, few significant correlations were found between simple characteristics (dry matter, NH_4^+, C:N) and N mineralisation parameters; only $C:N_{org}$ was positively correlated (p<0.05) to the length of net N immobilisation. Gross biochemical properties did not show much greater promise for explaining C and N mineralisation characteristics, although lignin seemed to be both highly negatively correlated to maximum C-mineralisation and positively correlated to net N mineralisation after 80 days.

In conclusion, the variability in deep litter chemical composition could not be predicted from age or dry matter content. Only weak predictions of C and N mineralisation patterns of the deep litters could be obtained from simple characteristics, and were not improved markedly by biochemical properties.

Field experiments to determine N accumulation under fertility building crops

A. Joynes[1], D.J. Hatch[1], A. Stone[1], S. Cuttle[1], G. Goodlass[2] and S. Roderick[3]
[1]*IGER, North Wyke Research Station, Okehampton, Devon, EX20 2SB, UK*
[2]*ADAS, High Mowthorpe, Malton, N Yorks, UK*
[3]*Duchy College, Stoke Climsland, Callington, Cornwall, UK*

Introduction

In recent years there has been a rapid growth in organic farming in the UK. The government's policy is to encourage organic farming because of the expected environmental and socio-economic benefits. Defra are providing support in the form of grants and subsidies for farmers who convert from conventional to organic forms of production and have funded a thee year project to evaluate the use of fertility building crops. Practical help in the form of leaflets and a booklet are planned to summarise the findings of this project. IGER, together with ADAS and Duchy College, are undertaking field experiments to investigate N_2 fixation by legumes, under various conditions of cutting, mulching and different levels of soil fertility. The first aim of the experiment is to see if the level of soil fertility has a significant effect on N_2 fixation. It is anticipated that soils with a high level of available N will reduce N_2 fixation. A second aim is to see whether the practice of mulching legumes by returning herbage to the sward, also causes a reduction in N_2 fixation. Additional measurements are being made on selected commercial organic farms over two growing seasons.

Methods

Forty eight paired plots (1.5 x 10m), were prepared for planting either with, or without composted FYM in autumn 2002. The plots were randomised and half were sown with a red clover/ryegrass mixture, while the other half was sown with ryegrass only. The plots were cut three times during the growing season. The following treatments were applied:

Treatments
A. Red clover/ryegrass mixture (herbage cut and removed)
B. Ryegrass (herbage cut and removed)
C. Red clover/ryegrass mixture (herbage cut and returned to plot)
D. Ryegrass (herbage cut and removed: herbage from treatment A spread on the plot)

The cut herbage was returned to the plots, spread with a fork and then mown again to produce a fine mulch. There were 6 replicates of each treatment. Herbage samples were taken after each cut and were analysed for dry matter (DM), total C and total N. Clover from the red clover/ryegrass plots was separated from the grass, weeds and dead material and analysed separately from the other material.

The experiment will be continued over two growing seasons. Nitrogen accumulation below ground is determined by analysing for soil mineral N and the total N contained in roots from

N yields from second harvest

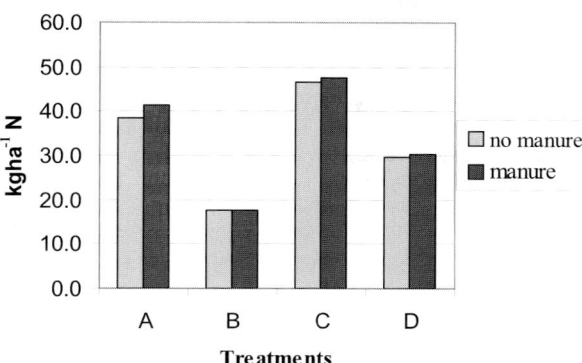

Figure 1. *Preliminary results from IGER plot experiments, second harvest 1st July 03.*

soil cores taken at the beginning and end of each season. Estimates of N_2 fixation are made by subtracting the N in nil-N grass controls from the treatments with red clover.

Results and discussion

The first harvest showed that N_2 fixation did not appear to have been suppressed by the addition of FYM (data not shown). The situation did not change with the results from the second harvest (Figure1): probably because the difference in fertility between the manured and non-manured plots was insufficient to depress N_2 fixation. This may be due to the readily available N in the manure being low (approximately only 10% of the soil N supply) and the remainder likely to have been mainly recalcitrant organic material (i.e. well-composted).

Table 1. *Preliminary results from IGER plot experiments second harvest.*

N yield of treatments less grass controls	Fixation N (kg ha^{-1})
Manure, clover cut and removed (A-B)	23.9
No manure clover cut and removed (A-B)	21.0
Manure clover cut and returned to plot (C-D)	17.3
No manure clover cut and returned to plot (C-D)	16.8

Table1 shows that the clover/ryegrass plots with herbage cut and removed (treatment A) had increased N_2 fixation when compared with the clover/ryegrass plots that had the mulch returned (treatment C). One explanation, may be that the extra available N from the mulch caused a reduction in N_2 fixation.

If these results are confirmed in future harvests then, in order to maximise N fixation, it would be more advantageous to cut and remove clover herbage and apply this as a mulch to a fallow, or non-leguminous sward to improve the overall farm soil fertility.

A comparison of different indices for nitrogen mineralisation

A. Kokkonen and M. Esala
MTT Agrifood Research Finland, Environmental research / Soils and environment, FIN-31600 Jokioinen, Finland

Introduction

This preliminary study was conducted to evaluate some of the methods used for predicting the N supply in soil for N mineralisation. We tested the methods that are simple, easy to perform and cheap for daily routine soil laboratories and could provide a rapid combined index for the directly available mineral N and for the mineralisable N.

Materials and methods

Three chemical indices for N mineralisation were tested on oven-dried (40°C, 24h) and field-moist clay, loam and peat soils by 1) 0.01M $CaCl_2$ solution, 2) cold water (deionized) and 3) neutral phosphate buffer solution consisting of $Na_2HPO_4 \cdot 2H_2O$ and KH_2PO_4. Soil N mineralisation capacity was analysed with a modification of the anaerobic incubation method described by Stengberg et al. (2000) both from fresh and dried soil samples. Finally, soil mineral N was extracted by 2 M KCl.

Results and discussion

The amount of Norg extracted by the different extraction methods ranged between 9-165 mg kg^{-1} in mineral soils (Figs. 1 and 2). In dried peat soil more than 1260 mg kg^{-1} N was extracted by the 0.01M $CaCl_2$ solution. Neutral phosphate buffer solution extracted more than 6% of the

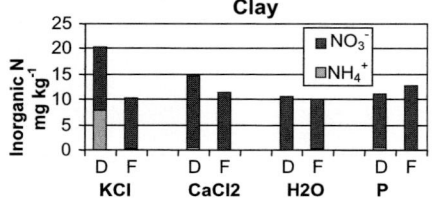
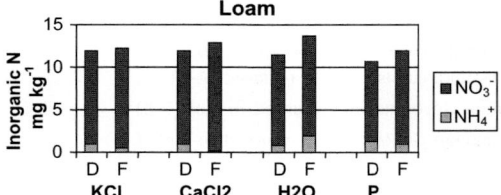
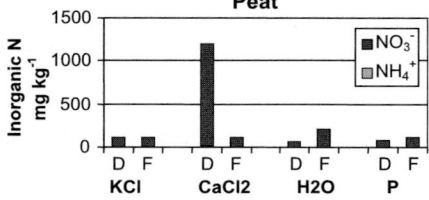

Figure 1. Nitrogen extracted by different methods.

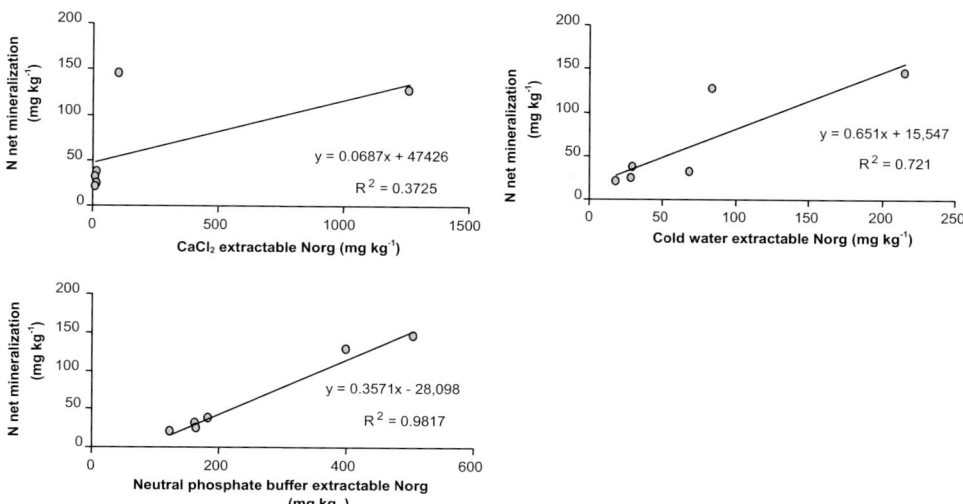

Figure 2. Relationship between extracted organic N and net N mineralisation obtained by an anaerobic incubation experiment.

total soil N and less than 2% of the total N was extracted by cold water or by 0.01M $CaCl_2$ solutions. The phosphate buffer extract showed a higher capability to exchange sorbed organic compounds than those with a low affinity (e.g. the chloride based extracts). Overall, the neutral phosphate buffer extraction provided the best indicator for N mineralisation. Phosphate extractable Norg gave the highest correlation ($R^2 = 0.91$) with net N mineralisation obtained by the anaerobic incubation experiment.

Conclusions

This study is a preliminary part of a larger project for predicting soil N mineralisation as well as modelling of N for N fertilisation optimation. The present results provide an important basis for improving understanding and in future work we will continue to test these methods and for N mineralisation indices.

References

Houba, V.J.G., Novozamsky, I., Huybregts, A.W.M. and Van Der Lee, J.J. (1986). Comparison of soil extractions by 0.01 M $CaCl_2$, by EUF and by some conventional extraction procedures, Plant and Soil, 96, 433-437.

Matsumoto, S., Ae, N. and Yamagata, M. (2000). Extraction of mineralisable organic nitrogen from soils by a neutral phosphate buffer solution. Soil Biology and Biochemistry, 32, 1293-1299.

Sippola, J. and Suonurmi-Rasi, R. (1985). Simple extraction methods as indicators of available soil-nitrogen supply. Annales Agriculturae Fenniae, 24, 125-129.

Stenberg, B., Johansson, M., Pell, M., Sjödahl-Svensson, K., Srenström, J. and Torstensson, L. (1998). Microbial biomass and activities in soil as affected by frozen and cold storage. Soil Biology and Biochemistry, 30, 393-402.

A novel approach to regulate nitrogen mineralisation in soil

K. Kumar[1], C.J. Rosen[1] and M.P. Russelle[1,2]
[1]Department of Soil, Water, & Climate, University of Minnesota, Saint Paul, MN 55108, USA
[2]USDA-ARS, Saint Paul, MN 55108, USA

Introduction

Protease enzymes play an important role in transforming organic N in soils and other organic wastes to inorganic forms of N, which are then utilised by plants (Figure 1).

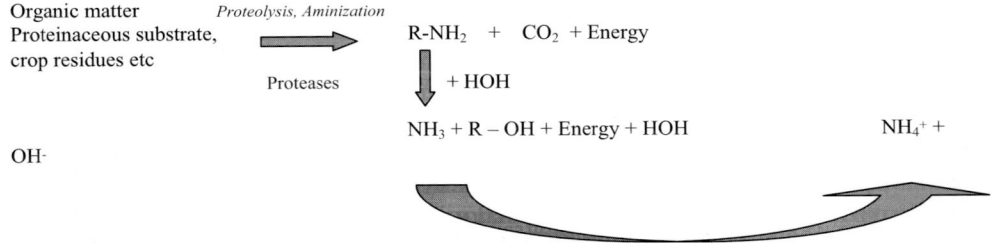

Figure 1. Nitrogen mineralisation process.

There has recently been a strong interest in protease inhibitors in treating HIV patients who have developed AIDS. Inhibitors of proteases are also present naturally in plants and recently their role in providing defense against insects and pests has been recognised (Ryan, 1990). Transgenic modifications have enhanced protease inhibitor expression in several species to develop crop cultivars resistant to insects. It is important to note that these plant protease inhibitors have specificities for animal and microbial enzymes, which are similar to the protease enzymes in soils. These inhibitors may also affect the activity of soil proteases, which are responsible for soil N mineralisation processes.

We hypothesise that addition of protease inhibitors to soil or organic amendments will alter the pattern of N release from organic N sources. The specific objectives of our research were to: (i) evaluate the effect of specific protease inhibitors on N mineralisation in soils with and without organic amendments and (ii) assess the potential for transgenic plants containing protease inhibitors to alter N mineralisation.

Materials and methods

Soil amended with alfalfa residues and un-amended soil was incubated in the laboratory with and without different protease inhibitors. The soil from these microcosms was extracted periodically with 2M KCl and inorganic N ($NH_4^+ + NO_3^-$) was determined in these extracts. Release of inorganic-N was taken as a measure of net N mineralisation (Equation 1).

Net N mineralisation = inorganic N at end of incubation - inorganic N at start of incubation.

Results and discussion

In a short-term incubation, some protease inhibitors were more efficient in reducing N mineralisation than others, and there were interactions between soils and protease inhibitors (Figure 2). The effects of protease inhibitors in reducing net N mineralisation were also seen after 50 days of incubation (Figure 3). The complete protease inhibitor cocktail was most efficient in reducing net N mineralisation in both the control soil and soil amended with alfalfa. The reduction in net N mineralisation for alfalfa-amended soils was at the same magnitude as the non-amended control soil. The additions of alfalfa residues probably lead to greater protease activity in soil and a single dose of inhibitor was not sufficient to markedly decrease the protease activity. However, another dose of inhibitor, 25 d after incubation, reduced N mineralisation during the next 25 d (Figure 3). The N mineralisation from transgenic plants with enhanced protease activity is being compared to N mineralisation of isogenic lines without the protease inhibitor gene.

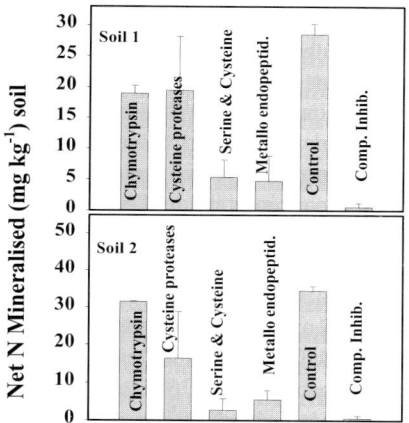

Figure 2. Effect of specific protease inhibitors on Net N mineralisation after 7 days of treatment.

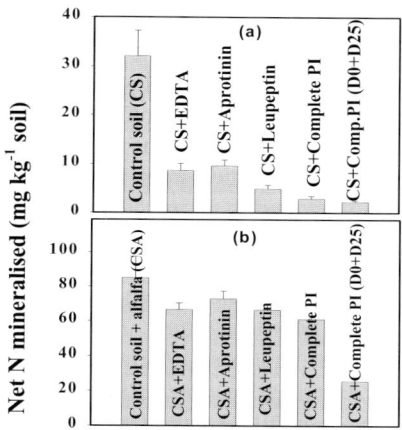

Figure 3. Effect of various protease inhibitors on net N mineralisation in 50 days (a) control soil, and (b) soil amended with alfalfa @ 11.2 Mg ha^{-1}.

Conclusions

Our results indicate that protease inhibitors can reduce protease enzyme activity in soil. Using either direct protease inhibitor additions or transgenic techniques, it should be possible to exploit this discovery to better synchronise N mineralisation with crop N demand.

Acknowledgements

This research is supported with funds from the USDA-CSREES (Soil and Soil Biology Programme) Grant 2002-35107-12436.

References

Ryan, C.A. (1990). Protease inhibitors in plants: genes for improving defenses against insects and pathogens. Annual Review of Phytopathology, 28, 425-449.

Controlling nitrogen flows and losses

Variable nitrogen fertilisation by tractor-mounted remote sensing

A. Link, J. Jasper and H.-W. Olfs
Hydro Agri, Centre for Plant Nutrition Hanninghof, Duelmen, Germany

Introduction

It is well known that the crop N demand during the growing season varies widely within individual fields. Previous research has shown that this in-field variation in soil and crop N dynamics might be as great as between-field variation (Dampney & Goodlass, 1997). Because of the importance of N in plant nutrition, the costs of N fertiliser and its environmental impacts, site-specific N fertiliser policy is one of the main objectives in precision agriculture. However, precision agriculture requires accurate and efficient tools to determine the actual N fertiliser demand for every spot in the field. Traditional methods of soil and plant analysis have proven to be costly and time consuming in delivering information on the actual N fertiliser demand at the required high spatial and temporal resolution (Wollring et al., 1998). Remote sensing techniques (e.g. canopy reflectance measurements) offer the opportunity to deliver this information quickly, precisely and cost efficiently (Baret & Fourty, 1997).

Materials and methods

Technical Set-up of the N-Sensor
The N-Sensor (Hydro, Germany), a tractor-mounted multispectral scanner, consists of two diode-array spectrometers, fiberoptics and processing electronics mounted together in one box on top of the tractor cabin. A special viewing geometry and an integrated irradiance correction guarantees an accurate measure of canopy reflectance on both sides of the tramline and a high stability of the calculated spectral index (Link et al., 2002). A terminal, which logs crop and GPS data on a chip card and displays the status of the crop, controls the system.

Calibration and Performance Trials
N rate trials with winter wheat and winter barley on traditional small plots were established to determine the relationship between the derived spectral index and the N status of the crop and in turn, to calculate the optimum N fertiliser rate based on the N-Sensor values measured at the same time. In 1999-2002, trials with cereals on 146 large fields were carried out in several countries (CZ, DE, DK, FR, GB, HU, IT, SE, US) to test the performance of the obtained calibration for variable N fertilisation, according to the N-Sensor readings. Treatments with a uniform application at the same N rate were established as a reference that allowed us to determine effects solely due to N redistribution within a field. In addition to grain yield, grain protein content was also measured in 91 out of the 146 field trials.

Results and discussion

In 77% of all trials, grain yield in the N-Sensor treatment was higher compared with the control (Figure 1). Variable N application increased grain yield on average by 0.17 t ha^{-1} compared with the uniform fertiliser application of the same total N amount. Higher grain yield in the variable N treatment in general had no negative effect on grain protein content. Grain protein content in the variable N treatment (11.86%) was increased, on average, by 0.08% compared with the uniform N treatment (11.78%).

Thus, the efficiency of N fertiliser was improved by site-specific N application. These results are in line with Goodwin et al. (2003) who found even bigger improvements from variable rate application when using a 'standard' uniform N rate according to normal farm practice as a reference.

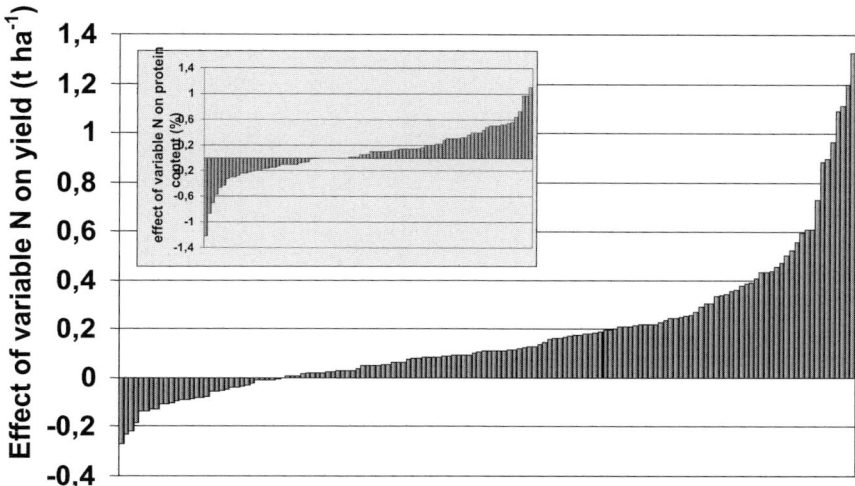

Figure 1. Yield (146 trials) and protein content (insert, 91 trials) from variable N application compared with uniform application.

Conclusion

With the N-Sensor it was possible to detect differences in the crop N status within a field and to apply N fertiliser accordingly. Varying N fertiliser application rates according to spatially variable crop N requirements has positive effects: over, or under-fertilisation within a field can be avoided. This increases yield, N fertiliser efficiency and reduces lodging. Other positive effects are a more homogeneous ripening of the crop, which reduces harvesting costs and grain losses.

References

Baret, F. and Fourty, T. (1997). Radiometric estimates of nitrogen status of leaves and canopies. In: G. Lemaire (ed). Diagnosis of the nitrogen status in crops, Springer-Verlag, Berlin/Heidelberg, 201-227.

Dampney, P.M.R. and Goodlass, G. (1997). Quantifying the variability of soil and plant nitrogen dynamics within arable fields growing combinable crops. In: J.V. Stafford (ed). Precision Agriculture '97: Proceedings of the 1st European Conference on Precision Agriculture, BIOS Scientific Publishers Ltd, Oxford, UK, 219-226.

Goodwin, R.J., Wood, G.A., Taylor, J.C. Knight, S.M. and Welsh, J.P. (2003). Precision farming of cereal crops: a review of a six year experiment to develop management guidelines. Biosystems Engineering, 84, 375-391.

Link, A., Panitzki, M. and Reusch, R. (2002). Hydro N-Sensor: Tractor-mounted remote sensing for variable nitrogen fertilization. Proceedings of the Sixth International Conference on Precision Agriculture, 932-937.

Wollring, J., Reusch, S. and Karlsson, C. (1998). Variable Nitrogen Application Based on Crop Sensing. The International Fertilizer Society Proceedings No. 423, 3-27.

A dynamic version of the predictive balance sheet method for fertiliser N advice

J.M. Machet[1], S. Recous[1], M.H. Jeuffroy[2], B. Mary[1], B. Nicolardot[1] and V. Parnaudeau[1]

[1]*INRA, Unité d'Agronomie de Laon-Reims-Mons, Rue Fernand Christ, 02007 Laon Cedex, France*
[2]*INRA-INA PG, UMR d'Agronomie, 78850 Thiverval-Grignon, France*

Introduction

The research conducted over the past ten years on the dynamics of organic matter in soils and on the fate of fertiliser-N on the one hand, and the increasing demand for high quality crops and protection of the environment on the other hand, led us to develop a new version of the predictive balance sheet method. Until now, this method for calculating fertiliser-N rates was the basis of the AZOBIL model (Machet et al., 1990), the software most commonly used in France by soil laboratories and advisors. The objectives of N fertiliser advice is to supplement the soil supply in order to ensure the optimal total N supply for crops, this being achieved not only by accurately calculating the total rate of fertiliser-N application, but also by matching the date of application with the crop N requirement. Moreover, environmental objectives must be also achieved: it is necessary to control the amount of residual mineral N in the soil after harvest, and to minimise gaseous losses and nitrate leaching. The present work describes the main new characteristics of the model.

Modelling the balance sheet dynamically

At the opening of the balance sheet (end of winter for winter crops, at sowing for spring crops), the soil inorganic N pool is measured at the rooting zone depth. In order to take into account the various contributions of previously applied crop residues and organic wastes to the residual mineral N (varying with the characteristics of added organic matter and climate), the decomposition of the different organic sources are simulated (using observed climatic data) from harvest of the previous crop, until the start of the balance sheet. Temperature and moisture functions are used to calculate a 'normalised time' which takes into account climatic variations and determines a potential rate of decomposition. From this time to the harvest of the crop, the subsequent net contribution of the organic residues (or wastes) and the net mineralisation of the humified organic matter, are simulated using normalised days calculated from the past years' mean climatic data of the area.

Availability of fertiliser-N

Tracers from fertiliser experiments using ^{15}N for different crops in the temperate climate of Northwest Europe showed that immobilisation of N by the soil heterotrophic microflora and gaseous losses compete with plant uptake for fertiliser-N (Recous et al., 1997). In order to better take into account these two processes in the calculation of fertiliser rates, the model includes new functions: volatilisation of NH_3, estimated by a simple model expressing the effects of soil factors (pH, cationic exchange capacity), fertiliser N form (physical and chemical) and of application method (at soil surface or into the soil), and crop status at date of fertiliser N application. The amount of fertiliser N immobilised is calculated for the upper 30 cm layer from the availability of C (calculated by the model at any time as C coming from rhizodeposition, crop residues and organic amendments), and an immobilisation ratio (immobilised N:decomposed C) that varies with N availability (Mary et al., 1996).

Net N mineralisation from soil and organic matter

Net N mineralisation in the soil is the sum of humified organic matter mineralisation and organic residue mineralisation. The mineralisation of humified organic matter is positive. It depends on the humified organic N pool, the soil texture (clay and limestone contents), the soil temperature and moisture (Mary et al., 1999). Mineralisation of organic residues (Figs. 1 and 2) can be either positive (net mineralisation) or negative (net immobilisation) and depends on decomposition rate, which itself depends on the nature of organic residues (chemical characteristics and C:N ratio) and temperature and moisture soil conditions (Nicolardot et al., 2001).

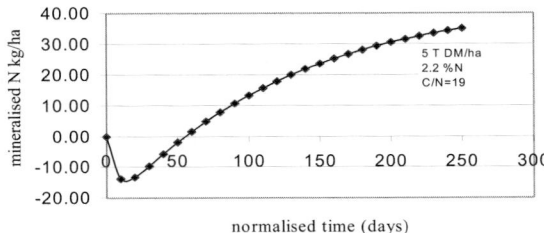

Figure 1. Decomposition of sugar beet leaves.

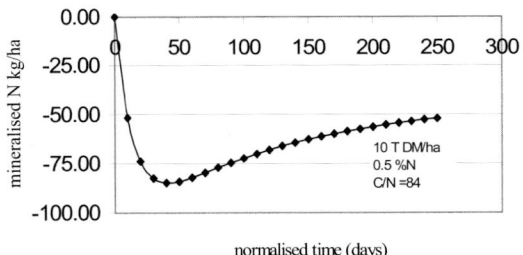

Figure 2. Decomposition of winter wheat straw.

Conclusions

This new version of the predictive balance sheet method constitutes the basis of AZOFERT® software which is a new decision support tool for fertiliser recommendations (Dubrulle et al., 2003). The calculation of gaseous loss and residual inorganic N in soil at harvest also allows the assessment of environmental risks associated with the use of a fertiliser rate and a chemical form in a given situation.

References

Dubrulle, P., Machet, J.M. and Damay, N. (2003). Azofert : a new decision support tool for fertiliser N recommendations. (This volume.) 12th Nitrogen Workshop, 21st - 24th September, Exeter, Devon, UK.

Machet, J.M., Dubrulle, P. and Louis, P. (1990). Azobil : a computer program for fertiliser N recommendations based on a predictive balance sheet method. Proceedings of 1st Congress of the European Society of Agronomy, s2, p21.

Mary, B., Recous, S., Darwis, D. and Robin, D. (1996). Interactions between decomposition of plant residues and nitrogen cycling in soil. Plant and Soil, 181, 71-82.

Mary, B., Beaudoin, N., Justes, E. and Machet, J.M. (1999). Calculation of nitrogen mineralisation and leaching in fallow soil using a simple dynamic model. European Journal of Soil Science, 50, 549-566.

Nicolardot, B., Recous, S. and Mary, B. (2001). Simulation of C and N mineralisation during crop residue decomposition: A simple dynamic model based on the C:N ratio of the residues. Plant and Soil, 228, 83-103.

Recous, S., Loiseau, P., Machet, J.M. and Mary, B. (1997). Transformations et devenir de l'azote de l'engrais sous cultures annuelles et sous prairies. In: G. Lemaire et B. Nicolardot (eds) Maîtrise de l'azote dans les agrosystèmes. Les colloques de l'INRA, 83, 105-120.

Residual effects of maize nitrogen fertilisation on winter barley crop: N content, yield, yield components and agronomic factors

M. Maiorana, F. Montemurro, G. Convertini and F. Fornaro
Istituto Sperimentale Agronomico, Via C. Ulpiani, 5, 70125 Bari, Italy

Introduction

The reduction in N fertiliser rates is very important in Mediterranean conditions. In fact, residual N could cause denitrification and NO_3^- leaching, with possible pollution of natural waters (Shepherd et al., 1993). Therefore, alternative agronomic practices, such as inter-cropping, are increasingly employed to reduce these risks and to increase N use efficiency. Our aim was to verify if growing two or more rotated crops would allow us to utilise N more efficiently than continuous cropping and to exploit residual fertility (Guillard et al., 1995). In this paper, the effects of residual N fertility of a fertilised maize crop have been evaluated in a following winter barley crop without N fertilisation.

Materials and methods

The research was carried out at Foggia (Southern Italy) in the Experimental Farm of the Institute. The soil is a silty-clay Vertisol of alluvial origin, classified as a fine, mesic, Typic Chromoxerert (Soil Taxonomy-USDA). The climate is 'accentuated thermomediterranean' (FAO-UNESCO classification), with low rainfall. On plots of 45 m^2, laid out in a randomised design with three blocks, winter barley (cv. Aida) was sown after a maize crop during the previous summer and fertilised with three N levels: 0, 100, 200 kg N ha^{-1}. The treatments of barley were indicated as: RF0, RF100, RF200 (RF = Residual Fertility), respectively.

Results and discussion

Table 1 shows that N applied to maize had positive effects as residual fertility on barley; in fact, there were significant differences for yield and N content of grain and straw and for total N uptake. No significant differences between the two N levels were found, except for N content of straw.

Table 1. Yield and N content of grain and straw and total N uptake of barley.

		RF0	RF100	RF200
Yield (t ha^{-1})	Grain	3.05b	3.86a	3.82a
	Straw	2.93b	4.76a	4.82a
N content (%)	Grain	1.29b	2.53a	2.39a
	Straw	0.82b	0.86b	1.26a
Total N uptake (kg ha^{-1})		62.72b	137.75a	152.84a

Values with different letters in each line are significantly different at P≤0.05 (DMRT test).

Residual N also significantly affected plant height, ear length, seed number per m^2 and leaf area index (LAI) at both flowering and waxy maturity stages, while it did not influence the other variables examined (Table 2). Table 3 shows that the harvest index of barley was significantly higher in the unfertilised treatment (RF0), thus confirming a good ability to produce barley with a low level of soil fertility in Mediterranean conditions (Delogu et al., 1998). Leaf N status significantly increased from RF0 to RF100 and RF200 in both physiological stages examined but, once more, there was no significant difference between fertilised treatments of maize. A similar trend was found in the indexes of N use efficiency: NAE (N Agronomic Efficiency) and NPE (N Physiological Efficiency), were not significantly affected by fertilised treatments of maize.

Table 2. Main yield components and agronomic characteristics of barley.

	RF0	RF100	RF200
Plant height (cm)	53.47b	67.63a	67.53a
Ear length (cm)	6.50b	7.01ab	7.43a
Seeds (n per m^2)	6657b	8231a	8565a
1000-seed weight (g)	45.77	47.07	44.77
Test weight (g)	65.73	65.00	64.77
Fertile ears (n per m^2)	502	532	583
LAI Flowering	1.65b	2.42ab	3.14a
Waxy maturity	1.94c	2.66b	3.54a

Values with different letters in each line are significantly different at P≤0.05 (DMRT test).

Table 3. Harvest index, leaf nitrogen status and N use efficiency of barley.

	RF0	RF100	RF200
Harvest index (%)	50.90a	45.12b	44.33b
Leaf N status Flowering	33.82b	39.69a	42.74a
Waxy maturity	40.63b	46.22a	47.99a
NAE (kg kg^{-1}) -	7.71	8.09	
NPE (kg kg^{-1}) -	7.62	11.62	

Values with different letters in each line are significantly different at P≤0.05 (DMRT test).

Conclusion

The results of this work seem to indicate that maize, a soil-depleting crop, did not affect negatively the following crop of winter barley. In fact, N applied to maize exerted a significant, positive residual effect on yield, quality (N grain content) and N uptake of barley. The levels of N (100 and 200 kg N ha^{-1}) as residual effects, did not have significant effects on barley performance. Considering that this behaviour was observed also in the maize crop, the lowest N dose (100 kg) seems to be the best choice, whether for crop yields, or for the environment. Therefore, the evaluation of N fertilisation effects on cropping systems can be considered to be a useful tool to reduce agronomic input and to preserve the agronomic and ecological systems.

References

Delogu, G., Cattivelli, L., Pecchioni, N., De Falcis, D., Maggiore, T. and Stanca, A.M. (1998). Uptake and agronomic efficiency in winter barley and winter wheat. European Journal of Agronomy, 9, 11-20.

Guillard, K., Griffin, G.F., Allinson, D.W., Rafey, M.M., Yamartino, W.R. and. Pietrzyk, S.W. (1995). Nitrogen utilisation of selected cropping systems in the U.S. Northeast. I. Dry matter yield, N uptake, apparent N recovery and N use efficiency. Agronomy Journal, 87, 193-199.

Shepherd, M.A., Davis, D.B. and Johnson, P.A. (1993). Minimising nitrate losses from arable soils. Soil Use and Management, 9, 94-99.

Turnover of grain legume N rhizodeposits and effect of rhizodeposition on the turnover of crop residues

J. Mayer[1*], F. Buegger[2], E.S. Jensen[3], M. Schloter[2] and J. Heß[1]

[1]*University of Kassel, Dept of Organic Farming and Cropping, Nordbahnhofstr. 1a, 37213 Witzenhausen, Germany*
[2]*GSF, Institute of Soil Ecology, PO Box 1129, 85758 Oberschleißheim, Germany*
[3]*Plant Research Department, PRD-301, Risø National Laboratory, 4000 Roskilde, Denmark*
(Present address: Swiss Federal Research Station for Agroecology and Agriculture, Reckenholzstrasse 191,CH-8046 Zürich, Switzerland)

Introduction

An important percentage of N taken up by grain legumes is deposited in the rhizosphere. For grain legumes, N rhizodeposition constituted 12% - 16% of total N and 35% - 45% of residual N (= total N - grain N). At maturity, a smaller percentage of N derived from rhizodeposition (NdfR), about 40%, was found to be recovered in the following pools: mineral N, microbial biomass and micro roots. However, about 60% could not be recovered in these pools and was probably immobilised as microbial residues (= dead microbial biomass, exoenzymes, mucous substances) (Mayer et al., 2003a). Plant roots also interact with the soil organic matter and influence the nutrient availability and aggregate stability, cause priming effects and alter the activity and composition of the microbial population. These interactions might affect the turnover of incorporated crop residues. The aim of the study was to investigate (i) the turnover of N rhizodeposition of grain legumes and (ii) the effects of rhizodeposition on the subsequent turnover of its crop residues.

Materials and methods

A sandy loam soil for the experiment was either stored at 6°C, or planted with faba bean (*Vicia faba* L.), pea (*Pisum sativum* L.), white lupin (*Lupinus albus* L.) in pots. Legumes were ^{15}N stem labelled *in situ* during growth and visible roots were removed at maturity. The remaining plant derived N in soil was defined as N rhizodeposition. The turnover of N derived from rhizodeposition of the respective grain legumes and the effects of the rhizodeposition on the subsequent C and N turnover of its crop residues were investigated in an incubation experiment (168 days, 15°C, 50% WHC). The turnover of C and N were compared in soils with and without previous growth of the three legumes and with or without incorporation of crop residues. During

Table 1. Net-N mineralisation and percentage of mineralised rhizodeposition N (NdfR) in soils without crop residues after 168 days of incubation (n=3 ±SD, different letters show significant differences of means, p<0,05).

Soil cropped with preceding legume	Net-mineral N (µg g^{-1} soil)	% NdfR mineralised
Faba bean	31^{bc} ±0.8	26 ±0.4
Pea	29^{b} ±0.1	27 ±1.1
Lupin	33^{c} ±0.7	21 ±0.7
Control	27^{a} ±1.0	-

incubation, CO_2, microbial C and N, mineral N, total N and respective ^{15}N contents were determined. The newly formed microbial residue N derived from crop residues was estimated by an approximate calculation (Mayer et al., 2003b).

Results and discussion

After 168 days of incubation, the net N mineralisation in the treatments, without crop residues, constituted between 33 and 29 µg g^{-1} mineral N g^{-1} soil compared with 27 µg g^{-1} in the control soil; 21% of lupin, 26% of faba bean and 27% of pea N rhizodeposits were mineralised (Table 1). Between 52% and 55% of mineralised rhizodeposition N was supplied by the microbial residue pool and a smaller amount (15% to 17%) was supplied by the microbial biomass (Figure 1). The effect of rhizodeposition on the C and N turnover of crop residues was inconsistent. Rhizodeposition increased the crop residue C mineralisation and the microbial C in only the lupin treatment, whereas the microbial N was increased by rhizodeposition in all treatments (Figure 2). The formation of microbial residue N (derived from crop residues) was accelerated in the soils previously cropped with grain legumes compared with the control soil. Thus, the effect of rhizodeposition seems to enhance the immobilisation of crop residue N either as microbial biomass or microbial residues. The recovery of residual ^{15}N in the microbial and mineral N pool was similar between the treatments containing only labelled crop residues, and labelled crop residues plus labelled rhizodeposits. This indicates a similar decomposability of both rhizodeposition N and crop residue N, and may be attributable to an immobilisation of both N sources (rhizodeposits and crop residues) as microbial residues, and a subsequent remineralisation mainly from this pool.

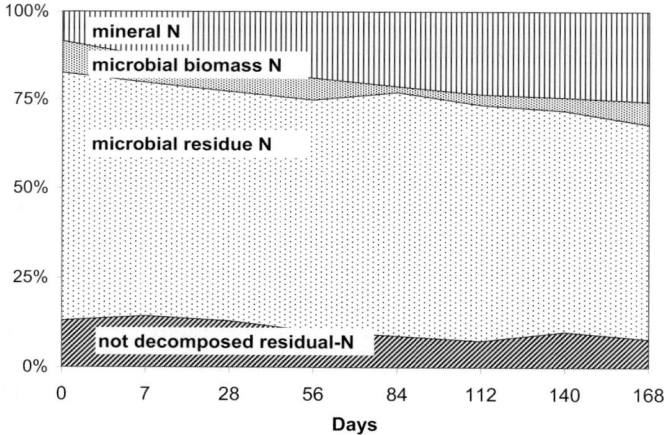

Figure 1. Distribution of rhizodeposition N of faba bean in differing soil pools during 168 days of incubation.

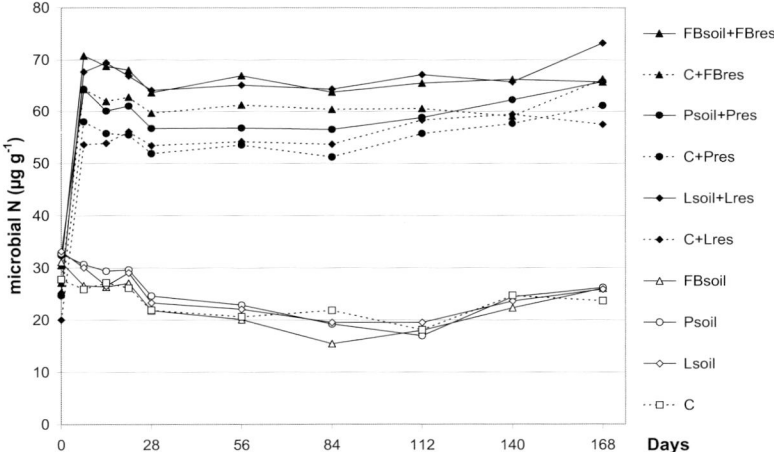

Figure 2. Microbial N in treatments without (FBsoil, Psoil, Lsoil, C) and with crop residue addition (FBsoil+FBres, C+FBres, etc.).

References

Mayer, J., Buegger, F., Jensen, E.S, Schloter, M. and Heß, J. (2003a). Estimating N rhizodeposition of grain legumes using a ^{15}N in situ stem labelling method. Soil Biology and Biochemistry, 35, 21-28.

Mayer, J., Buegger, F., Jensen, E.S., Schloter, M. and Heß, J. (2003b). Turnover of grain legume N rhizodeposits and effect of rhizodeposition on the turnover of crop residues. Biology and Fertility of Soils (in press).

Decomposition of soluble compounds obtained after fractionation of different animal wastes

T. Morvan[1] and B. Nicolardot[2]
[1]INRA, UMR SAS, 4, rue de Stang Vihan, 29000 Quimper, France
[2]INRA, Unité d'Agronomie de Laon-Reims-Mons, BP 224, 51686 Reims cedex 2, France

Introduction

In most decomposition models, exogenous organic matter added to the soil is described using several pools. Some attempts have been made to use chemical fractionation methods (e.g. method proposed by Van Soest, 1963) to fractionate organic matter; significant lack of fit between observed and simulated values was observed by using this approach in models (Corbeels et al., 1999, Henriksen & Breland, 1999). Some discrepancies might result from the different degradability of fractions obtained by the Van Soest method, when considering different types of organic matter. Thus, Morvan et al., (2001) suggested that the degradability of the soluble fraction was different between animal composts and slurries. An experiment was set up to test this hypothesis.

Materials and methods

Four animal effluents were considered: pig slurry (PS), cattle slurry (CS), farmyard manure (FYM) and composted farmyard manure (Comp). Organic fractions were obtained using two fractionation methods: water soluble fraction (SOLw) at 20°C and fractions obtained using the Van Soest method (SOLvs, soluble compounds; HEM, hemicellulose-like; CEL, cellulose-like; LIG, lignin-like). Decomposition of non-fractionated wastes, the insoluble water fraction and the insoluble Van Soest fraction (HEM + CEL + LIG) were measured in a loamy soil during a 107-day incubation (15°C, optimal soil water content). C mineralisation of the soluble fractions (SOLw and SOLvs) was estimated by difference between C mineralisation of the insoluble and non-fractionated materials, assuming a negligible priming effect.

Results

Some biochemical properties show very different organic profiles between the studied animal wastes (Figure 1). The SOLvs contained more than 50% of the organic C for PS, CS and Comp, whereas the SOLw at 20°C represented only 10 to 22% of organic C. The study of C mineralisation of non-fractionated cattle slurry and its insoluble compounds (Figure 2) shows that the soluble fractions greatly contributed to the first stages of the decomposition process. The relationship between the part of organic C present in the form of the soluble compounds (water or Van Soest soluble fraction) and the contribution of these compounds in the decomposition of organic products measured at day 107, is presented in Figure 3. It clearly shows that soluble fractions of composted cattle manure were not degradable, whereas the decomposition of soluble fractions of slurries was important, and higher than those of FYM.

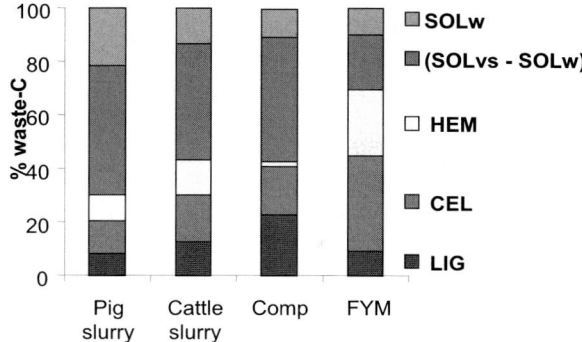

Figure 1. Biochemical characteristics of animal wastes.

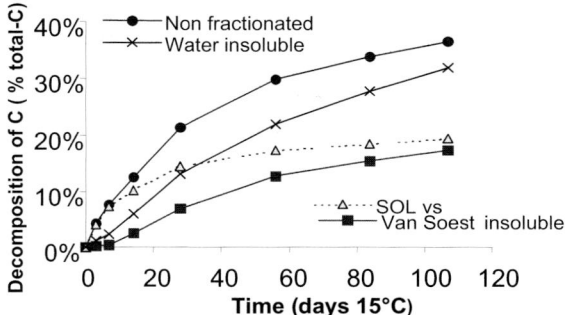

Figure 2. C decomposition of non-fractionated and fractions of cattle slurry (CS).

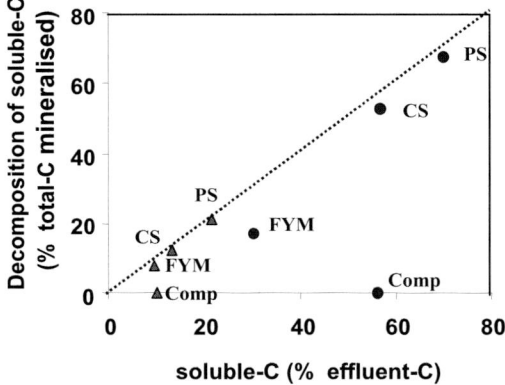

Figure 3. Proportion of soluble water (triangles) and Van Soest soluble (circles) fraction in the organic C of animal wastes and their decomposition in soil at day 107.

Conclusion

This study confirms that it is important to characterise the biodegradability of soluble fractions of organic products (slurries, composts, farmyard manures, etc.) when fractionation methods are used in decomposition models. Finally, these results show that it is first necessary to perform a typology study in order to put together organic products which have similar characteristics in terms of composition and decomposition behaviour.

References

Corbeels, M., Hofman, G. and Van Cleemput, O. (1999). Simulation of net N immobilisation and mineralisation in substrate amended soils by the NCSOIL computer model. Biology and Fertility of Soils, 28, 422-430.

Henriksen, T.M. and Breland, T.A. (1999). Evaluation of criteria for describing crop residue degradability in a model of carbon and nitrogen turnover in soil. Soil Biology and Biochemistry, 8, 1135-1149.

Morvan, T., Dach, J., and Parnaudeau, V. (2001). Relationships between biochemical composition of manures and composts and their carbon and nitrogen transformation in soil. 11th Nitrogen Workshop, Reims, 9-12 Sep 2001, 153-154.

Van Soest, P.J. (1963). Use of detergents in the analysis of fibrous feeds. II A rapid method for the determination of fiber and lignin. Journal of the A O A C, 46, 825-835.

Is good nutrient management possible by applying more than 170 kg nitrogen per hectare from manure?

A. Mulier[1,2], G. Hofman[1] and I. Verbruggen[2]

[1]*Ghent University, Department of Soil Management and Soil Care, Coupure Links 653, 9000 Gent, Belgium*

[2]*Policy Research Centre for Sustainable Agriculture (Stedula), Potaardestraat 20, 9090, Gontrode, Belgium*

Introduction

During the past decade, the European Nitrates Directive (91/676/EEC) has influenced the development of farm nutrient management significantly. To meet the objectives of this Directive, the Flanders Government designated vulnerable zones where the amount of N applied as manure may not exceed 170 kg ha^{-1} yr^{-1} (De Clercq et al., 2001). The area comprised by these vulnerable zones increased recently from 7 to 46% of the total utilised agricultural area (Press communiqué, 2002). Because of this expansion, many Flemish cattle farmers lost the opportunity of using all their manure on their own fields. However, farm gate N balances show that low N surpluses can also be realised on farms where more than 170 kg manure-N ha^{-1} is used in a judicious way.

Materials and methods

In our study (Mulier et al., 2002), we calculated farm gate N balances for 11 dairy cattle farms and 6 beef cattle farms in Flanders during 2 subsequent years (2000 and 2001). We also determined total N supply surpluses as the difference between (1) the N-demands of crops according to average fertiliser advice and (2) the actual use of manure- and fertiliser-N on the farm. Furthermore, this total surplus was split up into a manure-N surplus (= actual manure-N use - advised manure-N use) and a fertiliser-N surplus (= actual fertiliser-N use - advised fertiliser-N use). These extra calculations enabled us to assess the influence of the nutrient management of the farmer on the extent of his farm N surplus.

Results and discussion

Farm N surpluses, total applied manure-N, total fertilisation surpluses, manure- and fertiliser-N surpluses over the two years are summarised in Table 1. In 2000, the mean N surplus for the dairy farms was 222 kg N ha^{-1}, in 2001 it was reduced to 188, mainly from efforts in manure management. For the beef cattle farms, mean N surpluses were higher (289 kg N ha^{-1} in 2000, 244 kg N ha^{-1} in 2001). Here also, a considerable reduction was realised. We found a good positive linear correlation between the farm N surplus and the estimated total surplus (R^2=0.56, P<0.01). Among the farmers with a N surplus higher than the mean in 2000, some farmers applied too much manure (nos. 5, 6, 27), others applied too much fertiliser (nos. 35, 39), others applied too much of both (nos. 18, 2). Within the group with a N surplus below the mean (in 2000) we could also distinguish different types of farmers: nutrient managers with a preference for the use of manure (nos. 8, 41, 28, 40), and those with preference for the use of fertilisers (nos. 36, 9, 37). In 2001 the same trends were confirmed.

Table 1. Calculated surpluses in kg N ha⁻¹ (2000 / 2001).

Farm number	Farm type	Farm-N-surplus		Total applied manure-N		Total fertilisation		Manure surplus		Fertiliser surplus	
		2000	2001	2000	2001	2000	2001	2000	2001	2000	2001
4	Dairy	121	96	164	179	-10	-3	-3	10	-7	-13
8	Dairy	158	157	249	265	62	35	66	64	-4	-29
36	Dairy	167	121	200	186	36	-11	-9	-15	45	4
41	Dairy	199	168	322	328	73	93	123	135	-50	-42
Mean	Dairy	222	188	238	239						
39	Dairy	245	127	202	167	30	14	-11	-14	41	28
35	Dairy	255	183	171	218	84	36	-31	3	115	33
6	Dairy	266	235	314	274	125	55	92	53	33	2
5	Dairy	309	221	260	272	56	62	39	46	17	16
18	Dairy	385	288	307	249	241	178	144	62	96	116
37	Beef	205	195	162	174	37	63	-43	-31	80	94
9	Beef	225	236	238	212	131	102	44	16	87	86
28	Beef	249	113	465	355	206	52	261	140	-55	-88
40	Beef	283	330	332	344	197	217	142	136	54	81
Mean	Beef	289	244	313	280						
2	Beef	363	275	344	274	235	98	150	71	85	26
27	Beef	411	317	339	320	128	64	107	89	21	-25

Conclusions

We concluded that low N surpluses were found on farms with different types of nutrient management, including those applying >170 kg manure-N ha⁻¹, who were practising good management. Farmers, preferring the use of fertilisers to manure, also achieved farm N surpluses greater than the mean. That leads to a final question: should farmers not have the opportunity to use >170 kg manure N ha⁻¹, under conditions explicitly managed to minimise losses?

Acknowledgment

This project was financed by the Flemish Land Agency (VLM)

References

De Clercq, P., Gertsis, A.C., Hofman, G., Jarvis, S.C., Neeteson, J.J. and Sinabell, F. (2001). Nutrient Management Legislation in European Countries. Wageningen Pers. The Netherlands, 347p.

Mulier, A., Hofman, G., Carlier, L., De Brabander, D., De Vliegher, A., Fiems, L., Janssens, G., Van Cleemput, O., Van Huylenbroeck, G. and Willekens, K. (2002). Emission prevention in agriculture through the use of mineral balances: continuing project. Final report for the Flemish Land Agency, August 2002.

Regulation of N release from polyphenol-protein complexes by fungi in different tropical production systems

R. Mutabaruka and G. Cadisch
Department of Agricultural Sciences, Imperial College London, Wye Campus, Ashford Kent, TN25 5AH, UK

Introduction

Polyphenols are water soluble phenolic compounds that are capable of binding to plant proteins and form polyphenol-protein complexes, thus reducing the release of N from decomposing plant materials, and hence availability for plant uptake. The objective of this paper was to test if, in tropical systems with polyphenol rich vegetations, specialised fungal communities had evolved capable of breaking down these recalcitrant polyphenol-protein complexes.

Materials and methods

Fungi were isolated from soils under different agricultural systems (maize, sugarcane and *Gliricidia sepium* or *Peltophorum dasyrrachis* woodlots) and natural systems (secondary forest and *Imperata cylindrica* grassland) using potato dextrose agar (PDA) as a non-specific media, a TA-BSA (Tannic acid, hydrolysable tannin) or QUEB-BSA (Quebracho, condensed tannin) polyphenol-protein (BSA) complex media.

Fungi were identified and tested for their efficiency in breaking down the compounds by incubating 50 mg of TA-BSA and QUE-BSA complexes.

Results

The largest number of fungal polyphenol-protein complex degraders were isolated from natural forest (Figure 1). TA-BSA complex degraders were more abundant than QUE-BSA complex degraders in all systems, the latter being particularly low in cultivated or grassland systems. Isolated fungi degraded up to 60% of C of TA-BSA complexes, but only 30% of C of QUE-BSA

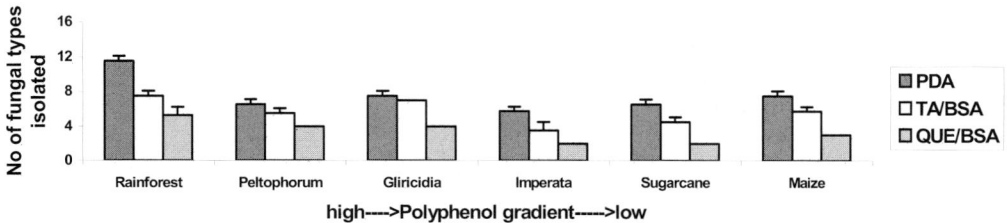

Figure 1. Number of types of fungi isolated from soils from under different vegetations growing on PDA, TA/BSA or QUE/BSA. Error bars are standard deviations.

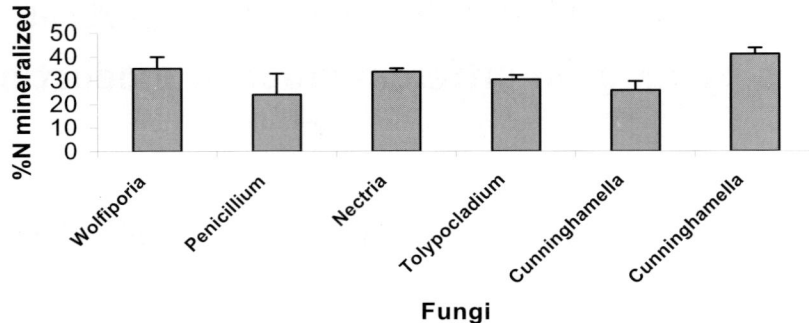

Figure 2. Percent N mineralised by different fungal species after incubation with QUE-BSA complexes for 96 days.

complexes over 96 days (data not presented). The fungi species differed in the efficiency of N release from the most recalcitrant polyphenol-protein complex (Figure 2).

Discussion

The natural forest system contained the highest number of fungal types of polyphenol-protein complex degraders, presumably because of its high diversity of plants with their varied litter quality as well as fewer disturbances. The number of QUE-BSA degraders was positively associated with the polyphenol content of the above ground vegetation. However, the efficiency of N release from this complex was not clearly linked to the litter quality input history, but to fungi species. TA-BSA complexes were degraded two times faster than QUE-BSA, because condensed tannins are more resistant to degradation than hydrolysable tannins, as are those with higher molecular weight (Cadisch & Giller, 2000).

Conclusion

The hypothesis that polyphenol rich plant communities led to the evolution of specialised degrader communities was proven in the case of the more recalcitrant QUE-BSA complex, but not for the hydrolysable tannin-protein (TA-BSA) complex.
 However, the efficiency of N release from these complexes was not clearly linked to the litter quality input history.

Acknowledgements

This project was partly funded by Airey Neave Trust.

References

Cadisch, G. and Giller, K.E. (2000). Soil organic matter management: The role of residue quality in carbon sequestration and nitrogen supply. In: B. Rees, B. Ball, C. Campbell and C. Watson, (eds) Sustainable Management of Soil Organic Matter. CAB International, Wallingford, 97-111.

Modelling nitrogen mineralisation from mature bio-waste compost applied to vineyard soils

C. Nendel[1,4], S. Reuter[1], K.C. Kersebaum[2], R. Nieder[3] and R. Kubiak[1]

[1]*State Education and Research Centre, Department of Ecology, Breitenweg 71, 67435 Neustadt an der Weinstraße, Germany*

[2]*Centre for Agricultural Landscape and Land Use Research, Institute for Landscape Systems Analysis, Eberswalder Straße 84, 15374 Müncheberg, Germany*

[3]*Braunschweig Technical University, Institute of Geoecology, Langer Kamp 19c, 38106 Braunschweig, Germany*

[4]*Institute of Vegetable and Ornamental Crops, Department of Modelling and Knowledge Transfer, Theodor-Echtermeyer-Weg 1, 14979 Großbeeren, Germany (present address)*

Introduction

The ongoing discussion about utilisation of bio-waste on agricultural sites reveals a lack of knowledge of the nutritional value of bio-waste compost under the special conditions of viticulture. In order to increase the knowledge of advisors on optimised fertilisation of vineyards, a simulation model was developed at the State Education and Research Centre, Neustadt. The Centre is the main institution for viticultural fertiliser advice in the state of Rhineland-Palatinate, an important vine-growing area in Germany.

The model

The model presented is based on a simulation model for N-dynamics in arable soils (Kersebaum & Richter, 1991). Parameters for plant growth of the perennial *Vitis vinifera* L. were taken from the literature. Some structural changes in the plant growth submodel (as well as the introduction of a new empirical equation for root growth to link plant and soil processes) enabled the simulation of continuous plant growth over a period of several years, including the main effects of viticultural management (Nendel, 2002).

Mineralisation parameters

Nitrogen mineralisation patterns of a typical, mature bio-waste compost were investigated in a laboratory incubation experiment according to Stanford & Smith (1972). Based on the results of the experiment, the compost was characterised as having 5% rapidly decomposable, 60% slowly decomposable, and 35% (in a medium time-frame) non-decomposable constituents. The parameters were derived from non-linear regressions and subsequently verified in a 2 year field incubation experiment using micro-lysimeters. For validation of the model, data from a 3 year Nmin monitoring experiment in four typical Rhineland-Palatinate vineyards were used.

Simulation results

Water and Nmin contents of the investigated soils could be simulated quite satisfactorily. Reproduction of the measured Nmin values by the model resulted in 50% of the simulated values to be within the standard deviation range of the field data for vineyard soils amended with bio-waste compost. For unamended soils, 70% of the simulated Nmin values accomplished this level of accuracy. On average, the Nmin content of a 10 cm-soil layer was over-estimated by the model by 1.7 kg ha^{-1} with a standard deviation of 10.4 kg ha^{-1}. Further improvement was prevented by a frequently occurring priming effect, as a consequence of the compost amendment, which the model was not able to reproduce (Figure 1). The simulation results revealed a higher N release from mature bio-waste compost under viticultural conditions as compared with known figures from arable soils. A similar pattern was observed for the micro-lysimeter experiment (Nendel, 2002).

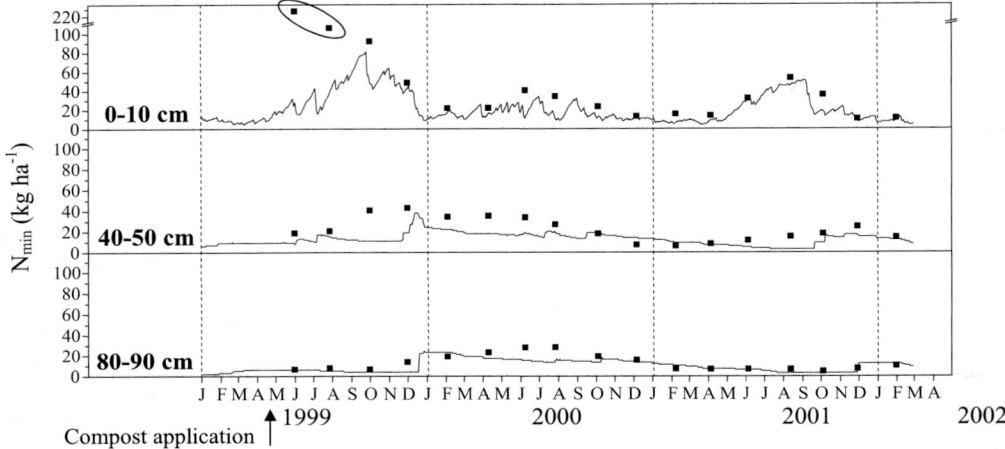

Figure 1. Simulation of Nmin dynamics in different soil layers at the Bad Kreuznach intensive investigation site. The vineyard was fertilised with 50 Mg ha^{-1} compost. Presumed priming effects encircled.

Conclusion

The N mineralisation potential of mature bio-waste compost in viticultural soils is considerably higher than in agricultural soils, mainly because of higher soil temperatures in vineyards. This has to be considered when more fertiliser is used in addition to the compost.

Acknowledgements

The authors wish to thank the German Viticulture Research Group (FDW) for financial support.

References

Kersebaum, K.C. and Richter, J. (1991). Modelling nitrogen dynamics in a plant-soil system with a simple model for advisory purposes. Fertilizer Research, 27, 273-281.

Nendel, C. (2002). Die Wirkung von Bioabfallkompost auf den Stickstoffhaushalt in Rebflächen (The effect of bio-waste compost on the nitrogen budget in vineyards). Shaker Publishing, Aachen, 184 pp.

Stanford, G. and Smith, S.J. (1972). Nitrogen mineralization potentials of soils. Soil Science Society of America Journal, 36, 465-472.

Inorganic N dynamics in vineyard soils under different covering strategies

B. Nicolardot [1], P. Thiébeau [1], C. Herre [1], A.F. Doledec [2], A. Perraud [2] and B. Mary [3]

[1]*INRA, Unité d'Agronomie de Laon-Reims-Mons, BP 224, 51686 Reims cedex 2, France*
[2]*Comité Interprofessionel du Vin de Champagne, BP 135, 51204 Epernay, France*
[3]*INRA, Unité d'Agronomie de Laon-Reims-Mons, rue Fernand Christ, 02007 Laon, France*

Introduction

There is a lack of knowledge for the management of organic inputs and N fertilisation in vineyard soils of the Champagne area. Most available data were obtained for vineyards located in southern France or Europe, and may not be applicable to northern conditions. Environmental concerns led wine-producers to set up experimental programmes to evaluate improved agricultural practices. In Champagne vineyards, inputs of organic matter are mainly used to avoid erosion and to maintain or increase soil organic matter. Amounts and nature of organic inputs may have an impact on water quality, particularly NO_3^- concentration. The aim was to evaluate N dynamics in vineyards soils according to the management practice: soil maintained bare, soil covered with organic matter, or grass cover.

Materials and methods

A field experiment was set up in 1991 at Montbré (surroundings of Reims, Champagne area). It included 6 treatments (3 replicates): soil without input (control soil), soil with mineral fertiliser (50 kg N ha^{-1}), soil covered with oak bark or with coniferous bark or with dehydrated manure, soil under grass with mineral fertiliser (50 kg N ha^{-1} in May). Some of the experimental plots were equipped with porous cups at 50 cm depth. Soil water was periodically sampled during the drainage period and analysed for NO_3^- concentration. The N mineralisation potentials of soils, sampled in April 1999, were measured in the different plots in the upper layers (0-25 cm and 25-50 cm) during soil incubations under controlled conditions (15°C, field capacity). Soil inorganic N was periodically measured from 1998 to 2002 by soil coring to 75 cm (cores divided into 3 equal layers) where vine roots were present. Each soil sample analysed was obtained by mixing 6 cores taken in three adjacent rows of vine (3 samples in, and 3 samples between rows). Soil was sampled every month in the control soil and at 3 dates in other treatments (bud, flowering, vintage). The amounts of drained water and N leached below 75 cm and N mineralised were calculated in the control soil using the LIXIM model (Mary et al., 1999). They were also calculated during the non-vegetative period of vine (15 October to 15 April), since LIXIM is applicable to bare soil or to soil with non-growing plants.

Results and discussion

During the year, the amounts of NO_3^--N in the control soil varied between 30 and 65 kg N ha^{-1} (Figure 1). The lowest values were after winter and the maximal values were observed in June when plant uptake increased considerably. The NO_3^- contents were generally higher for soil + 50 kg N ha^{-1}, soil covered with dehydrated manure and grass cover + 50 kg N ha^{-1}. In the latter treatment, the amount of inorganic N was similar to the control soil when grass ceased to take up inorganic N. The N release in soil covered by oak or coniferous bark was comparable to that of the control soil. The mineralisation rates (Vp) calculated by LIXIM in the control soil for the 3 non-vegetative periods (Figure 2) were 0.34 kg N ha^{-1} day^{-1} (day at 15°C and field capacity). The Vp values obtained in controlled conditions (Table 1) varied from 0.37 to

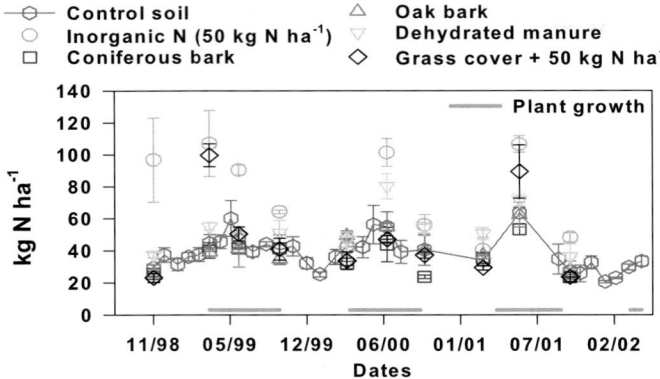

Figure 1. Soil inorganic N dynamics (0-75 cm layer) for the different soil cover.

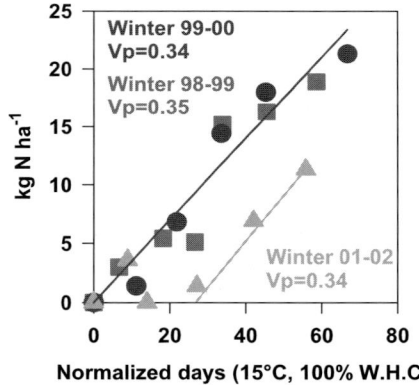

Figure 2. Cumulative N mineralisation and Vp values for the control soil (LIXIM model calculation).

Table 1. Mineralisation rates (kg N ha^{-1} normalised day^{-1}) calculated from soil incubations (15°C, 100% W.H.C.) for the different soil cover.

Soil cover	Soil layer		
	0-25 cm	25-50 cm	0-50 cm
Control soil	0.29	0.19	0.48
Inorganic N (50 kg N ha^{-1})	0.24	0.15	0.39
Coniferous bark	0.20	0.16	0.37
Oak bark	0.27	0.14	0.41
Dehydrated manure	0.29	0.17	0.46
Grass cover + 50 kg N ha^{-1}	0.31	0.12	0.43

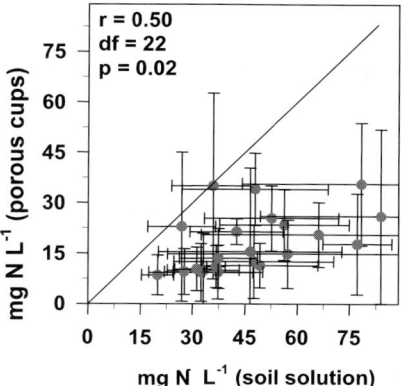

Figure 3. Relationships between NO_3^--N concentrations measured for control soil (porous cups) and the soil solution (25-75 cm layer).

0.48 kg N ha^{-1} day^{-1} in the upper two layers (the contribution of the 25-50 cm layer representing 30-40% of the total mineralisation). They did not differ significantly between treatments and were comparable to the value obtained for the control soil in field conditions.

LIXIM outputs indicated that NO_3^- concentration in the drained water was < 50 mg l^{-1} for all treatments, except for the treatment 'soil + 50 kg N ha^{-1}'. Nitrate concentrations measured in porous cups at 50 cm were generally lower than those measured at 25-75 cm. For these experimental sites, the results showed that water collected by porous cups did not represent water present in the soil layer where the porous cups were located. Finally, simulations of N mineralisation during the vegetative period (using Vp values calculated during the non-vegetative period) showed that soil N supply from mineralisation of soil organic matter was sufficient for the N nutrition of vines. Fertiliser N applications are not necessary in these situations; grass cover or applications of organic matter with high C:N ratios may be used to limit erosion with moderate effects on the quality of the drainage water.

Acknowledgements

This work was supported by Région Champagne-Ardenne and Région Bourgogne.

References

Mary, B., Beaudoin, N., Justes, E. and Machet, J.M., (1999). Calculation of nitrogen mineralisation and leaching in fallow soil using a simple dynamic model. European Journal of Soil Science, 50, 549-566.

Effect of rate and timing of nitrogen fertiliser on yield and grain protein content in winter wheat. The chlorophyll meter as a nutritional diagnostic tool

M.A. Ortuzar[1], A. Alonso[1], A. Aizpurua[1], A. Castellón[1] and J.M Estavillo[2]
[1]NEIKER Basque Institute for Agricultural Research and Development, Berreaga 1, 48160 Derio, Spain
[2]Dpto Biología Vegetal y Ecología, Facultad de Ciencia y Teconología, UPV/EHU, Apdo. 644, 48080 Bilbao, Spain

Introduction

Grain protein is an important factor in determining milling and baking quality of wheat, and the N fertiliser management has a large influence on it (Cassman et al., 1992). The Hydro N-Tester chlorophyll meter offers a possible alternative technique to estimate crop N status to determine the need for additional N fertiliser (Follet et al., 1992). The aim of this work was to study the effect of rate and timing of N applications on yield and grain protein content in the wheat variety, Soissons and to test the chlorophyll meter as a diagnostic tool.

Materials and methods

Fertiliser experiments were carried out in Álava (779 mm and 11.5 ºC, average annual rainfall and temperature) from November 2001 to July 2002 at 3 sites: Betolaza, Aranguiz 1 and Aranguiz 2. Previous crops were winter wheat for Betolaza and Aranguiz 1, and beans for Aranguiz 2. The soil mineral N values (0-60 cm depth) at the end of the winter (February) were 35, 28 and 56 kg N ha^{-1} for Betolaza, Aranguiz 1 and Aranguiz 2, respectively. Two N application rates (140 and 180 kg N ha^{-1}) were applied in two or three amendments at Z20, Z30 and Z37 Zadoks growth stages (Zadoks et al., 1974), as shown in Table 1. The first amendment was applied as sulphate ammonium nitrate and the rest as calcium ammonium nitrate. A completely randomised block experimental design was used with four replications. Hydro N-Tester measurements were taken at Z32 and Z37 at the uppermost fully extended leaf, being expressed as relative values (RN32 and RN37) with respect to those obtained for an over-fertilised plot (220 kg N ha^{-1}), to which the 100% value was assigned.

Results

There were no significant interactions between N application rates and timing. The high application rates only produced a significant increase in yield in Aranguiz 1, which was predicted from the Hydro N-Tester values in Z32 (Table 2). Both in Betolaza and Aranguiz 2, Hydro N-Tester values indicated a better plant N status than in Aranguiz 1. The lack of response in Aranguiz 2 was attributed to the higher soil N availability, as a consequence of the N fixed by the previous bean crop, leading to the highest yields. In relation to timing, late N additions increased grain protein content in Aranguiz 1 and Betolaza, although in Aranguiz 1 the yield was decreased. Both at Z32 and Z37, the relative Hydro N-Tester value diagnosed the poorer N nutritional stage that led to the decrease in yield in Aranguiz 1, which was because only

Table 1. N fertilisation rates and timing.

Rate (kg N ha^{-1})	Timing of N application		
	Tillering starts (Z20)	Jointing starts (Z30)	Flag leaf (Z37)
140	40	100	-
140	40	60	40
180	60	120	-
180	40	60	80

Table 2. Treatment means for grain yield, grain protein content (Prot) and N Tester values at Z32 and Z37 Zadocks growth stages (RN32 and RN 37).

N rate	Betolaza				Aranguiz 1				Aranguiz 2			
(kg N ha^{-1})	Yield	Prot	RN 32	RN 37	Yield	Prot	RN32	RN 37	Yield	Prot	RN32	RN 37
	(kg ha^{-1})	(%)	(%)	(%)	(kg ha^{-1})	(%)	(%)	(%)	(kg ha^{-1})	(%)	(%)	(%)
140	4390	12.4	93.2	91.6	5104	12.4	83	86.2	9139	13.0	97.5	91.8
180	4457	11.4	95.2	95.4	5623	11.4	92	87.2	9011	13.1	102	95.4
	ns	ns	ns	ns	*	ns	*	ns	ns	ns	ns	ns
Timing												
2 applications	4345	11.0	96.9	97	5588	10.6	91	89	9003	13.2	102	92.9
3 applications	4502	12.4	91.6	90	5139	13.3	84	84	9147	13.0	96.9	94.3
	ns	*	ns	*	*	*	*	*	ns	ns	ns	ns

Figures followed by * are significantly different at $P < 0.01$.   ns = not significant.

100 kg N ha^{-1} had been applied before Z32 and Z37, the third application being made later (Table 1). In Betolaza, the same trend was observed in N-Tester values at Z37, but these values were greater and so they did not correspond to a decrease in yield. In Aranguiz 2, no effect of timing was observed as influenced by the previous bean crop.

Conclusion

Late N additions can increase grain protein content, but fertiliser strategy should include enough N in the first two additions for the yield not to diminish. The chlorophyll meter seems to be a useful tool for adjusting N fertiliser rates in our climatic conditions.

Acknowledgements

This work was financially supported by the Spanish Government (MCyT n° AGL-2001-2214/C06/06)

References

Cassman, K. G., Bryant, D. C., Fulton, A. E. and Jackson, L. F. (1992). Nitrogen supply effects on partitioning of dry matter and nitrogen to grain of irrigated wheat. Crop Science, 32, 1252-1258.

Follet, R.H. and Follet, R. F. (1992). Use of a chlorophyll meter to evaluate the nitrogen status of dryland winter wheat. Communications in Soil Science and Plant Analysisl, 23, 687-697.

Zadoks, J. C., Chang, T. T. and Konzak, C. F. (1974). A decimal code for the growth stages of cereals. Eucarpia Bulletin, 7, 10.

Controlling nitrogen flows and losses

Assessing N dynamics of organic wastes in field conditions using a calculation model

V. Parnaudeau[1], P. Robert[2], C. Herre[1], F. Millon[1], B. Mary[1] and B. Nicolardot[1]
[1]*INRA, Unité d'Agronomie Laon-Reims-Mons, BP 224, 51686 Reims cedex 2, France*
[2]*ASAE, 2 esplanade R. Garros, 51686 Reims cedex 2, France*

Introduction

Recycling organic wastes by land spreading may increase the risk of NO_3^- leaching when N contained in the wastes mineralises, or when they contain inorganic N. To limit environmental impacts, it is necessary to get better knowledge on the behaviour of wastes after spreading onto soil (i.e. N mineralisation dynamics) in order to provide more accurate recommendations and guidelines for farmers. For this purpose, we characterised N release of sewage sludges (SS) and untreated waste waters (WW) from food-processing industries in field conditions.

Materials and methods

The field trial was set up during August 2000 for 18 months. Seven treatments (3 replicates) were considered: control soil without spreading, and soil after spreading with six different wastes selected from a previous laboratory experiment study: effluents, including waste water (WW), from alfalfa dehydrating industry or distillery or sugar refinery, municipal sludge or sludge from a distillery. Each experimental plot was 16 m wide and 35 m long, and the wastes were applied with standard spreading equipment. Distribution and application rates of wastes were precisely measured using rectangular trays put on the soil surface along four transects during spreading. Wastes were also sampled during spreading for their characterisation. All plots were maintained bare fallow during the whole experiment using chemical herbicides. Soil cores were sampled every 2-3 weeks to 120 cm depth (divided into 4 equal layers), moisture and inorganic N (NH_4^+ and NO_3^-) contents being determined in soil from each layer. Meteorological data were collected automatically every day. The daily values of water evaporation and drainage, N mineralisation and leaching were calculated using the measured data and the LIXIM model (Mary et al., 1999). LIXIM is a capacity-type model applied to a layered soil. Input data are: meteorological data (rainfall, PET, temperature), basic soil properties (bulk density, moisture at field capacity and wilting point), and moisture and inorganic N contents in each soil layer at each sampling date. Nitrogen mineralised and N leached after waste application were calculated as the difference between amended and control soils.

Results and discussion

Analyses and field trial results confirmed the diversity of the wastes (Table 1). The C:N ratios varied from 5.3 to 21.2. LIXIM was able to reproduce satisfactorily the water and inorganic N contents measured in all treatments (e.g. Distillery Waste Water, Figure 1). The wastes also induced very different N mineralisation rates and dynamics: wastes no. 2 and no. 5 (with a low C:N ratio) induced net N mineralisation just after spreading, whereas wastes no. 3 and no. 6 led to N immobilisation (for one month) and subsequent N release. Nitrogen was immobilised for more than one year after spreading WW no. 1 and no. 4 (sugar refinery WW was almost stable with little N immobilised, while alfalfa WW had a high immobilisation rate). Nevertheless, in this last case, amounts of N immobilised were low since this effluent contained a low amount of organic N. N leached after application of wastes nos. 1, 2, 5 and 6, (which represented from 24 to 39% of waste total N), came from the mineralisation of organic N and inorganic N contained in the wastes. These results justify the growth of catch crops when these wastes are spread in summer or autumn. WW no. 4 did not contain inorganic N and did not release N, thus it induced less leaching than the control soil. Waste no. 3 caused no leaching during the first winter, but it did in the following year, because of net N mineralisation during this period (results not shown), which should be taken into account for fertilisation of the following crop.

Table 1. Amounts of applied N, N mineralisation rates and N leaching following spreading (SS = sewage sludge, WW = waste water).

	Organic N applied (kg N ha⁻¹)	NH₄⁺-N applied (kg N ha⁻¹)	Organic C:N ratio	Waste N mineralised (% of added organic N)			N leached (% of added total N)		
Days after spreading				29	85	374	29	85	374
1. Alfalfa dehydrating WW	42	36	13.4	-74	-55	-49	0	0	+29
2. Liquid distillery SS	117	38	5.3	+4	+28	+38	0	-6	+29
3. Distillery WW	141	2	15.5	-3	+8	+26	0	-3	+3
4. Sugar refinery WW	230	0	21.2	-4	-3	-2	0	+1	-6
5. Solid distillery SS	52	21	7.4	0	+60	+58	0	-13	+39
6. Municipal SS	113	20	14.6	-2	+15	+46	0	-2	+24

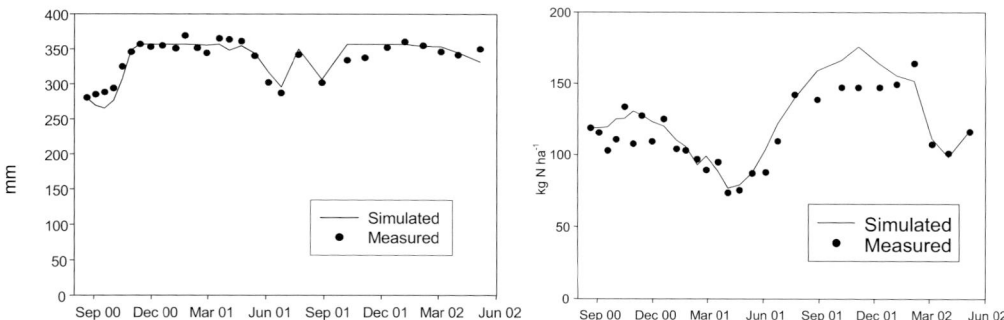

Figure 1. Observed and simulated evolution of water (left) and inorganic N (right) in soil profile after spreading of distillery waste water.

Acknowledgements

This work was granted by Agence de l'Eau Seine-Normandie (AESN) and Agence de l'Environnement et de la Maîtrise de l'Energie (ADEME).

References

Mary, B., Beaudoin, N., Justes, E. and Machet, J. M. (1999). Calculation of nitrogen mineralization and leaching in fallow soil using a simple dynamic model. European Journal of Soil Science, 50, 549-566.

^{15}N labelling and use of dairy manure components for N cycling studies

J.M. Powell[1] and K.A. Kelling[2]
[1]USDA-ARS Dairy Forage Research Center, 1925 Linden Drive West, Madison, WI 53706, USA
[2]Dept. Soil Science, University of Wisconsin, 1525 Observatory Drive, Madison, WI 53706, USA

Introduction

Indirect estimates of manure N availability to crops are highly variable, but can be improved by using ^{15}N-enriched manure (Dittert et al., 1998). However, labeling and feeding sufficient forage to dairy cows in order to obtain uniformly labelled manure for application to field plots is laborious and expensive. The objective of this study was to compare total corn ^{15}N uptake and the amount and forms of soil ^{15}N remaining in field plots amended with manure from dairy cows fed costly ^{15}N-labelled forage or less costly ^{15}N-labelled urea.

Materials and methods

Two methods were used to differentially enrich dairy manure components with ^{15}N. Forage manure (FM) was produced by feeding dry cows ^{15}N-enriched alfalfa (*Medicago sativa* L.) hay and corn (*Zea mays* L.) silage to label urine N, faecal endogenous N (consisting mostly of microbial products), and faecal undigested feed N. Urea manure (UM) was produced by direct feeding ^{15}N-enriched urea to cows with unlabelled forage. This effectively labelled only urinary N and faecal endogenous N. Over a two-year period, we applied ^{15}N-enriched FM or UM (Table 1) to 1.5 m wide x 2.3 m long field plots, incorporated it within 4-6 hours, and determined corn ^{15}N uptake and residual soil nitrate- and total-^{15}N.

Table 1. Quantities of manure N applied using two methods of manure ^{15}N enrichment.

Manure Enrichment Method	Manure Component	Manure Application Rate Year of Application	
		1999	2000
Forage	Wet weight (Mg ha^{-1})[a]	63	63
	Total N (kg ha^{-1})	284	235
	Urine N[a]	135	94
	Faecal microbial N[b]	88	89
	Faecal undigested feed N[b]	23	20
	Straw bedding	38	32
Urea	Wet weight (Mg ha^{-1})	63	63
	Total N (kg ha^{-1})	280	291
	Urine N[b]	131	119
	Faecal microbial N[b]	93	81
	Faecal undigested feed N	18	27
	Straw bedding	38	32

[a] Manure dry matter content was approximately 170 g kg^{-1} in both study years.

[b] Manure components ^{15}N enriched. Other components at ^{15}N natural abundance.

Results and discussion

Of the total labelled manure N applied, 14 to 18% was accounted for in corn harvested during the first year, and 4 to 8% in the second year after manure application. Our estimates of manure ^{15}N uptake by corn corresponded well to manure ^{15}N uptake by other cereal crops (Jensen et al., 1999). No significant differences were noted in total corn N uptake, and total soil ^{15}N levels relating to year of application or manure enrichment method (Table 2). On average, 67% of applied manure ^{15}N was accounted for in crop uptake (22%) and the soil (45%).

Table 2. [15]N recovery from field plots amended with ^{15}N-enriched dairy manure.

Manure Enrichment Method	Manure ^{15}N recovery (% of applied)		
	Corn[a]	Soil (0-90 cm) [b]	Total
Forage	21 (2.7) [c]	41 (2.9)	62
Urea	23 (4.9)	48 (4.0)	71

[a]Sum of average first and second year total corn N uptake. [b]Soils were sampled in autumn of 2000.
[c] standard errors in parentheses.

Except for plots amended with UM in 2000, which received the greatest amount of manure N (Table 1), most (72 to 98%) [15]NO_3^--N and total ^{15}N was found in the upper 30-cm of soil. This suggests either relatively little downward movement of applied manure N, or that leached N may have moved out of the 0- to 90-cm layer. Averaged across years, the relative amount of manure N recovered (i.e. % of total manure ^{15}N applied) in the 0-30, 30-60 and 60-90 cm soil depths were similar for each manure type, averaging 32, 6 and 3% in the FM-amended plots and 38, 7 and 3% in the UM-amended plots, respectively. Most of the ^{15}N unaccounted for (33%) was likely to have been lost via NH_3 volatilisation, and to a lesser extent, denitrification.

Conclusion

Our results suggest that the less laborious and less costly urea method of labelling the most labile dairy manure N components (urinary and faecal endogenous N) would be adequate for evaluating short-term N dynamics in manure-amended soils. The N contained in the undigested feed in faeces did not apparently affect crop N uptake during this 2-year study to a measurable extent. However, this manure N component should be labelled using the forage method to produce manure for use in long-term N cycling studies.

References

Dittert, K., Goerges T. and Sattelmacher, B. (1998). Nitrogen turnover in soil after application of animal manure and slurry as studied by the stable isotopes [15]N: a review. Zeitschrift fur Pflanzenernahrung Bodenkunde, 161, 453-463.

Jensen, B., Sørensen, P., Thomsen, I. K., Jensen, E. S. and Christensen, B. T. (1999). Availability of nitrogen in ^{15}N–labeled ruminant manure components to successively grown crops. Soil Science Society of America Journal, 63, 416-423.

Predicting soil mineral nitrogen in spring, based on soil mineralisable N in autumn

M. Quemada

Universidad Politécnica de Madrid, Depto. Producción Vegetal: Fitotecnia, 28008 Madrid, Spain

Introduction

Optimal rates of N application help to minimise residual levels of N in the soil after the crop is harvested and reduce potential fertiliser losses to the environment. The mineral N present in the top metre of soil in late winter or early spring has been used successfully to adjust the N requirements of winter cereal. However, sampling in late winter can be impossible in many wet soils, and laboratory results may only be available to advisers when a decision for fertiliser application should have already been taken. The problem could be eased by measuring soil mineral N in autumn, and using computer simulation to predict changes during winter and early spring (Addiscott & Whitmore, 1987). The problem lies in estimating the amount of soil organic N that is going to be mineralised from sampling to the period in which N-fertiliser is usually applied to cereal crops. Different attempts (incubation methods or chemical extractions) have been proposed to provide indices of soil mineralisable nitrogen (SMN), but in all cases, there are difficulties in extrapolating results to the field scale where many factors affect N mineralisation. The objective was to develop a computer tool that allows prediction of soil mineral N in spring at a field scale, based on indices of SMN determined by laboratory, or field methods. Estimates of crop uptake and nitrate leached are also obtained as outputs.

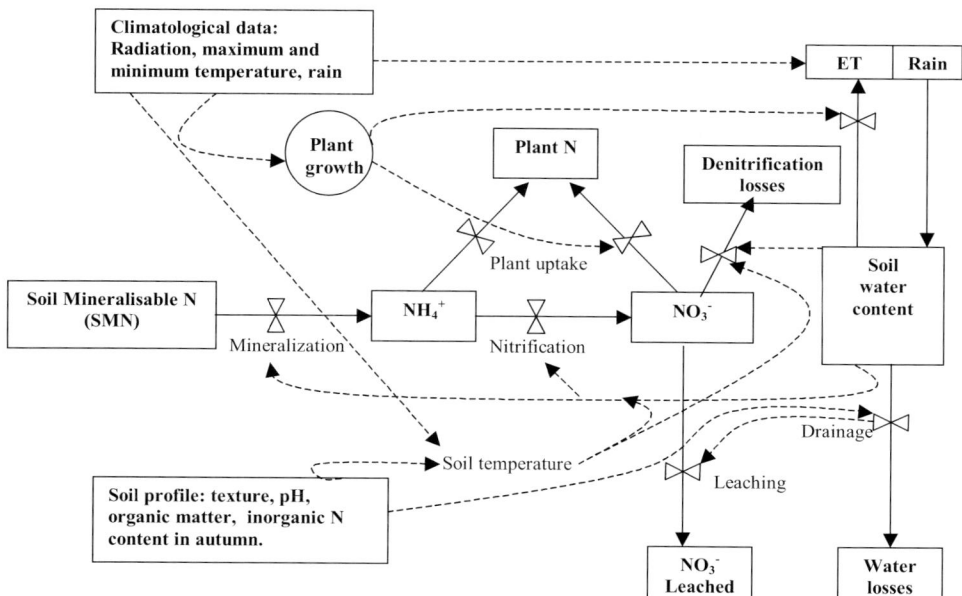

Figure 1. Flow diagram of the model.

Materials and methods

A dynamic soil-crop model that simulates changes in soil mineral N by computing (on a daily basis) the amount of N leached, mineralised, denitrified, nitrified, and uptake by the crop is shown in Figure 1.

The mineralisation submodel simulates the decay of a single pool of organic matter ('soil mineralisable N':SMN). This pool needs to be defined by the user, and it can be obtained from an aerobic incubation of soil samples, or from a function based on soil characteristics. The model also includes a subroutine that allows estimation of SMN from previous field studies in which the main components of the N balance were estimated. The mineralisation and nitrification submodels assume first-order kinetics, being the rate constant in the topsoil modified daily by the effect of moisture and temperature. The leaching submodel is uni-dimensional along a vertical axis and divides the soil profile into layers, each characterised by specific water retention levels. For each time step, surface infiltration, drainage, evaporation and NO_3^- redistribution in the soil solution are computed. The dry matter production and N uptake submodel is based on radiation-use efficiency and transpiration, adjusted for water and N availability. The model needs daily climatic data, and a description of the soil profile. Soil mineral N and water content of the soil profile in autumn must be initialised. The model was tested by using data from two different field studies in Navarra (Spain). Soil mineral N in the soil profile (0-90 cm) was measured before sowing (October) and at the beginning of tillering (February or March). Soil mineralisable N was determined from the model calibration using field results, from aerobic incubation of soil samples (Stanford & Smith, 1972), and from soil characteristics (Marion et al., 1981).

Results and conclusions

Model simulations based on field calibration and aerobic incubation were very close to observed data (Table 1). When SMN was obtained from soil characteristics, the model over-estimated mineral N in spring. Developing better relationships may lead to more accurate predictions.

Table 1. Observed soil mineral N content in autumn and late winter in fields cultivated with winter cereal in Navarra (Spain), and simulated soil mineral N content in late winter using the model when SMN was determined from model calibration using field results, from aerobic incubation of soil samples and from soil characteristics.

Year	Location	Crop	Soil mineral N in October (kg N ha^{-1})	Soil mineral N in February or March (kg N ha^{-1})			
				Observed	Simulated		
					Field calibration	SMN from incubation	SMN from soil characteristics
99/00	Beriain	Wheat	100± 5.3	80 ± 5.5	83	77	98
00/01	Beriain	Barley	90 ± 8.0	66 ± 7.9	71	59	85
01/02	Beriain	Wheat	72 ± 5.2	119 ± 24.6	115	107	145
00/01	Tafalla	Barley	116 ± 17.8	150 ± 27.3	–	143	186
01/02	Tafalla	Barley	66 ± 4.1	72 ± 9.9	76	68	98

Acknowledgement

This project was funded by the Ministerio de Ciencia y Tecnología, Spain. Ref. nº AGL2001-2214-C06-01

References

Addiscott, T.M. and Whitmore, A.P. (1987). Computer simulation of changes in soil mineral nitrogen and crop nitrogen during autumn, winter and spring. Journal of Agricultural Science, 109, 141-157.

Marion, G.M., Kummerow, J. and Miller, P.C. (1981). Predicting nitrogen mineralization in chaparral soils. Soil Science Society of American Journal 45, 956-961.

Stanford, G., and Smith, S.J. (1972). Nitrogen mineralization potentials of soils. Soil Science Society of America Proceedings, 109, 190-196.

Effect of seasonal split of N dosage on the fate of ^{15}N fate in citrus trees

A. Quiñones, J. Bañuls, E. Primo-Millo and F. Legaz
Departamento de Citricultura y otros Frutales. Instituto Valenciano de Investigaciones Agrarias (I.V.I.A.). Postal address 46113. Moncada (Valencia), Spain

Introduction

On the Mediterranean coast, where the cultivation of citrus fruits predominates, a severe increase in contamination by leaching of NO_3^- has been observed in subterranean waters as a consequence of the excessive use of N fertilisers. The use of the ^{15}N technique allows a better understanding of the final destination of applied N in the plant-soil system (Feigenbaum et al., 1987; Martínez et al., 2002). For this reason, the objective of this work was to study the influence of different variables such as irrigation system (flooding and drip), timing of N application and N fractionation on the ^{15}N uptake efficiency in citrus production.

Materials and methods

The fate of N derived from labelled potassium nitrate (7.0%^{15}N excess) was determined in eight year old Navelina orange trees (*C. Sinensis*) grafted on Carrizo cintrange (*C. sinensis x P trifoliata*). Each tree was grown outdoors in 5 m^3 lysimeters filled with a sandy-loam soil and fed with 125 g N from fertiliser, plus 50 g supplied by the irrigation water using 2 irrigation systems: flooding (6,498 l tree^{-1}) and drip irrigation (5,649 l tree^{-1}). A group of 3 trees received the N dose in 5 applications of 20% each, in March, May, June, August and September by flooding (Treatment A). The other group of 3 trees received the same N dose in 66 equal splits from March to October by drip (Treatment B). The trees were harvested at the end of the vegetative cycle (December) and the isotopic ratio (^{15}N/^{14}N) was examined in the soil-plant system.

Results

Figure 1 shows that the relative percentages of N absorbed in the new organs was higher in the trees subjected to treatment A (flood with low N fractionation); whereas the N accumulated fron fertiliser in the old organs and roots was lower in the trees subjected to treatment B (drip with high N split). The ^{15}N recovery was higher for the drip-irrigated trees, (70.8%) than for the flood-irrigated trees (63.2%). The yield and fruit quality was not affected by treatments, except for the external colour index of the fruit that was significantly lower for drip irrigation compared with flooding (Table 1).

Table 1. Yield and fruit quality at harvest.

Treatment	A	Significance[a]	B
Yield (kg per tree)	39.0		43.9
Colour Index	15.5	*	13.2

[a] Statistics analysis. t-Student $P<0,05$

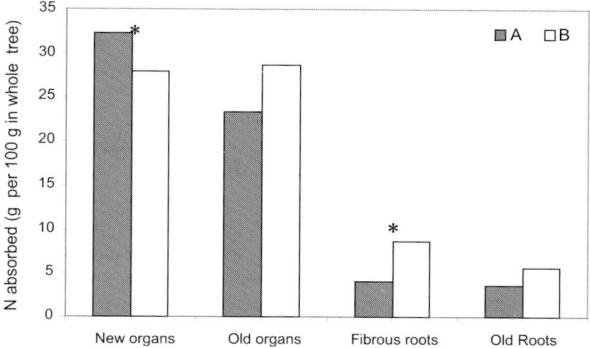

Figure 1. Relative distribution among the different tree organs of the N absorbed (g) from N applied.

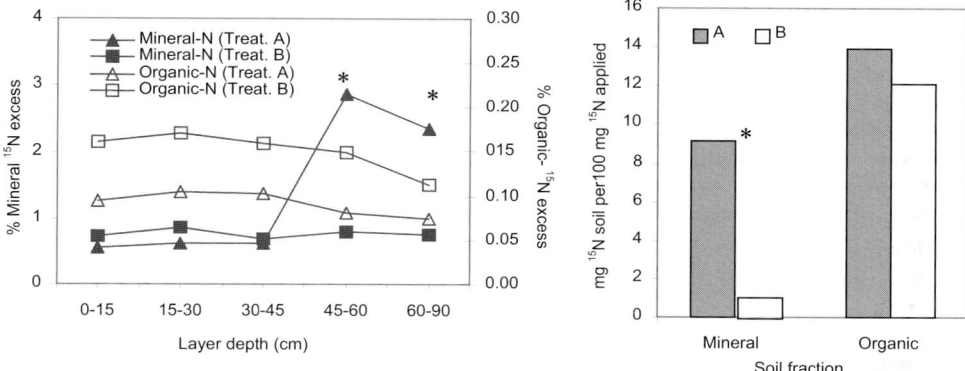

*Figure 2. Effect of the irrigation system on NO_3^--^{15}N and organic-^{15}N in the soil profile at the end of the trial. *denotes significant differences between treatments (Student's t test, P<0.05).*

Nitrate was accumulated in similar concentrations in the upper soil layers for both irrigation systems (Figure 2). However, a higher NO_3^- concentration was found in the deeper soil layers for flooding. The percentages of mineral-N retained in the soil profile were 9.2 and 1.1% of the ^{15}N applied for the flooding and drip irrigation, respectively. Moreover, the ^{15}N retained in organic form was higher in the flood-irrigated soil.

Conclusions

The N use efficiency was higher for the drip-irrigated trees. This would lead to a reduction in the N dose required. The NO_3^- concentration was lower for the drip-irrigated soil. This would tend to reduce the potential contamination of groundwaters.

References

Feigenbaum, S., Bielorai, H., Erner, Y. and Dasberg S, (1987). The fate of ^{15}N labelled nitrogen applied to mature citrus trees. Plant and Soil, l97, 179-187.

Martínez, J.M., Bañuls, J., Quiñones, A., Martín B., Primo-Millo E., and Legaz, F. (2002). Fate and transformations of ^{15}N labelled nitrogen applied in spring to Citrus trees. Journal of Horticultural Science and Biology, 77, 361-367.

Methodology of soil incubation studies -comparison between laboratories

T. Salo[1], L.S. Jensen[2], F. Palmason[3], T.A. Breland[4], B. Stenberg[5], A. Pedersen[2], C. Lundström[5] and M. Esala[1]

[1]MTT, Environmental Research, 31600 Jokioinen, Finland
[2]The Royal Veterinary and Agricultural University, Dept. of Agricultural Sciences, 1871 Frederiksberg C, Denmark
[3]Agricultural Research Institute, Keldnaholt, 112 Reykjavik, Iceland
[4]Agricultural University of Norway, Department of Horticulture and Crop Sciences, 1432 Aas, Norway
[5]Swedish University of Agricultural Sciences, Department of Agricultural Research, 53223 Skara, Sweden

Introduction

Laboratory incubations are a popular method to study soil and plant litter C and N mineralisation and to estimate parameters for empirical or mechanistic models. It is documented that the results of such incubations depend on soil pre-treatments, experimental conditions during incubation, length of incubation and soil characteristics (Benbi & Richter, 2002). We have not, however, found studies that discuss the reproducibility of laboratory incubations and their comparability between different laboratories. We present results from laboratory incubations of a soil with and without a reference plant material, conducted with an almost identical methodology in five different laboratories.

Materials and methods

Laboratory incubations were made in two batches in each of five different laboratories in 2001-2002. Together we made 10 laboratory incubations with the same control soil, either unamended or amended with a reference plant material (Timothy: *Phleum pratense*). Incubations were carried out with similar protocols and lasted for ca. 220 days. Treatments were carried out in triplicate at 15°C and 50-60% of soil water-holding capacity and they were sampled 13-14 times for CO_2 production by alkali traps. Soil mineral N was extracted by 1M KCl on 8-9 occasions. Thus, CO_2 production and soil mineral N concentrations of control and reference pots in different incubations could be compared both between and within laboratories. CO_2 production of the reference material was calculated by subtracting CO_2 production of control soil treatment from the total CO_2 production of the reference treatment. In one laboratory, true alkali blanks proved unreliable and thus, only the CO_2 production of reference material could be calculated, hence control soil CO_2 production is available only for 8 laboratory incubations.

Results and discussion

The average CO_2 production of the control soil over the 220 days was 2.70 mg C g^{-1} dry soil (SE = 0.15 mg, CV=16%, n=8). The mean value for the reference material was 4.64 mg C g^{-1} dry soil (SE = 0.15 mg, CV=10 %, n=10). Statistical analysis with mixed models showed that there were more differences in CO_2 production within laboratories than between laboratories.

With the control soil, in two laboratories out of four, similar amounts of CO_2 were produced in their batches (Table 1), and there was a significant batch x time interaction for CO_2 production. Total CO_2 production of the control soil during 220 days incubation varied between laboratories from 2.29 to 3.19 mg C g^{-1} dry soil. Least significant difference (5% level) for between laboratory comparisons was 0.65 mg C g^{-1} dry soil. The corresponding value for within laboratory comparisons was 0.31 mg C g^{-1} dry soil. For the reference material there was no significant difference between laboratories, although the highest difference in total CO_2 production between laboratories was 0.89 mg C g^{-1} dry soil. There were within-laboratories differences for the reference plant material in two out of five cases (LSD=0.24 mg C g^{-1} dry soil).

Table 1. Statistics for the comparison of means of batches within laboratories. Different superscript letters indicate statistically significant difference (p<0.05). N/A = not available.

Laboratory, batch	Sum of CO_2-C production (mg g^{-1})		Mean of soil mineral N (mg kg^{-1})	
	Control	Reference	Control	Reference
1,1	3.08	4.39	171[a]	124
1,2	3.29	3.95	145[b]	122
2,1	2.11[a]	4.90[a]	146[a]	106[a]
2,2	2.67[b]	4.65[b]	153[b]	115[b]
3,1	N/A	5.37	147[a]	124[a]
3,2	N/A	4.65	132[b]	98[b]
4,1	3.09[a]	5.20[a]	136	89[a]
4,2	2.78[b]	4.91[b]	144	105[b]
5,1	2.36	4.20	141	111
5,2	2.22	4.20	136	95

The mean value for soil mineral N content of control soil was 145 mg N kg^{-1} dry soil (SE=4 mg kg^{-1}, CV=8%, n=10) and for the reference material 109 mg N kg^{-1} dry soil (SE=4 mg kg^{-1}, CV=11%, n=10). Soil mineral N contents were not statistically different between laboratories. Within laboratories, soil mineral N contents differed in three out of five cases (Table 1). The difference between soil mineral N contents of batches was usually low but systematic, which caused the lowest LSD values of within laboratory comparisons to be 7 mg N kg^{-1} dry soil.

Conclusions

This study shows that although great efforts have been made to keep the methodology constant, there can be considerably larger variation in results from incubation studies than indicated by the experimental uncertainty of the individual incubation batch. However, this variability is most often completely ignored in the interpretation of laboratory incubation data when used for parameterisation of simulation models.

References

Benbi, D. K. and Richter, J. (2002). A critical review of some approaches to modelling nitrogen mineralization. Biology and Fertility of Soils, 35, 168-183.

References

Chapman, S.C. and Barreto, H.J. (1997). Using a chlorophyll meter to estimate specific leaf nitrogen of tropical maize during vegetative growth. Agronomy Journal. 89, 557-562.

Fotyma, E., Fotyma, M. and Bezduszniak, D. (1998). Chlorophyll meter (SPAD-502, Minolta) a new tool for evaluating the nitrogen nutritional status of cereals. Short Communications. Volume II. Fifth ESA Congress, Nitra, The Slovak Republic, 304-305.

Fotyma, E. (2002). Variety differentiation of chlorophyll content in the leaves of winter cereals. Pamiętnik Puławski, 130, 171-178.

Wollring, J., Reusch, S. and Karlsson, C. (1998). Variable nitrogen application based on crop sensing. Proceedings of the International Fertiliser Society. 423, pp. 28.

Zadoks, J.C., Chang, T.T. and Konzak, G.F. (1974. A decimal code for growth stages of cereals. Weed Research. 14, 415-421.

Organic nitrogen loads at different scale levels in Spain

J. Soler-Rovira and J.M. Arroyo-Sanz

E.U.I.T. Agrícola, Departamento de Producción Vegetal: Fitotecnia, Universidad Politécnica de Madrid, Ciudad Universitaria s/n, 28040 Madrid, Spain

Introduction

Recently, Spanish livestock production has become more specialised and intensive (Soler-Rovira & Arroyo-Sanz, 2003). Several indicators have been developed in order to study N environmental issues and the European Commission has selected organic N load per ha as an indicator in nutrient policy, with a target value of 170 kg N ha^{-1} (De Clercq et al., 2001). The aim of this work is to study the N load at different scale levels in order to identify zones with high N production and generate indicators that characterise them.

Materials and methods

Agricultural censuses were used as the statistical source (INE, 1991; INE, 2003). Nitrogen loads have been calculated with the number of livestock units (of cattle, sheep, goats, pigs, horses and poultry), utilised agricultural area (i.e. arable land plus permanent grassland and pastures) and N excretion factors (Soler-Rovira et al., 2001). Several statistical indices have been calculated (i.e. median, 75% percentile, etc. Table 1). Gini indices have been calculated with Lorenz curves for each province with municipal data. Sigma convergence has been calculated as the standard deviation of the values as Neperian logarithms.

Results and discussion

In a large country like Spain, the national N load is low (i.e. 39 kg N ha^{-1}): the lowest in EU-15. However, two regions (Cataluña and Galicia) show a N load higher than 100 kg N ha^{-1}, but less than 170 kg N ha^{-1}, which is exceeded in some European regions (De Clercq et al., 2001). Divergence between regions exists but the sigma coefficient has decreased between 1989 and 1999 (0.70 and 0.62, respectively). Divergence between provinces is higher than between regions and has also decreased, due to an increment in N production (0.86 and 0.84). Thirteen provinces have been selected with different N loads and location (Table 1). Pig livestock produces more than 50% of N in six provinces and cattle in two provinces.

In all provinces (except Badajoz) there are municipalities with N loads higher than the target value, but the range is very variable. Nitrogen production is more spatially concentrated in those provinces that show higher values of the Gini index and Sigma coefficient (e.g. Barcelona, Valencia or Almería). Seventeen districts show N loads higher than 170 kg N ha^{-1} in provinces with high, but also low N loads (i.e. Castellón and Almería), because of the unequal distribution among municipalities in the province. As transport of manure is not economically feasible over long distances, these districts have a structural N surplus over the target value. Nitrogen surplus amounts to 52,557 t, i.e. 36% of the total organic N produced in those districts. Among these 17 districts there are six NO_3^- vulnerable zones, with 74 municipalities designated in Barcelona, Lérida and Castellón (Soler-Rovira et al., 2001).

Table 1. Values of the statistical indices calculated for each province at different scales.

Provinces	Province		District					Municipality				
	Mean[a]	Species[b]	Median[a]	P75[a]	Range[a]	N°> 170[c]	Range[a]	%> 170[c]	VC[d]	Gini I.	σ-C[e]
Barcelona	230	P64-C21	238	309	74-526	6	0-42,220	25	647	0.76	1.81
Ponte-vedra	221	C37-Po-29-P28	231	250	138-260	3	29-1,267	47	85	0.35	0.64
La Coruña	161	C73	158	165	157-171	1	48-1,137	36	70	0.25	0.43
Lérida	156	P51-Po35	137	197	4-364	4	0-3,236	37	146	0.57	1.38
Las Palmas	105	Po31-G31-C21	120	126	34-131	0	9-1,017	32	119	0.54	1.17
Guipúzcoa	102	C58	102	102	102	0	12-402	8	50	0.24	0.48
Castellón	100	P58-Po30	62	153	34-246	2	0-494	21	118	0.59	1.90
Huesca	63	P56	59	94	17-163	0	0-617	11	128	0.56	1.27
Segovia	61	P58-C23	57	73	28-89	0	0-434	10	114	0.52	1.13
Valencia	42	P48-Po31	30	66	okt-84	0	0-2,249	9	319	0.81	2.05
Almería	33	P59	17	27	7-195	1	0-1,101	7	269	0.92	1.52
Toledo	33	P42-C27-Po16	47	50	13-60	0	0-370	6	135	0.58	1.39
Badajoz	30	P35-S34-C24	24	36	14-58	0	0-141	0	71	0.37	0.89

[a] kg N ha^{-1}; [b] % of contribution to N load: C: cattle, G: goat, P: pig, Po: poultry, S: sheep. [c] Number or % of entities that exceed target value. [d] Variation coefficient (%). [e] Sigma coefficient.

Conclusions

Although at national level the organic N load is low, there is a clear divergence between regions, with provinces and districts that show high N production levels. Some districts have structural N surpluses over 170 kg N ha^{-1}. In NO_3^- vulnerable zones there should be clear implications for farmers. Nitrogen loads and convergence indices are efficient indicators of N availability or surplus. Other indicators should complete the study of N environmental issues.

References

De Clercq, P., Gertsis, A.C., Hofman, G., Jarvis, S.C., Neeteson, J.J. and Sinabell, F. (2001). Nutrient Management Legislation in European Countries. Department of Soil Management and Soil Care, Wageningen Pers, The Netherlands, 347 pp.

INE (1991). Censo Agrario 1989. INE, Madrid.

INE (2003). Censo Agrario 1999. INE, Madrid.

Soler-Rovira, J. and Arroyo-Sanz, J.M. (2003). Spatial and Temporal Trends of Environmental Indicators Relating to Nitrogen and Phosphorous in Spain. In: E. Schnug, J. Nagy, T. Németh, Z. Kóvacs and T. Dovéngy-Nagy. (eds). Fertilizers in Context with Resource Management in Agriculture. 14th International Symposium of Fertilizers. pp: 321-329.

Soler-Rovira, J., Soler-Rovira, P. and Soler-Soler, J. (2001). Environmental Pressures and National Environmental Legislation with Respect to Nutrient Management: España. In: P. De Clercq, A.C. Gertsis, G. Hofman, S.C. Jarvis, J.J. Neeteson and F. Sinabell. (eds). Nutrient Management Legislation in European Countries. Department of Soil Management and Soil Care, Wageningen Pers, The Netherlands, pp: 114-137.

Organic nutrient management indicators at district level in Spain

J. Soler-Rovira and J.M. Arroyo-Sanz

E.U.I.T. Agrícola, Departamento de Producción Vegetal: Fitotecnia, Universidad Politécnica de Madrid, Ciudad Universitaria s/n, 28040 Madrid, Spain

Introduction

Nutrient management planning should include a set of indicators in order to evaluate the effects of actual practices. Spatial differences within a large country like Spain could be studied at different scale levels, due to concentration, specialisation and intensification of agricultural production (Soler-Rovira & Arroyo-Sanz, 2003). The aim of this work is to select and study organic nutrient management indicators at the district level in 13 Spanish provinces.

Materials and methods

Thirteen provinces (with 89 districts) have been selected as having different N loads and location. The statistical source was from the last Agricultural Census (INE, 2003). A primary set of 48 indicators has been built up as follows: number of farms (with land, landless and total), livestock units (LU's) per farm (cattle, sheep, goats, pigs, horses and poultry) and livestock density per ha of utilised agricultural area (UAA); organic N and P loads per ha of UAA, using excretion factors for each species; N and P surpluses over the target value (i.e. 170 kg N ha^{-1} and 22 kg P ha^{-1}); slurry produced by cattle and pigs per ha of UAA, considering that all the excreta were managed as slurry; animal excreta per ha of UAA from cattle, sheep, goats, poultry and horses, considered the solid and the liquid excreta. These indicators were studied by principal component analyses (PCA), using STATITCF software. The coordinates of the observations with the principal axes were studied by cluster analyses.

Results and discussion

The first principal axis of the PCA is highly correlated with indicators related to number of farms, excreta, cattle and sheep (Table 1). The second axis is correlated with N and P load and surplus, slurry, pig and poultry and total livestock density. Indicators of each group are well correlated amongst themselves (Table 2). The number of farms is not well correlated with livestock density per farm or per ha, except cattle density. Pig livestock density indicator is highly correlated with N and P load and surplus and slurry production (Table 2). The 89 districts can be divided by cluster analyses into seven classes (Figure 1). Indicators correlated with the first axis can differentiate classes 1, 4, 5 and 7. Class 1, located near to the origin of the plane, includes a high proportion of districts with average values of indicators. Class 4 includes six districts from the Galicia region with a large number of farms, based mainly on cattle production with grasslands (Soler-Rovira & Arroyo-Sanz, 2003). Classes 2, 3 and 6 are situated along the second axis. N, P and pig density indicators characterise them, as they are located in the Mediterranean coast and Ebro Valley, where intensive rearing farms are situated near to the points of introduction of raw materials for feed production (Soler-Rovira & Arroyo-Sanz, 2003).

Table 1. Indicators correlated with the two principal axes of the PCA.

Axis 1 (30% of variance absorption)				Axis 2 (20% of variance absorption)			
Indicator	r	Indicator	r	Indicator	r	Indicator	r
Poultry farms	+0.91	Excreta per ha	+0.71	P load per ha	+0.91	Pigs per ha	+0.85
Pig farms	+0.89	Cattle per ha	+0.68	P surplus per ha	+0.89	N surplus per ha	+0.80
Cattle farms	+0.87	Horses per ha	+0.55	Total LU's per ha	+0.89	Poultry per ha	+0.59
Sheep farms	+0.87	Sheep per farm	-0.62	N load per ha	+0.88	Poultry per farm	+0.58
Horses farms	+0.86	Cattle per farm	-0.54	Slurry per ha	+0.85	Pigs per farm	+0.48

Table 2. Determination coefficients (r^2) between some indicators.

	Cattle farms	Sheep farms	Pig farms	Poultry farms	Cattle/ ha	Pigs/ ha	Poultry/ ha	Total LU's/ ha
LU's per farm	0.11	0.20	0.09	0.02	0.02	0.13	0.35	-
LU's per ha	0.53	0.01	0.00	0.07	1.00	1.00	1.00	1.00
Total LU's per ha	0.05	0.04	0.06	0.06	0.40	0.83	0.34	1.00
N load per ha	0.06	0.06	0.08	0.07	0.40	0.76	0.40	0.99
P load per ha	0.03	0.04	0.04	0.04	0.28	0.81	0.45	0.98
Excreta per ha	0.44	0.31	0.40	0.40	0.85	0.21	0.26	0.58
Slurry per ha	0.05	0.01	0.04	0.03	0.44	0.92	0.10	0.90
N surplus per ha	0.00	0.01	0.00	0.00	0.23	0.71	0.22	0.76
P surplus per ha	0.01	0.03	0.02	0.02	0.23	0.81	0.41	0.91

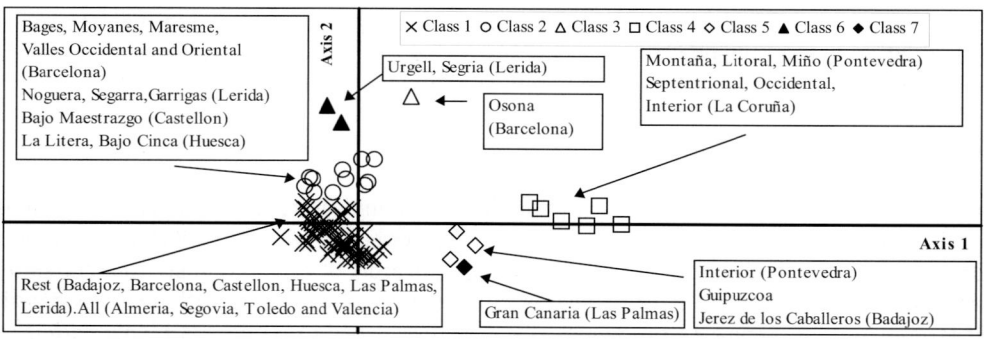

Figure 1. Representation of the districts on the plane generated by the two principal axes.

Conclusions

Twenty indicators have been selected in order to characterise organic nutrient management in 89 districts. Six districts show a high number of farms of low definition. Another 14 districts are characterised by pig density and N and P load and surplus. This study needs to be completed with the generation of other pressure, state and response indicators.

References

INE (2003). Censo Agrario 1999. INE, Madrid.

Soler-Rovira, J. and Arroyo-Sanz, J.M. (2003). Spatial and Temporal Trends of Environmental Indicators Relating to Nitrogen and Phosphorous in Spain. In: Schnug, E., Nagy, J., Németh, T., Kóvacs, Z. and Dovéngy-Nagy, T. (eds). Fertilizers in Context with Resource Management in Agriculture. 14th International Symposium of Fertilizers. pp: 321-329.

Physical separation of pig slurry has a small effect on the overall utilisation of nitrogen

P. Sørensen, I.K. Thomsen and B.T. Christensen
Department of Agroecology, Danish Institute of Agricultural Sciences, Tjele, Denmark

Introduction

A high utilisation of N in animal manure is a prerequisite for reducing losses to the environment. Physical separation of animal slurry has been suggested as a tool for improving the distribution of nutrients and avoiding over-fertilisation on farms with a high livestock density. Slurry is divided into a dry-matter-rich (DMR) and a liquid (LIQ) fraction by centrifugation. The DMR fraction contains most of the slurry P (Møller et al., 2002), and it has a higher C/N ratio than unseparated slurry. The liquid fraction has a low C/N ratio and provides better soil infiltration, reducing the risk of NH_3 volatilisation. Thus, slurry separation may influence the turnover of slurry N in soil as well as loss of volatile N after application. We measured the mineral fertiliser equivalent (MFE) of separated and unseparated slurry and quantified losses of ^{15}N from ^{15}N-labelled pig slurry fractions, after application to spring barley and winter wheat.

Materials and methods

Labelled and unlabelled pig urine and faeces were produced by feeding pigs on ^{15}N-labelled and unlabelled diet (mainly barley and pea). The uniformity of labelling was evaluated and found satisfactory. Two similar pig slurries were established by mixing ^{15}N-labelled urine and faeces with their unlabelled counterparts. After a few weeks storage, the slurries were separated by repeated centrifugation. The average composition of the slurry fractions is shown in Table 1. Slurry fractions and unseparated slurry were applied to confined field plots (30 cm diameter) by incorporation (simulated ploughing) before sowing spring barley and by surface-banding in growing winter wheat (100 kg N ha^{-1} in DMR, 100 kg NH_4^+-N ha^{-1} in LIQ and slurry + extra unlabelled fertiliser N). The DMR (100 kg N ha^{-1}) was also incorporated in the autumn before sowing winter wheat. The recovery of urinary ^{15}N and faecal ^{15}N was measured in the crop and soil (0-40 cm) at harvest. The recovery of total manure N was the weighted recovery of labelled faecal and urinary N. The MFE of separated and untreated slurries (labelled and unlabelled) was calculated from the measured N uptake in crops (grain + straw) using N fertiliser response curves. Slurry fractions from full-scale separation by a decanting centrifuge were compared with the fractions produced in the laboratory and found to have a similar composition and MFE. In a parallel experiment, the residual effect of slurry and fertiliser ^{15}N in the year after application was calculated from the uptake of labelled N in a barley crop, and the extra residual effect, compared with fertiliser application, was calculated.

Table 1. Characteristics of experimental slurry fractions (g kg^{-1}).

Slurry fraction	Dry matter	Total N	NH_4^+-N	Total C	Total P	C/N
Dry-matter-rich fraction (DMR)	300	12.5	4.36	104	8.2	8.3
Liquid fraction (LIQ)	19	6.29	5.08	6.1	0.36	1.0
Unseparated pig slurry	57	7.22	5.41	16.9	1.4	2.3

Results and discussion

The MFE of slurry fractions following incorporation to spring barley and after surface banding in winter wheat is shown in Table 2. The DMR fraction contained a high proportion of NH_4^+-N (30-40% of total N), and the MFE was 44% when it was incorporated before sowing spring barley. After incorporation in spring, losses of labelled N were similar from all sources. A significant part of the N in the DMR was lost (38%), presumably by leaching and denitrification, when the fraction was incorporated in the autumn and by NH_3 volatilisation when the fraction was applied to the growing winter wheat. The MFE of slurry was generally higher after incorporation with spring barley than after surface banding in winter wheat, probably because the NH_3 loss was reduced by incorporation. In winter wheat, the LIQ fraction infiltrated the soil better than untreated slurry, resulting in a lower NH_3 loss, but this was counterbalanced by the higher loss from the DMR fraction (Table 2). Thus, the overall plant availability of slurry N was not affected by separation, either under conditions with minimal N loss (barley) or under conditions with high NH_3 volatilisation (winter wheat). The overall utilisation was lowest when the DMR fraction was applied in the autumn. During storage of the DMR fraction, N may also be lost by NH_3 volatilisation if the fraction is not covered, resulting in a reduced overall N utilisation.

Table 2. Plant utilisation and loss of N from physically separated pig slurry fractions following application to spring barley and winter wheat (n=4). The mineral fertiliser equivalent (MFE) of pig slurry was calculated from the crop N uptake, and the 'loss' of ^{15}N was calculated from the total recovery of labelled N in crops and soil (0-40 cm) after harvest.

[15]N-labelled amendment	Spring barley (incorporated)		Winter wheat (surface-banded)	
	MFE (% of total N)	[15]N 'loss' (% of input)	MFE (% of total N)	[15]N 'loss' (% of input)
DMR fraction, autumn			14 [b]	38 [b]
DMR fraction, spring	44	14	27	38
LIQ fraction	95	20	83	15
DMR+LIQ fractions[a]	85	19	69	21
Unseparated pig slurry	82	15	69	25
Mineral fertiliser N	100	20	100	5
LSD (P<0.05)	13	NS	9	7.3

[a] DMR+LIQ: the weighted effect of the two fractions after application in spring. [b] The DMR fraction was incorporated before sowing winter wheat. NS = not significant.

The residual N effect in the year after application of separated and unseparated pig slurry was calculated from the uptake of labelled N in a barley crop and expressed as MFE. The residual N effect as percentage of the total N amendment was 4.5% for the DMR fraction, 1.2% for the LIQ fraction and 2.9% for unseparated slurry. Thus, the overall residual N effect of slurry fractions was similar to that of the untreated slurry. We conclude that after centrifugation of pig slurry the overall crop uptake and loss of N were similar for unseparated pig slurry and for corresponding dry-matter-rich + liquid fractions when these were applied in spring to the same crop.

References

Møller, H.B., Sommer, S.G. and Ahring, B.K. (2002). Separation efficiency and particle distribution in relation to manure type and storage conditions. Bioresource Technology, 85, 189-196.

The use of anaerobic incubations to predict net N immobilisation after the application of organic residues to soils

J.R. Sousa[1], R. Lagoa[1], F. Cabral[2] and J. Coutinho[1]
[1]*Univ Tras-os-Montes, Dep Soil Science - CCEA, ap. 1013, 5000-911 Vila Real, Portugal*
[2]*Inst Sup Agronomia, Dep Química Agrícola e Ambiental, 1349-017 Lisboa, Portugal*

Introduction

Sound management of organic residues needs to be based on a reliable prediction of the N mineralisation-immobilisation turnover (MIT) processes as one consequence of the organic recycling of on- and off-farm by-products. However, most of the laboratory methods used to assess net N mineralisation potential were developed by taking into account only the soil itself and its native organic N. Thus, the aim of the present study was to evaluate anaerobic incubations as an adequate methodology to predict different N mineralisation patterns after the application of fresh organic matter to soil. A special focus was given to the treatment with straw (a high C:N ratio residue) which was expected to immobilise soil mineral N.

Materials and methods

The study was carried out with a coarse textured umbric Cambisol (pH 5.6, 60.5 g organic matter kg^{-1}, C:N ratio 14), using poultry manure (28.9 g N kg^{-1} DM, C:N ratio 10) and wheat straw (2.3 g N kg^{-1} DM, C:N ratio 166) as organic residues, applied at a rate equivalent to 75 mg N kg^{-1} dry soil. A control treatment (soil without amendment) was conducted to calculate apparent net N fluxes. Anaerobic incubations were performed during 7 days at 40 °C according to the method proposed by Keeney (1982). A standard aerobic incubation was also conducted during 28 days at 25 °C as a reference method, following the procedure described by Hart et al. (1994) with weekly sampling periods and adjusting first order kinetic equations. Soil inorganic N (NH_4^+, or NH_4^+ and NO_3^- for anaerobic and aerobic incubations, respectively) was extracted with 2M KCl and the extracts were analysed in a segmented-flow system by molecular absorption spectroscopy. Apparent net N mineralisation values were calculated by subtracting the control from the residue treatment results.

Results and discussion

Soil mineral N content of the treatments obtained with both methods is presented in Table 1. As expected, both methods predicted positive net N mineralisation fluxes either for soil alone or for soil with poultry manure, this flux being particularly intense for the latter treatment. Anaerobic incubation also demonstrated N immobilisation after straw application. This fact was revealed by the complete depletion of mineral N after 7 days of waterlogged conditions, but also occurred in the aerobic incubation.

Table 1. Soil mineral N at the start and at the end of the aerobic (30d) or anaerobic (7d) incubation periods.

treatment	aerobic method (mg N kg^{-1})		anaerobic method (mg N kg^{-1})	
	start	end	start	end
soil	5.7	26.0	2.0	20.5
soil + poultry manure	12.9	67.0	12.3	64.0
soil + straw	7.9	0.2	3.1	2.8

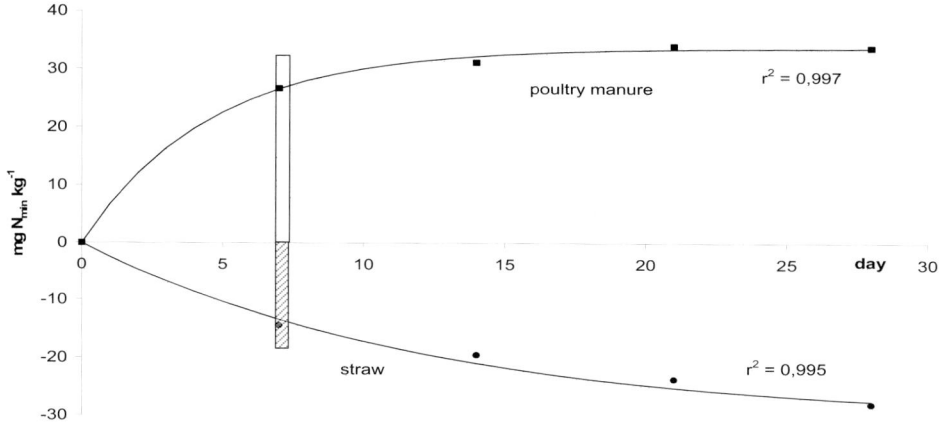

Figure 1. Apparent net N mineralisation in soil after the application of organic residues measured by the anaerobic (bars) or the aerobic incubation (lines and dots): (straw = ● and ▨, poultry manure = ■ and ▢).

Under waterlogged conditions, the incubation values also seem to reflect an anaerobic microbial activity with specific N assimilation requirements, since the application of straw to the soil led to a negative value for the apparent net N mineralisation (Figure 1). The present set of data does not support the previous remarks made by Boone (1990), who suggested that waterlogged incubations would most likely estimate the N pool mineralised from aerobic organisms killed by anaerobic conditions and would not represent an adequate balance between N mineralisation and N immobilisation.

Conclusions

The available data suggest that anaerobic incubation may be used to predict soil N immobilisation fluxes after the application of fresh organic matter. Thus, the present observation is useful to advance research in order to develop and validate quantification of MIT net N fluxes using the short-term anaerobic incubation methodology.

Acknowledgements

This work was partially funded by FCT through the project POCTI/AGG/46559

References

Boone, R.D. (1990). Soil organic matter as a potential net nitrogen sink in a fertilized cornfield, South Deerfield, Massachusetts, USA. Plant and Soil, 128, 191-198.

Hart, S.C., Stark, J.M., Davidson, E.A. and Firestone, M.K. (1994). Nitrogen mineralization, immobilization, and nitrification. In: R.W. Weaver (ed). Methods of Soil Analysis. Part 2, Soil Science Society of America, Madison, Wisconsin, USA, 985-1018.

Keeney, D.R. (1982). Nitrogen availability indexes. In: A.L. Page, R.H. Miller and D.R. Keeney (eds). Methods of Soil Analysis. Part 2, Soil Science Society of America, Madison, Wisconsin, USA, 711-733.

Nitrogen uptake by ryegrass (*Lolium perenne*) as affected by the decomposition of apple leaves and pruning wood in soil

M. Tagliavini[1], G. Tonon[1], D. Solimando[1], P. Gioacchini[2], M. Toselli[1], P. Boldreghini[1] and C. Ciavatta[2]

[1]*Dipartimento di Colture Arboree, Bologna University, V. Fanin 46 , 40127 Bologna, Italy*
[2]*Dipartimento di Scienze e Tecnologie Agroambientali, V. Fanin 46, 40127 Bologna, Italy*

Introduction

In deciduous tree orchards, significant amounts of N return to the soil by leaf abscission and pruned wood. The knowledge of the fate of these N sources and their availability for subsequent uptake is of great importance in sustainable fruit production, and will allow a better use of within-orchard resources and less dependence on external N supply (Tagliavini et al., 1996).

Materials and methods

^{15}N enriched leaves and pruning wood (both < 2 mm size) from apple trees (cv. Gala) were incorporated on February 22, 2002 into a clay loam soil (0.12% total N) and used to fill 10 l pots sown with perennial ryegrass (*Lolium perenne* L.). Litter material had the following characteristics: total N = 1.76 (leaves) and 1.15% (wood); total C = 46.2 (leaves) and 44.9% (wood). Sufficient amounts of labelled N per unit of soil (ratios of plant material:soil (w:w)=1:250, leaves and 1:330, wood), were added to be able to follow its fate in soil and plants. Destructive harvests of three replicates were performed on April 22, May 24, June 28, August 12 and October 28. The dry weight, total N and ^{15}N abundance of ryegrass roots and leaves (including the mown grasses) were determined. Microbial N (Brooks et al., 1985) and N_{min} (Keeney & Nelson, 1982) were determined on the same dates. Soils remaining from October 28 harvest were extracted with 0.1 M NaOH plus 0.1 M Na-pyrophosphate and then the extract fractionated into humic (HA) and fulvic acids (FA) and non-humified (NH) fractions (Francioso et al., 2000).

Results

Plants grew more in control soils than in those enriched with residues (especially leaf litter) until the May harvest (data not shown), while growth in residue amended soils was comparable to the control soil from May 24 to June 28: at the end of the experiment, treated and control plants produced a similar biomass. Severe reduction in total N uptake occurred as a consequence of litter addition (Table 1) during the first two months from the beginning of the incubation. At the April sampling, the N_{min} concentration was lower in soil receiving leaf litter (2.5±0.1 µg N g^{-1}, mean ± SE) than in control soil (4.9±0.1 µg N g^{-1}) and both substrates increased the microbial pool, compared with the control (40 µg N g^{-1}, on average, compared with 18 µg N g^{-1}). The N uptake in control soil was completed by May 24 (Table 1), while plants grown in soil enriched with litter continued to absorb N until the end of the experiment

(Table 1). During the first two months after addition, there was practically no uptake of N derived from leaf litter (Table 1), which only occurred from April 22 to June 28. The uptake of N derived from decomposition of wood (Table 1) followed a pattern similar to that of leaf litter, except during the first two months, when N from wood litter was slightly more available for uptake. The fraction of total labelled soil N that was extracted, averaged 83% and 72% in leaf- and wood-amended soils, respectively; some 45-50% of the extracted labelled N was incorporated in the humic fraction (HA+FA).

Discussion and conclusions

Severe reduction in availability of native soil N occurred during the first two months after litter addition. In this period, leaf residues caused stronger immobilisation of native soil N than wood litter. This effect occurred in spite of the higher C:N ratio of wood (39:1) than leaf (24:1) residues and may be explained by the higher amount (+33%) of leaf litter than wood litter added per unit of soil. The native soil N, initially immobilised in amended soils, was made available afterwards for plant uptake as indicated by no further N content increase in control plants (Table 1) after May 24, while those growing in residue enriched soils continued to absorb N and their N content was similar to that of control plants at the end of the experiment. Plant availability of N derived from decomposition of litter was faster for wood than for leaves and plant uptake was completed for both plant residues after three months from addition. Ryegrass took up similar amounts of N derived from leaf and wood residues, which corresponded to 7 and 13% of the N added by leaves and wood, respectively. In conclusion, during the 8 months from addition, no net increase in N availability occurred as a result of decomposition of apple residues in soil, while short term immobilisation of native soil N took place. The majority of the N added in residues underwent decomposition and significant amounts were recovered in the humified fraction. If short-term availability of N from the return of apple residues in soil is limited, medium-long term benefits in the ecosystem will be likely.

Table 1. Total amount of N and N deriving from leaf and wood litter present in ryegrass plants.

Sampling date	Total plant N (mg N g^{-1} soil)			N from Litter (mg N kg^{-1} soil)	
	Control	Leaf litter	Wood litter	Leaf litter	Wood litter
April 22	24.5 ± 1.4	3.6 ± 0.6	10.5 ± 0.2	0.16 ± 0.03	1.47 ± 0.29
May 24	30.3 ± 2.0	14.4 ± 1.0	21.5 ± 0.9	2.52 ± 0.49	3.99 ± 0.49
June 28	29.5 ± 2.4	19.7 ± 2.0	27.3 ± 1.8	4.17 ± 0.69	5.29 ± 0.26
August 12	29.0 ± 3.6	22.4 ± 1.5	24.2 ± 0.4	3.51± 0.36	3.81 ± 0.39
October 28	30.4 ± 1.4	28.6 ± 4.1	30.8 ± 3.7	4.23 ± 0.62	4.99 ± 0.46

Data are the average of three replicates ± SD

Acknowledgements

Project "Ciclo dell'azoto in ecosistemi arborei da frutto e da legno" founded by University of Bologna, Programme "Ricerca Fondamentale Orientata ex quota 60% 2001-2003".

References

Brookes, P.C., Landman A., Pruden, G. and Jenkinson, D.S. (1985). Chloroform fumagation and the release of soil N: a rapid direct extraction method to measure microbial biomass in soil. Soil Biology and Biochemistry, 17, 837-842.

Francioso, O., Ciavatta, C., Sanchez-Cortes, S., Tugnoli, V., Sitti, L. and Gessa, C. (2000). Spectroscopic characterization of soil organic matter in long term-amendment trials. Soil Science, 165, 495-504.

Keneey, D.R. and Nelson, D.W. (1982). Nitrogen - inorganic forms. In: A.L. Page (eds) Methods of Soil Analysis, 2nd edition, Agronomy monographs 9, ASA Madison, WI, 643-698 pp.

Tagliavini, M., Scudellari, D., Marangoni, B. and Toselli, M. (1996). Nitrogen fertilization management in orchards to reconcile productivity and environmental aspects. Fertilizer Research, 43, 93-102.

Use of the NDICEA model analysing nitrogen efficiency

G.J. van der Burgt
Louis Bolk Instituut, Hoofdstraat 24, NL-3972 LA Driebergen, The Netherlands

Introduction

In organic arable farming, N management is a complicated matter. Farming systems based on solid manure application in late summer offer a particular challenge in meeting the crop N demand while maintaining low, acceptable levels of N losses. In order to meet these goals, a good understanding of the N dynamics is necessary and modelling can be of great help.

Materials and methods

The experimental farm OBS in Nagele, The Netherlands, is located in the relatively new 'polder' on a clay topsoil and loamy subsoil with optimal drainage conditions. It has a conventional and an organic section. The organic section has a six-year crop rotation including one and a half years of a grass-clover ley. A data base of all available data of the six fields during ten years has been used for this study. The data have been used as an input for the NDICEA model (Koopmans & Bokhorst, 2000). This two-layer model, with a time-step of one week, integrates sub-models for the mineralisation of soil-bound organic matter, water-balance and crop growth (Habets & Oomen, 1993). The validation of the model has been described by Koopmans & Bokhorst (2002).

The site-specific calibration of the model has been made for all six fields separately. The resulting calibrated soil parameters of the six fields showed slight differences only, except that for denitrification (data not shown). This supports the assumption of homogeneous soil conditions over the six fields. For the analysis of the N dynamics, the average of the six field-specific soil data sets were used as well as the average agronomic data (sowing date, harvest date, yield, manure etc). In the same way, the average data of the conventional section of the OBS farm have been used to compare the two systems. In the NDICEA model every organic matter input is treated separately in the calculation procedure, so the N dynamics can be studied in detail. Because of this, 'virtually labelled' manure or fertiliser N can been followed in both systems throughout a complete rotation cycle.

Results and discussion

Soil organic matter content and N mineralisation from organic matter are greater in the organic section. The average crop N uptake is 196 and 190 kg ha^{-1} yr^{-1} for the organic and the conventional section, respectively. Total N uptake (including green manure) is 247 and 238 kg ha^{-1} yr^{-1}, respectively, and 122 and 138 kg ha^{-1} yr^{-1} is taken out of the systems in products. These data do not differ much between the two sections. The reason for this is that the generally lower yields in the organic section are compensated by the high N uptake of peas and clover.

The N input of both systems is completely different. The input in the organic section is 98 kg ha^{-1} yr^{-1} by manure application and 62 kg by N$_2$ fixation (clover, pea). For the conventional section this is 140 kg ha^{-1} yr^{-1}, all as fertilisers. N efficiency, calculated as the ratio between N output (by products) and input (by manure and fertiliser), is 1.24 and 0.98 for organic and conventional, respectively. If N$_2$ fixation and N deposition (29 kg ha^{-1} yr^{-1}) are taken into account as input, the N efficiency is 0.64 (organic) and 0.81 (conventional). The experiment with the 'virtually labelled' N shows that in the organic system about 33% of the initial N application is still in the system after a six year rotation (Figure 1), compared with 15% in the conventional system after a four year rotation (not shown).

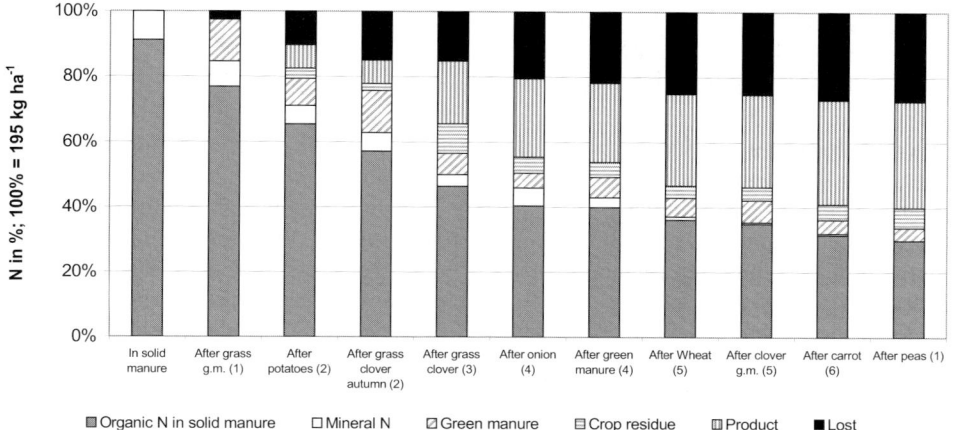

Figure 1. Modelled fate of N (%) during crop rotations under organic management in summer after peas and before grass (green manure) farmyard manure is applied (195 kg N): numbers in brackets on the horizontal axis are the year-numbers of the crop in the rotation.

Conclusions

The conventional section has a higher N efficiency, but both sections meet the legal requirements of maximum N loss. The organic section has more rapid internal N dynamics and an internal N supply (fixation) with a higher level of soil organic matter and a delayed N release. The delayed N release and more rapid internal N dynamics might support yield stability and suppression of soil-born diseases, which could be the subject for further research. The relatively high N efficiency is a combined result of good yields, use of green manures, adequate timing of agronomic activities and a good soil structure.

References

Habets, A.S.J. and Oomen, G.J.M. (1993). Modelling nitrogen dynamics in crop rotations in ecological agriculture. In: J.J. Neeteson and J. Hassink (eds), Nitrogen mineralization in agricultural soils: Proceedings symposium at the Institute for Soil Fertility Research, Haren NL, 19-20 April 1993. AB-DLO, Haren, The Netherlands. P. 255-268.

Koopmans, C.J. and Bokhorst, J. (2000). Optimising organic farming systems: nitrogen dynamics and long-term soil fertility in arable and vegetable production systems in the Netherlands. In: T. Alföldi, W. Lockeretz and U. Nigli (eds.). IFOM-2000 - The World Grows Organic. Proceedings of the 13th International IFOAM Scientific Conference, Basel, 20 to 31 August 2000. Hochschulverlag AG an der ETH Zürich. P. 69-72.

Koopmans, C.J. and Bokhorst, J. (2002). Nitrogen mineralization in organic farming systems: a test of the NDICEA model. Agronomie, 22, 855-862.

Microbial biomass measurements in soil of the central highlands of Mexico

M.S. Vásquez-Murrieta and L. Dendooven
Laboratory of Soil Ecology, Department of Biotechnology and Bioengineering, CINVESTAV, Mexico City

Introduction

Soils in the central highlands of Mexico are extremely important for agricultural production, so a drop in soil fertility would have an important impact on crop production. Microbial biomass is the key factor in the dynamics of nutrients in soil, but little information exists for the soils of Mexico. We sampled soils in the region, with a gradient of organic C and belonging to the three most important soil texture classes, and measured microbial biomass C with the chloroform fumigation-incubation method (CFIM) (Jenkinson & Powlson, 1976) and related this to the release of ninhydrin-positive compounds (NPC) (Amato & Ladd, 1988) and organic C released by $CHCl_3$.

Materials and methods

Twenty-four sites in the State of Guanajuato (Mexico) (ten clayey, five clay loamy, nine sandy clay loam soils) were sampled. Three sub-samples of 25 g of each site and plot were extracted with 100 ml $0.5M$ K_2SO_4 and compared with the unfumigated soil. Three sub-samples of 25 g soil were fumigated for 24 and 240 h (Sparling et al., 1993) with ethanol-free chloroform after 1 day and compared with that released after 10 days, ($CHCl_3$) in the dark at $22^{\circ}C$ and extracted with $0.5M$ K_2SO_4 (Ec) (Vance et al., 1987; Joergensen & Brookes, 1990). The CFI technique was according to the original method of Jenkinson & Powlson (1976).

Results and discussion

The soils from this area ranged in organic matter C content from 4.0 g to 24.5 g C kg^{-1} soil and pH from 6.2 to 7.8 (Table 1). The soil microbial biomass C (SMBC), as measured with the CFI method, ranged from 138 to 2195 mg C kg^{-1}. The release of NPC and extractable C increased with the time of exposure to $CHCl_3$ (Table 2). The ratio of microbial biomass C related to the NPC was 32.5 after 1 day and 22.1 after 10 days, while the relationship with extractable C was 2.3 and 1.9, respectively. The efficiency of the extraction for organic C or the k_{EC} value (extractable C after fumigation/microbial biomass C) ranged from 0.21 to 0.86 (mean 0.47). The ratio E_C/F_C value (extractable C after fumigation/flush of CO_2) ranged from 0.47 to 1.90 (mean 1.10). We found that relationships between the SMBC, as measured by the CFI, and the Ec and NPC extractable after 1 and 10 days of $CHCl_3$ fumigation (for agricultural soils of the central highlands of Mexico), were comparable to values reported for soils in other regions of the world. The values reported here could be used to calculate SMBC from the NPC, or Ec released in a fumigation of one day without the need for a fumigation incubation of 10 days.

Table 1. Characteristics of soils.

Textural classification	pHH$_2$0	WHC [a]	Carbon (g kg^{-1} soil)		Total Nitrogen (g kg^{-1} soil)
			Organic	Inorganic	
Clay	6.2 - 7.5	550 - 1130	6.8 - 21.8	0.4 - 1.1	0.5 - 1.4
Clay loam	7.0 - 7.4	590 - 870	6.3 - 18.9	0.4 - 1.1	0.5 - 2.6
Sandy clay loam	6.6 - 7.8	410 - 1320	4.0 - 24.5	0.4 - 3.8	0.5 - 3.1

[a] WHC : Water holding capacity

Table 2. Production of CO_2, microbial biomass C, and Ninhydrin-N and soluble-C (mg C kg^{-1}) and the proportion (%) released after 24 or 240 h.

Textural classification	Ninhydrin-N			Soluble-C			CO_2 production (mg C kg^{-1})		Microbial
	24h	240h	%	24 h	240 h	%	NF [a]	FU [b]	C[c] (mg C kg^{-1})
Clay	2-12	5-24	18-100	66-194	100-247	27-100	22-136	98-282	138-442
Clay loam	4-20	12-32	36-100	87-186	103-228	55-100	35-84	154-190	224-335
Sandy clay loam	2-62	6-82	10-75	63-467	77-541	63-100	25-442	110-1347	161-2195

[a] CO_2 produced from unfumigated soil.

[b] CO_2 produced from fumigated soil.

[c] Microbial C = (CO_2 FU soil - CO_2 from NF soil)/0.45 (Jenkinson, 1988).

Acknowledgements

We thank I. Migueles-Garduño and A. Meza-Serrano for technical support, INIFAP for access to the experimental sites. The research was funded by the department of Biotechnology and Bioengineering and CONACyT, Mexico.

References

Amato, M. and Ladd, J.N. (1988). Assay for microbial biomass based on ninhydrin-reactive nitrogen in extracts of fumigated soils. Soil Biology and Biochemistry, 20, 107-114.

Jenkinson, D.S. and Powlson, D.S. (1976). The effects of biocidal treatments on metabolism in soil. V. A method for measuring soil biomass. Soil Biology and Biochemistry, 8, 209-213.

Joergensen, R.G. and Brookes, P.C. (1990). Ninhydrin-reactive nitrogen measurements of microbial biomass in 0.5 K$_2$SO$_4$ soil extracts. Soil Biology and Biochemistry, 22, 1023-1027.

Sparling, G.P., Gupta, V.V.S.R. and Zhu C. (1993). Release of ninhydrin-reactive compounds during fumigation of soil to estimate microbial C and N. Soil Biology and Biochemistry, 25, 1803-1805.

Vance, E.D., Brookes, P.C. and Jenkinson, D.S. (1987). Microbial biomass measurements in forest soils: determination of k$_C$ values and tests of hypotheses to explain the failure of the chloroform fumigation-incubation method in acid soils. Soil Biology and Biochemistry, 19, 689-696.

Natural ^{15}N abundance as an indicator of the effect of management intensity on nitrogen cycling in montane grasslands

M. Watzka and W. Wanek
Institute of Ecology and Conservation Biology, University of Vienna, Althanstr. 14, A-1090 Vienna, Austria

Introduction

Landscape patterns of different grassland management practices have changed widely in the last few decades in the montane regions of Austria which are dominated by dairy farming (Pötsch, 1998). Trends to either management intensification or to abandonment of grassland sites, as well as to shifting from conventional to organic farming, are of concern with respect to their effect on N cycling. As the natural abundance of the stable isotope ^{15}N in different compartments of ecosystems can be used as an integrator of N cycle processes (Robinson, 2001), we hypothesised that patterns of δ^{15}N in grassland soil and plants are related to management intensity and could provide information on turnover, accumulation and loss of N.

Materials and methods

Investigations were conducted on 8-50 year old mown permanent grassland plots located at the Federal Research Institute for Agriculture in Alpine Regions (BAL) in Gumpenstein (700m asl), Austria. The 36 treatments (3 replicates each) differed in the amount and quality of applied fertiliser (0-200 kg N ha^{-1}yr^{-1}; mineral fertiliser (Nitramoncal), cattle slurry, stable manure) and/or cutting frequency (1-6 times per season). Samples of fertiliser, soil from different depths, hay and aftermath were analysed for N content and δ^{15}N. Pools of mineral N in surface soils were measured twice and gross mineralisation, N$_2$O emission and transfer of mineral N in the soil profile were determined once in the growing season 2001.

Results and discussion

Fertiliser quality and the amount of fertiliser input were the most important management factors influencing δ^{15}N patterns in the investigated grassland ecosystems. Higher δ^{15}N values of organic fertilisers (δ^{15}N=6-11‰) compared with mineral fertiliser (δ^{15}N=0‰) were reflected by higher δ^{15}N values in soil and harvested plant material. No direct influence of cutting frequency on δ^{15}N was detected. The δ^{15}N signature of surface soils and plants increased with the amount of applied fertiliser-N, independent of fertiliser quality. Correlation between δ^{15}N of harvested plant biomass and surface soil (δ^{15}N$_{plants}$=1.023, ^{15}N$_{soil}$-3.72; r^2=0.87**) indicated that fertiliser N, as well as a considerable portion of soil N, were subject to fast turnover processes, leading to an 'isotopic equilibration' of N available for plants with total soil N, irrespective of fertilisation regime. Taking fertiliser quality into account, we found positive correlations between δ^{15}N in surface soil and the N-balance calculated from known N inputs and output in harvest (Figure 1). For organic fertilised plots these correlations may be attributed partly to N accumulation in soil. However, at least in mineral fertiliser plots, soil $\delta^{15}N$ also indicated that processes which discriminate against ^{15}N (e.g. nitrification, denitrification, NH_3 volatilisation) were stimulated by an increased supply of readily available N, leading to loss of the lighter, and accumulation of the heavier N isotope in soil. These findings were supported by higher NO_3^- concentrations and a tendency to higher transport rates of NO_3^- in the soil profile at fertilisation levels above 100 kg N ha^{-1}yr^{-1}.

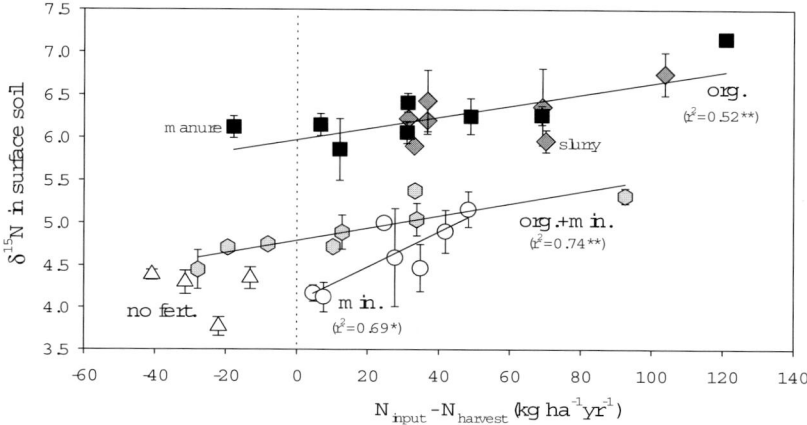

Figure 1. Correlations of $\delta^{15}N$ in surface soil (0-5cm) with N balance calculated from known inputs (fertiliser + atmospheric deposition + symbiotic N_2 fixation) and output in harvest (org.: organic fertiliser, min.: mineral fertiliser, no fert.: no fertiliser applied).

Conclusions

As the natural ^{15}N abundance in soil and vegetation of montane grasslands reflected the type of applied fertiliser, it may be used as an indicator for the medium term fertilisation regime of meadows. The strong correlation between $\delta^{15}N$ values of plants and soils points to fast incorporation of fertiliser N into the soil N pool and to involvement of a considerable portion of total soil N in turnover. This confirms previous results of rapid N turnover in managed grassland (Ledgard et al., 1998). Taking fertiliser quality into account, soil $\delta^{15}N$ was correlated with input-output balances of investigated plots, thus indicating different rates of N accumulation and/or losses. Incorporation of $\delta^{15}N$ patterns into a N cycle model should make it possible to use them as an easily measureable integrator of the N balance of differently managed permanent meadows in the investigated region.

Acknowledgements

We wish to thank the 'Federal Research Institute for Agriculture in Alpine Regions' (BAL), for the use of their long term experimental plots, and Dr. Buchgraber and his team for scientific and logistic support. This project was funded by the Austrian Academy of Sciences as part of the 'Man and Biosphere' Programme: 'Grassland in the mountain region of Austria'.

References

Ledgard, S.F., Jarvis, S.C., and Hatch D.J. (1998). Short-term nitrogen fluxes in grassland soils under different long-term nitrogen management regimes. Soil Biology and Biochemistry, 30, 1233-1241.

Pötsch, E. M. (1998). About the influence of the fertilising intensity on the N-cycle of alpine grassland. Bodenkultur, 49, 19-27.

Robinson, D (2001). $\delta^{15}N$ as an integrator of the nitrogen cycle. Trends in Ecology and Evolution, 16, 153-162.

SECTION 3
CONTROLLING LOSSES TO AIR

Controlling losses to air

E.A. Davidson[1] and A.R. Mosier[2]
[1]*The Woods Hole Research Center, Woods Hole, MA, 02543, USA*
[2]*Agricultural Research Service, Fort Collins, CO, 80521, USA*

Abstract

Agriculture contributes about two-thirds of global NH_3 emissions to the atmosphere, over one-third of N_2O emissions, and about one-fourth of nitric oxide (NO) emissions. These gases have numerous deleterious effects on the environment, including eutrophication of downwind ecosystems (NH_3 and NO), noxious odours (NH_3), production of atmospheric aerosols that reduce air quality (NH_3 and NO), tropospheric ozone (O_3) production (NO), stratospheric O_3 depletion (N_2O), and global warming (N_2O and tropospheric O_3 derived from reactions involving NO). Because nitrogen (N) amendments are commonly needed for optimal agricultural productivity, some leakage of applied N out of agricultural fields and into aquatic and atmospheric reservoirs is probably inevitable. The magnitude of that leakage, however, can be controlled by sound management practices. Our knowledge of the processes of gaseous production and transport in soils should allow us to minimise this leakage, thereby protecting air and water quality while maintaining food security. We contend that, although new technological developments in fertiliser products would be helpful, sufficient understanding of the processes of gaseous N losses already exists to allow substantial decreases in N losses. Rather than technological and scientific know-how, the most important limiting factors to progress in this area appear to be socio-economic mechanisms to provide farmers with incentives and/or requirements to reduce emissions.

Keywords: ammonia, denitrification, nitric oxide, nitrification, nitrous oxide

Introduction

During the last few decades, the introduction of reactive N into the biosphere by food and energy production has been greater than rates of N fixation in native terrestrial ecosystems (Table 1), and this anthropogenic input has been steadily increasing (Galloway & Cowling, 2002). Increased inputs of N to the atmosphere can increase tropospheric O_3 formation, reduce atmospheric visibility, and increase acid deposition (Driscoll et al., 2003). Increased N deposition can acidify soils, streams, and lakes and can alter forest productivity (Matson et al., 2002). Increased inputs of N to aquatic ecosystems from atmospheric deposition, sewage, and agricultural runoff can cause eutrophication, including damage to fisheries in coastal ecosystems (Rabalais, 2002). The formation of N_2O during nitrification and denitrification in all systems results in tropospheric warming and stratospheric O_3 depletion (Prather et al., 2001). These undesirable 'cascading effects', as Galloway et al. (2003) call them, of reactive N moving through aquatic and terrestrial ecosystems and the atmosphere, do not stop until the reactive N is eventually converted back to N_2 through the process of denitrification.

The objectives of this paper are: (1) to provide an overview of the magnitude of the problem of gaseous N losses from agriculture; and (2) to identify current process-level understanding that can be used operationally to reduce gaseous emissions.

Table 1. Budget of current (1990) inputs of N to the biosphere (compiled from data in Galloway & Cowling, 2002).

	10^{12} g N yr^{-1}
Natural terrestrial ecosystems	
Biological N fixation	90
Lightning	5
Total	95
Human-controlled inputs	
Haber-Bosch N fertiliser & industry	85
Biological N fixation in agriculture	33
Combustion in industry and transportation	21
Total	140
Natural N fixation in oceans	140

Overview of global gaseous N budgets

Ammonia

The largest term in the global NH_3 emissions budget is excreta from domestic animals, followed by direct emissions from agricultural soils (Table 2). Ammonia emissions from biomass burning, much of which is related to tropical slash-and-burn agriculture, are also important. These three primarily agricultural sources account for about two-thirds of the global total. While there is some indication of a levelling-off of NH_3 emissions from agricultural soils, emissions from animal feed operations continue to increase globally because of increasing demand for meat in human diets in many developed and developing nations. However, trends of increased N use efficiency are expected to partially offset the expected growth in demand for meat, thus causing NH_3 emissions to increase less rapidly (Bouwman & Van der Hoek, 1997). Ammonia is usually deposited within a few kilometres to perhaps hundreds of kilometres downwind of the source, so local and regional budgets may be more relevant than global budgets. Local N emissions may be dominated by NH_3 near industrial-scale, intensive animal feeding operations.

Nitrous oxide

Natural sources of N_2O from oceans and non-agricultural soils contribute a little more than half of the total global emissions (Table 2). Agriculture contributes about 80% of anthropogenic emissions of N_2O and nearly 40% of total global emissions. Direct emissions from soils, produced by soil nitrifying and denitrifying microorganisms, are the dominant agricultural sources. Unlike NH_3 and NO, N_2O is stable in the troposphere, has a mean residence time of >100 years in the atmosphere (Prather et al., 2001), and is accumulating in the atmosphere. Hence the heat-trapping and stratospheric O_3-depleting implications of N_2O emitted by modern agriculture will be borne by several future generations.

Nitric oxide

Agriculture contributes ca. 25% (±10%) of total global NO emissions (Table 2). Industrial and transportation sectors are more important globally and in industrial and densely populated regions of developed countries. However, emissions of NO from agricultural soils can be locally and regionally important for photochemical production of O_3 in rural areas (Davidson et al., 1998; Hall et al., 1996). Biomass burning can be a very significant local and regional source of NO, particularly where fire is used as an agricultural management tool in many tropical countries.

Table 2. Global budgets for the 1990s of NH$_3$, N$_2$O , and NO emissions to the atmosphere in Tg N yr^{-1} (1Tg = 10^{12}g), compiled from data in Bouwman & Van der Hoek (1997), Davidson & Kingerlee (1997), Holland et al. (1999) and Prather et al. (2001). Best estimates to no more than two significant places are followed by ranges of reported literature estimates in parentheses.

Source	NH$_3$	N$_2$O	NO
Ocean	8.2 (7-13)	3.0 (1-5)	
Atmospheric NH$_3$ oxidation		0.6 (0.3-1)	2 (0.5-3)
Tropical soils			
forests		3.0 (2-4)	0.4 (0.3-0.9)
savannas		1.0 (0.5-2)	4.6 (1.8-4.6)
Temperate soils			
forest/woodland/chaparral		1.0 (0.1-2)	3.1 (0.1-3.1)
grasslands		1.0 (0.5-2)	0.7 (0.4-0.7)
All soils with natural vegetation	2.4 (2-5)		9 (3-9)
Lightning			5 (3-25)
Natural subtotal	*11 (9-18)*	*10 (5-16)*	*16 (7-37)*
Agricultural soils	9.0 (6-9)	4.2 (1-15)	3.9 (2-4)
Crops and crop decomposition	3.6 (3-5)		
Biomass burning	5.9 (2-6)	0.5 (0.2-1)	6.0 (3 - 13)
Industrial/transportation	0.3 (0.1-2)	1.3 (1-2)	21 (20 - 25)
Cattle and feedlots	22 (22-32)	2.1 (1-3)	
Human and pet excreta	2.6 (2-4)		
Anthropogenic subtotal	*44 (34-58)*	*8 (2-21)*	*30 (25-42)*
Total Sources	54 (45-75)	18 (7-37)	46 (32-79)
Stratospheric sink		12 (9-16)	
Atmospheric accumulation		4 (3-5)	
Implied source		16	

Process-level controls on gaseous emissions of N

Emissions of NO and N$_2$O from soil have been conceptualised as leaks of N flowing through metaphorical pipes of nitrification and denitrification (Fig. 1): this concept has been adopted as a theme for this conference. Leaks in the pipe are significant only when there is ample N flowing through the pipe, as occurs in naturally fertile soils or soils that have received significant N inputs from atmospheric deposition or from fertilisers. Hence, emissions of either gas tend to be low where inputs of N or rates of N mineralisation are low relative to plant demand (Davidson et al., 2000). The relative sizes of the holes in the pipes illustrate the importance of environmental factors such as soil water content, acidity, and ratios of electron donors to electron acceptors (Firestone & Davidson, 1989). Soil water content is a particularly strong controller of the sizes of the 'holes in the pipe' by regulating the redox conditions of the soil (Davidson et al., 2000; Davidson & Verchot, 2000). Under well-drained, aerobic soil conditions, NO is the dominant gas, usually from nitrification. Under wetter conditions, where anaerobic microsites become increasingly abundant, N$_2$O from nitrification and denitrification becomes the dominant gas. Under very wet conditions, with little O_2 diffusion, N_2 from denitrification is the dominant end product. This conceptual model is simplistic in that it ignores heterotrophic nitrification and details of whether autotrophic nitrifiers produce these gases from oxidative or reductive pathways (Firestone & Davidson, 1989). Nevertheless, this model captures the two most important regulating factors that distinguish conditions for large and small fluxes of NO and N_2O. First, reducing overall emissions requires that N availability be managed so that it seldom exceeds plant demand (e.g. less N entering the nitrification and denitrification 'pipes'). Avoiding emissions trading (e.g. decreasing N_2O emissions by reducing excess water, but thereby increasing NO emissions) requires an understanding of the dominant factors affecting the ratios of the gaseous products (the sizes of the holes in the pipes).

Ammonia emissions are not presented as leaks from a pipe in the revised 'hole-in-the-pipe' (Fig. 1), because NH_3 is not a by-product that 'leaks' out of an enzymatic process that is less than 100% efficient, as is the case for NO and N_2O emissions from nitrification and denitrification. Rather, the $NH_3:NH_4^+$ equilibrium is a pH-dependent, abiological process. Indeed, NH_3 can be the dominant fate of NH_4^+ if the soil pH is sufficiently high. Biological processes of urea hydrolysis, plant uptake of NH_4^+, and proton production and consumption certainly influence the $NH_3:NH_4^+$ equilibrium, but these interactions are best conceptualised independently of the 'hole-in-the-pipe' metaphor. Reducing NH_3 emissions from soils, composts, and animal wastes requires understanding of the factors regulating NH_4^+ production and consumption and soil acidity. Sommer et al. (2003) provide an up to date review of the processes involved.

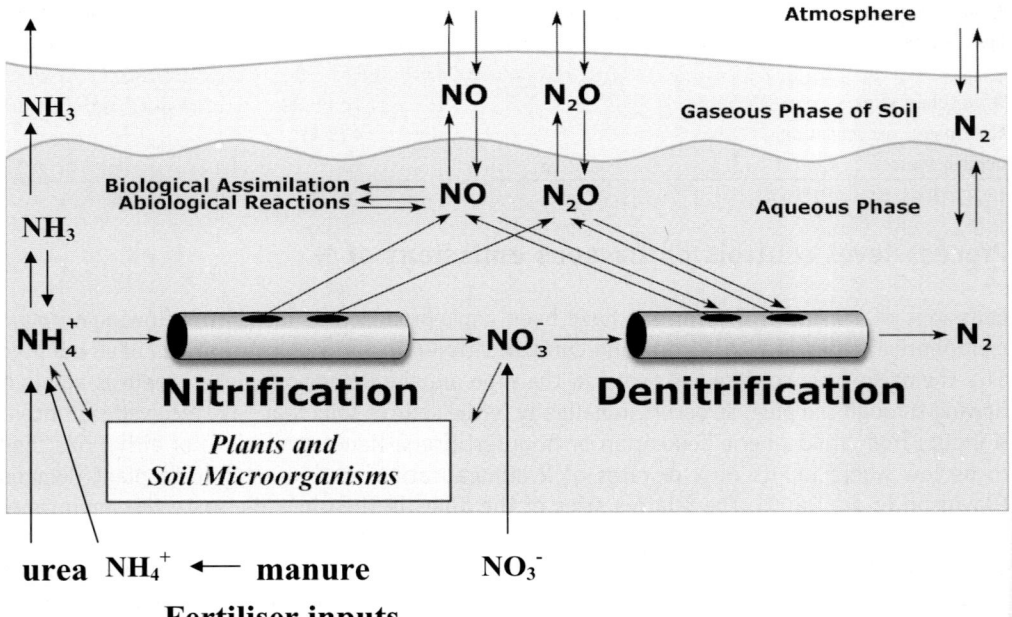

Figure 1. The 'Hole-in-the-Pipe' conceptual model from Firestone & Davidson (1989) and from Davidson et al. (2000), modified to include NH_3 emissions and fertiliser inputs.

The depth of production and the gaseous diffusivity of the soil between the site of production and the soil surface are important for all three of these gases. When diffusivity is low and/or the diffusion path is relatively long, the probability of efflux is lower and the probability increases that $NH_3/NH_4^+/NO_3^-$ immobilisation can first occur, or that NO and N_2O can be reduced to N_2. This conceptual understanding can be applied to the placement of N fertilisers, thus explaining why emissions are often higher following surface applications and lower when fertiliser-N is worked into deeper soil layers by various means (Peoples et al., 1995).

The conceptual basis of the 'hole-in-the-pipe model' has been codified into numerical models, including the well-known CASA (Potter et al., 1996), CENTURY (Del Grosso et al., 2000), and DNDC (Li et al., 1992; Li, 2000) models. Of course, the models also include other important ecosystem functions, and they vary in their complexity and their applicability to specific agronomic systems.

Emission factors

Another approach to estimating N gas emissions from agricultural soils is the application of emission factors to specific fertiliser types. In general, fertilisers containing NO_3^- have higher N_2O emission factors than do other types (Table 3), presumably because the substrate is already available and denitrification rates can be high if there is sufficient available C and if the soils are wet. The ranges of observed emission factors are large because other factors besides fertiliser type, such as soil water content, are often important but not accounted for by emission factors. Nevertheless, for broad-brush accounting, such as that prescribed for national greenhouse gas inventories by the Intergovernmental Panel on Climate Change (IPCC, 1997; Mosier et al., 1998), the use of emission factors may be adequate and may be necessary, where other data needed to drive more sophisticated models are not widely available. As with N_2O emission factors, it is possible to broadly generalise about NH_3 emissions based on fertiliser types, although as has already been noted, numerous other factors are also important. In general, NH_3 emissions seldom exceed 10% of ammonium-based fertilisers and range from 10-40% of surface-applied urea (Peoples et al., 1995). Incorporating urea into the soil usually reduces NH_3 emissions to below 10% of the urea-N.

Table 3. Emission factors for N_2O from fertilised soil from a review of 35 studies published since 1994 (Mosier & Kroeze, 1999).

Fertiliser type	Range of emission factors (% of fertiliser-N emitted)
Ammonium nitrate	1.2-3.6
Ammonium sulfate	0.2-1.0
Calcium nitrate	5.2-12
Manure	1.0-3.2
Urea	0.2-3.8
Urea/Ammonium nitrate	0.1-6.8

Management practices to lower gaseous-N losses from agriculture

Reviews of N budgets for numerous cropping systems reveal that crops seldom take up more than 50% of applied N, and uptake of only 10-30% is common (Cassman et al., 2002, Peoples et al., 1995). The remaining 50-90% of the applied N is available for leaching to surface and ground waters and for gaseous losses. Obviously, strategies that get more of the N into the crop would reduce the amount that could potentially be lost to the aquatic biosphere and the atmosphere. Therefore, the main challenge is to match fertiliser supply with crop demand. Most of the strategies listed in Table 4 apply to minimising both gaseous and hydrologic losses of N. Perhaps the only strategy that may affect gaseous N losses more than hydrologic export is the incorporation of applied N below the soil surface. As already noted, surface applications of urea often lead to much greater losses of NH_3 (Peoples et al., 1995) and NO (Matson et al., 1996).

We understand the underlying processes sufficiently well to postulate with some confidence that each of these practices listed in Table 4 is likely to significantly lower gaseous N losses. However, we have less confidence in making quantitative estimates of how much loss will be prevented under specific field conditions. Equally importantly, estimates of the costs of implementing these practices (dollars or euros per hectare) relative to the reduction in gaseous emissions (kg N ha^{-1}) and the benefits to society (both monetary and unpriced, Davidson, 2000), remain extremely difficult to estimate. The costs of implementing these strategies could be lowered by technological developments, such as production of cheaper and more effective controlled release fertilisers and nitrification inhibitors. Regionally appropriate parameterisation of numerical models could also improve estimates of both costs, and efficacy of N loss reductions, of these proposed changes in management practices.

Table 4. Strategies to minimise N losses (summarised from Peoples et al., 1995).

Management practice	Mechanism for reducing emissions
Match timing of fertilisation with crop demand	N is applied at times and in doses that promote more efficient crop uptake, thus minimising availability of inorganic-N for nitrification, denitrification and NH_3 volatilisation.
Utilise soil testing	
Multiple applications of small fertiliser doses (e.g., fertigation)	
Precision fertiliser application	
Lower annual fertiliser application rates	
Improved management of manures	
Control release fertilisers	Delays release of inorganic-N into the nitrification "pipe," thus increasing the time for crop uptake before N loss.
Nitrification inhibitors	
Conserve residues	Immobilises N for later, more gradual mineralisation.
Minimise fallow periods	Avoids accumulation of inorganic-N when plant uptake is minimal.
Integrate animal and crop production	Minimises accumulation and concentration of animal wastes that promote ammonification, nitrification, and denitrification.
Manage soil acidity	Circumneutral conditions minimise NH_3 formation and promote N_2O reduction to N_2, although nitrification is also promoted.
Incorporate N below soil surface	Increases path length for diffusional losses of gases, thus increasing the probability of NH_3 conversion to NH_4^+ and reduction of NO and N_2O to N_2

Although advances in scientific and technological know-how would be welcome, this is not necessarily the limiting factor for implementing many of these practices in the near future. Lowering annual application rates of N fertilisers, for example, could be implemented now, if farmers had the appropriate incentives or regulatory requirements. The dose-response curves of most crops are well known (e.g. Fig. 2), and diminishing returns of crop yield for additional N application beyond an optimal rate is well characterised (Cassman et al., 2002). When additional N is applied where the yield response curve levels out, most of that additional N is lost as gaseous and hydrologic export (Howarth et al., 2002). Hence, a relatively small reduction in N application would usually have little impact on crop yield (where the curve is relatively flat), but could significantly reduce the excess N that is likely to be lost. The reason that farmers often apply somewhat more than the optimal amount of N is due to uncertainty in the amount of N that will be mineralised from the soil, which depends on climatic variation that is also difficult to predict (Cassman et al., 2002). Adding a little extra fertiliser N is an effective form of insurance for good crop yield: in case less N is mineralised from the soil organic matter than expected. If this insurance could be provided to the farmer through other sorts of agricultural policies, similar to other crop failure protection programmes, then hydrologic and gaseous N losses could be significantly reduced with relatively modest reductions in fertiliser-N use and agricultural productivity. There are also opportunities to make changes in manure management in order to reduce emissions of NH_3 and N_2O.

In some cases, farmers may also reap economic benefit from more efficient use of fertilisers.

Matson et al. (1998) demonstrated that large-scale corn farmers in Mexico could increase profits by 17% and reduce denitrification losses by 90% by changing the timing of their fertiliser application and lowering the annual application rate by about 25%. Merging the economic analysis with the N cycling measurements provided a novel demonstration of a win-win situation. Because N_2O and CO_2 are greenhouse gases, and C-trading schemes designed to reduce atmospheric accumulation of greenhouse gases are beginning to emerge, it is possible to imagine financial incentives for farmers both to conserve soil C and to avoid N_2O emissions. Greenhouse gas credits that have economic value could be awarded for improved nutrient

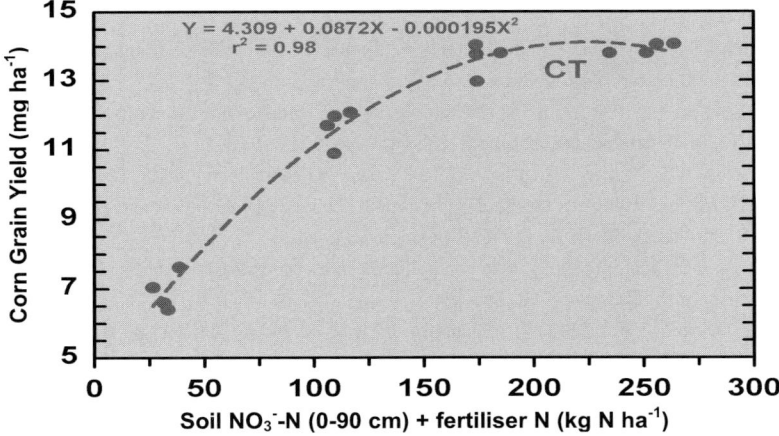

Figure 2. A typical dose-response curve for corn, adapted from Halvorson et al. (2003).

management practices. Nutrient management plans might also be a requirement for qualification for enrolment in agricultural programmes such as low-interest loans, price supports, or conservation set-asides. Effective nutrient management plans could be based on current understanding of N cycling processes, but the limiting factors to implementing such programmes are the social, economic, and political perceptions that such policies are needed and worth their costs.

Conclusions

As the cascading negative effects of excess reactive N in the atmosphere and biosphere become increasingly apparent to society, the costs of not implementing improved nutrient management practices in agriculture will also become more apparent. Where subsistence agriculture is practised in the developing world, avoiding N losses would increase much needed agricultural productivity. Although there is always room for improvement in our understanding of the basic science and in the development of technologies to address the problem of N losses, our current understanding and management tools are adequate to do a much better job at preventing losses of NH_3, N_2O, and NO. Rather than focussing on our desire to gain more knowledge, which scientists are understandably inclined to emphasise, agronomists and ecologists might best serve society at this moment by emphasising what we could do if the economic and political willpower existed to seriously address the problem of N losses from agriculture.

References

Bouwman, A.F. and Van der Hoek, K.W. (1997). Scenarios of animal waste production and fertilizer use and associated ammonia emission for the developing countries. Atmospheric Environment, 31, 4095-4102.

Cassman, K.G., Dobermann, A. and Walters, D.T. (2002). Agroecosystems, nitrogen-use efficiency, and nitrogen management. Ambio 31, 132-140.

Davidson, E.A. (2000). You Can't Eat GNP. Perseus Publishing, Cambridge, Massachusetts.

Davidson, E.A., Keller, M., Erickson, H.E., Verchot, L.V. and Veldkamp, E. (2000). Testing a conceptual model of soil emissions of nitrous and nitric oxide. BioScience, 50, 667-680.

Davidson, E.A. and Kingerlee, W. (1997). A global inventory of nitric oxide emissions from soils. Nutrient Cycling in Agroecosystems, 48, 37-50.

Davidson, E.A., Potter, C.S., Schlesinger, P. and Klooster, S.A. (1998). Model estimates of regional nitric oxide emissions from soils of the southeastern United States. Ecological Applications, 8, 748-759.

Davidson, E.A. and Verchot, L.V. (2000). Testing the hole-in-the-pipe model of nitric and nitrous oxide emissions from soils using the TRAGNET database. Global Biogeochemical Cycles, 14, 1035-1043.

Del Grosso, S.J., Parton, W.J., Mosier, A.R., Ojima, D.S., Kulmala, A.E. and Phongpan, S. (2000). General model for N_2O and N_2 gas emissions from soils due to dentirification. Global Biogeochemical Cycles, 14, 1045-1060.

Driscoll, C.T., Whitall, D., Aber, J., Boyer, E., Castro, M., Cronon, C., Goodale, C.L., Groffman, P., Hopkinson, C., Lambert, K., Lawrence, G. and Ollinger, S. (2003). Nitrogen pollution in the north eastern United States: Sources, effects, and management options. Bioscience, 53, 357-374.

Firestone, M.K. and Davidson, E.A. (1989). Microbiological basis of NO and N_2O production and consumption in soil. In: M.O. Andreae and D.S. Schimel (eds). Exchange of trace gases between terrestrial ecosystems and the atmosphere. John Wiley & Sons, New York, pp. 7-21.

Galloway, J.N. and Cowling, E.B. (2002). Reactive nitrogen and the world: 200 years of change. Ambio, 31, 64-71.

Galloway, J.N., Aber, J.D., Erisman, J.W., Seitzinger, S.P., Howarth, R.W., Cowling, E.B. and Cosby, B.J. (2003). The nitrogen cascade. BioScience, 53, 341-356.

Hall, S.J., Matson, P.A. and Roth, P.M. (1996). NO_x emissions from soil: Implications for air quality modeling in agricultural regions. Annual Review of Energy and the Environment, 21, 311-346.

Halvorson, A.D., Mosier, A.R., Reule, C.A. (2003). Irrigated crop management effects on productivity, soil nitrogen and soil carbon. Proceedings of 2003 Fertilizer Industry Round Table, October 28-30, 2003, Winston-Salem, North Carolina.

Holland, E.A., Dentener, F.J., Braswell, B.H. and Sulzman, J.M. (1999). Contemporary and pre-industrial global reactive nitrogen budgets. Biogeochemistry, 46, 7-43.

Howarth, R.W., Boyer, E.W., Pablich, W.J. and Galloway, J.N. (2002). Nitrogen used in the United States from 1961-2000 and potential future trends. Ambio, 31, 88-96.

IPCC. (1997). Houghton, J.T., Meria Filho, L.G., Lim, B., Trennton, K., Mamaty, I., Bonduki, Y., Griggts, D.J. and Callander, B.A. (eds). Revised 1996 Intergovernmental Panel on Climate Change (IPCC) Guidelines for National Greenhouse Gas Inventories (Vol. 1-3).

Li, C. (2000). Modelling trace gas emissions from agricultural ecosystems. Nutrient Cycling in Agroecosystems, 58, 259-276.

Li, C., Frolking, S. and Frolking, T.A. (1992). A model of nitrous oxide evolution from soil driven by rainfall events: 1. Model structure and sensitivity. Journal of Geophysical Research, 97, 9759-9776.

Matson P.A., Billow, C. and Zachariassen, J. (1996). Fertilization practices and soil variations control nitrogen oxide emissions from tropical sugar cane. Journal of Geophysical Research, 101, 18533-18545.

Matson, P.A., Lohse, K.A. and Hall, S.J. (2002). The globalization of nitrogen deposition: consequences for terrestrial ecosystems. Ambio, 31, 113-119.

Matson P.A., Taylor, R. and Ortiz-Monasterio, I. (1998). Integration of environmental, agronomic, and economic aspects of fertilizer management. Science, 280, 112-115.

Mosier, A.R. and Kroeze, C. (1999). Contribution of agroecosystems to the global atmospheric N_2O budget. In Proceedings of the International N_2O Workshop on Reducing Nitrous Oxide Emissions from Agroecosystems. R.L. Desjardins, J.C. Keng and K. Haugen-Kozyra (eds). Banff, Alberta, Canada, March 3-5, 1999, pp. 3-15.

Mosier, A.R., Kroeze, C., Nevison, C., Oenema, O., Seitzinger, S. and van Cleemput, O. (1998). Closing the global N_2O budget: nitrous oxide emissions through the agricultural nitrogen cycle. Nutrient Cycling in Agroecosystems, 52, 225-248.

Peoples, M.B., Freney, J.R. and Mosier, A.R. (1995). Minimizing gaseous losses of nitrogen. In: P.E. Bacon (ed). Nitrogen fertilization in the environment. Marcel Dekker, Inc., New York, pp 565-602.

Potter, C.S., Matson, P.A., Vitousek, P.M. and Davidson, E.A. (1996). Process modeling of controls on nitrogen trace gas emissions from soils world-wide. Journal of Geophysical Research, 101, 1361-1377.

Prather, M., Ehhalt, D., Dentener, F., Derwent, R., Dlugokencky, E., Holland, E., Isaksen, I., Katima, J., Kirchhoff, P., Matson, P., Midgley, P. and Wang, M. (2001). Atmospheric Chemistry and Greenhouse Gases. In: J.T. Houghton, Y. Ding, D.J. Griggs, M. Noguer, P.J. van der Linden, X. Dai, K. Maskell and C.A. Johnson (eds). Climate Change 2001: The Scientific Basis. Contribution of Working Group I to the Third Assessment Report of the Intergovernmental Panel on Climate Change. Cambridge University Press, New York, p 240-287.

Rabalais, N.N. (2002). Nitrogen in aquatic ecosystems. Ambio, 31, 102-112.

Sommer, S.G., Géneremont, S., Colllier, P., Hutchings, N.J., Olesen, J.E. and Morvan, T. (2003). Processes controlling ammonia emission from livestock slurry in the field. European Journal of Agronomy, 19, 465-486.

Pitfalls in measuring nitrous oxide production by nitrifiers

N. Wrage[1], G.L. Velthof[2], H.L. Laanbroek[3] and O. Oenema[1,2]

[1]*Wageningen University and Research Centre, Soil Quality, PO Box 8005, 6700 EC Wageningen, The Netherlands*
[2]*Wageningen University and Research Centre, Alterra, PO Box 47, 6700 AA Wageningen, The Netherlands*
[3]*NIOO Centre of Limnology, Rijksstraatweg 6, Nieuwersluis, The Netherlands*

Abstract

Nitrifiers and denitrifiers are the most important sources of the greenhouse gas N_2O. Nitrifiers produce N_2O in nitrification during the oxidation of hydroxylamine (NH_2OH) and in nitrifier denitrification by reducing NO_2^- via N_2O to N_2. The experimental method for differentiation between nitrification, nitrifier denitrification, denitrification and other soil sources of N_2O is based on incubations with combinations of acetylene (C_2H_2) and O_2. Small partial pressures of C_2H_2 (0.02 kPa) should inhibit nitrification and nitrifier denitrification, and large partial pressures of O_2 (100 kPa) nitrifier denitrification and denitrification. In a survey of N_2O production in four soils under a range of conditions, the inhibitors sometimes increased the N_2O production, resulting in the calculation of negative fluxes, especially for nitrifier denitrification. Pure culture experiments revealed that small partial pressures of C_2H_2 did not inhibit the N_2O production by the nitrifier *Nitrosospira briensis*. Large concentrations of O_2 partly inhibited NH_3 oxidation in pure cultures. A sensitivity analysis showed that nitrifiers have probably been underestimated as sources of N_2O in studies using C_2H_2 or both C_2H_2 and O_2 as inhibitors. This is presumably due to i) inhibition of the N_2O reductase of denitrifiers even at low concentrations of C_2H_2, ii) only partial inhibition of nitrification and nitrifier denitrification by C_2H_2, and iii) partial inhibition of nitrification by large concentrations of O_2. Evidently, the quantification of specific sources of N_2O in soils remains a challenge for further research.

Keywords: inhibitors, nitrifier denitrification, nitrous oxide

Introduction

Soils are the main source of the greenhouse gas N_2O (Bouwman, 1990). In soils, nitrifiers and denitrifiers are the principal producers of the gas (Granli & Bøckman, 1994). Nitrifiers produce N_2O by nitrification and by nitrifier denitrification. In nitrification, N_2O develops during the oxidation of hydroxylamine (NH_2OH). In nitrifier denitrification, nitrifiers reduce NO_2^- via N_2O to N_2 (Wrage et al., 2001). Not much is known about this latter pathway yet. It is supposed to be similar to denitrification, where NO_3^- or NO_2^- is reduced via NO and N_2O to N_2.

In studies of the different sources of N_2O, 0.02 kPa acetylene (C_2H_2) and 100 kPa O_2 have been used as inhibitors. Small partial pressures of C_2H_2 are assumed to inhibit nitrification and nitrifier denitrification, and large partial pressures of O_2 are assumed to inhibit nitrifier denitrification and denitrification (Robertson & Tiedje, 1987; Webster & Hopkins, 1996). To study the respective role of each pathway, soil samples are incubated with combinations of these inhibitors (Table 1).

A recent survey of the sources of N_2O in four soils under a range of conditions showed that the incubation method with C_2H_2 and O_2 was not suitable for all soils (Wrage et al., 2003a). In several cases, the addition of inhibitors increased the N_2O production. Sometimes, negative values were calculated for fluxes of N_2O, especially for N_2O from nitrifier denitrification. These problems were further studied in pure culture experiments with the NH_3 oxidisers *Nitrosomonas europaea* and *Nitrosospira briensis*. In these experiments, C_2H_2 did not inhibit the N_2O production by *N. briensis* (Wrage et al., 2003b). *Nitrosospira* species are the prominent NH_3 oxidisers in most soils (Kowalchuk & Stephen, 2001). Since C_2H_2 is used frequently to differentiate between nitrification and denitrification (e.g. Kester et al., 1997, Ambus, 1998, Priemé & Christensen, 2001), an insensitivity of some *Nitrosospira* species to C_2H_2 could have large consequences for our understanding of N_2O production in soils. The pure culture experiments also showed that large concentrations of O_2 (100 kPa) might inhibit NH_3 oxidation (Wrage et al., 2003b). An additional study with mutants of *N. europaea* that were deficient in nitrite reductase (NirK, Beaumont et al., 2002) or nitric oxide reductase (NORB, Beaumont, pers.comm.) was carried out. NirK and NORB are two enzymes needed for the nitrifier denitrification pathway. A negative effect of 100 kPa O_2 on NH_3 oxidation was in this study (Wrage, 2003).

Table 1. Effects of inhibitors on pathways of N_2O production in soil (+: can take place; -: is blocked). After Webster & Hopkins, (1996).

Affected process	Control	with 0.02 kPa C_2H_2	with 100 kPa O_2	with 0.02 kPa C_2H_2 in 100 kPa O_2
Nitrification	+	-	+	-
Nitrifier denitrification	+	-	-	-
Denitrification	+	+	-	-
Other sources[1]	+	+	+	+

[1] Other sources are, for example, chemodenitrification or heterotrophic nitrification.

What are the implications of such problems with the inhibition of sources of N_2O by C_2H_2 and O_2 for our understanding of the pathways of N_2O production in soils? Have sources of N_2O been systematically over- or underestimated? We used a sensitivity analysis to estimate the impact of different inhibition problems on investigations of soil sources of N_2O.

Methods

For the sensitivity analysis, we carried out the following four steps. First of all, possible artifacts caused by 0.02 kPa C_2H_2 and 100 kPa O_2 were compiled. The effect of these artifacts was then estimated by using on expert judgement and literature. Next, hypothetical soils were defined, where the N_2O production was dominated by different, well defined sources. Finally, the effects of the artifacts on sources of N_2O production in these soils were investigated. These four steps will now be explained in more detail.

We considered the following possibilities for the effect of 0.02 kPa C_2H_2 and 100 kPa O_2 in pathways of N_2O production in soils:

a. the inhibitors act as expected with respect to repression of nitrification, nitrifier denitrification and denitrification;
b. NH_3 oxidation is partially inhibited by 100 kPa O_2;
c. nitrifier denitrification is incompletely suppressed by 100 kPa O_2;
d. denitrification is incompletely suppressed by 100 kPa O_2;
e. nitrification and nitrifier denitrification are incompletely inhibited by 0.02 kPa C_2H_2;
f. N_2O reductase in denitrifiers is partially inhibited by 0.02 kPa C_2H_2;
g. a combination of partial inhibition of NH_3 oxidation by 100 kPa O_2, incomplete suppression of nitrification and nitrifier denitrification by 0.02 kPa C_2H_2, and inhibition of N_2O reductase of denitrifiers by 0.02 kPa C_2H_2.

For a partial inhibition of NH_3 oxidation by 100 kPa O_2 (possibility b), it is assumed that one third of the 'normal' N_2O production by nitrification was suppressed by O_2. In pure culture studies, inhibition of NH_3 oxidation by such large O_2 concentrations could not be excluded, but did not lead to significant effects (Wrage et al., 2003b). In studies with NirK- and NORB-deficient *N. europaea*, 100 kPa O_2 caused a 20-75% reduction of NH_3 oxidation (Wrage, 2003). In cases of incomplete suppression (possibilities c, d and e), it was assumed that the inhibitor did not stop the N_2O production by the pathway in question, but only inhibited one third of it. In pure culture studies, 0.02 kPa C_2H_2 did not affect the N_2O production by *N. briensis* (Wrage et al., 2003b). There are no data yet that allow us to quantify a possibly incomplete suppression of nitrifier denitrification and denitrification by large partial pressures of O_2 (possibilities c and d).

Partial inhibition of the N_2O reductase of denitrifiers by C_2H_2 (possibility f) was assumed to increase the N_2O production of denitrifiers by a factor of three. Large concentrations of C_2H_2 (1-10 kPa), which are known to block N_2O reductase, have increased N_2O production by a factor of 2 to 6 compared with controls without C_2H_2 (e.g. Duxbury & McConnaughey, 1986; Colbourn, 1992; Dendooven et al., 1999; Estavillo et al., 2002). Partial inhibition of N_2O reductase of denitrifiers can occur in soils at C_2H_2 concentrations as small as 0.01 kPa (Ryden et al., 1979). We postulated no interaction between inhibitors. Thus, the inhibitors' effect was assumed to be the same whether used in combination with another inhibitor or alone. For a combination of incomplete inhibition of nitrification by C_2H_2 plus partial inhibition of nitrification by O_2 (possibility g), N_2O production by nitrification was expected to be fully inhibited in incubations with C_2H_2 plus O_2.

To estimate the importance of the different possibilities, four hypothetical soils were considered with N_2O production i) dominated by nitrification, ii) dominated by nitrifier denitrification, iii) dominated by denitrification, and iv) not dominated by a single source, but equally caused by nitrification, nitrifier denitrification and denitrification (Table 2). The amounts of N_2O produced by the different sources were calculated following Equations 1-4, where the subscripts A, O and AO stand for incubations with 0.02 kPa C_2H_2, 100 kPa O_2, and both inhibitors, respectively (see also Table 1):

Equation 1: $N_2O_{Nitrification} = N_2O_O - N_2O_{AO}$
Equation 2: $N_2O_{Denitrification} = N_2O_A - N_2O_{AO}$
Equation 3: $N_2O_{Nitrifier\ Denitrification} = N_2O_C - N_2O_O - N_2O_A + N_2O_{AO}$
Equation 4: $N_2O_{Other} = N_2O_{AO}$

Table 2. N_2O production rates in four hypothetical soils: a nitrification-dominated one, a nitrifier-denitrification-dominated one, a denitrification-dominated one and a balanced[1] one. Since the production rates add up to 100, the production per source can be interpreted as a percentage of the total production.

	Nitrification dominated	Nitrifier denitrification dominated	Denitrification dominated	Balanced[1]
Nitrification	60	20	5	33
Nitrifier denitrification	20	60	10	33
Denitrification	15	15	80	33
Other sources[2]	5	5	5	1
Total	100	100	100	100

[1] Balanced: nitrification, nitrifier denitrification and denitrification contribute equally to the N_2O production.

[2] Other sources are for example chemodenitrification or heterotrophic nitrification.

Results and discussion

The results of the sensitivity analysis indicated that problems with the inhibitors C_2H_2 and large partial pressures of O_2 can lead to both over- and underestimations of the sources of N_2O (Table 3). Frequently, two sources were linked due to the calculations, so that overestimations of one led to underestimations of the other. The sum of N_2O produced by the linked sources remained unchanged. For example, problems caused by 100 kPa O_2 (Table 3, columns b-d) did not change the sum of N_2O produced by nitrification plus nitrifier denitrification or by denitrification plus other sources compared with the ideal situation (Table 3, column a). However, internal shifts of N_2O production occurred between nitrification and nitrifier denitrification, as well as between denitrification and other sources. When problems with 0.02 kPa C_2H_2 were assumed (Table 3, columns e and f), the sum of N_2O produced by nitrification plus other sources, or by nitrifier denitrification plus denitrification, was the same as in the ideal situation (Table 3, column a). Again, internal shifts occurred; this time between nitrification and other sources, as well as between nitrifier denitrification and denitrification. A combination of artefacts by C_2H_2 and O_2 (Table 3, column g) caused an underestimation of nitrification, a large underestimation of nitrifier denitrification and a large overestimation of denitrification.

The different sources of N_2O varied in their sensitivities to problems caused by the inhibitors. The calculated amount of N_2O produced by nitrification deviated by a factor 0.3 to 3 from the correct value. For denitrification, this factor was 0.3 to 6.5, showing that problems with the inhibitors lead to over- rather than underestimations of N_2O production by denitrification. Overestimations of N_2O production by denitrification were highest in nitrification- and nitrifier denitrification-dominated soils. Other sources of N_2O, such as chemodenitrification of heterotrophic nitrification, were not underestimated in the cases considered, but could be overestimated by up to 23 times. Nitrifier denitrification deviated by a factor -15.8 to 2 from the correct values. Thus, N_2O production by nitrifier denitrification was susceptible to large underestimations, especially in denitrification-dominated soils. The large underestimations also led to negative values. In the considered cases, negative values were only obtained for nitrifier denitrification. This fits in well with the results from a soil survey (Wrage et al., 2003a), where most negative values were also derived for nitrifier denitrification.

Table 3. Sensitivity analysis with hypothetical soils where the N_2O production is dominated by different processes. The considered artefacts caused by the inhibitors are shown in the footnotes. It is assumed that the effect of each inhibitor was independent of its use alone or in combination with the other inhibitor. N: nitrification, ND: nitrifier denitrification, D: denitrification, O: other sources of N_2O. Changes relative to the ideal situation (column a) are indicated in bold. In the last two columns, the largest range and the largest under- and overestimation found for each process are indicated in bold italics.

Soils	Ideal a	Artefacts with 100 kPa O_2 b	c	d	Artefacts with 0.02 kPa C_2H_2 e	F	Combination g
Nitrification dominated							
N	60	40	73	60	20	60	40
ND	20	40	7	20	7	-10	-43
D	15	15	15	5	28	45	98
O	5	5	5	15	45	5	5
Nitrifier denitrification dominated							
N	20	13	60	20	7	20	13
ND	60	67	20	60	20	30	-17
D	15	15	15	5	55	45	98
O	5	5	5	15	18	5	5
Denitrification dominated							
N	5	3	12	5	2	5	3
ND	10	12	3	10	3	-150	-158
D	80	80	80	27	87	240	250
O	5	5	5	58	8	5	5
Nitrification, nitrifier denitrification and denitrification contribute equally							
N	33	22	55	33	11	33	22
ND	33	44	11	33	11	-33	-66
D	33	33	33	11	55	99	143
O	1	1	1	23	23	1	1

In studies of the sources of N_2O in soils, 0.02 kPa C_2H_2 is used much more frequently as an inhibitor than 100 kPa O_2. In such studies, N_2O produced in incubations with C_2H_2 is assumed to be produced by denitrification. The reciprocal inhibited part, is ascribed to nitrification. Table 1 shows that the inhibited part comprises both nitrification and nitrifier denitrification, i.e. all N_2O produced by nitrifiers. The part which is not inhibited consists of N_2O from denitrification plus that from other sources such as chemodenitrification. Thus, N_2O production by denitrification is already overestimated from the contribution of other sources. We have also seen that problems caused by C_2H_2 lead to an overestimation of denitrification at the cost of nitrifier denitrification. Nitrification is underestimated to the advantage of other sources (Table 3, column e, but not column f). Pure culture studies have shown that C_2H_2 does not inhibit N_2O production by all nitrifiers (Wrage et al., 2003b). An inhibition of N_2O reductase of denitrifiers by small concentrations of C_2H_2 has also been demonstrated (Ryden et al., 1979). Therefore, it is very likely that the N_2O production by denitrification has been overestimated in studies using low partial pressures of C_2H_2, while the N_2O production by nitrifiers has been underestimated.

Table 4. Total N₂0 production by different soils and percentage derived from nitrifiers (i.e. nitrification plus nitrifier denitrification).

Site description	Method	Total N$_2$O emission (n.b. different units)	N$_2$O emission from nitrifiers (% of total emission)	References
• Early successional forest	C$_2$H$_2$ inhibition	0.024-0.074 (g N kg$_{soil}^{-1}$ a^{-1})	3-40	Robertson & Tiedje, 1987
• Freely drained sandy loam with ryegrass-chickweed (greenhouse)	dicyandiamide inhibition	0.3-2.3 (kg N ha^{-1} a^{-1})	40	Skiba et al., 1993
• Douglas fir stand	nitrapyrin inhibition	0.005-0.04 (g N kg$_{soil}^{-1}$ a^{-1})	40-96	Martikainen & de Boer, 1993
• Agricultural acid brown earth	^{15}N tracer method	0.005-0.051 (g N kg$_{soil}^{-1}$ a^{-1})	70	Stevens et al., 1997

What do the results of the sensitivity analysis and the observed problems with the inhibitors C$_2$H$_2$ and O$_2$ mean for our understanding of N$_2$O production by different sources? So far, problems with the inhibitors have been studied in pure culture experiments. Results from pure culture studies cannot be directly translated to soils. However, it is likely that an inhibitor that fails to work in pure culture experiments will also fail to inhibit the same organisms in soil incubations. Table 4 shows the results of studies investigating N$_2$O production by nitrifiers with different methods. Most of the studies used nitrification inhibitors to differentiate between sources of N$_2$O. We have already elaborated on problems with C$_2$H$_2$, which lead to underestimations of N$_2$O production by nitrifiers. Nitrapyrin has been shown to either stimulate or partially inhibit denitrification (Henninger & Bollag, 1976; Klemedtsson et al., 1988). Therefore, nitrification plus nitrifier denitrification might have been either over- or underestimated in the study with nitrapyrin (Table 4). The study using the tracer method (Stevens et al., 1997) probably gives the most accurate idea of N$_2$O production by nitrification, without accounting for nitrifier denitrification. Stevens et al. (1997) ascribed 70% of the total N$_2$O emission from an agricultural acid brown soil to nitrification. Thus, we see that nitrifiers potentially produce substantial parts of the total N$_2$O emission.

Clearly, it is still difficult to measure the contribution of different soil sources to the production of the greenhouse gas N$_2$O. The contribution of nitrifiers to N$_2$O production has probably been underestimated, especially in studies with C$_2$H$_2$. Stable isotope analyses provide a means to differentiate at least between N$_2$O produced from NH$_3$ and N$_2$O from NO$_2^-$ or NO$_3^-$. Both tracer experiments and investigations of the natural abundance of stable isotopes in developing N$_2$O seem promising. Unfortunately, many studies still use inhibitors, rather than stable isotope analysis, to investigate sources of N$_2$O. Furthermore, measurements are often only carried out during a few days or at best months. Continuous measurements over extended periods are still rare. Therefore, both the overall amount of N$_2$O produced by soils and the distribution to its sources pose key challenges for further research. Only a profound knowledge of N$_2$O production and its sources will allow us to develop sensible reduction programmes for N$_2$O production from soils.

Conclusions

It is very likely that nitrifiers have been underestimated as producers of N_2O in studies using C_2H_2 or both C_2H_2 and O_2 as inhibitors. The most probable reasons for this are i) only partial inhibition of nitrification and nitrifier denitrification by small partial pressures of C_2H_2, ii) an inhibition of the N_2O reductase of denitrifiers, even at small concentrations of C_2H_2, and iii) partial inhibition of nitrification by 100 kPa O_2. Other inhibitors are also known to have side-effects which might lead to under- and overestimations of sources of N_2O. Therefore, it is concluded that inhibitors should not be used to assess the strength of different sources of N_2O in soils.

References

Ambus, P. (1998). Nitrous oxide production by denitrification and nitrification in temperate forest, grassland and agricultural soils. European Journal of Soil Science, 49, 495-502.

Beaumont, H.J.E., Hommes, N.G., Sayavedra-Soto, L.A., Arp, D.J., Arciero, D.M., Hooper, A.B., Westerhoff, H.V. and van Spanning, R.J.M. (2002). Nitrite reductase of *Nitrosomonas europaea* is not essential for production of gaseous nitrogen oxides and confers tolerance to nitrite. Journal of Bacteriology, 184, 2557-2560.

Bouwman, A.F. (1990). Exchange of greenhouse gases between terrestrial ecosystems and the atmosphere. In: A.F. Bouwman (ed). International Conference, Soils and the Greenhouse Effect, John Wiley and Sons Ltd., Chichester, 61-127.

Colbourn, P. (1992). Denitrification and N_2O production in pasture soil: the influence of nitrogen supply and moisture. Agriculture, Ecosystems and Environment, 39, 267-278.

Dendooven, L., Murphy, D.V. and Catt, J.A. (1999). Dynamics of the denitrification process in soil from the Brimstone Farm experiment, UK. Soil Biology and Biochemistry, 31, 727-734.

Duxbury, J.M. and McConnaughey, P.K. (1986). Effect of fertilizer source on denitrification and nitrous oxide emissions in a maize-field. Soil Science Society of America Journal, 50, 644-648.

Estavillo, J.M., Merino, P., Pinto, M., Yamulki, S., Gebauer, G., Sapek, A. and Corré, W.J. (2002). Short term effect of ploughing a permanent pasture on N_2O production from nitrification and denitrification. Plant and Soil, 239, 253-265.

Granli, T. and Bøckman, O.C. (1994). Nitrous oxide from agriculture. Norwegian Journal of Agricultural Sciences, Supplement 12, 7-128.

Henninger, N.M. and Bollag, J-M. (1976). Effect of chemicals used as nitrification inhibitors on the denitrification process. Canadian Journal of Microbiology, 22, 668-672.

Kester, R.A., Meijer, M.E., Libochant, J.A., de Boer, W. and Laanbroek, H.J. (1997). Contribution of nitrification and denitrification to the NO and N_2O emissions of an acid forest soil, a river sediment and a fertilized grassland soil. Soil Biology and Biochemistry, 29, 1655-1664.

Klemedtsson, L., Svensson, B.H. and Rosswall, T. (1988). A method of selective inhibition to distinguish between nitrification and denitrification as sources of nitrous oxide in soil. Biology and Fertility of Soils, 6, 112-119.

Kowalchuk, G.A. and Stephen, J.R. (2001). Ammonia-oxidizing bacteria: a model for molecular microbial ecology. Annual Reviews of Microbiology, 55, 485-529.

Martikainen, P.J. and de Boer, W. (1993). Nitrous oxide production and nitrification in acidic soil from a Dutch coniferous forest. Soil Biology and Biochemistry, 25, 343-347.

Priemé, A. and Christensen, S. (2001). Natural perturbation, drying-wetting and freezing-thawing cycles, and the emission of nitrous oxide, carbon dioxide and methane from farmed organic soils. Soil Biology and Biochemistry, 33, 2083-2091.

Robertson, G.P. and Tiedje, J.M. (1987). Nitrous oxide sources in aerobic soils: nitrification, denitrification and other biological processes. Soil Biology and Biochemistry, 19, 187-193.

Ryden, J.C., Lund, L.J. and Focht, D.D. (1979). Direct measurement of denitrification loss from soils: I. Laboratory evaluation of acetylene inhibition of nitrous oxide reduction. Soil Science Society of America Journal, 43, 104-110.

Skiba, U., Smith, K.A. and Fowler, D. (1993). Nitrification and denitrification as sources of nitric oxide and nitrous oxide in a sandy loam soil. Soil Biology and Biochemistry, 25, 1527-1536.

Stevens, R.J., Laughlin, R.J., Burns, L.C., Arah, J.R.M. and Hood, R.C. (1997). Measuring the contribution of nitrification and denitrification to the flux of nitrous oxide from soil. Soil Biology and Biochemistry, 29, 139-151.

Webster, E.A. and Hopkins, D.W. (1996). Contributions from different microbial processes to N_2O emission from soil under different moisture regimes. Biology and Fertility of Soils, 22, 331-335.

Wrage, N. (2003). Pitfalls in measuring nitrous oxide production by nitrifiers. PhD Thesis, Wageningen University and Research Centre, 141 pp.

Wrage, N., Velthof, G.L., Laanbroek, H.J. and Oenema, O. (2003a). Nitrous oxide production in grassland soils: Assessing the contribution of nitrifier denitrification. Soil Biology and Biochemistry (accepted for publication).

Wrage, N., Velthof, G.L., Oenema, O. and Laanbroek, H.J. (2003b). Acetylene and oxygen as inhibitors of nitrous oxide production in *Nitrosomonas europaea* and *Nitrosospira briensis*: a cautionary tale. FEMS Microbiology Ecology (in press).

Wrage, N., Velthof, G.L., van Beusichem, M.L. and Oenema, O. (2001). Role of nitrifier denitrification in the production of nitrous oxide. Soil Biology and Biochemistry, 33, 1723-1732.

The effect of organic and mineral nitrogen fertilisers on emissions of NO, N_2O and CH_4 from cut grassland

R. Rees[1], S. Jones[1,4], R.E. Thorman[2], I. McTaggart[3], B. Ball[1] and U. Skiba [4]

[1]*Scottish Agricultural College, Bush Estate, Penicuick, EH26 0PH, UK*
[2]*now at ADAS Boxworth, Battlegate Road, Boxworth, Cambridge, UK*
[3]*now at Department of Global Agricultural Sciences, University of Tokyo, 1-1 Yayoi, 1-Chome, Bunkyo-ku, Tokyo 113-8657, Japan*
[4]*Centre for Ecology and Hydrology, Edinburgh Research Station, Bush Estate, Penicuik, EH26 0QB, UK*

Abstract

Grasslands are a major source of greenhouse gases in the UK, and can be significantly influenced by management. Here we report the results from two separate experiments from a grassland near Edinburgh in which the influence of inorganic N fertilisers and manures on the emissions of N_2O and CH_4 and the O_3 precursor NO were studied. The site management involved three annual cuts, but no grazing. Plots fertilised with NH_4NO_3, cattle slurry, sewage and sludge pellets were compared with a non-fertilised control. In 1998/1999, the emissions of NO and N_2O were measured and in 2002/2003, emissions of N_2O and CH_4 were monitored. Flux measurements were made using static chambers at weekly to fortnightly intervals throughout the growing seasons. The results from both experiments showed that the use of organic manures can increase the emissions of NO, N_2O and CH_4. In the period July 1998 to September 1999, NO emissions ranged from -0.50 to +5.94 g N ha^{-1} day^{-1} and N_2O fluxes from 0.10 to 4908 g N ha^{-1} day^{-1}. During the growing season of 1999, cumulative NO emissions from plots fertilised with cattle slurry, and sewage sludge pellets were 3 and 2 times larger than from plots fertilised with NH_4NO_3. Cumulative N_2O emissions in 1999 were not significantly different from plots treated with NH_4NO_3 or organic N, but during the growing season of 2002, N_2O emissions from plots fertilised with sewage sludge pellets were 26 fold (107 kg N ha^{-1}) larger than from NH_4NO_3 fertilised plots. Emissions from plots fertilised with cattle slurry were 50% smaller. Cattle slurry did, however, triple the emissions of CH_4 compared with fertilisation with NH_4NO_3. Nitrous oxide release from the NH_4NO_3 mineral fertiliser was short-lived with emissions returning to near background levels within 2 weeks. Release of N_2O from the manure treatments followed a very different pattern with significantly higher emissions occurring throughout the growing season.

Keywords: grassland, manure, methane, nitric oxide, nitrous oxide

Introduction

Many intensively managed grasslands in western Europe receive large inputs of fertiliser N. Average N inputs to UK grasslands used for silage between 1993-1997 were 177 kg N ha^{-1} yr^{-1} (Chalmers, 2001), with mineral fertilisers often supplemented by inputs of organic manures (Smith et al., 2001). It has become apparent that these large inputs can be associated with significant nutrient losses both in the form of gaseous pollutants and losses in drainage water

(Jarvis et al., 1996). The addition of organic manures to these grasslands is inevitable, where grazing is a component of the management, however, their use is known to contribute to large but uncertain losses of gaseous pollutants (Allen et al., 1996). Together, grasslands and the livestock industry contribute to about 50% of the total UK N_2O emission (Skiba, 2000), which was estimated to be 137 kt in 2001(Watterson, 2003). This estimate is based on the IPCC methodology, which suggests that 1.25% of the fertiliser input is emitted as N_2O (Houghton, 2001). The regression works less well when applied to N supplied with organic manures. Recent studies have suggested that the fraction of N released from manures as NO can be considerably higher than that for mineral fertilisers (Chadwick et al., 1998), although this fraction is highly dependent upon the prevailing weather conditions. Nitrous oxide is produced in soils as a consequence of both the oxidative process of nitrification, and the reductive process of denitrification. These processes may go on simultaneously within soils but, nitrification is more likely to occur in lighter textured soils and in drier conditions and denitrification is most prevalent in heavier soils and wetter conditions. The UK emissions of NO are 750 kt N yr^{-1}, however, the contribution of soils to this total is highly uncertain and is thought to vary between 6 and 66 (Simpson et al., 1999; Skiba & Sozanska, 2001). Emission of NO is associated primarily with nitrification.

In developing more sustainable systems of grassland management, an integrated approach to pollutant loss needs to be used in order to modify management practices. In this study, our objective was, therefore, to assess three gaseous pollutant losses (N_2O, NO and CH_4) within contrasting management options. The work has concentrated on comparisons between N inputs with mineral fertilisers and organic manures over four years.

Materials, methods and sites

Studies were carried out on a grassland soil at the Bush Estate 15 km to the south of Edinburgh in Scotland between 1998 and 2003. The site (Cowpark) was at an altitude of 200 m on an imperfectly drained clay loam soil. Fuller site details and a detailed description of the experimental design are described by (Ball et al., 2004). The treatments considered here were; (1) zero N control, (2) ammonium nitrate, (3) sewage sludge pellets, and (4) cattle slurry. Treatments were applied in 1998 to 2000, (results from 2000 are not reported, and plots were untreated in 2001) and in 2002 and 2003. Treatments were applied to a blocked and replicated field experiment in field plots measuring 12 m by 6 m. Each treatment was replicated three times. Organic manures were applied in late April and early July in each treatment year. Mineral fertilisers were also applied at these times, but in addition, a mineral fertiliser application was made during mid August in 2002 and 2003. In 1998 and 1999 the application of organic manures was adjusted to provide 240 kg N ha^{-1} yr^{-1}. In 2002 and 2003 application of organic manures was adjusted to provide a target of 300 kg available N ha^{-1} yr^{-1}. The rates and timings of manure applications to the different treatment plots are given in Table 1. Gas flux measurements were made using static chambers located randomly within each of the plots using the method described by (Ball et al., 1997). A sampling period of approximately 60 min was used, and samples were collected in portable evacuated aluminium tubes. Gases were analysed by electron capture and flame ionisation gas chromatography. Soil samples were collected randomly from within the plots to a depth of 0.2 m periodically throughout the experimental period. Soil mineral N (NH_4^+ and NO_3^-) concentrations were determined following extraction in 1M KCl, followed by colorimetric analysis to provide estimates of available N.

Nitric oxide emissions were measured only in 1998-99 using a closed chamber technique (Thorman, 2002). Polypropylene cylindrical chambers (0.2 m high, 0.4 m diameter) were inserted into the soil to a depth of not less than 0.3 m. An ambient air sample was withdrawn for 1 min from inside the open chamber, via tubing fitted with a three-way tap and inserted through a hole in the chamber wall. Subsequently, the chamber was sealed for a 10 min period with a collapsible, clear, plastic, dome shaped lid. A gas sample was taken through a PTFE line and analysed immediately using a chemiluminescent NO_x analyser (Thermo Environmental Instruments, model 42C) in 1998 and a chemiluminescent NO_2 analyser fitted with a $CrCO_3$ converter (Scintrex) in 1999. Meteorological data were collected using recording equipment located within 500 m of the field plots.

Table 1. *Applications and N content of manure treatments in 1998/1999 and 2002/2003.*

Organic manure	Application date	Dry Matter (%)	Rate m^3 ha^{-1}	Total N kg ha^{-1}	Ammoniacal N kg ha^{-1}	Available N kg ha^{-1}
Cattle slurry	July 1998	5.9	60	220	100	122
	April 1999	10.0	100	430	154	160
	July 1999	3.7	100	190	68	110
	April 2002	9.2	100	300	127	Nd
	July 2002	7.2*	100	300*	78	Nd
	April 2003	7.2*	100	380	181	Nd
	July 2003	7.2*	100	150	62	Nd
Sludge pellets	July 1998	97.4	15	510	95	118
	April 1999	97.4	17.5	508	72	128
	July 1999	97.4	17.5	508	72	128
	April 2002	99	37.5	1533	83	150
	July 2002	99	30	1533	60	150
	April 2003	99	30	1533	93	150
	July 2003	99	30	1533	71	150

*Estimated value based upon mean of previous applications. Nd = not determined.

Results

The amounts and distribution of rainfall within the sampling period varied widely. The long- term average rainfall (1955-1995) is 849 mm, however, in 1998 the total annual rainfall was above 1200 mm, whilst in 2003 the total rainfall between Jan-Sept was only 388 mm (Table 2).

The different distribution of rainfall within years also showed large variability, thus the total rainfall in the period between 1 June-30 September (when soil temperatures and therefore microbial activity are highest) ranged between 69 mm in 2003 and 386 mm in 2002.

The emissions of NO and N_2O were generally lower in 1998 than in 1999, (Figures 1 and 2a), coinciding with considerably reduced monthly rainfall in 1999 (Table 2). In both years the

Table 2. Rainfall (mm) at Cowpark in 1998-99 and 2002-2003.

	1998	1999	2002	2003
January	142.5	132.1	45.4	101.0
February	73.9	53.1	154.6	36.0
March	58.5	56.9	13.6	27.0
April	106.2	57.9	48.1	44.0
May	100.9	63.8	18.5	111.0
June	102.0	102.7	107.2	37.0
July	119.8	56.2	54.5	28.0
August	84.6	48.6	179.5	2.0
September	62.7	83.0	44.7	2.0
October	152.0	50.4	231.2	
November	117.2	107.8	96.32	
December	83.0	140.5	76.44	
Total	1203	953	1070	388

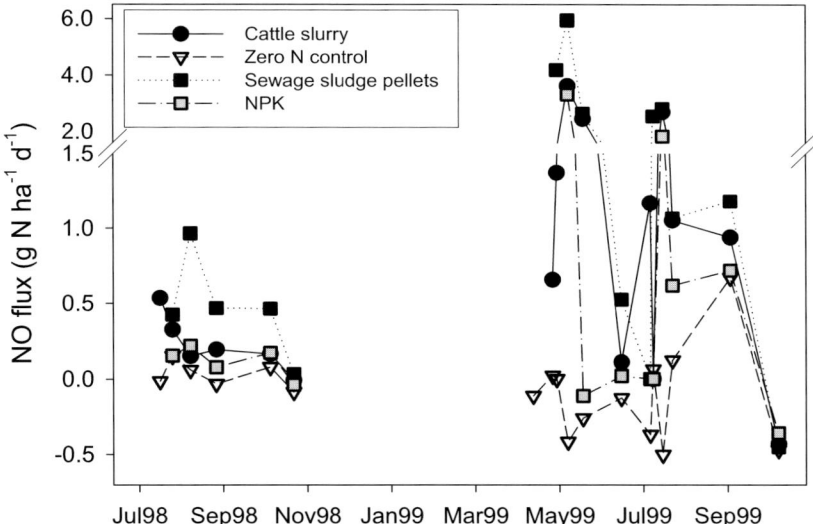

Figure 1. Nitric oxide emissions from grassland fertilised with manures (black symbols) mineral N (grey symbols) or a zero N control (open symbols).

highest NO flux was observed from plots receiving sewage sludge pellets with maximum peaks of 1.2 g N ha^{-1} day^{-1} in August 1998 and 5.9 g N ha^{-1} day^{-1} in May 1999. In 1999, cumulative NO emissions from plots fertilised with cattle slurry and sewage sludge pellets were 3 and 2 times larger than from plots fertilised with NH_4NO_3 (p < 0.05) (Table 3). Nitrous oxide emissions were approximately three orders of magnitude larger than NO emissions. In 1998-99, the largest emission peaks of N_2O were associated with the NH_4NO_3 fertiliser application in both years (4.9 and 0.87 kg N_2O-N ha^{-1} d^{-1}, respectively). Apart from these very short-lived emission peaks, N_2O emission from plots fertilised with NH_4NO_3, sewage sludge pellets, or cattle slurry were

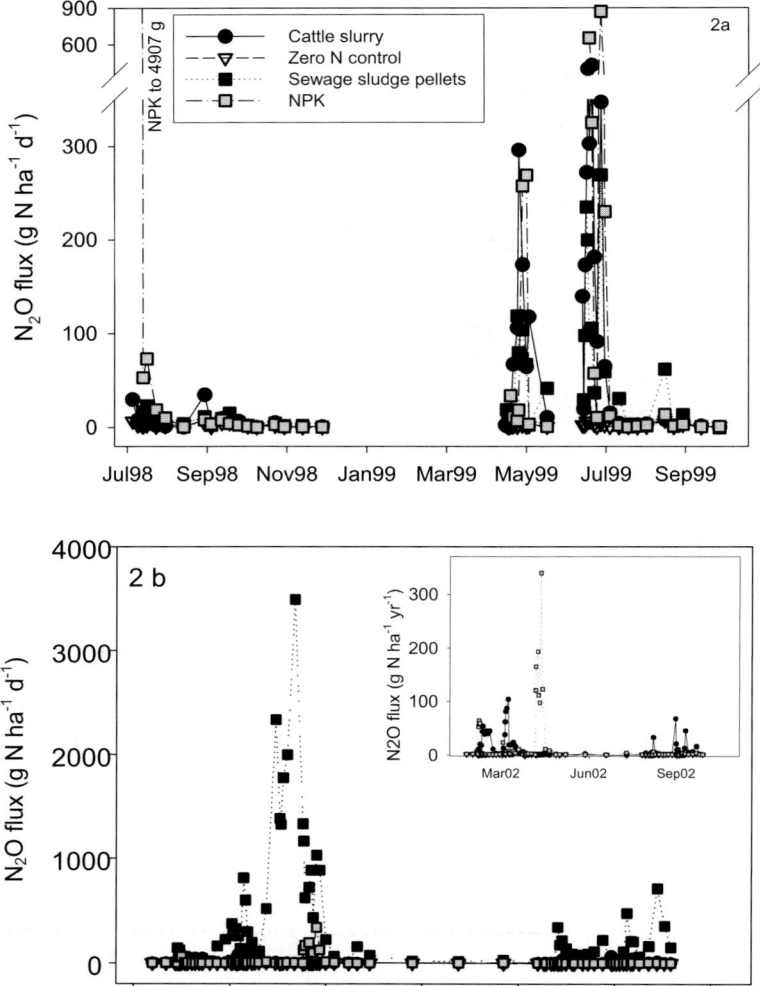

Figure 2. Nitrous oxide fluxes from grassland fertilised with manures (black symbols), mineral N (grey symbols) and a zero N control (open symbols) for the periods July 1998 to October 1999 (2a) and March 2002 to August 2003 (2b). The insert in Fig 2b magnifies the y-axis of the main graph and demonstrates emission peaks for cattle slurry and NH_4NO_3 fertilisers, but symbols for the zero N control still remain hidden.

not significantly different (Figure 2a). For the growing season of 1999, cumulative emissions of N_2O were 33% smaller from the sewage sludge pellets compared with the NH_4NO_3 fertiliser, but this difference was not significant (Table 3).

In the treatment years 2002/2003, N_2O emission peaks induced by the NH_4NO_3 fertiliser were considerably smaller than in 1998 and 1999 (Figs. 2a and b). The only sizeable emission peaks were measured after the April and August fertilisation in 2002 (64 and 339 g N_2O-N ha^{-1} yr^{-1}, respectively). Emissions from plots fertilised with cattle slurry were also lower in 2002/2003 compared with the previous years. Cumulative emissions for the same periods (April - October) in 1999 and 2002 were 3.5 times larger in 1999, in spite of very similar rates of total and ammoniacal N applied (Tables 2 and 3). In contrast, the total N of the dried sewage sludge pellets applied in 2002/2003 was three times larger than in 1998/1999, but the ammoniacal N remained unchanged (Table 2). Consequently, rates of N_2O emission were much larger than in previous years and peaked in August 2002 at 3.5 kg N ha^{-1} day^{-1} and at 0.7 kg N ha^{-1} day^{-1} in August 2003 (Figure 2b). Emission factors for N_2O calculated for the growing seasons of 1999 and 2002 varied significantly between fertiliser type, month and year of application and ranged from 0.23 to 3.67% of the total N applied (Table 4).

Methane emissions were generally small (< 0.1 kg C ha^{-1} d^{-1}) in 2002-03 in all treatments, except for periods of 2-4 days following applications of manures. Cattle slurry was associated with largest emission peaks (2.8 kg C ha^{-1} d^{-1}), which was triple the emission resulting from fertilisation with NH_4NO_3.

Table 3. Cumulative fluxes of NO, N_2O and CH_4 for the growing seasons (April - October) 1999 and 2002.

	1999		2002	
	g NO N ha^{-1}	kg N_2O N ha^{-1}	kg N_2O N ha^{-1}	kg CH_4 C ha^{-1}
Control	0.23	0.33	0.34	0.35
NH_4NO_3	3.45	7.22	4.10	0.33
Sewage sludge pellets	10.0	4.80	107.31	0.19
Cattle slurry	7.30	8.91	2.41	0.78

Table 4. The loss of N_2O in the growing seasons April to October 1999 and 2002 expressed as a percentage of total N input.

Treatment	Total N applied lost as N_2O (%)	
	1999	2002
NH_4NO_3	3.67	1.25
Cattle slurry	1.20	0.34
Sludge pellets	0.23	3.43

Discussion

Nitrous oxide was the dominant greenhouse gas emitted, however, patterns of loss from mineral and organic fertilisers differed significantly in the different years. In each year, N_2O emissions following mineral fertiliser application peaked very rapidly, returning close to background levels within 7-10 days. Emission factors for the mineral N fertilisers for the growing seasons of 1999 and 2002 (Table 4) are within the range reported by previous authors (Dobbie & Smith, 2003). Losses of N_2O, following manure applications, were more variable, but normally extended over a longer period than those from the mineral fertilisers (Table 4 and Figure 2). In 2002, very high emissions of N_2O (with a cumulative emission of 107 kg N ha^{-1} yr^{-1}) were measured from the sewage sludge treated plots, making a very significant contribution to the global warming potential from these plots. By contrast, in 1998 and 1999, N_2O emissions from the plots receiving sewage sludge pellets were significantly lower than those from plots receiving NH_4NO_3 fertiliser. Although in this later period there were higher rates of application of total and available N in the manure treated plots, this cannot fully explain the differences. Nitrous oxide emitted as a percentage of the total N in the sewage

sludge pellets was 0.23 in 1999 and increased to 3.4 in 2002. This latter value is higher than the IPCC default emission factor, but again well within the range of previously reported values (Dobbie & Smith, 2003). However, this variability also indicates a weakness in the concept of emission factors when applied to manure treated soils. Inter-year variation in precipitation also results in variation about the emissions factor mean. Although N_2O loss may be reported as a proportion of N applied, there is no certainty that the N released originates from the applied N. In this experiment, treatments were maintained on the same plots over a period of six years, allowing an accumulation of organic matter and N. It is possible that this, in conjunction with the relatively warm and wet soil conditions in 2002, led to a significant release of N_2O from labile organic matter pools in the manure treated plots. Long-term manure applications have been shown to lead to an accumulation of labile organic matter pools (Jensen et al., 1999; Hao et al., 2003). Ginting et al. (2003) found that potentially mineralisable N in manure treated plots was 40-70% higher than in soils treated with mineral fertilisers after just four years of treatments. Smith et al. (2000) suggested that application of sewage sludge would result in an increase in soil organic C at a rate of 0.49% yr^{-1} at sludge application rates of 1 t ha^{-1} yr^{-1}. If applied to Cowpark, this would indicate a 16% annual increase in SOM storage. Smith's calculations were derived from soils with lower organic matter contents that those used in the current study, but indicate the large potential for increases in SOM content.

The variations in magnitude of the fertiliser induced emission peaks and the cumulative emissions over the growing seasons can be largely explained by differences in the timing and rate of rainfall in the weeks before, during, and after fertiliser application. The largest N_2O emission peak from plots receiving NH_4NO_3 (almost equivalent to 5 kg N ha^{-1} d^{-1}) was observed after the spring fertilisation in 1998, when rainfall was twice as high as that in April 1999, 2002, or 2003 (Table 1). However, application of cattle slurry or sewage sludge pellets in April 1998 stimulated a much smaller response than in the following years. It is likely that the additional input of C and N during the wet spring of 1998 promoted complete denitrification to N_2, and therefore reduced the rate of N_2O loss. Nitrous oxide emissions from manure treated plots may be more likely to occur during warm and wet summers, such as 2002, when enhanced mineralisation processes contribute to the release of large amounts of mineral N. The rainfall in 2002 was only 26% higher than the long-term average, however, during the summer of 2002 (June-Aug) rainfall was 64% above average. This period coincided with the highest temperatures and the periods in which manures were applied. The 2002 applications were also higher than those of preceding years and followed earlier applications, which would have allowed accumulation of organic matter.

Compared with the N_2O emissions, the potential contribution of NO and CH_4 fluxes to global warming were insignificant. Again, both gases were stimulated by fertiliser application, with organic manures significantly increasing the emissions of both. It is possible, that given the relatively high clay content of these soils, a significant proportion of the NO produced will have been consumed within the soil by denitrification, as suggested by Thorman (2002). Methane emissions from the treatments applied were also generally low, with some small uptake of CH_4 outside of the periods of manure application, a pattern which has been previously observed in other studies (Chadwick et al., 2000).

Conclusions

Nitrous oxide was the dominant greenhouse gas released from the grassland soils examined in this study. Emissions of N_2O in manure treated plots can be very much higher than those from mineral fertilisers, and show large variability that can be linked to climatic variability and previous site history. Rainfall distribution at times of fertiliser application appears to be more important than fertiliser type in influencing N_2O emissions. It is possible that repeated, large applications of manures lead to an increase in the potential for N_2O release from grassland soils, however further studies would be necessary to identify the source of these fluxes. Nitric oxide and methane are produced in relatively small quantities from these managed grasslands and are unlikely to pose a significant pollution threat.

Acknowledgement

Financial support for this work was provided by the Scottish Executive Environment, and Rural Affairs Department, Department for Environment Food and Rural Affairs and the EU (Greengrass).

References

Allen, A.G., Jarvis, S.C. and Headon, D.M. (1996). Nitrous oxide emissions from soils due to inputs of nitrogen from excreta return by livestock on grazed grassland in the UK. Soil Biology and Biochemistry, 28, 597-607.

Ball, B.C., Horgan, W., Clayton, H. and Parker, J.P. (1997). Spatial variability of nitrous oxide fluxes and controlling soil and topographic properties. Journal of Environmental Quality, 26, 1399-1409.

Ball, B.C., McTaggart, I.P. and Scott, A. (2004). Mitigation of greenhouse gas emissions from soil under silage production by use of organic or slow release fertilisers. (Pers. Comm.)

Chadwick, D.R., Pain, B.F. and Brookman, S.K.E. (2000). Nitrous oxide and methane emissions following application of animal manures to grassland. Journal of Environmental Quality, 29, 277-287.

Chadwick, D.R., van der Weerden, T., Martinez, J. and Pain, B.F. (1998). Nitrogen transformations and losses following pig slurry applications to a natural soil filter system (Solepur process) in Brittany, France. Journal of Agricultural Engineering Research, 69, 85-93.

Chalmers, A.G. (2001). A review of fertilizer, lime and organic manure use on farm crops in Great Britain from 1983 to 1987. Soil Use and Management, 17, 254-262.

Dobbie, K.E. and Smith, K.A. (2003). Nitrous oxide emission factors for agricultural soils in Great Britain: the impact of soil water-filled pore space and other controlling variables. Global Change Biology, 9, 204-218.

Ginting, D., Kessavalou, A., Eghball, B. and Doran, J.W. (2003). Greenhouse gas emissions and soil indicators four years after manure and compost applications. Journal of Environmental Quality, 32, 23-32.

Hao, X.Y., Chang, C., Travis, G.R. and Zhang, F.R. (2003). Soil carbon and nitrogen response to 25 annual cattle manure applications. Journal of Plant Nutrition and Soil Science-Zeitschrift fur Pflanzenernahrung und Bodenkunde, 166, 239-245.

Houghton, J.T., Meira Filho, L.G., Lim, B., Treanton, K., Mamaty, I. and Bonduki, Y. (2001). Revised 1996 IPCC Guidelines for National Greenhouse Gas Inventories, Reference Manual Vol 3, D.J. Griggs and B.A. Callender (Eds). IPCC/OECD/IEAUK Meteorological Office, Bracknell.

Jarvis, S.C., Wilkins, R.J. and Pain, B.F. (1996). Opportunities for reducing the environmental impact of dairy farming managements: A systems approach. Grass and Forage Science, 51, 21-31.

Jensen, B., Sorensen, P., Thomsen, I.K., Jensen, E.S. and Christensen, B.T. (1999). Availability of nitrogen in N-15-labeled ruminant manure components to successively grown crops. Soil Science Society of America Journal, 63, 416-423.

Simpson, D., Winiwarter, W., Borjesson, G., Cinderby, S., Ferreiro, A., Guenther, A., Hewitt, C.N., Janson, R., Khalil, M.A.K., Owen, S., Pierce, T.E., Puxbaum, H., Shearer, M., Skiba, U., Steinbrecher, R., Tarrason, L. and Oquist, M.G. (1999). Inventorying emissions from nature in Europe. Journal of Geophysical Research-Atmospheres, 104, 8113-8152.

Skiba, U. (2000). Nitrous oxide, report for the SNIFFER project UK Air Pollution Information System (APIS).

Skiba, U. and Sozanska, M.M.S.a.F.D. (2001). Spatially disaggregated inventories of soil NO and N_2O emissions for Great Britain. Water Air and Soil Pollution, 1, 109-118.

Smith, K.A., Brewer, A.J., Crabb, J. and Dauven, A. (2001). A survey of the production and use of animal manures in England and Wales. III Cattle manures. Soil Use and Management, 17, 77-87.

Smith, P., Powlson, D.S., Smith, J.U., Falloon, P. and Coleman, K. (2000). Meeting the UK's climate change commitments: options for carbon mitigation on agricultural land. Soil Use and Management, 16, 1-11.

Thorman, R. (2002). Nitric oxide emissions from agricultural soils. PhD University of Edinburgh.

Watterson, D. UK Greenhouse Gas Inventory, 1990 to 2001. (2003). Annual Report for submission under the Framework Convention on Climate Change.

A new agricultural ammonia emission inventory for Switzerland based on a large scale survey and model calculations

B. Reidy[1], S. Pfefferli[2] and H. Menzi[1]
[1]*Swiss College of Agriculture, Zollikofen, Switzerland*
[2]*Swiss Federal Research Station for Agricultural Economics and Engineering, Tänikon, Switzerland*

Abstract

The Gothenburg Protocol will require member countries to report regularly on the evolution of NH_3 emissions from agriculture. Existing emission inventory approaches are not sufficient to detect relatively small changes of emissions reproducibly, because they are mainly based on expert assumptions. A new emission inventory approach has therefore been developed, which is based on a representative survey and model calculations with a newly developed empirical NH_3 emission model (DYNAMO). The stratified survey considered 36 different classes for geographical and farm type criteria. Based on individual emission calculations for each farm analysed in the survey, average emissions per animal were calculated for 24 livestock categories and each class of the survey. Multiplying these results with the corresponding number of animals at district level resulted finally, in the new Swiss NH_3 emission inventory. The new emission inventory approach allows a more detailed analysis of the regional distribution of NH_3 emissions, as well as a detailed monitoring of the development of emissions in time, taking into account a large number of relevant management options. Thanks to the stratified sampling of the survey, it is also possible to compare emissions of different regions and altitudes, as well as emissions from different main farm types.

Keywords: ammonia, emissions, inventory, manure, survey

Introduction

Agricultural production is well recognised as the major contributor of NH_3 to the atmosphere. In Switzerland, NH_3 losses from agriculture were estimated at 53.5 and 51.3 kt N for 1990 and 1995, respectively, (Menzi et al., 1997). Agriculture was responsible for nearly 90% of the total NH_3 emissions. Ammonia emissions have serious ecological consequences. Beside eutrophication and acidification of natural ecosystems, NH_3 is known to be responsible for the formation of secondary aerosols, leading to adverse effects on human health (Dockery et al., 1993). In the 1999 Gothenburg Protocol to Abate Acidification, Eutrophication and Ground-level Ozone within the framework of the Convention on Long-range Transboundary Air Pollution, NH_3 was therefore newly introduced as an air pollutant for which member countries of the convention have to achieve emission ceiling values.

Emission ceiling values imply that the development of emissions has to be closely monitored and that emission inventories must reliably show relatively small changes of emissions in time. Various European countries have published detailed national NH_3 emission inventories in recent years (e.g. Pain et al., 1998; Stadelmann et al., 1998; Misselbrook et al., 2000; Hutchings et al., 2001; Döhler et al., 2002; Hyde et al., 2003). As most European inventories published so far (e.g. Buijsmann et al., 1987; Klaassen, 1992) and the 'simple methodology' of EMEP/CORINAIR (2001), many of these inventories were based on the principle of multiplying defined emission factors (e.g. emissions per cow, or per pig) with the corresponding activity data (e.g. animal numbers; Menzi et al., 2002). Together with ECETOC (1994), the Swiss emission inventory for 1990 and 1995 (Menzi et al., 1997; Stadelmann et al., 1998) was one of the first inventories introducing a new calculation approach based on N flux (Menzi & Katz, 1997). By calculating emissions as %N passing the different stages of emissions (grazing, housing, manure storage, manure application), this approach respects interdependencies between emissions at different emission stages. This is a prerequisite for taking into account differences and changes in production technique without having to adjust emission factors. Nitrogen flux approaches were recently also developed by Cowel & ApSimon (1998) and Dämmgen et al. (2002).

In spite of the N-flux approach, the Swiss emission inventory is considered insufficient for detecting relatively small changes of emissions in time and space reliably, as this will be necessary for emission monitoring and reporting under the Convention on Long-range Transboundary Air Pollution. This is mainly because farm management characteristics, considered in the emission calculation model, were based on estimates and assumptions of national experts, as statistical information was generally lacking. No expert is capable of reliably and reproducibly recognising small changes over a short period of two to five years on a national or regional scale. Furthermore, the consistency of the estimates of experts from different regions (or in the case of replacing an expert) cannot be guaranteed. An additional limitation of the inventory was the limited number of farm management properties considered in the calculations. Since NH_3 emissions are strongly influenced by a great number of management factors, which may vary between different farms as well as over time, it is important to consider as many of these factors as possible in the calculation, in order to obtain a reasonable resolution of the emissions in time and space.

To overcome these major drawbacks, we have developed a new emission inventory approach based on a large scale survey and calculations with a specially developed empirical model. The new calculation approach is independent of expert assumptions on farm management, and takes into account a great number of relevant farm management parameters. It thus allows a standard procedure for regular inventory calculations used for emission monitoring.

General approach

The new emission inventory calculation approach is based on a multi-stage procedure (Figure 1). Firstly, a new empirical model for the calculation of NH_3 emissions (DYNAMO) was developed at the whole farm level. In order to obtain the model input data for the emission calculation, step two consisted of a representative farm survey of a great number of management factors relevant to NH_3 emissions. The survey questionnaire was sent to a stratified random sample of approximately 6% of Swiss farms. In a subsequent step, emissions were calculated individually using the DYNAMO model for each farm that participated in the survey with the input data obtained from the survey. Step four resulted in the calculation of weighted average emission

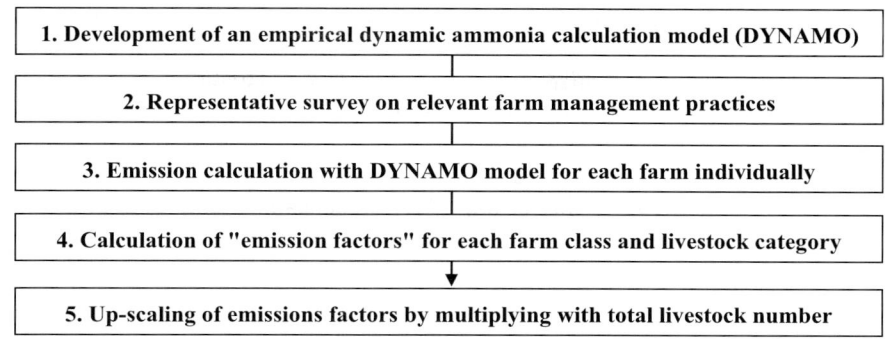

1. Development of an empirical dynamic ammonia calculation model (DYNAMO)

2. Representative survey on relevant farm management practices

3. Emission calculation with DYNAMO model for each farm individually

4. Calculation of "emission factors" for each farm class and livestock category

5. Up-scaling of emissions factors by multiplying with total livestock number

Figure 1. Overview of the new ammonia emission inventory calculation approach.

factors for specific farm classes and livestock categories. These emission factors were finally up-scaled by multiplying them with the relevant livestock number at district level. The aggregation of the emissions from livestock production and manure management with other emission sources resulted in the new Swiss agricultural NH$_3$ emission inventory.

Dynamic ammonia emission calculation model (DYNAMO)

DYNAMO is an Excel based empirical calculation model, basically following the N flux approach presented by Menzi & Katz (1997), but substantially extended with respect to the number of variables included and its user-friendliness. As an N-flux model, the program calculates the N flow from the excretion over the different emission stages on a farm: for different livestock categories and manure types. Ammonia emissions are calculated by emission rates expressed in percent of the amount of volatile N present at each emission stage. The same model can be used in the 'single farm' as well as in the 'regional mode', where it is possible to calculate the NH$_3$ emissions for whole geographical regions.

The model calculates NH$_3$ emissions separately for 24 different animal categories: cattle (dairy cows, dairy followers, calves, beef cattle and suckler cows), pigs (dry sows, farrowing sows, boars, fattening pigs and weaners), poultry (laying hens, growers, broilers, turkeys, geese and other poultry), horses (mares, fillies, mules and donkeys), goats and sheep. Considered emission stages include grazing, hardstandings, animal houses, manure storage and manure application, mineral fertiliser, as well as emissions from crops and meadows. Depending on the animal type and housing system, the model can integrate three different manure types (slurry, solid manure from tide housing systems, deep litter solid manure) in the calculation. The amount of NH$_3$ emitted depends on the number and categories of animals present on a farm and on the relevant farm management practice. The model therefore considers ca. 300 farm management practice variables which are known to have an influence on NH$_3$ emissions. Examples of such variables are the composition of the feed ration for milking cows and pigs, grazing time and number of grazed animals, the use and surface type of hardstandings, covering of slurry tanks and manure application techniques. The management practice variables are used to generate correction factors which modify the relevant standard emission rates.

Results are provided in a detailed, as well as in a summarised form, by means of tables and graphs and can also be easily exported to other software packages. In the detailed form, absolute NH_3 emissions are given per animal category and emission stage, and also in percent of the N excreted. The summarised form provides results on the total NH_3 emission per farm or region and on the relative contribution of each animal category to the total NH_3 emissions.

Representative survey on farm management practice

To collect data on farm management practice, a mail survey was conducted with a representative sample of farms. The chosen farms received an extensive questionnaire containing ca. 300 questions on farm management systems and practice such as grazing, hardstandings, animal housing, manure management systems, manure utilisation practice, manure application techniques, feeding and mineral fertiliser use. The questionnaire was designed to allow an automated data input. To account for differences in farm management between different farm types and geographical regions, the stratified sample of farms was grouped into 36 classes (Table 1).

Table 1. *Thirty-six farm classes representing three geographical regions, three altitude zones and four farm types of Switzerland.*

Geographical region	Altitude zone	Farm type
eastern Switzerland	lowlands	cattle farms
west/south Switzerland	hill region	pig/poultry farms
central Switzerland	mountain region	arable farms
		mixed farms

To allow a reliable statistical analysis of the results, at least 2% of the farms (or a minimum of about 20 farms per class) were considered necessary. This assumption was based on the experience of the Federal Research Station for Agricultural Economics and Engineering and the Swiss Federal Statistical Office. Based on these assumptions and an expected participation rate of 40%, the total sample size was 3880 farms out of the existing ca. 60000 Swiss farms meeting the defined minimum farm size criteria.

The stratified random sampling of the farms was done by the Swiss Federal Statistical Office. Parallel to sending out the questionnaire to the farms, an information campaign on the project was launched in farmer journals. As an incentive to participate, a lottery for the participating farms was organised. After two reminders, 50.3% of the farms (1950 farms) had returned the questionnaire. To reduce the work for the participants, the questionnaire did not include data on livestock numbers and farm surfaces, which are regularly collected by the Statistical Office. To be able to use the statistical data without breaching statistical confidentially, the survey data and the statistical data were combined by the Statistical Office in a database that did not allow the identification of the individual farm. Subsequent data analysis was performed with this anonymous database. Extensive plausibility checks were performed to check the reliability of the data. They revealed surprisingly good data quality and the conclusion that farmers had answered honestly. Furthermore, the comparison of the structure of the farms included in the calculation with that of the total of the farms in each class gave no reason to assume a biased response.

Emission calculations and up-scaling of results

Using the DYNAMO model, emissions were calculated for each of the 1950 farms participating in the survey. Emission calculation was done automatically by using a software interface that allowed the automatic import of the input data from the database and the subsequent saving of the results for each farm. Emissions for each livestock category present on a farm and for each emission stage were calculated, as well as the total emissions of the farm. As major inputs used for up-scaling to the emission inventory, mean emissions factors per animal and per emission stage (grazing, hardstandings, housing, manure storage, manure application,) were derived for each of the 24 livestock categories considered and for each of the 36 different farm classes. To obtain the total NH_3 emissions from livestock production, the differentiated emission factors were subsequently multiplied with the total animal numbers for each of the farm classes on a district level. They were then aggregated at the national level to obtain national emissions from livestock production and manure management (Table 2). In the next step, these results will be combined with the emissions from mineral fertiliser application and other agricultural emission sources (i.e. emissions from crops and grassland and from application of sewage sludge), as well as non-agricultural sources, to produce the new Swiss NH_3 emission inventory.

Table 2. New Swiss ammonia emission inventory for livestock. Preliminary data.

Animal category	kt NH_3-N
Cattle	28.6
Pigs	6.5
Poultry	1.7
Horses, ponies, donkeys	0.7
Sheep, Goats	0.9
Livestock total	38.4

As compared with the 1990 inventory (Menzi et al., 1997; Stadelmann et al., 1998), emissions from livestock production decreased by approximately 19%. This change was mainly due to a decrease of over one third of emissions from pig production and of approximately 17% from cattle production. This development resulted largely from a decrease in animal numbers. The cattle share of total emissions from livestock production increased from 73 to approximately 75%.

The share of emissions from animal houses, relative to total emissions from livestock production, increased from 25 to approximately 29% because of a strong increase in loose housing systems for cattle (Figure 2). Emissions from manure applications decreased from 59 to 54% because of the increasing use of band application systems for slurry while emissions from manure storage decreased slightly from 11 to 10%. In spite of the introduction of hardstandings on nearly 80% of the cattle farms within the last eight years, emissions from hardstandings were only ca. 1% of total emissions from livestock production, because the utilisation time of these hardstandings is low, especially on farms with tied housing systems.

To better reveal differences in the regional distribution of NH_3 emissions, the results at district level were combined with land use data to produce emission maps (Figure 3).

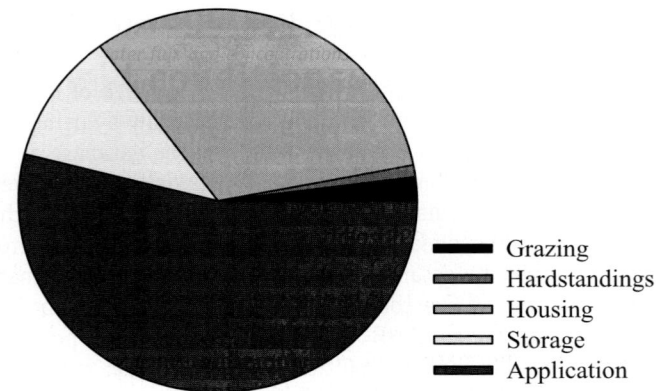

Figure 2. Contribution of each emission stage (%) of the total NH_3 emissions from livestock production and manure management. Preliminary data.

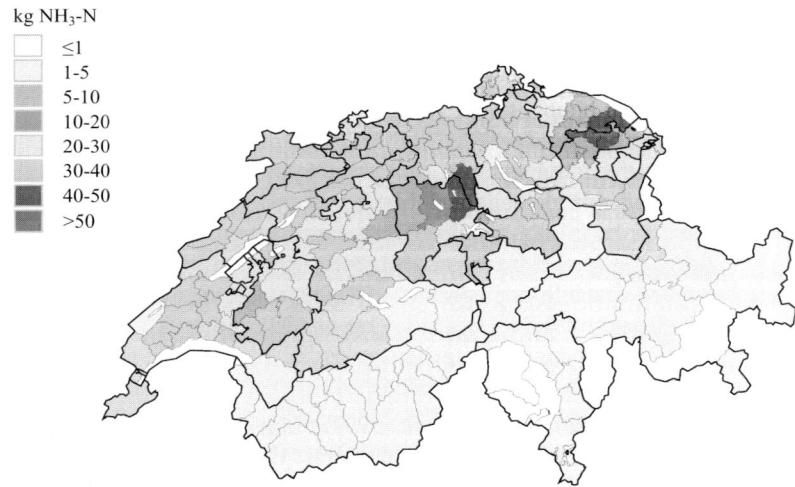

Figure 3. Emission map showing the regional distribution of agricultural NH_3 emission in Switzerland at the district level. Emissions are expressed in kg NH_3-N ha^{-1} yr^{-1} and are preliminary data.

Conclusions

As compared with the previous Swiss NH_3 emission inventory, the new inventory calculation approach has several main advantages. (1) As the input data on farm management practice is based on a survey, rather than on expert assumptions, the results are independent from the personal view of individuals. The data can therefore, be considered to reliably reflect actual conditions. Moreover, the standardised approach will allow a good comparison of future inventories with the 'baseline' for 2000. (2) The differentiation of different regions and farm types allows a much more detailed and reliable analysis of the factors influencing the emissions. Furthermore, the maps on the spatial distribution of emissions are more detailed and reliable. (3) Because emissions are calculated for each farm, it is possible to obtain information on the variability of emissions between farms (e.g. Table 3). This, together with the information on farm management practice obtained from the survey, will allow a much more detailed assessment of the emission abatement potential of different geographical regions and farm types. It also presents the possibility to derive more specific practical recommendations on how farmers can reduce NH_3 emissions. The interest of farmers for such recommendations is considerable, because of the restrictive limits on N use imposed by the Swiss agricultural policy.

Nevertheless, the new calculation approach also bears certain risks or disadvantages. The results will only reliably reflect actual conditions if the sample of farms selected for the survey is representative, and if the information provided by the farmers is reliable. A good survey design with questions that allow cross-checking and feasibility analysis is the best precaution to check the reliability of the information received. The survey approach is only possible as long as there is no serious direct opposition to the inventory from the side of farmers. At present, no such problems were encountered, but the attitude of farmers might change in the future, e.g. if significant legal restrictions are issued to mitigate NH_3 emissions. It can also be assumed that this aspect could be a greater problem in other countries, because the Swiss agricultural policy with its strict limits on N use is a good incentive for farmers to reduce N losses.

Table 3. Calculated mean emission factors per dairy cow for the three altitude zones. The emission factors are shown for each emission stage and in total. The variation coefficient (v.c.) for each value is indicated in italics. Data are preliminary.

Geographical region	Farms	(kg NH_3-N yr^{-1})					
	(N)	grazing	hardstandings	housing	storage	application	total
Lowland	574	0.5	0.6	8.5	3.8	19.6	33.0
v.c. (%)		*194*	*95*	*35*	*57*	*35*	*23*
hill region	456	0.5	0.6	7.6	3.5	19.3	31.5
v.c. (%)		*213*	*87*	*31*	*57*	*34*	*22*
mountain region	435	0.3	0.8	6.7	3.4	20.5	31.7
v.c. (%)		*200*	*82*	*29*	*52*	*47*	*32*

We hope that the new approach will not only allow a more reliable and detailed emission inventory but also a more profound assessment of the abatement potential and corresponding practical recommendations in Switzerland. Such results could be of tremendous help to improve the awareness of farmers to the problem of N losses through NH_3 volatilisation and to support their willingness to adopt voluntary measures to reduce such losses.

References

Buijsman, E., Maas, H.F.M. and Asman, W.A.H. (1987). Antropogenic NH_3 emissions in Europe. Atmospheric Environment, 21, 1009-1022.

Cowell, D.A. and ApSimon, H.M. (1998). Cost effective strategies for the abatement of ammonia from European agriculture. Atmospheric Environment, 32/33, 573-580.

Dämmgen, U., Lüttich, M., Döhler, B., Eurich-Menden, B. and Osterburg, B. (2002). GAS-EM - a procedure to calculate gaseous emissions from agriculture. Landbauforschung Völkenrode, 52, 19-42.

Döhler, H., Dämmgen, U., Eurich-Menden, B., Osterburg, B., Lüttich, M., Berg, W., Bergschmidt, A. and Brunsch, R. (2002). Anpassung der deutschen Methodik zur rechnerischen Emissionsermittlung an internationale Richtlinien sowie Erfassung und Prognose der Ammoniak-Emissionen der deutschen Landwirtschaft und Szenarien zu deren Minderung bis zum Jahre 2010. Abschlussbericht im Auftrag von BMVEL und UBA, UBA-Texte 05/02, Umweltbundesamt, Berlin, 192 pp.

Dockery, D.W., Pope, C.A., Xu, X.P., Spengler, J.D., Ware, J.H., Fay, M.E., Ferris, B.G. and Speizer, F.E. (1993). An association between air-pollution and mortality in 6 United-States cities. New England Journal of Medicine, 329, 1753-1759.

ECETOC (European Centre for Ecotoxicology and Toxicology of Chemicals) (1994). Ammonia emissions to air in western Europe. Technical Report No. 62, 196 pp.

EMEP/CORINAIR (2001): Atmospheric Emission Inventory Guidebook, 3rd Ed., Draft of Chapter 100900: Manure Management Regarding Nitrogen Compounds.

Hutchings, N.J., Sommer, S.G., Andersen, J.M. and Asman, W.A.H. (2001). A detailed ammonia emission inventory for Denmark. Atmospheric Environment, 35, 1959-1968.

Hyde, B.P., Carton, O.T., O'toole, P. and Misselbrook, T.H. (2003). A new inventory of ammonia emissions from Irish agriculture. Atmospheric Environment, 37, 55-62.

Klaassen, G. (1992). Discussion and conclusion on emission inventories and emission coefficients for ammonia. In: G. Klassen (ed.). Ammonia emissions in Europe: emission coefficients and abatement costs. Proceedings of workshop February 4-6 1991. IIASA, Laxenburg. pp 159-168.

Menzi, H. and Katz, P.E. (1997). A differentiated approach to calculate ammonia emissions from animal husbandry. In: J.A.M. Voermans and G.J. Monteny (eds). "Ammonia and odour emissions from animal production facilities", Proceedings of International Symposium, Vinkeloord, NL, 6-10 October 1997, 35-42.

Menzi, H., Frick, R. and Kaufmann, R. (1997). Ammoniakemissionen in der Schweiz: Ausmass und technische Beurteilung des Reduktionspotentials. Schriftenreihe der Forschungsanstalt für Agrarökologie und Landbau (FAL), 26, 107 pp.

Menzi, H., Dämmgen, U. and Döhler, H. (2002). Emissionsinventare für Ammoniak, Methan und Lachgas. B. Eurich-Menden und H. Döhler (eds). Proceedings of Symposium "Emissionen aus der Tierhaltung und Beste Verfügbare Techniken zur Emissionsminderung", Kloster Banz, Deutschland, 3-5 Dezember 2001, 215-230.

Misselbrook, T.H., Van Der Weerden, T.J., Pain, B.F., Jarvis, S.C., Chambers, B.J., Smith, K.A., Phillips, V.R. and Demmers, T.G.M. (2000). Ammonia emission factors for UK agriculture. Atmospheric Environment, 34, 871-880.

Pain, B.F., Van Der Weerden, T.J., Chambers, B.J., Phillips, V.R. and Jarvis, S.C. (1998). A new inventory for ammonia emissions from UK agriculture. Atmospheric Environment, 32, 309-313.

Stadelmann, F.X., Achermann, B., Lehman, H.J., Menzi, H., Pfefferli, S., Sieber, U. and Zimmermann, A. (1998). Ammonia emissions in Switzerland: Present situation, development, technical and economic assessment of abatement measures, recommendations. Institut für Umweltschutz und Landwirtschaft Liebefeld (IUL) und Forschungsanstalt für Agrarwirtschaft und Landtechnik (FAT). 56 pp.

N$_2$O emissions from a field trial as influenced by N fertilisation and nitrification inhibitors

D. Báez, J. Coutinho, N. Moreira and H. Trindade
Department of Plant Science and Agricultural Engineering, UTAD, Ap. 1013, 5001-911 Vila Real, Portugal

Introduction

As a greenhouse gas, N$_2$O contributes to global warming. Animal manure, slurries and inorganic N fertiliser application may cause important emissions of N$_2$O. Ammonium from fertilisers is oxidised to NO$_3^-$ by nitrifying micro-organisms and may be the major source of N$_2$O emissions. In recent years, it has been proved that the use of nitrification inhibitors (NIs) may increase N use efficiency by crops and also decrease N$_2$O losses. The objectives of our study were (i) to determine in a field trial (after autumn N fertilisation) the effect of N fertiliser source (cattle slurry and inorganic N fertilisers) on the N$_2$O emissions, and (ii) to evaluate the influence of two NIs on N$_2$O emissions, dicyandiamide (DCD) and 3,4-dymethyl pyrazole phosphate (DMPP), when applied together with fertilisers.

Materials and methods

The experiment was carried out at Vila Real (Portugal) from October 2002 to January 2003 on a poorly drained silty loam soil, classified as Dystric Cambisols (FAO classification) with a (0-30 cm) pH value of 6.4, OM content of 29 g kg^{-1} and 4.9 cmol(+) kg^{-1} of CEC. Oat (*Avena sativa* L.) was sown at 130 kg ha^{-1} on 11 November. The trial was laid out as a two factor factorial with a split-block layout and three replications. One factor consisted of N source with 9 treatments, ammonium sulphate nitrate (ASN, 19.5% NH$_4^+$-N 6.5% NO$_3^-$-N), 2 NIs-stabilised N fertilisers Entec®26 (COMPO, 1% DMPP relative to NH$_4^+$-N in the ASN) and Nitrotop (ADP, 8% NH$_4^+$-N 16% Amide-N, 4 % DCD-N relative to total N) and 6 combinations of surface banded (SS) or injected (SI, 30 cm depth) cattle slurry, either with or without DMPP or DCD as a NI. There was also a nil N treatment control. The second factor consisted of two tillage systems, no-till and conventional tillage practice. Slurry was applied on 30 October and mineral fertilisers on 8 November. Fluxes of N$_2$O were measured weekly using the closed chamber method. The gas samples were analysed for N$_2$O using a gas chromatographic system with ^{63}Ni electron capture detector (ECD). At each gas sampling, soil samples were taken to a depth of 10 cm, and the water filled pore space (WFPS) and mineral N content (NH$_4^+$-N and NO$_3^-$ -N) were determined. Cumulative N$_2$O losses were calculated assuming a mean flux rate between a one date of gas sampling and the next date.

Results and discussion

No tillage effect was observed on N$_2$O emissions. Higher N$_2$O losses could be expected in the no-tillage system because of higher denitrification activity (Rodriguez & Giambiagi, 1995) compared with conventional tillage system. Probably the tillage effect on N$_2$O emission is related with the time since no-tillage system has been established (Jacinthe & Dick, 1997) and the data in our experiment correspond to the first winter season after no-tillage establishment.

During the study, the values of WFPS were >70%, suggesting that denitrification was the major process responsible for emissions and N_2 was probably the final reaction product. Cumulative N_2O-N emissions during autumn (Table 1, 11 Nov - 28 Jan) were highest on ASN treatment and smallest on the cattle slurry injected together with DCD with similar values to the control plot. In relation to the amount of N applied, emissions decreased in the following order: ASN> Entec>Nitrotop> SI+DMPP>SS>SI>SS+DMPP>SS+DCD>SI+DCD. Inorganic N fertilisers produced higher losses than slurry but, because of the large variability in the measurements, were only significantly different when the slurry was injected with DCD.

Higher NH_4^+-N contents (0-10 cm soil layer) in both NIs treatments, in comparison with ASN and SS treatment without NI, were observed 25 (inorganic fertilisers) and 34 days (slurry fertilisers) after fertilisation (data not shown), which showed a clear inhibition of NH_4^+ oxidation after the application of DMPP and DCD. However, differences in cumulative N_2O emissions between treatments were only observed in slurry injected treatments, showing that the use of DCD led to smaller N_2O emissions. Apparently, although the differences were statistically not significant, both NIs also decreased cumulative N_2O emissions when added to slurry surface applied.

Table 1. Cumulative N_2O emissions (g N_2O-N ha^{-1}) from the field trial during autumn for cattle slurry and inorganic fertiliser treatments and percentage of N fertiliser lost as N_2O (% N-fert).

Treatment	N-fert. kg N ha^{-1}	N_2O emission 11 Nov- 28 Jan	N_2O emission 31 Oct- 28 Jan	% N-fert
Slurry surface banded (SS)	88	1636 ab	1926 a	1.0
SS + DMPP	93	1377 bcd	1549 a	0.5
SS + DCD	92	1281 bcd	1447 a	0.4
Slurry injected (SI)	121	1660 ab	1826 a	0.6
SI + DMPP	105	1976 ab	2234 a	1.1
SI + DCD	100	953 d	1103 b	0.1
ASN	30	2256 a		4.2
Entec 26	30	2179 ab		4.1
Nitrotop	30	1705 ab		2.5
Control	0	953 cd	1044 b	

N_2O-N emission values followed by different letters are significantly different (p<0.05, LSD test)

Conclusions

During the experiment, no tillage effect was observed on N_2O emissions. The application of inorganic N fertilisers showed a higher percentage of N lost as N_2O than after cow slurry applications.

Acknowledgements

This study was funded by project Nº 177-Acção 8.1 DE & D-Programa AGRO and D. Báez held a post-Doc. grant from the Spanish government. We thank COMPO Agricultura and Adubos de Portugal for supplying DMPP and DCD.

References

Jacinthe P. A. and Dick, W. A. (1997). Soil management and nitrous oside emissions from cultivated fields in Southern Ohio. Soil Tillage Research, 41, 221-235.

Rodriguez, M. B. and Giambiagi, N. (1995). Denitrification in tillage and no tillage pampean soils:relationships among soil water, available carbon, and nitrate and nitrous oxide production. Commun. Soil Science and Plant Analysis, 26, 3205-3220.

Denitrification and interactions between nitrification and methane oxidation under elevated atmospheric CO_2

E.M. Baggs[1], H. Blum[2], M. Richter[2], G. Cadisch[1] and U.A. Hartwig[2]
[1]*Imperial College London, Wye Campus, Department of Agricultural Sciences, Wye, Ashford, Kent TN25 5AH, UK*
[2]*Institute of Plant Sciences, ETH Zürich, 8092 Zürich, Switzerland*

Introduction

During the past few decades, atmospheric concentrations of CO_2, CH_4 and N_2O have been increasing at rates of 0.5, 0.8 and 0.3% yr^{-1}. There is uncertainty about the effect of increasing atmospheric concentrations of CO_2 on the processes of denitrification, nitrification and CH_4 oxidation that determine net emissions of N_2O, N_2 and CH_4 from soils. Emissions of N_2O have been shown to increase from fertilised (56 g N m^{-2} yr^{-1}) *Lolium perenne L.* swards under elevated pCO_2 (60 Pa) (Baggs et al., 2003a), which is hypothesised to result from the greater belowground C allocation under elevated pCO_2 driving denitrification. Such fertiliser application may inhibit CH_4 oxidation in soil, and it is likely that N_2O is produced during NH_3 oxidation by methylotrophs during this inhibition, resulting in an interaction between NH_3 oxidation (nitrification), N_2O production and CH_4 oxidation. We report here two short-term experiments undertaken at the Swiss FACE experiment in which these processes, and the interactions between them, were examined in *L. perenne* swards under ambient (36 Pa) and elevated (60 Pa) pCO_2.

Materials and methods

Experiment 1: Effect of elevated pCO2 on loss of N by denitrification
Fertiliser was applied at 11.2 g N m^{-2} as (a) $^{14}NH_4^{15}NO_3$ or (b) $^{15}NH_4^{15}NO_3$ (1 atom% excess ^{15}N) to different replicates. ^{15}N-N_2O and ^{15}N-N_2 fluxes measured from (a) replicates were attributed to denitrification, and ^{15}N-N_2O fluxes from (b) replicates minus ^{15}N-N_2O from (a) replicates were attributed to nitrification, as tests had shown that there was no dissimilatory $^{15}NO_3^-$ reduction or immobilisation and subsequent re-mineralisation of $^{15}NO_3^-$ (Baggs et al., 2003b).

Experiment 2: Effect of elevated pCO2 on CH4 oxidation and N2O emission during nitrification
^{15}N labelled fertiliser was applied to *L. perenne* swards at 11.2 g N m^{-2} as in Experiment 1. ^{13}C-CH_4 (11 µl l^{-1}; 11 atom% excess ^{13}C) was applied to a closed chamber headspace and oxidation rates estimated from the decrease in ^{13}C-CH_4 concentrations as a function of time. N_2O emitted during nitrification was determined as described above.

Results and discussion

In experiment 1, total denitrification (N_2O + N_2) was increased under elevated pCO_2 with emissions of 6.2 and 19.5 mg ^{15}N m^{-2} over 22 days from ambient and elevated pCO_2 swards, respectively, supporting the hypothesis that increased belowground C allocation in *L. perenne* swards under elevated pCO_2 provided the energy for denitrification. Nitrification was the predominant N_2O producing process under ambient pCO_2, whereas denitrification was predominant under elevated pCO_2. The N_2:N_2O ratio during denitrification was often higher under elevated pCO_2 (Figure 1a).

Elevated pCO2 had no significant effect on net CH4 emission, but CH4 oxidation was more rapid under elevated pCO2 (Figure 1b). It is possible that elevated pCO2 reduced the inhibitory effect of high N on CH4 oxidation.

Gross nitrification and emissions of N_2O during nitrification were greater under ambient pCO_2. This may have been due to N_2O production by methylotrophs following inhibition of CH_4 oxidation.

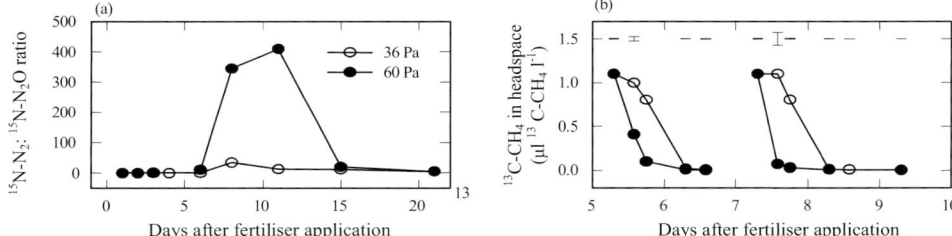

Figure 1. (a) ^{15}N-N_2-to-^{15}N-N_2O ratio during denitrification following application of 11.2 g N m^{-2} (1 atom% excess ^{15}N) in Experiment 1, (b) Concentrations of 13C-CH4 in microplot chamber headspace following application of 11 ppm 13C-CH4 (11 atom% 13C) on days 5 and 7 in Experiment 2.

Conclusions

Our results indicate significant interactions between increasing atmospheric levels of CO_2, N_2O emissions and CH_4 oxidation that need to be considered before appropriate strategies can be proposed that mitigate both N_2O and CH_4 emissions. Our findings are of global significance as increases in atmospheric concentrations of CO_2 may, depending on fertiliser management, increase greenhouse gas emissions of N_2O and CH_4, thereby exacerbating the forcing effect of elevated CO_2 on global climate.

Acknowledgements

This work was funded in part by a Research Fellowship under the OECD Co-operative Research Programme: Biological Resource Management for Sustainable Agriculture Systems, and by a BBSRC Wain Research Fellowship.

References

Baggs, E.M., Richter, M., Hartwig, U.A. and Cadisch, G. (2003a). Nitrous oxide emissions from grass swards during the eighth year of elevated atmospheric pCO_2 (Swiss FACE). Global Change Biology, 9, 1214-1222.

Baggs, E.M., Richter, M., Cadisch, G. and Hartwig, U.A. (2003b). Denitrification in grass swards is increased under elevated atmospheric CO_2. Soil Biology and Biochemistry, 35, 729-732.

Soil water content as a factor that controls N_2O production by denitrification and autotrophic and heterotrophic nitrification

E.J. Bateman, G. Cadisch and E.M. Baggs
Imperial College London, Wye Campus, Department of Agricultural Sciences, Wye, Ashford, Kent, TN25 5AH, UK

Introduction

Emissions of N_2O from fertilised agricultural soils are a major contributor to the increase in atmospheric concentration of N_2O (currently 0.8 ± 0.2 ppb per year, IPCC, 2001). This is of concern because N_2O contributes to global warming and the destruction of stratospheric O_3. The biological processes of nitrification and denitrification are the major source of N_2O from fertilised soils. Denitrification is an anaerobic process whereby NO_3^- is reduced to N_2 with N_2O as a regular intermediate. Nitrification, the aerobic oxidation of NH_4^+ to NO_3^-, is primarily carried out by a monophyletic group of autotrophic bacteria. There is also evidence that heterotrophic bacteria and fungi are capable of nitrifying NH_4^+ as well as some other organic N compounds in culture and soil (Papen et al., 1989). However, there is uncertainty about the relative contributions of these processes to N_2O emissions from soils under different conditions, which therefore limits the development of strategies to control them.

Materials and methods

Because nitrification (autotrophic and heterotrophic) and denitrification may proceed simultaneously within different microsites of the same soil, the development of techniques to distinguish the source of N_2O produced in soils is essential. In this study, acetylene (C_2H_2), a specific inhibitor of autotrophic nitrification, was shown to significantly reduce N_2O emissions from soils fertilised with NH_4NO_3 for up to seven days. Using a stable isotope (^{15}N) technique, ^{15}N-N_2O from soils fertilised with $^{14}NH_4^{15}NO_3$ was attributed to denitrification, whilst ^{15}N-N_2O from soils fertilised with $^{15}NH_4^{15}NO_3$ was attributed to both nitrification and denitrification (Baggs et al., 2003). The difference in ^{15}N-N_2O between these treatments was attributed to nitrification, as tests had shown that there was no dissimilatory $^{15}NO_3^-$ reduction or immobilisation and subsequent re-mineralisation of $^{15}NO_3^-$. Soil water content and aeration are known to affect the magnitude of N_2O emissions from soils following fertiliser application (Bollmann & Conrad, 1998). Therefore laboratory soil incubations were carried out using the C_2H_2 and ^{15}N techniques in combination to determine how soil water filled pore space (WFPS: a function of soil water content and porosity) influences denitrification, and autotrophic and heterotrophic nitrification contributions to N_2O production.

Results

N_2O emissions increased with soil water content in the range 20 to 70% WFPS (Figure 1). Nitrification and denitrification both contributed to N_2O emissions at 60% WFPS and below

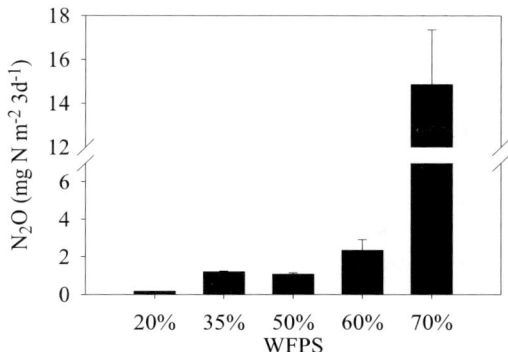

Figure 1. Total N₂O emissions (+SE) in 3 days after N application to soils at different WFPS

Table 1. Total ^{15}N-N$_2$O emissions (SE) in 3 days after N application to soils at different WFPS, and the percentage contribution of nitrification and denitrification.

% WFPS	Total ^{15}N-N$_2$O μg m^{-2} (3 d)	% Denitrification	% Autotrophic nitrification	% Heterotrophic nitrification
50%	32 (10.6)	22.1	64.7	13.2
70%	825 (295)	100	0	0

although autotrophic nitrification was the dominant source, whilst denitrification was predominant at 70% WFPS. A potential role for heterotrophic nitrification was indicated at 50% WFPS (Table 1).

Discussion and conclusions

Inhibition and ^{15}N techniques provide useful estimates of how soil WFPS affects the different processes of N$_2$O production. Our results suggest that traditional understanding of nitrification and denitrification does not fully account for N$_2$O emissions from soils. We found that denitrification is not strictly anaerobic since it contributed at a low water content, and a role for heterotrophic nitrification, as well as autotrophic nitrification, was indicated. These methods may be used to investigate other soil factors that are influenced by agricultural practices. This would allow the development of individual profiles for each N$_2$O$^-$ producing process. Consideration of the separate processes involved and their different response to environmental conditions is necessary to fully understand variation in measured N$_2$O fluxes, and to propose mitigation practices that could reduce emissions.

Acknowledgements

This work was funded by a BBSRC Research Committee studentship. We thank Jon Fear for the stable isotope analyses and Trudi Krol for assisting with the soil mineral N analyses.

References

Baggs, E.M., Richter, M., Cadisch, G. and Hartwig, U.A. (2003). Denitrification in grass swards is increased under elevated CO_2. Soil Biology and Biochemistry, 35, 729-732.

Bollmann, A. and Conrad, R. (1998). Influence of O_2 availability on NO and N_2O release by nitrification and denitrification in soils. Global Change Biology, 4, 387-396.

IPCC (2001). Climate Change 2001. The Scientific Basis. Houghton, J.T., Ding, Y., Griggs, D.J., Noguer, M., van der Linden, P.J., Xiaosu, D.(eds). CUP, UK.

Papen, H., von Berg, R., Hinkel, I., Thoene, B. and Rennenberg, H. (1989). Heterotrophic nitrification by *Alcaligenes Faecalis* - NO_2^-, NO_3^-, N_2O, and NO production in exponentially growing cultures. Applied and Environmental Microbiology, 55, 2068-72.

Short-term CO$_2$ and N$_2$O emission after application of manure and maize residues to three different soil types: a laboratory study

D. Beheydt, P. Boeckx, L. Geers, A. Goossens and O. Van Cleemput
Ghent University, Laboratory of Applied Physical Chemistry - ISOFYS, Coupure Links 653, B-9000 Gent, Belgium

Introduction

In the EU-15, agriculture is responsible for about 10% of the annual greenhouse gas (GHG) emission. The Kyoto Protocol (1997) to the 1992 UN Framework Convention on Climate Change (UNFCCC) determines international emission reduction targets for six greenhouse gases. By demonstrating an increase of C sequestered in terrestrial ecosystems, countries are allowed to offset their emission reduction targets against this increase. The application of crop residues or manure can be proposed as a management option to raise the C content in arable soils. An increase in soil C, however, may enhance the denitrification rate and as a result, the N$_2$O emission from soil. Because N$_2$O has a global warming potential which is 296 times greater than CO$_2$, N$_2$O emission may offset the amount of C sequestered. The aim of this experiment was twofold. In the first place, the emissions of N$_2$O and CO$_2$ were measured after the introduction of two management options to sequester C. The second was the determination of the role of the newly added C to the total CO$_2$ emission, using stable isotope measurements.

Materials and methods

The upper 10 cm of three different arable soils ($\delta^{13}C_{soil}$ varying between -25.9 and -26.4‰) in Belgium, with different textures (according the Belgian texture triangle), were sampled. A pre-incubation period of 7 days with a water-filled-pore-space of 50% and a temperature of 15°C was used. This temperature is common in spring during manure application, and in autumn, when maize stubbles remain on the field after harvesting. The soils (ca. 200 g) were incubated in plastic cylindrical containers at the same bulk density as in the field. Three treatments, each in triplicate, were carried out. In the first treatment, the soil was mixed with ground maize stubble (Table 1). In the second treatment, cow manure was used and the third treatment was a control without amendments. During 40 days, GC analyses, measuring CO$_2$ and N$_2$O fluxes (Shimadzu GC 14-B, ECD ^{63}Ni detector), and isotopic composition of the emitted CO$_2$ (ANCA-SL 20-20, IRMS, PDZ Europa, UK) were carried out. Results were compared using a double-sided t-test with a significance level of 0.05.

Results

Integration of the flux measurements during 40 days (Figure 1) shows that maize and manure application increased the total GHG emission compared with their controls. There was, however, no statistical difference between the total greenhouse gas (GHG) emission from either treatments when a specific soil texture is considered, although the C-addition in the maize and manure treatment differed (0.41 g C kg^{-1} and 0.62 g C kg^{-1}, respectively). Comparing,

Table 1. Properties of the soil amendments.

	Manure	Maize
C/N	17.4	139.5
$\delta^{13}C$	-18.2	-11.8
Application rate	30,000 l ha^{-1}	equivalent to stubble from 100,000 plants ha^{-1}

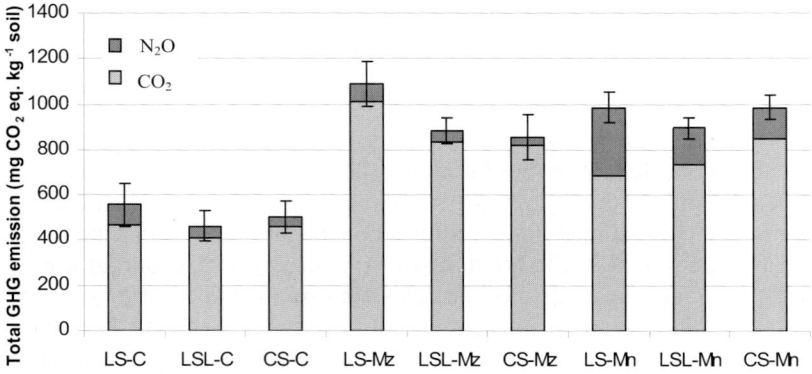

Figure 1. Emission of CO_2 and N_2O after 40 days from a loamy sand (LS), a loam sandy loam (LSL) and a clayey sand (CS) soil after application of maize (Mz) and manure (Mn) in comparison with a control (C).

instead, the two treatments for a certain soil texture, only the loamy sand soil showed a statistical difference in emitted CO_2. The N_2O emission after manure application, however, was for each texture, statistically higher than the incubations with maize or the control.

During the incubation period of 30 days, samples for $\delta^{13}C$-CO_2 analysis were also taken. Based on the $\delta^{13}C$ signature of the soil, the C amendments and the emitted CO_2, it is possible to calculate the contribution of the newly added C to the total CO_2 emission (Högberg & Ekblad, 1996). Results (not shown) showed that CO_2 emission is only partly based on newly added C but because of large standard deviations; no further conclusions could be drawn at this point.

Conclusions

Similar applications in cattle manure or ploughing in harvest residues led to the same amount of GHG emission, when measured during a short term (40 d). Cattle slurry, however, was a more important source of N_2O.

Acknowledgements

We wish to thank the Prime Minister's Office - Belgian Federal Science Policy Office for their financial support.

References

Högberg, P. and Ekblad, A. (1996). Substrate-induced respiration measured in situ in a C$_3$-plant ecosystem using additions of C$_4$-sucrose. Soil Biology and Biochemistry, 28, 1131-1138.

Influence of urinary N concentration on N_2O and CO_2 emissions and N transformation in a temperate grassland soil

R. Bol[1], S.O. Petersen[2], K. Dittert[3] and M.N. Hansen[2]

[1]*Institute of Grassland and Environmental Research, North Wyke Research Station, Okehampton, Devon, EX20 2SB, UK*

[2]*Danish Inst Agricultural Sciences, Dept Agroecology, Research Centre Foulum, DK-8830, Tjele, Denmark*

[3]*Inst of Plant Nutrition and Soil Science, Kiel University, D-24118 Kiel, Germany*

Introduction

Carbon storage in grasslands has been proposed as a CO_2 mitigation option, but this depends on N inputs, since associated N_2O emissions can off-set the removal of atmospheric CO_2. Petersen et al. (1998) found that a higher cattle urine N content was associated with higher urea content and increased NH_3 volatilisation. The amount of N excreted in urine depended on the protein intake of grazing animals, hence volatilisation can be (partly) controlled by management of the feed provided. The present study focused on the influence of enhanced urea-N in urine on soil CO_2, N_2O and NH_3 emissions, soil N transformations.

Materials and methods

The field experiment was carried out between 20th September and 4th October 2001 on an 8 year old grass-clover pasture at the Research Centre Foulum, Denmark (55°52'N, 9°34E). The sandy loam soil contained 2.7% C and 0.18% N and is classified as a Typic Hapludult. On day 0, cattle urine (of known urea-N content) was added to give application rates equivalent to 179 (UL) and 538 kg urea-N ha^{-1} (UH). There were three blocks of three treatments, i.e. a control (Co; no amendments) and two urine treatments (UL and UH). Urea was dual-labelled with 13.4 at % ^{13}C and 13.4 at % ^{15}N. Adjacent plots were established to estimate NH_3 volatilisation from the same rates of unlabelled urea using Ferm chambers. The N_2O and CO_2 emissions, plant uptake of urea-C and -N, and the urea, NH_4^+ and NO_3^- pools, were monitored over 14-d. Isotopic composition was used to estimate the proportion of N derived from the urinary-urea.

Results

The total recovery of ^{15}N was 41%, two thirds of which were found at 0-5 cm soil depth. Plant N derived from urea ranged from 5 to >20% during the experiment, but urea-derived C was only <0.5% after 3 h and decreased rapidly. Ammonia volatilisation was estimated at 19 (UL) and 9% (UH) of applied urea-N. Ammonium concentrations increased during the first few days, peaking at concentrations corresponding to 116 and 69% of added urea-N added in the UL and HL treatments, respectively. Nitrification was delayed for about two days, but then proceeded almost linearly until day 14. Nitrous oxide emissions increased during the 14-d period, as did the proportion derived from urea-N. On day 14, the contribution from urea was 58 and 24% in the UL and UH treatments, respectively. Accumulated losses of N_2O during this period amounted to 0.12 and 0.06% of urea-N level applied in UL and UH. Carbon dioxide fluxes derived from urea were only detectable within the first 24 h. The CO_2 emissions were ca. 1000 larger than for N_2O. However, expressed as global warming potential equivalents (1 $N_2O \sim 310\ CO_2$), they were only 2-3 times higher.

Discussion

The amounts of ^{13}C (20%) and ^{15}N (40%) recovered in the soil were lower than the 83% observed for ^{15}N in the associated laboratory experiment (Petersen et al., this volume). An explanation could be that a significant part of the urine applied was intercepted by plants, as was indicated by the apparent direct uptake of ^{13}C and ^{15}N. The initial concurrent increase in plant ^{13}C and ^{15}N content in the first 24 hours indicated that direct uptake was the predominant process of N and C acquisition. Urea-derived C remained a significant component of the plant until day 6, while the level of urea-derived N in the plant material was maintained throughout the experiment. Surface application maximised urine-plant contact, and hence, the potential for urine retention and NH_3 volatilisation from plants. The NH_4^+ and NO_3^- pool changes were similar to those observed in the laboratory (Petersen et al., this volume) and in the field (Monaghan & Barraclough, 1992). The amount and proportion of urea-derived C in respired CO_2 was greatest in the first 24 hours after urea application to the soil, and reflected degassing of CO_2 released during hydrolysis. In contrast, for N_2O, emission rate and proportion of urea-derived N increased during the 14 day experiment, and was highest in UH. Hence, the effects of urea on the C and N cycling in grassland soils were decoupled. The accumulated emissions of N_2O were low (ca. 0.1% of urea-N applied) during the two week experiment, indicating losses below the emission factor (EF) of 2% recommended by IPCC. The time courses of N_2O and NO_3^- release were correlated and indicated that N_2O emissions were associated with nitrification (Koops et al., 1997). However, net NO_3^- accumulation amounted to only 30 mg N kg^{-1} soil, i.e. < 5% of the urea-N added, and the large NH_4^+ pool suggested that nitrification would continue beyond the 14-d experimental period. In a three-year grazing study at the experimental site, urinary N deposition ranged from 175 to 331 kg ha^{-1} yr^{-1}. With an N_2O EF of 2%, this would offset a C storage of 464 to 879 kg C ha^{-1} yr^{-1}. Carbon accumulation in grassland is typically in the order of 3-400 kg C ha^{-1} yr^{-1} (Post & Kwon, 2000).

In conclusion, this study has indicated a decoupling of the turnover and fate of urea-C and N in urine deposited on pasture. Even though an EF well below the value currently recommended by IPCC was indicated, N_2O emissions from urine deposition alone, may off-set the effect of C storage in the greenhouse gas budget.

Acknowledgements

R. B. thanks the BBSRC for the ISIS grant. This project was conducted as part of EU project EVK2-2000-22045 (MIDAIR).

References

Koops, J.G., van Beusichem, M.L. and Oenema, O. (1997). Nitrous oxide production, its source and distribution in urine patches on grassland soils. Plant and Soil, 191, 57-65.

Monagahan, R.M. and Barraclough, D. (1992). Some chemical and physical factors affecting the rates and dynamics of nitrification in urine-affected soil. Plant and Soil, 143, 11-18.

Petersen, S.O., Sommer, S.G., Aaes, O. and Søegaard, K. (1998) Ammonia losses from urine and dung of grazing cattle: Effect of N intake. Atmosphere and Environment, 32, 295-300.

Post, W.M. and Kwon, K.C. (2000) Soil carbon sequestration and land-use change: processes and potential. Global Change Biology, 6, 317-327.

A model-based evaluation of options for the mitigation of agricultural nitrous oxide emission

L. Brown[1], L. Cardenas[1], P. Bellamy[2], S.C. Jarvis[1], J. Hollis[2], R.W. Sneath[3], S. Yamulki[1] and K.W.T. Goulding[4]
[1]*Institute of Grassland and Environmental Research, Okehampton, Devon, UK*
[2]*National Soil Resources Institute, Silsoe, Bedford, UK*
[3]*Silsoe Research Institute, Silsoe, Bedford, UK*
[4]*Institute of Arable Crops Research, Harpenden, Herts., UK*

Introduction

As signatory States to the Kyoto Protocol, European countries are required to reduce their overall emission of greenhouse gases by 8% by the period 2008-2012, relative to 1990 levels. It was estimated that in 1990, 47% of the total UK N_2O emission was due to agriculture (Salway et al., 1999). As industrial sources have declined in recent years, the relative contribution of the agricultural sector to the total emission has increased. This would suggest that strategies implemented in the agricultural sector may have a large impact on the total N_2O emission in the UK. While the agricultural sector emissions have shown small reductions in recent years in the UK, this has been due to incidental changes, which may vary from year to year (e.g. animal numbers, fertiliser use). If we are to achieve more significant and consistent reductions, effective strategies will be required for application at farm level.

Methodology

A range of proposed farm-level mitigation options were evaluated according to their availability, practicality and public acceptability. The efficacy of some selected strategies was evaluated using the UK-DNDC model (Brown et al., 2002). This is a development of the original DNDC model (Li et al., 1992), a model of N_2O evolution from soil. Here we present some results at field scale using representative scenarios for grassland and arable (winter wheat) farming. For each farming type, two representative soils and county locations were selected. Soils for grassland were Brickfield and Denbigh series (referred to as 1 and 2, respectively) and those for arable were Andover and Beccles series (3 and 4, respectively). Locations for grassland were Devon (D) and Cheshire (C) and for arable crops Norfolk (N) and Wiltshire (W). For grassland, the N input from ammonium nitrate fertiliser was 200 kg N ha^{-1} yr^{-1}, applied on 5 dates. Slurry was applied on 3 occasions, (totalling 180 kg N ha^{-1} yr^{-1}) and grazed with 2.2 cattle ha^{-1}, from 1 April to 31 August. For the arable scenario, 185 kg N ha^{-1} was applied as ammonium, on 3 dates. Farmyard manure was applied at 240 kg N ha^{-1} yr^{-1}.

Results

There was a wide range in effectiveness of grassland strategies in reducing emission (Figure1). Reducing fertiliser input by 20% resulted in only a modest decrease in emission. The two strategies relating to grazing animals (decreased stocking rate and shortened grazing season) were most effective in reducing emission. Strategies should be compared with caution, however,

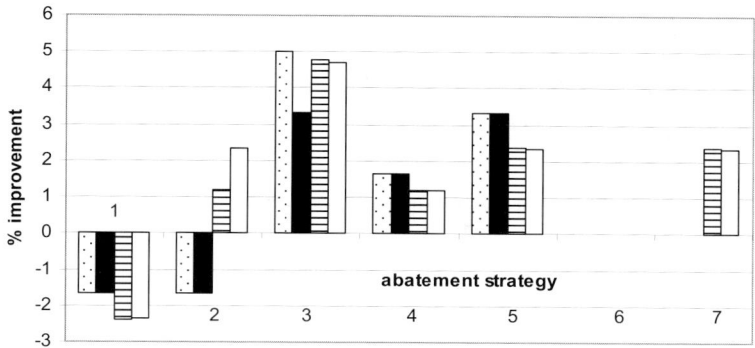

Figure 1. Effect of abatement strategies for N_2O emission for grassland on 4 sites (C1 (stippled), C2 (black), D1 (striped), D2 (white)). Abatement strategies: 1=urea, 2=injected urea, 3=reduced stocking rate, 4= reduce fertiliser by 20%, 5=shorten grazing season, 6=reduce C:N of manure, 7= synchronise fertiliser to plant demand

because they are not standardised. For the winter wheat scenarios (not shown), many of the strategies (e.g. no-till, use of urea fertiliser, injection rather than surface application of urea or ammonium nitrate) increased N_2O emission. Reducing N fertiliser input reduced emission by almost 5%.

Conclusion

The use of a model such as DNDC for investigating farm-scale mitigation strategies for the reduction of N_2O emission has great advantage over more simple models because it facilitates evaluation of the extent to which the adoption of strategies for mitigation of one pollutant may lead to increased emission of another. In addition to consideration of other losses, abatement strategies should be evaluated according to their potential impact on production. Most of the strategies that were successful in reducing N_2O emission would have an associated production penalty, or would involve changes in farming practice that may incur further losses (e.g. increasing the period of housing animals, or increasing the area of land required for agricultural production).

Acknowledgements

This research was funded by DEFRA under contract CC0243. We are grateful to Changsheng Li (University of New Hampshire, USA) for his assistance with DNDC.

References

Brown, L., Syed, B., Jarvis, S. C., Sneath, R. W., Phillips, V. R., Goulding, K. W. T. and Li, C. (2002). Development of a mechanistic model to estimate emission of nitrous oxide from UK agriculture. Atmospheric Environment, 36, 917-928.

Li, C., Frolking, S., and Frolking, T. A. (1992). A model of nitrous oxide evolution from soil driven by rainfall events. 1. Model structure and sensitivity. Journal of Geophysical Research, 97, 9759-9776.

Salway, A. G., Dore, C., Watterson, J. Murrells, T. (1999). Greenhouse Gas Inventories for England, Wales, Scotland and Northern Ireland: 1990 and 1995. National Environmental Technology Centre.

Variations in aerobic respiratory and denitrifying activities in the vadose zone: laboratory and field experiments

P. Cannavo[1], F. Lafolie[1], A. Richaume[2], B. Nicolardot[3] and P. Renault[1]

[1]*INRA Unité Climat Sol et Environnement, 84914 Avignon cedex, France*
[2]*LEM, UMR CNRS 5557, Lyon University, 69622 Villeurbanne cedex, France*
[3]*INRA Unité d'Agronomie, 51686 Reims cedex, France*

Introduction

Understanding microbial mechanisms which regulate soil chemical constituents, and notably NO_3^-, are crucial for understanding the fate of C and N in the deep vadose zone (DVZ). The objective of this work was to characterise, on a yearly basis, the aerobic respiratory and denitrifying activity dynamics and to link them to environmental factors. A model predicting C and N transformations in the vadose zone was used in order to estimate gas transport and the importance of microbial activity in the DVZ.

Materials and methods

A field experiment (43°91' N; 5°06' E) was carried out over 7 months in a bare soil, a fluvic hypercalcaric cambisol (FAO classification). The experiment started just after maize harvest and residues were incorporated into the soil. *In situ* measurements consisted of recording (1) climate variables (temperature, rainfall), (2) soil water content and temperature, (3) soil NO_3^- and dissolved organic C (DOC), and (4) soil CO_2 and N_2O in the entire 2.5 m-thick vadose zone. Four layers were defined in which heterotrophic and denitrifying bacterial counts were performed using the Most Probable Number (MPN) method. Aerobic respiratory and semi-potential denitrifying activities (i.e. ARA and SPDA, respectively) were also estimated in autumn, winter and spring, from the depletion of O_2 and the accumulation of N_2O in the presence of C_2H_2, respectively, for soils in sealed flasks incubated under optimal conditions. C and N transformation predictions were carried out with the PASTIS model (Garnier et al., 2003), in which a gas transport module using the *dusty gas theory* was added.

Results

Both ARA and SPDA decreased with increasing depth (Figure 1). Significant seasonal ARA variations were observed in the DVZ (ANOVA, $P<0.95$). ARA suggested the existence of aerobic respiration in the whole profile, whereas SPDA distribution suggested that denitrifying activity mainly occurred at the soil surface. The PASTIS model correctly simulated CO_2 concentration whatever the depth, taking into account both DOC adsorption and death of zymogenous biomass during the winter period (Figure 2B). N_2O peak concentrations during important rainfall events were overestimated by the model (Figure 2C). The model correctly simulated NO_3^- leaching during important rainfall events which occurred in February 2002 (Figs. 2A and 3A). However,

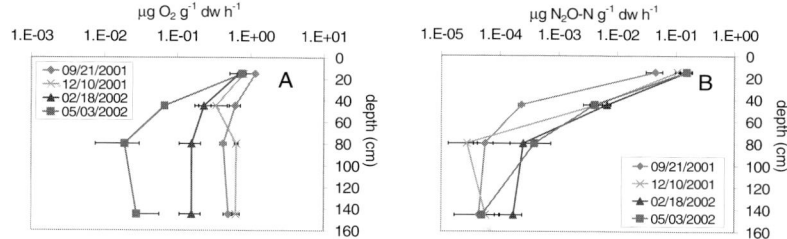

Figure 1. Temporal variations in aerobic respiratory (A) and semi-potential denitrifying (B) activities in the soil profile. Bars represent standard deviation values.

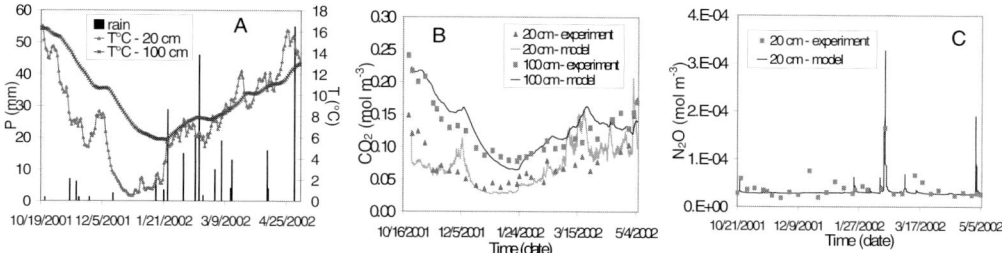

Figure 2. Soil temperature and rainfall during the experiment (A). Simulation of CO_2 (B) and N_2O concentrations using the dusty gas theory.

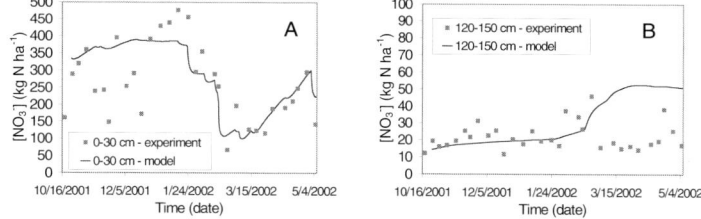

Figure 3. NO_3^- concentration simulation in the 0-30 (A) and 120-150 cm (B) layers.

the model simulated some NO_3^- leaching in the DVZ, which was in contradiction with experimental data.

Discussion

Potential activities and *in situ* microbial activities, as revealed by CO_2 and N_2O concentration measurements, were in good agreement. Continuous N_2O production from denitrification under aerobic conditions was observed during the experiment (i.e. 1.5 kg N_2O-N ha^{-1} yr^{-1}). Low DOC concentrations and variations, coupled with high ARA seasonal variations in the DVZ, suggested that DOC quality should have influenced such variations. With the model, it was possible to estimate that 15% of CO_2 produced during the experiment came from deeper than 60 cm, strengthening the importance of taking into account microbial activity in the DVZ. The substantial NO_3^- leaching from the 0-30 cm layer (i.e. 300 kg N ha^{-1}) suggested that denitrification with N_2 as terminal product and/or NO_3^- assimilation within the biomass should have occurred, avoiding leaching in the DVZ.

Conclusion

This work confirmed the importance of taking into account both aerobic respiration and denitrification in the DVZ. Taking DOC quality into account in models should improve the prediction of NO_3^- groundwater contamination.

Acknowledgements

The authors wish to thank T. Flacher, for use of his land, N. Guillemaud for helping in microbial assessments, B. Mary and O. Delfosse for ^{15}N measurements, and M. Bourlet for pedological analyses.

References

Garnier, P., Néel, C., Aita, C., Recous, S., Lafolie, F. and Mary, B. (2003). Modelling carbon and nitrogen dynamics in a bare soil with and without straw incorporation. European Journal of Soil Science, 54, 1-14 (*in press*).

Use of a 2-pool model to evaluate the effect of fertiliser application on nitrogen emissions from grassland soils

L.M. Cárdenas, J.M.B. Hawkins, D. Chadwick and D. Scholefield
Institute of Grassland and Environmental Research, North Wyke Research Station, Okehampton, Devon, EX20 2SB, UK

Introduction

Soils are an important source of atmospheric N via production of NO, N_2O and N_2 in nitrification and denitrification processes controlled by the activity of soil microbes. However, there are many uncertainties concerning the global budget of these gases because of the large variability in emission encountered in the field. Application of fertiliser to grassland soils is the main source of N gases. We present the results of the application of a 2-pool model (Dhanoa et al., 1985) to simulate N_2O and N_2 emissions in experiments in a fully automated laboratory system. The model was originally developed to simulate the excretion patterns in ruminant faeces. It calculates the rates of increase and decrease of the fluxes k_2 and k_1, the lag time TT (time it takes for the system to respond to the amendment addition), and the mean retention time MRT (the total time it takes for the process to occur). The model was used to statistically evaluate the fluxes.

Materials and methods

An automated laboratory system was used for the measurement of gaseous emissions from soils (Cárdenas et al., 2003) and the original atmosphere of incubated intact soil cores from a grassland soil was replaced by flushing with a mixture of He and O_2. Once N_2 had reached baseline levels, the effects of inorganic N and C additions on the production of N_2O and N_2 were measured; NO_3^--N was added at the equivalent of 50 kg N ha^{-1} and sucrose at 363 kg C ha^{-1}.

Results

The 2-pool model was able to simulate the fluxes from soil receiving N only (Cárdenas et al., 2003). It was less effective when C was applied in addition to N. This suggested higher complexity in the processes that ultimately result in the production/emission of gaseous N. We later applied a more complex model comprising several pools, and found that the N_2O fluxes, as well as the N_2 fluxes, from the N + C treatment were well simulated. Figs. 1 and 2 show examples of the models fitted to N_2O and N_2 fluxes from N + C treatment. Statistical analysis of the data was carried out by using Anova on the model parameters. Table 1 shows the resulting parameters of the 2-pool model applied to replicates with N only, and the multi-pool model applied to replicates that received N + C.

Table 1. Parameters obtained from the 2-pool (NO_3^-) and multi-pool model (NO_3^- + glucose).

Treatment	k_1, h^{-1}	k_2, h^{-1}	TT,h	MRT, h
NO_3^-	-0.0077	-0.0913	5.46	146.36
NO_3^-	-0.013	-0.1683	3.20	86.13
NO_3^-	-0.0167	-0.0984	2.96	73.07
NO_3^- + glucose	0.0268	0.1401	8.76	53.17
NO_3^- + glucose	0.0508	0.0815	30.8	62.76
NO_3^- + glucose	0.0581	0.1125	14.13	40.22

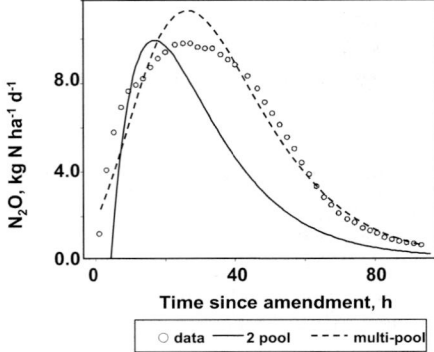

Figure 1. N_2O fluxes from $+NO_3^-$ + C treatment.

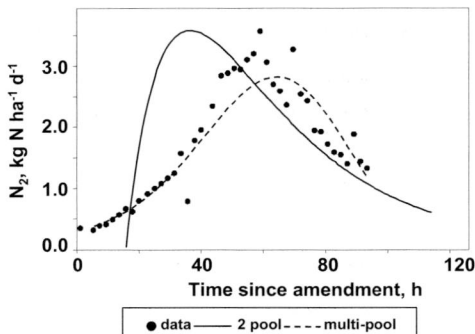

Figure 2. N_2 fluxes from $+NO_3^-$ + C treatment.

Discussion

The 2-pool model fits the data well when only N was applied, suggesting that the system behaves as a 2-pool system with NO_3^- (in the first pool), being converted to N_2O (second pool), some of which diffuses out of the system. When both N + C were applied, the existence of a third pool is suggested, containing N_2 reduced from the N_2O in the second pool. As a result, emissions of N_2O and N_2 from both pools occurred. The Anova analysis showed that TT and MRT did not differ for both treatments, suggesting that the initiation of the processes does not depend on the addition of C and that the original C content in the soil is sufficient to start the process. The other model parameters, however, were significantly different.

Conclusions

The application of a model developed for use in ruminant digestion to gaseous emissions from incubated soil cores was successful. A 2-pool model fitted the data well when only N was applied. A multi-pool model was needed to fit the data when N and C were applied.

Acknowledgements

This project was funded by a BBSRC, UK, grant.

References

Cárdenas, L.M., Hawkins, J.M.B., Chadwick, D. and Scholefield, D. (2003). Biogenic gas emissions from soils measured using a new automated laboratory incubation system, Soil Biology and Biochemistry, 35, 867-870.

Dhanoa, M.S., Siddons, R.C., France, J. and Gale, D.L. (1985). A multicompartmental model to describe marker excretion patterns in ruminant faeces. British Journal of Nutrition, 53, 663-671.

Scholefield, D., Hawkins, J.M.B. and Jackson, S.M. (1997). Development of a helium atmosphere soil incubation technique for direct measurement of nitrous oxide and dinitrogen fluxes during denitrification. Soil Biology and Biochemistry, 29, 1345-1352.

Opportunities for reducing ammonia emissions from pig housing in the UK

T.G.M. Demmers[1], R. Kay[2] and N. Teer[1]
[1]*Silsoe Research Institute, Environmental Engineering, Wrest Park, Silsoe, MK45 4HS, UK*
[2]*ADAS Terrington, Terrington St Clement, King's Lynn, Norfolk PE34 4PW, UK*

Introduction

This paper reports the potential to reduce the NH_3 emission from pig housing using a number of simple techniques compared with a fully-slatted system. The EU Directive on Integrated Pollution Prevention and Control (IPPC) Directive 96/61/EC, requires that control measures are put in place to reduce NH_3 emissions from pig units with more than 2000 finishing pig places or 750 sow places. Work from Hoeksma et al. (1994) suggests that NH_3 emissions from pig housing may be reduced by 30-60% using partly (50%) slatted floors compared with fully-slatted. A key factor in the use of part-slatted systems is the reduction of the area of manure contact with air.

Materials and methods

In the Environment House facilities at ADAS Terrington, the NH_3 emissions were quantified from 5 different slurry-based housing systems. These were a typical UK, fully-slatted house (1); rubber mats over 50 % of the slats in an existing fully slatted house (2); structural modification, i.e. covering over 50% of the slats with a domed floor, plus air barriers preventing air movement under the new solid area, (3); the current UK, part-slatted system with 33% slats (4) and the novel Dutch part-slatted system, (IC-V) comprising 50% slatted area (including a narrow slatted area at the front of the solid lying area to collect water spillage) a convex solid area to avoid pooling of excreta and a slurry channel with sloping sides (Zeeland, 1997) (5). The study was replicated three times over time, in a randomised block design. 120 pigs per replicate, were housed in pen groups of 12 pigs with 2 pens per room. Ammonia concentration was measured continuously at the air inlet and outlet of the rooms throughout a 10 week trial using stainless steel catalytic converters combined with a NO_x-analyser (Demmers et al., 1999). The ventilation rate was measured continuously in each room using calibrated fan wheel anemometers. The emission rate was calculated from the NH_3 concentration and average ventilation rate for that period.

Results

The emission factors for all systems given in Table 1 are based on the liveweight measured at the start and end of the trail and assuming a linear growth curve.

The food consumption ratio and daily weight gain were average for all trials. No problems occurred regarding animal welfare. The NH_3 concentration did not exceed 20 ppm during the trials. The ventilation rate was not significantly different in any of the rooms. The NH_3 emission factors obtained during the replicates are substantially lower than those used in the current

Table 1. Ammonia emission factors (g N LU⁻¹ d⁻¹) for the monitored systems in the Environment Facility at ADAS Terrington.

Season	Fully-slatted	Rubber matting	Structural modifications	UK part-slatted	Dutch IC-V part-slatted
winter	41	37	35	22	19
spring	42	40	35	21	23
summer	60	75	68	73	29
average	48	51	46	39	24

UK ammonia inventory (80 g N LU^{-1} d^{-1}; Misselbrook et al., 2000). The very tightly controlled environment in the rooms will have contributed to this.

Discussion

Both part-slatted options had a significantly lower emission compared with the conventional fully slatted finishing building during winter and spring conditions. However, the UK part-slatted system failed in the summer, due to pigs fouling the solid area in the pen. The Dutch IC-V system proved to be the best option to reduce NH$_3$ emission overall. The altered airflow pattern ensured proper dunging behaviour and greatly reduced contact between air and manure in the pit. The modifications from fully-slatted to part-slatted, using either rubber mats or more structural modifications, proved considerably less effective in reducing NH$_3$ emission, especially during summer, when emissions increased compared with the fully-slatted system. For the rubber mats, this was expected as the problem of fouling the solid area persisted and the air movement under the mats had not been altered. Hence, the NH$_3$ emission from the manure pit remained the same. For the structural modification option, however, air movement under the new solid area was blocked. Therefore a 'traditional' part-slotted system was created with a relatively large manure store. Hence, a significantly lower emission was expected similar to the 'standard' UK part-slatted system. No suggestion to explain the poor results for this option can be offered.

Conclusions

Part-slatted housing systems can reduce NH$_3$ emissions from pig housing in the UK. Compared with fully-slatted systems, the UK part slatted and the Dutch IC-V system reduce NH$_3$ emissions by 19% and 50%, respectively. Neither simple structural modifications to convert fully-slatted systems to conventional part-slatted, nor covering 50% of the slats with rubber mats appeared to reduce NH$_3$ emission significantly.

Acknowledgement

This work was funded by Defra with support from the Environment Agency and the pig industry.

References

Demmers, T.G.M., Burgess, L.R., Short, J.L., Phillips, V.R., Clark, J.A. and Wathes, C.M. (1999). Ammonia emissions from two mechanically ventilated UK livestock buildings. Atmospheric Environment, 33, 217-227.

Hoeksma, P., Daanen, M., Oosthoek, J., and Voermans, J. (1994). Reduction of ammonia emission from pig houses. pp 81-91 In : Environmental Challenges and Solutions in Agricultural Engineering. International Commission of Agricultural Engineering.

Misselbrook, T.H., van der Weerden, T.J., Pain, B.F., Jarvis, S.C., Chambers, B.J., Smith, K.A., Phillips, V.R. and Demmers, T.G.M. (2000). Ammonia emission factors for UK agriculture. Atmospheric Environment, 34, 871-880.

Zeeland van, A.J.A.M. (1997). 'Sloped plates in the slurry channel in a fattening pig pen', Report P4.22, (Research Institute for Pig Husbandary: Rosmalen: The Netherlands).

A comparison of nitrous oxide and ammonia fluxes from managed grassland

C.Di Marco[1,2], M. Anderson[1], C. Milford[1], U. Skiba[1], M.A. Sutton[1] and K. Weston[2]

[1]*Centre for Ecology and Hydrology, Edinburgh Research Station, Bush Estate, Penicuik EH26 0QB, UK*

[2]*School of Earth, Environmental and Geographical Sciences, University of Edinburgh, West Mains Road, Edinburgh EH9 3JZ, UK*

Introduction

Agricultural activities are an important source of NH_3, which contributes to the acidification and eutrophication of soils and waters, with consequent changes in plant and animal communities. These changes influence N_2O emissions. Nitrous oxide is a major greenhouse gas showing large spatial and temporal variations (Christensen et al., 1996). In order to assess the relative importance of NH_3 and N_2O emissions from typical intensive grassland management in northern Britain, we have measured fluxes of both gases almost continuously throughout the growing season.

Materials and methods

The experimental site consisted of grassland situated 10 km south of Edinburgh on a sandy clay loam soil. It was cut for silage 1-2 times per year, followed by grazing with cattle and sheep for the latter part of the growing season. Nitrogen fertiliser (as NH_4NO_3) was applied three times per year, with a total N application of 270 kg N ha^{-1} yr^{-1}. Fluxes of N_2O were measured using eddy covariance from June 2002 to June 2003. This technique requires a fast response gas analyser with high sensitivity. Tuneable Diode Laser (TDL) absorption spectroscopy has been used to determine N_2O concentrations. Ammonia measurements were made in the growing seasons of 1998, 1999, 2001 and 2002 using the continuous flow denuder (AMANDA) (Milford et al., 2001) and the aerodynamic gradient technique. Using these micrometeorological techniques, it is possible to obtain a very detailed picture of the fluxes of N_2O and NH_3 at the field scale (10^3-10^4 m^2), which are very valuable for extrapolation to regional scales.

Results

Peak emissions of N_2O were measured immediately after fertiliser was applied on 7 June 2002. Maximum emission was observed and peaked at 2565 ng N_2O-N m^{-2} s^{-1}, about 24 hours after fertiliser application. Daily emissions then decreased to an average of about 200 ng N_2O-N m^{-2} s^{-1} for 10 days after the event. About 20 days after fertiliser application, N_2O losses decreased to a daily average of 17 ng N_2O-N m^{-2} s^{-1} (Figure 1).

Measurements of NH_3 fluxes at the site during 1998 and 1999 showed a consistent pattern with increased emissions following cutting and N fertilisation, with peak emissions of 1.7 µg NH_3-N m^{-2} s^{-1} and an annual net emission of 1.8 kg N ha^{-1} yr^{-1}. However, during 2002, NH_3 emissions were much smaller with peak emissions (following cutting) of 0.2 mg NH_3-N m^{-2} s^{-1}, with the

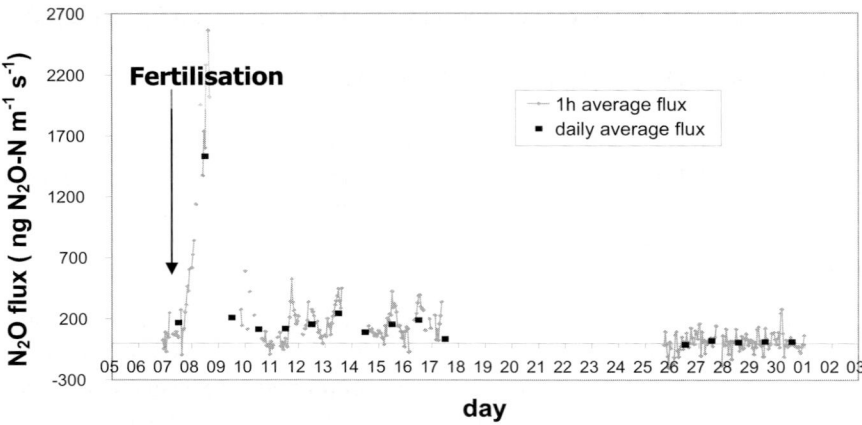

Figure 1. *N$_2$O flux measurements with eddy covariance at Easter Bush site in June 2002. Fertiliser was applied on the 7 June 2002.*

flux for the main exchange period (June-September inclusive) being a deposition of 0.5 kg N ha^{-1} (compared with 1.6 kg N ha^{-1} and 2.3 kg N ha^{-1} in 1998 and 1999, respectively).

Discussion

High time resolution analysis of the fluxes of both N$_2$O and NH$_3$ showed similar diurnal patterns with larger fluxes during daytime periods when temperature and turbulence is raised. The low emissions of NH$_3$, compared with previous years, may be related to wetter soil conditions during 2002.

Conclusions

These data demonstrate that intensively managed grasslands can be very important sources of both N$_2$O and NH$_3$. Gas emissions related strictly to environmental conditions, such as rainfall, soil moisture and temperature. Inter-annual differences point to the need for parallel long-term measurements of N$_2$O and NH$_3$ over several years.

Acknowledgements

This project was funded by Natural Environment Research Council.

References

Christensen, S., Ambus, P., Arah, J. R. M., Clayton, H., Galle, B., Griffith, D. W. T., Hargreaves, K. J., Klemedtsson, L., Lind, A. M., Maag, M., Scott, A., Skiba, U., Smith, K. A., Welling, M. and Wienhold, F.G. (1996). Nitrous oxide emission from an agricultural field: Comparison between measurements by flux chamber and micrometeriological techniques, Atmospheric Environment, 30, 4183-4190.

Milford, C., Theobald, M. R., Nemitz, E. and Sutton, M. (2001). Dynamics of ammonia exchange in response to cutting and fertilising in an intensively-managed grassland, Water, Air and Soil Pollution, 1, 167-176.

Gaseous losses of various nitrogeneous compounds from fertilisation of a wheat crop

S. Génermont[1], C. Hénault[2], P. Laville[1], M.H. Jeuffroy[1] and D. Flura[1]
[1]Institut National de la Recherche Agronomique, Thiverval-Grignon, France
[2]Institut National de la Recherche Agronomique, Dijon, France

Introduction

International agreements, such as the European convention on long-range transboundary air pollution, require a better knowledge of the processes leading to gaseous emissions from various anthropogenic activities. Agriculture is the main source of N gases, especially N_2O (up to 75% in France) which contributes to greenhouse gas effects, and NH_3 (~95% in NW Europe) which leads to acidifaction and eutrophication in ecosystems where it is deposited, but also NO_x (=$NO+NO_2$, 14%), mainly implicated in O_3 increase in the troposphere. This work aims at giving concurrent information on emissions of the three gasses after mineral fertiliser applications.

Methods

Experiments were carried out in Grignon, near Paris (France) in 2000, on a wheat canopy, (*Triticum aestivum* var. Soissons) sown at three densities (75, 150 and 300 plants m^{-2}) on a silty loam soil. Wheat was fertilised with 100 kg ^{15}N ha^{-1} as NH_4 NO_3 at four phenological stages, corresponding also to different climatic conditions (tillering, beginning of stem elongation, 2 nodes stage and flag leaf emergence). The control was fertilised bare soil. NH_3 and NO_x emissions, together with O_3 deposition, were continuously measured over periods of 14 days, after N application, using wind-tunnels. N_2O emissions were monitored using static chambers with eight replicates. Climatic conditions and micrometeorological conditions inside and outside of the tunnels were monitored. Wheat canopy structure and properties, and soil properties were characterised on the day of N application and at the end of gas emission measurements. Laboratory experiments were performed in order to identifiy microbiological processes involved in N_2O emissions.

Table 1. Main climatic and phenological conditions of the experiments.

	Experiments			
	R1	B3	B4	R2
Dates	22/02/00	14/03/00	04/04/00	25/04/00
	07/03/00	28/03/00	18/04/00	09/05/00
Phenology	Tillering	Beginning of stem elongation	2 node stage	Flag leaf emergence

Results and discussion

Gas measurements show that emissions increased with time over the year (Figs. 1 and 2): temperature seemed to be the most important factor for the three gases, combined with rain for N_2O. N_2O and NO emissions ranged from 9 to 45 g N-NO ha^{-1} d^{-1} which were in the same

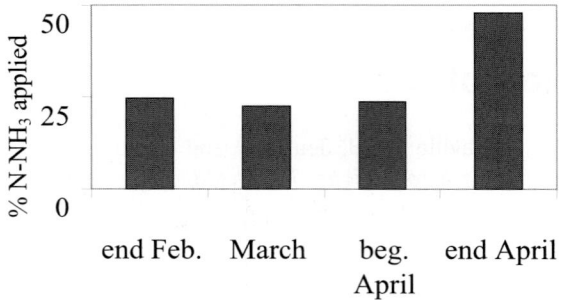

Figure 1. Cumulative N losses from NH_3 volatilisation, 14 days after fertiliser application on bare soil, on 4 occasions.

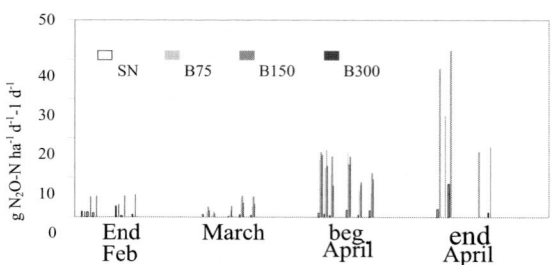

Figure 2. N_2O fluxes measured after fertilisation on bare soil (SN), and wheat (B) at several plant densities.

range as reported data. NH_3 losses were higher than commonly reported for NH_4^+-based fertilisers (from 5 to 17 kg N ha^{-1}).

NH_3 emissions were reduced by the wheat crop, because of the attenuation of the incoming solar radiation and wind at the soil surface inside the canopy. For each experiment, NH_3 losses were inversely related to LAI. This may be explained by better air circulation under the canopy which enhanced volatilisation at low densities.

NO emissions were similar for bare soil and wheat for weakly developed plants. Significant differences between bare soil and wheat (150 plants m^{-2}) only appeared for the 2 nodes stage (45 g N ha^{-1} d^{-1} and 15 g N ha^{-1} d^{-1}, respectively) (Figure 3): this may result from mobilisation of N by wheat at this stage of development. Differencies between densities only appeared at the flag leaf emergence stage: NO emissions were greater with the 150 plants m^{-2} than with 300 plants m^{-2}. N_2O emissions were enhanced on soil with wheat compared with bare soil. This cannot be explained either by mineral N in soil, or soil moisture, which were comparable in all situations, but could to be due to C availability in the soil. The plant density had no effect on N_2O emissions.

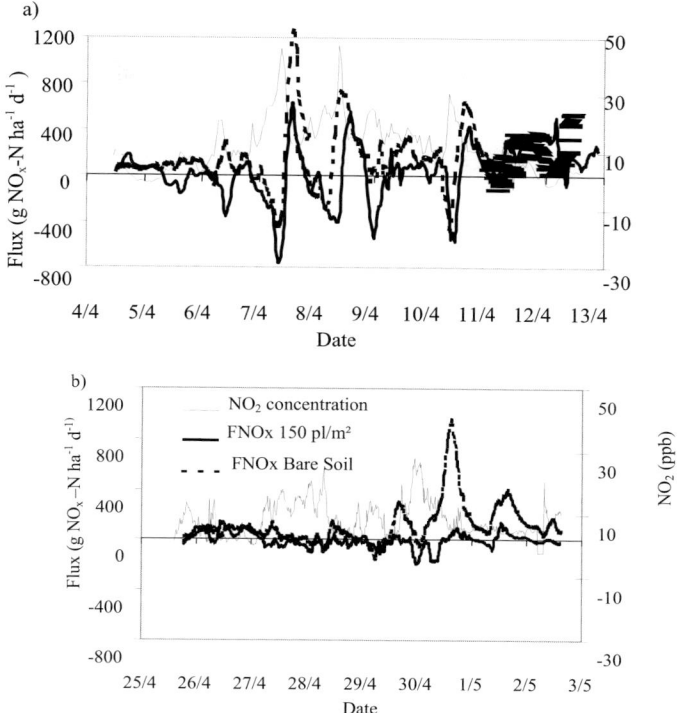

Figure 3. Background NO$_2$ concentration and NO emission measurements for a bare soil and a wheat cover at a) the two node stage and b) at the flag leaf emergence stage.

Conclusions

For each gas, the processes are different, and therefore the effect of the crop on the losses has different effects on the environment. It will thus be difficult to identify which abatement or fertilisation strategy to choose, if no objective criterium is available to compare their impacts on air pollution.

A simple denitrification model?
Review, sensitivity and application

M. Heinen
Alterra, Green World Research, P.O. Box 47, 6700 AA Wageningen, The Netherlands

Introduction

When modelling the N-cycle in soils, denitrification is one of the processes that needs to be considered. For this purpose, a literature review was carried out to determine if a simple description for denitrification is commonly used. To determine if such a model is practical to use, a sensitivity analysis was done, and it was applied to some available data sets on measured denitrification.

Consensus model

Denitrification models can be roughly divided in three groups: 1) microbial growth models, 2) soil structural models, and 3) simplified process models. Here we consider the third type. A review of literature resulted in more than fifty descriptions of simple denitrification models (Heinen, 2003). Although they are not fully independent, it indicates that in many studies they seem to suffice. Two major classes can be distinguished:
1. Denitrification is considered to be a first order decay process that can be reduced by governing soil conditions.
2. Actual denitrification rate is the potential denitrification rate reduced by governing soil conditions. Potential denitrification is defined as denitrification with an excess amount of NO_3^- under anaerobic conditions at a given reference temperature. The reduction functions pertain to NO_3^- content, degree of saturation, soil temperature, and soil pH. Written as an equation, this model is given as (disregarding pH)

$$D_a = D_p f_N f_S f_T = D_p \underbrace{\frac{N}{K+N}}_{f_N} \underbrace{\left(\frac{S-w_1}{1-w_1}\right)^{w_2}}_{f_S} \underbrace{Q_{10}^{0.1(T-T_{ref})}}_{f_T}$$

where D_a is the actual denitrification rate (mg N kg^{-1} d^{-1}), D_p is the potential denitrification rate (mg N kg^{-1} d^{-1}), f_N is the dimensionless reduction function for NO_3^--N content in the soil (N, mg N kg^{-1}), mostly given by a Michaelis-Menten function (parameter K), f_S is the dimensionless reduction function for degree of saturation S in the soil, mostly given by some power function (parameters w_1, w_2), and f_T is the dimensionless reduction function for soil temperature T (°C), mostly given by a Q_{10} function (parameters Q_{10}, T_{ref}). Oxygen content is the driving variable for the denitrification process. However, it is difficult to measure or to model. Therefore, use is made of the complementary variable degree of saturation S (volumetric water content divided by porosity). Although the analytical form of the reduction functions is mostly the same in the different models, the parameters used by the authors differ (Heinen, 2003).

Sensitivity analysis

The mathematical descriptions of the reduction functions are relatively simple, so that an analytical sensitivity analysis is possible by looking at the change in denitrification relative to the change in a parameter. From such analysis, it appears that the simple model is most sensitive to the water content reduction function. This means that volumetric water content, porosity and the curve shape parameters must be determined with great accuracy. The model applied to randomly chosen soil conditions at given reduction functions indicates that actual denitrification will always be much smaller than potential denitrification.

Application to six data sets

Six data sets were used to parameterise the reduction functions. Parameter optimisation was possible. In general, the order of magnitude of actual denitrification can be predicted by the model, but the spreading between measured and predicted data remains large (see Table 1, a summary taken from Heinen, 2003).

Table 1. Optimised parameters occurring in the reduction functions for six data sets: K (mg N kg^{-1}), w_1, w_2, and Q_{10}. The regression line without intercept through the data (measured versus predicted) is given by the slope RC and the goodness of this regression is given as R^2 (details in Heinen, 2003).

Soil type	K	w_1	w_2	Q_{10}	RC	R^2
Loam	15.50	0[d]	5.94	3.93	0.928	0.899
Dry sand	22.84	0[d]	6.39	3.11[b]	0.635	0.341
Wet sand	4.38	0.33	3.69	3.11[b]	0.798	0.757
Peat	135.3	0[d]	10.15	3.32	0.553	0.291
Sandy peat	69.9	[c]	[c]	2.44	0.933	0.743
Heavy loam	22.58	0[d]	1[d]	4.99	0.787	0.757

[a] fixed value; [b] determined from a separate parameter optimisation; [c] cannot be determined since data refer to anaerobic ($f_S = 1$) measurements; [d] lower bound during optimisation

Conclusion

Basically, two types of denitrification models have been described in the literature, however, they have many different parameter values. Calibration on Dutch and UK data sets revealed that site-specific calibration of parameter values is required. The order of magnitude of predicted denitrification by a calibrated model was correct; exact correspondence with measured data was not obtained. The model is most sensitive to the reduction function for degree of saturation.

Acknowledgements

This project was funded by the Dutch Ministry of Agriculture, Nature and Food Quality, research programme *"Onderzoek naar emissieroutes en effecten van maatregelen bij bepaalde bedrijfssystemen (Mest en Mineralenprogramma, 398-II)"*.

References

Heinen M. (2003). A simple denitrification model? Literature review, sensitivity analysis, and application. Alterra Report 690, Alterra, Wageningen, The Netherlands, 131 p.

Nitrous oxide emissions from soil - fertiliser and soil type effects

B. Hyde[1], A. Fanning[2], M. Ryan[2], M. Hawkins[3] and O.T. Carton[2]
[1]*Department of Environmental Resource Management, University College Dublin, Ireland*
[2]*Teagasc, Johnstown Castle, Wexford, Ireland*
[3]*Department of Statistics, University College Dublin, Ireland*

Introduction

Agriculture is the largest contributor to Irish greenhouse gas (GHG) emissions accounting for approximately 35% of the national total. Nitrous oxide from soils accounts for 35% of the total agricultural GHG emission and is the second largest source, after CH_4 emissions from enteric fermentation in ruminants.

A new Irish research project, with experiments at field and lysimeter scales, is nearing completion, the objectives of which are to determine the effects of fertiliser N, soil type and grassland management on N_2O emissions. The resulting values will be compared with the Intergovernmental Panel on Climate Change default emission factors. The potential for changing the default values will be evaluated in terms of their implications for national GHG inventory calculations.

Materials and methods

Field experiment. The objective of this experiment is to measure N_2O emissions, using chambers, from grassland grazed by steers and receiving 0, 225 and 390 kg N ha^{-1}. The N applications correspond to a basal or control level, an intermediate (the rate advised by Teagasc for a stocking rate equivalent to 170 kg organic N ha^{-1}) and a maximum rate, as advised by Teagasc for intensively stocked (~ 250 kg organic N ha^{-1}) grassland. The grassland is a permanent (> 10 years old) perennial ryegrass (*L. perenne*) sward. The soil is predominantly an imperfectly drained gley of loam over clay loam texture derived from Irish Sea till. The grassland management is based on rotational grazing (~ 21 days) using steers (300 - 350 kg live weight at spring turnout to grazing). The stocking rates are 1, 2.5 and 3.0 livestock units per ha for the 0, 225 and 390 kg N ha^{-1} treatments, respectively. Fertiliser N applications are synchronised so that the timing of the N application to the treatments throughout the grazing season occur at the same time. Nitrous oxide emissions are measured using the closed chamber technique. The chambers, which are 11.3 cm in diameter and 15.0 cm high, are inserted into the soil to a depth of 2 cm to leave an effective volume above the soil of 1300 cm^3. The chambers are left in place for 1 hour between 10.30 and 11.30 on each measurement date. Samples are then withdrawn from the chambers by a 10 ml syringe through a rubber septum and stored in pre - evacuated, 7 ml screw cap septum vials for analysis within 6 hours. The gas samples are analysed using a Gas Chromatograph. On each measurement date, 8 chambers are used in each of three replicate plots per N treatment (8 x 3 x 3 = 72 chambers).

Lysimeter experiment. The objective of this experiment was to measure N_2O emissions from grassed lysimeter units (0.6m diameter and 1m deep) with five soil types receiving N fertiliser application rates similar to those used in the field experiment. The soil drainage classes used in the lysimeters range from seriously impeded, through to moderately well drained, to very well drained soils (Table1). The range is considered to be representative of the majority of Irish soils. There were nine replicate monoliths of each soil type so that there were three replicates per fertiliser N treatment on each soil type. The grass (*L. perenne*) in the lysimeter unit was cut at six-week intervals from March through to November. Nitrogen was applied in four to five equal applications during the growing season. A single chamber, similar to those described above for the field experiment, was used per lysimeter column.

Table 1. Experimental soil types used in the lysimeter unit.

Castlecomer	Poorly drained organic clay loam over clay loam texture
Rathangan	Poorly drained loam overlying clay loam structure
Elton	Well drained gravelly loam overlying gravelly clay loam texture
Clonroche	Well drained loam to clay loam, overlying shaley loam textured brown earth
Oakpark	Gravelly brown earth of course sandy loam texture

Preliminary results

Field experiment. Two preliminary models of N$_2$O emissions from grazed grassland have been developed; (i) a multiple regression model based on average daily emissions per replicate plot in which the importance of causative factors, the nature of their influence, and their interrelationship can be identified and (ii) a mixture model in which individual flux chamber data is modelled (i.e. the raw data are modelled).

Lysimeter experiment. Preliminary results are in line with expected trends. The soil effect appears to be as great as the fertiliser effect with higher emissions associated with the seriously impeded soils (Figure 1).

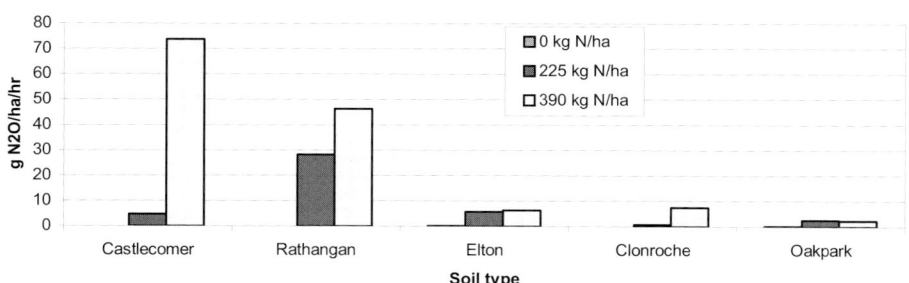

Figure 1. Effect of soil type and N application rate on N$_2$O emissions.

Conclusion

Higher N$_2$O emissions are associated in both experiments with higher N fertiliser inputs, wet rather than dry conditions and in the periods immediately following the fertiliser applications.

Acknowledgements

This work is funded by Teagasc and the EPA under the ERTDI programme as part of the Irish National Development Plan 2000 - 2006.

The Irish ammonia emission inventory - implications for compliance with the Gothenburg Protocol

B.P. Hyde[1], T. Misslebrook[2] and O.T. Carton[1]
[1]Teagasc, Johnstown Castle, Wexford, Ireland
[2]Institute of Grassland and Environmental Research, North Wyke Research Station, Okehampton, Devon EX20 2SB, UK

Introduction

An inventory of NH_3 emissions from Irish agriculture was recently constructed. The objectives were to produce a reliable inventory of NH_3 emissions from Irish agriculture, to identify information deficits, to determine the likelihood of achieving compliance with our national commitments under the Gothenburg Protocol (Protocol to the 1979 convention on long-range transboundary air pollution to abate acidification, eutrophication and ground level ozone - United Nations Economic Commission for Europe (UNECE, 1999), Geneva) and, if necessary, to identify the possible abatement strategies required. Ireland's commitments are to achieve a 9% reduction in emission levels by 2010 using 1990 as the base year.

Methods: the inventory construction

A number of inventory models were identified from the literature and compared in terms of their applicability to Irish circumstances, particularly in relation to information requirements. The UK NH_3 emission inventory (Misselbrook et al., 2000) was selected as being the most appropriate, based on its input requirements and the applicability of the emission factors used. Where applicable and where information was available, emission factors were adjusted to reflect Irish agricultural practices. Input data sets for 1990 and 2010 were compiled. The 2010 data were largely based on animal numbers and fertiliser use predictions for that year (Binfield et al., 2000) while data for 1990 were based on agricultural census data.

Results

The results of the inventory calculations for 1990 and 2010 are summarised in Table 1. Total emissions in 1990 were estimated to be 109 kt, with cattle accounting for in excess of 80% of the total.

The emission estimate for 2010 is 111 kt, which means that Ireland will not achieve its Gothenburg obligations to achieve a 9% reduction by 2010. Housing and land spreading of manure accounted for 39 and 40%, respectively, of total NH_3 emissions, therefore a reduction in emission levels from these sources would have a large impact on national emission totals. Abatement techniques, which can be used in animal buildings, are prohibitive because of excessive cost and are generally only applicable to new housing systems. In addition, the N that is conserved will remain in the manure with the potential for increased emission in the storage and landspreading processes. Therefore, a reduction in NH_3 emissions from the landspreading of animal manures offers the best potential in meeting reduction targets.

Table 1. Inventory estimates of NH_3 (kt) from Irish agriculture in 1990 and 2010.

Source	1990	2010
Cattle	84.8	81.5
Pigs	9.1	13.3
Sheep	4.5	3.7
Poultry	4.1	4.4
Horses	0.4	0.4
Conserved grassland	5.6	7.6
Tillage crops	0.7	0.6
Total	109.2	111.5

Discussion

One of the main factors contributing to the increase in emissions in the future is the growth in the size of the national pig and poultry herds during the 1990s which each increased by approximately 50%. Information in relation to some of the input data, particularly on manure types and manure management practices, was based on expert opinion. A national farm facility survey is now under way to address this information deficit. This will include data that will allow estimates of emissions from hard standings to be included in any future inventory. In Ireland, conventional vacuum tankers are used to apply nearly all livestock slurry. Even though effective at transporting and spreading large volumes of manure relatively cheaply, these machines leave a thin coating of manure on the surface of the soil or crop, which facilitates the volatilisation of NH_3. With the adoption of new landspreading technologies such as 'trailing shoe' application, bandspreading and the injection of liquid manure into the soil, Irish agriculture may go a long way towards meeting its commitments. Other emission reduction scenarios include reduced N fertiliser applications to grassland, longer grazing periods (and therefore shorter housing periods) and an increase in the utilisation of manure N as a valuable nutrient source.

Conclusion

Ireland has fallen behind many of its European counterparts in terms of NH_3 emission research. This paper and the results presented aim to provide a 'stepping stone' for future research. The main sources of NH_3 emissions within Irish agriculture have been identified with a view to assessing their suitability for reduction. The adoption of low emission manure spreading techniques appears to offer the best potential for achieving compliance with NH_3 emission reduction commitments under the Gothenburg Protocol. However, there are obvious cost implications in adopting such an approach.

References

Misselbrook, T.H., van der Weerden, T.J., Pain, B.F., Jarvis, S.C., Chambers, B.J., Smith, K.A., Philips, V.R. and Demmers, T.G.M. (2000). Ammonia emission factors for UK agriculture. Atmospheric Environment, 34, 871-880.

Binfield, J., Donnellan, T. and McQuinn, K. (2000). In: Outlook 2000: Medium Term Analysis for the Agri-Food Sector, Teagasc, Dublin, Ireland.
UNECE (1999). Protocol to the 1979 convention on long-range transboundsary air pollution
to abate acidification, eutrophication and ground-level ozone. United Nations Economic Commission for Europe (UNECE), Geneva.

Greenhouse gas emissions in meadow mesocosms exposed to elevated O_3 and CO_2

T. Kanerva[1], K. Koivisto[1], K. Karhu[1], K. Regina[2] and S. Manninen[1]
[1]Department of Ecology and Systematics, University of Helsinki, Helsinki, Finland
[2]MTT Agrifood Research Finland, Environmental Research, Jokioinen, Finland

Introduction

Concentrations of tropospheric gases, including CO_2 and O_3, have progressively increased over the past century. CO_2 has a major effect on global climate and a direct, largely beneficial effect on C3 plants (Bowes, 1991). Ozone also contributes to climate change, but is known to be highly phytotoxic (Ashmore & Bell, 1991). Most studies on the effects of CO_2 or O_3 have focused on responses of individual ecosystems, instead of the structure and function of the ecosystem as a whole (McGrady & Andersen, 2000). Interest in understanding how below ground microbial communities and their functions change along environmental gradients is increasing. We studied N cycling and metabolism in semi-natural grassland under elevated CO_2 and O_3 which were expected to have feedback effects on microbial trace gas fluxes (CH_4, N_2O, CO_2) from the soil. Elevated CO_2 may enhance N_2O production (Ineson et al., 1998), whereas O_3 may have an indirect harmful effect on soil microbiology (Islam et al., 2000). The objectives of the study were to evaluate effects on N_2O emissions from meadow mesocosms and on soil capacity to function as a sink or source of greenhouse gases.

Materials and methods

Seven perennial species (*Fragaria vesca* L., *Campanula rotundifolia* L., *Trifolium medium* L., *Vicia cracca* L., *Lathyrus pratensis* L., *Ranunculus acris* L., *Agrostis capillaris* L. and *Anthoxanthum odoratum* L.) were used to form mesocosms simulating a northern European dry meadow. The mesocosms were exposed to the following treatments in open-top chambers (3 replicates per treatment): NF (non-filtered air), NF+O_3 (1.5 x ambient O_3), NF+CO_2 (1.5 x ambient CO_2), and NF+O_3+CO_2. An open field plot (AA) served as a control during the summers of 2002-2004 in Jokioinen, SW Finland (60°49'N and 23°28'E). Greenhouse gas emissions were measured every second week during each growing season, using chambers (60 x 60 x 40 cm) placed on permanent frames for 40 min. to collect gases. Gas chromatography was used to analyse the samples. Emission rates were calculated from the linear increase in gas concentration in the chamber for every treatment.

Results and discussion

There were no statistically significant differences in the N_2O production between treatments in either of the years (Figs. 1 and 2), but the level of N_2O production in all of the treatments was lower during the second year. The reduction in the N_2O emission levels between 2002 and 2003 may be due to the increased vegetative cover in the mesocosms (data not shown). This suggests that enhanced growth of the plants in 2003 left less available N in the soil, which may explain the reduced denitrification activity. This assumption is supported by the fact that

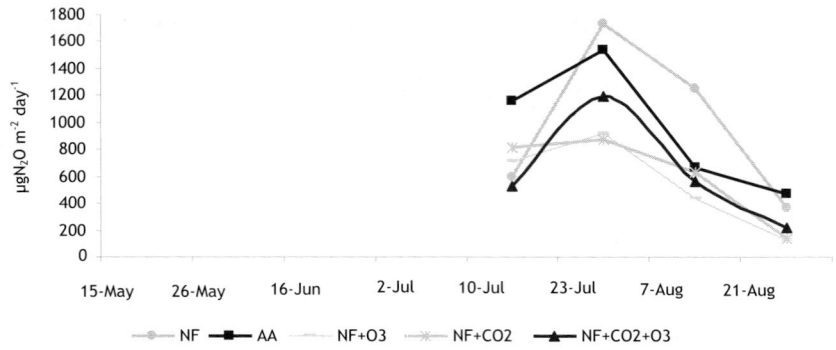

Figure1. *The release of N_2O from the mesocosms measured on four occasions during the growing season 2002.*

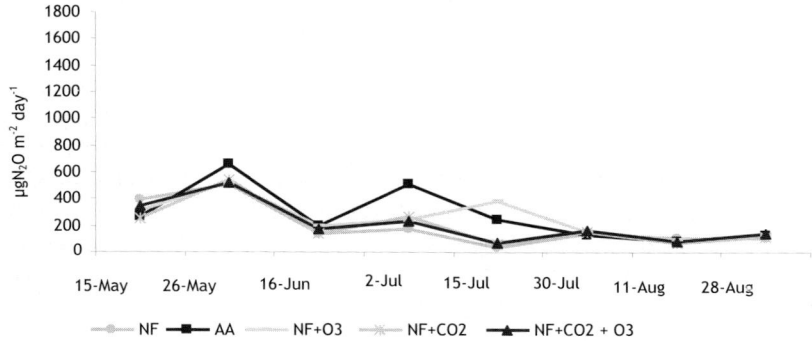

Figure 2. *The release of N_2O from the mesocosms measured on eight occasions during the growing season 2003.*

the highest N_2O production came from the AA treatments, which had the lowest vegetative cover in both years (data not shown). CH_4 and CO_2 emissions showed no significant differences between treatments in either of the years, but in 2003 there was a slight tendency for higher CO_2 production in the $NF+CO_2$ treatment and for lower CO_2 production in the $NF+O_3$ treatment compared with the control treatment.

The lack of changes in greenhouse gas production between the treatments may be attributed to several factors. Two years of exposure may not be sufficiently long for below-ground changes to take place, and prolonged study periods are therefore needed. Until now there has been no effect of CO_2 and/or O_3 on N_2O emissions from the meadow mesocosms and the treatments have not influenced the soil's capacity to function as a sink or source of greenhouse gases.

References

Ashmore, M.R. and Bell, J.N. (1991). The role of ozone in global change. Annals of Botany, 67, 39-48.

Bowes, G. (1991). Growth at elevated CO_2: photosynthetic responses mediated through Rubisco: commissioned review. Plant, Cell and Environment, 14, 795-806.

Ineson, O., Coward, P.A. and Hartwing, U.A. (1998). Soil gas fluxes of N_2O, CH_4 and CO_2 beneath *Lolium perenne* under elevated CO_2: The Swiss free air carbon oxide enrichment. Plant and Soil, 198, 89-95.

McGrady J. K. and Andersen C. P. (2000). The effect of Ozone on belowground carbon allocation in wheat. Environmental Pollution, 107, 465-472.

Refining the uncertainty in nitrous oxide emissions from New Zealand agricultural soils

F.M. Kelliher[1], A.S. Walcroft[2], S.F. Ledgard[3], H. Clark[4], G. Rys[5], H. Plume[6], M. Buchan[7] and R.R. Sherlock[7]

[1]*Landcare Research, P.O. Box 69, Lincoln, New Zealand*
[2]*Landcare Research, Private Bag 11052, Palmerston North, New Zealand*
[2]*AgResearch, Private Bag 3123, Hamilton, New Zealand*
[4]*AgResearch, Private Bag 11008, Palmerston North, New Zealand*
[5]*Ministry of Agriculture and Forestry, P.O. Box 2526, Wellington, New Zealand*
[6]*New Zealand Climate Change Office, P.O. Box 10362, Wellington, New Zealand*
[7]*Lincoln University, P.O. Box 84, Lincoln, New Zealand*

Introduction

At the beginning of the Kyoto Protocol base year, on 30 June 1990, New Zealand (NZ) had 57.9 million (M) sheep, 4.6 M beef cattle and 3.4 M dairy cattle (3-year-running arithmetic means). Year round grazing on most farms means pastoral soils are the principal source of N_2O emissions. We used the Intergovernmental Panel on Climate Change (IPCC) model to estimate N_2O emissions from NZ's agricultural soils. For the first time, this included process-based submodels to determine N intake and N excretion by grazing animals on a monthly basis, including lambs and calves. Here we consider the uncertainties for 1990 from statistical methods and Monte Carlo numerical simulation.

Methods

Emissions were determined by N inputs and emission factors that dictate the fraction emitted to the atmosphere as N_2O. Measurements were made on four plots, on 45 occasions over 227 d (n = 173) following application of 581 kg N in cow urine to pasture on Templeton silt loam soil at Lincoln, N.Z. on 16th October 2002. In 1990, 96% of the modelled N input was excreta N, deposited during grazing. Field measurements have focused on the excreta N emission factor (EF), known as EF_3. The effects of variability in the nine most influential parameters on the uncertainty of annual N_2O emissions were based on their statistics, deduced from the data available (Table 1, mean = arithmetic mean, excretion rates are kg N per head per year).

Results

A feature of N_2O emissions from grazed pasture is its episodic nature. To illustrate, we use a surrogate measurement, N_2O concentration in the soil. Following urine application to soil, the probability or frequency distribution of soil N_2O concentration was strongly and positively skewed (Figure 1). For ca. 60 days after urine application, there was 165 mm of rain on 20 days and N_2O emissions were significantly greater than from the (no urine application) control plots. The seven highest concentrations measured (maximum = 365 ppmv) are not included in Figure 1. The N_2O emission factors had this form of probability distribution (Table 1).

Table 1. Statistics of IPCC model parameters for NZ agricultural soils.

Parameter	Mean, minimum, maximum, probability distribution function
Fertiliser EF	0.0126, 0.002, 0.033, log normal (positively-skewed curve)
Excreta EF (EF_3)	0.011, 0.002, 0.033, log normal
Volatilised N EF	0.011, 0.002, 0.033, log normal
Leached N EF	0.0273, 0.01, 0.07, log normal
Fraction N volatilised	0.20, 0.10, 0.30, Gaussian (symmetrical, bell-shaped curve)
Fraction N leached	0.07, 0.03, 0.10, Gaussian
Sheep excretion rate	12.0, 8.0, 16.0, Gaussian
Beef cattle excretion rate	65.3, 46.9, 83.7, Gaussian
Dairy cattle excretion rate	105.0, 70.8, 139.2, Gaussian

For sheep and cattle, annual N excretion was 1380 ± 414 Gg N. This uncertainty was attributed mostly to pasture N content variance across NZ's 14 M ha of pasture, including seasonality. However, parameter EF_3 accounted for 91% of the uncertainty in N_2O emissions from NZ's agricultural soils. For 1990, we are 95% sure that N_2O emissions from agricultural soils were between 17 and 58 Gg (Figure 2).

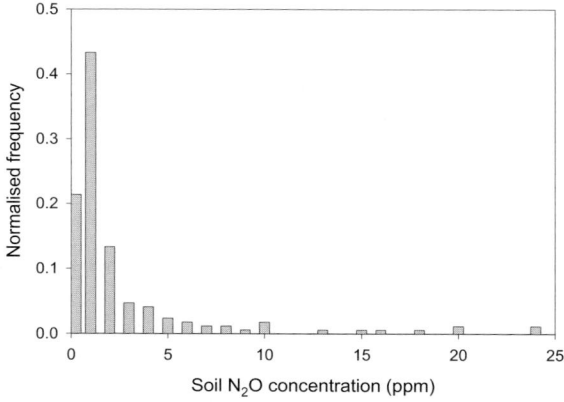

Figure 1. The normalised frequency distribution of N_2O concentration measured in soil (depth = 0.15 m) following cow urine application.

Conclusions

Wide variance is an inherent feature of EF_3, especially when it is estimated over annual and national scales. Consequently, an annual estimate of N_2O emissions from NZ agricultural soils included significant uncertainty. At such large scales, it is challenging to determine N_2O emissions, changes from year to year, and the efficacy of mitigation strategies with confidence. With sufficient data, spatial and temporal (e.g. seasonal) disaggregation of EF_3 may deliver more certain estimates of N_2O emissions from NZ's agricultural soils.

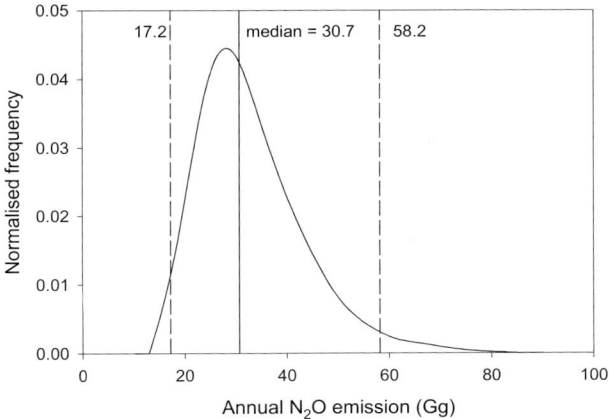

Figure 2. The normalised probability or frequency distribution of N_2O emissions from New Zealand's agricultural soils, for 1990. The vertical dashed and solid lines are 95% confidence limit and median emissions, respectively.

Dynamics of N_2O and NO production by *Alcaligenes faecalis parafaecalis*: effect of pH, temperature, substrate and oxygen supply

M. Kesik[1], S. Blagodatsky[1,2], H. Papen[1] and K. Butterbach-Bahl[1]

[1]*Institute for Meteorology and Climate Research, Karlsruhe Research Centre, Kreuzeckbahnstrasse 19, 82467 Garmisch-Partenkirchen, Germany*
[2]*Institute of Physicochemical and Biological Problems in Soil Science, RAS, 142290 Pushchino, Russia*

Introduction

Nitric oxide (NO) and N_2O are primarily and secondarily important atmospheric active trace gases. In view of the heterogeneity of N gas emissions from terrestrial ecosystems in time and space, mechanistic models are required to quantify the emission strength at regional and global scales in order to improve future predictions of climate change. Some of the most advanced models used for such purposes are the PnET-N-DNDC model (Li et al., 2000) for forest soils and the DNDC model (Li, 2000) for agricultural soils. These models explicitly describe the microbiological processes involved in N cycling and N-trace gas emissions, so that the rates of NO and N_2O emissions directly depend on the production, as well as the consumption, of NO and N_2O during nitrification and denitrification. However, the effect of environmental factors such as pH, temperature or substrate on the magnitude of N trace gas production and consumption by nitrification and denitrification is still not sufficiently understood. This knowledge gap can be at least partially closed, if nitrifying and/or denitrifying organisms are studied in a pure chemostat culture and parameterisation experiments are performed with regard to the effects of pH, substrate type, temperature or O_2 availability on N trace gas production/consumption.

Methods

The heterotrophic micro-organism, *Alcaligenes faecalis* subsp. *Parafaecalis*, is able to undertake nitrification and denitrification and therefore, is a good test organism for investigating the kinetics and mechanisms of N_2O and NO production. *A. faecalis p.* was grown under sterile controlled conditions in a Bioflo-III fermentor as a continuous culture on two different substrates (peptone-meat-substrate and ammonia-citrate-substrate) with different levels of dissolved O_2, pH and temperature. N_2O concentrations were measured by taking air samples with a gas-tight syringe and analysing them with a gas chromatograph. NO was automatically quantified with a chemiluminescent analyser and CO_2 by infra-red spectrometry. Biomass, NO_3, NO_2^- were measured hourly in culture samples. While measuring the temperature dependency of N_2O and NO production, the pH value was held constant at 7. The temperature was permanently at 28°C when different pH values were studied.

Results and discussion

Very high N_2O and NO production rates were found at pH 3 and 4, whereas at pH 5 and 6 NO production was remarkably low (Figure 1A, B, C). N_2O, as well as NO production, was positively correlated with temperature (Figure 1D, E). However, the N_2O to NO ratio (Figure F) declined, indicating that the temperature effect is more pronounced with regard to N_2O as compared with NO production. Figure 2 demonstrates the dynamics of NO production after changing for 5 h to either temperature (Figure 2A), or pH (Figure 2B), and shows that the chemostat culture of *A. faecalis p.* grown in peptone-meat substrate needed at least 5h to adapt to the new environmental conditions. Comparable results have been found also for ammonium-acetate substrate. However, N_2O, as well as NO production, was significantly lower for the ammonium-acetate substrate. O_2 was a major control of the magnitude of N trace gas production. Highest rates of N_2O, as well as of NO production, were always found during transition from aerobic to anaerobic conditions: such conditions stimulate the consequent production of NO and then N_2O by *A. faecalis p.* Observed dynamics of NO and N_2O production may be explained by a fast inactivation of N_2O-reductase by O_2, while NO_2^- and NO-reductase are still active at low O_2 concentrations. This mechanism could be responsible for a sharp increase in N_2O/NO emissions by soils during drying-rewetting and freezing-thawing cycles.

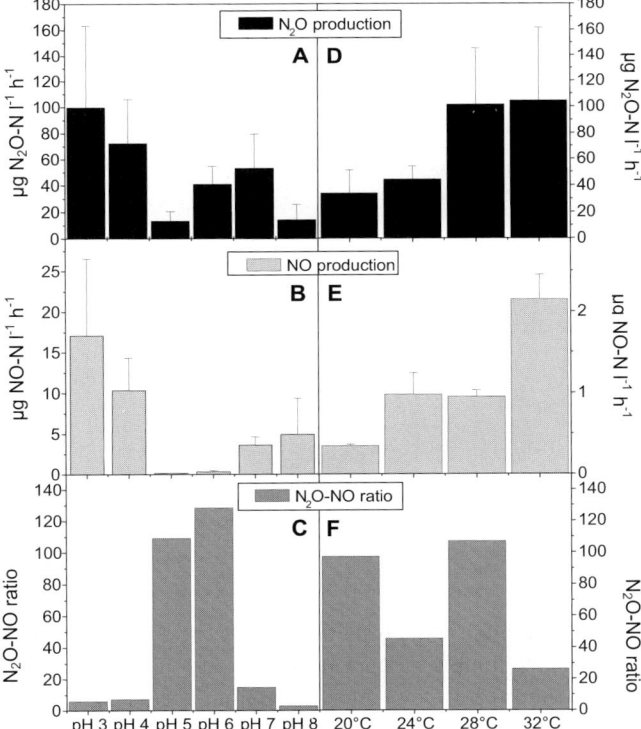

Figure 1. Influence of pH and temperature on N_2O and NO production.

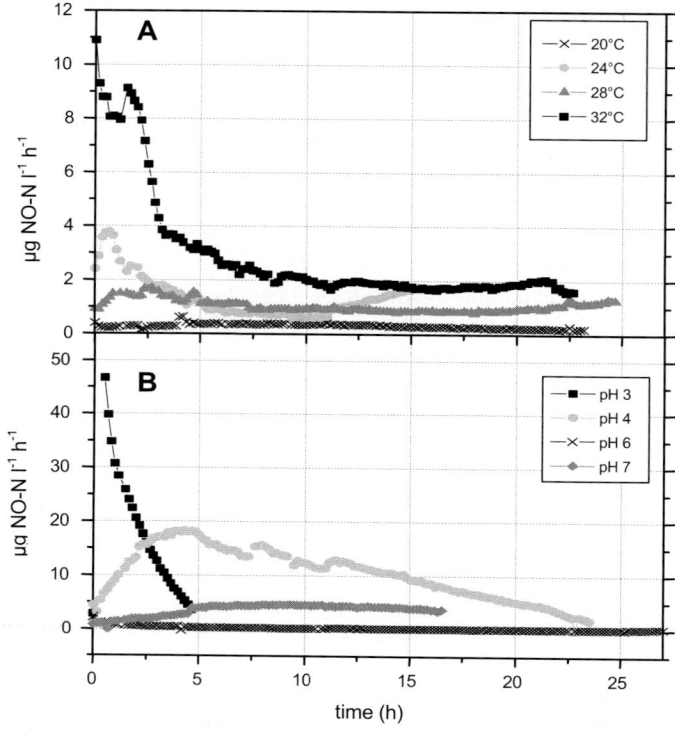

Figure 2. NO production as influenced by temperature and pH over time.

Acknowledgements

The work was supported by the European Commission in the NOFRETETE project (EVK2-CT2001-00106) of the Fifth Framework Programme.

References

Li, C. 2000. Modelling trace gas emissions from agricultural ecosystems. Nutrient Cycling in Agroecosystems, 58, 259-276.

Li, C., Aber, J.D., Stange, F., Butterbach-Bahl, K. and Papen, H. (2000). A process-orientated model of N_2O and NO emissions from forest soils. 1. Model development. Journal of Geophysical Research, 105, 4369-4384.

Denitrification and nitrate loss from organic agricultural soil at low temperatures

H.T. Koponen, A. Pärnä, H. Silvennoinen and P.J. Martikainen
Research and Development Unit of Environmental Health, Department of Environmental Sciences, University of Kuopio, P.O. BOX 1627, FIN-70211 Kuopio, Finland

Introduction

Agricultural soils contribute approximately 80% of the total N_2O in the atmosphere, and as such are the most important anthropogenic source of N_2O (Isermann, 1994). In recent studies it has been shown that wintertime's share of annual N_2O emissions can be up to 70% of the total (Röver et al., 1998). The high N_2O production at low temperature has been shown both in field (e.g. Christensen & Tiedje, 1990) and in laboratory experiments (e.g. van Bochove et al., 2000), but the underlying processes are still unknown. To control both gaseous N-losses and N-leaching from agriculture, more information is needed about the low temperature effects on the N cycle. In this experiment, we studied, in well-controlled laboratory conditions, the dependence of denitrification and fate of NO_3^- on temperature in an organic agricultural soil.

Materials and methods

The study site is located in southern Finland (60°49'N, 23°30E). Soil samples (0-20 cm) were collected in May 2001 from an exclusion area. The exclusion area has been kept uncultivated for 2 years by ploughing it regularly during the growing season. Field moist soil was homogenised by sieving (mesh size 5 mm). Sieved soil was stored at +4°C for 1 month before the experiment.

Denitrification and N_2O production were studied at temperatures of +15, +5, +2.5, +1.5°C, +0.5, 0°, -0.5 and -1.5°C. For each temperature, individual soil cores were packed. To ensure favourable conditions for denitrification, soil water content was adjusted to 70% WFPS. There were also N treatments (100 kg N ha^{-1} with 7 replicates of 10 kg N ha^{-1} to 4 replicates, as KNO_3). Four 100 kg N ha^{-1} replicates received N as $K^{15}NO_3$ (40 atom%). Soil cores were incubated at each temperature for five days. N_2O and CO_2 production was measured on incubation days 3, 4 and 5 by a closed chamber method. Nitrite, NO_3^- and NH_4^+ were measured before and after the incubations. Denitrification potential was also determined on the fifth incubation day using the acetylene inhibition technique.

Gas samples were analysed for CO_2 and N_2O using a Hewlett Packard 5890 Series II gas chromatographic system (Nykänen et al., 1995). Soil total N (N-tot and labelled ^{15}N-tot) was determined by a Roboprep-CN analyser (Europa Scientific Ltd, UK) linked to a VG Micromass 622 mass spectrometer (Esala, 1991). ^{15}N label of the mineral-N pool (NH_4, NO_3 and NO_2) was determined by a micro-diffusion method (Brooks et al., 1989).

Results and discussion

The production of N_2O increased with increasing in temperature. At temperatures near 0°C, emissions were negligible. Differences between the two added N levels were minor. Q_{10} was calculated for N_2O production using the whole temperature range, being 23.5 with 100 kg N ha^{-1} and 18.5 with 10 kg N ha^{-1}. When Q_{10} was calculated using temperatures above 0°C, the value was 2.3 with 100 kg and 10 kg N ha^{-1}. Though the N_2O production rate was affected by temperature, the denitrification potentials remained similar in soils incubated at different temperatures. CO_2 emissions decreased with decreasing temperature, being extremely low at less than 20 mg CO_2 m^{-2} h^{-1} below +2.5°C. Q_{10} was calculated for the whole temperature range: values for 100 kg and 10 kg N ha^{-1} were 3.6 and 3.0, respectively.

A decrease in temperature did not affect the NO_3^- loss from soil at temperatures above 0°C (Figure 1), NO_3^- loss remained constant down to +0.5°C and was minor at temperatures below 0°C. According to the ^{15}N analysis, in addition to N_2O, N_2 was also formed, especially at higher temperatures. N_2O emissions accounted for less than 1% of the NO_3^--N loss during the incubation. According to the ^{15}N analysis, NO_3^- immobilisation was more or less constant at temperatures above 0°C

Conclusions

Nitrate immobilisation near freezing point (0°C) can be high, and at +0.5°C close to that at +15°C. This result has a great importance when evaluating N losses and leaching of NO_3^- from agricultural soils, especially during autumn and spring. Immobilisation at low temperature evidently decreases the amount of NO_3^- susceptible to denitrification at periods of low uptake of N by vegetation.

Acknowledgements

Finnish Global Change Research Program FIGARE (S46554) funded this project. We thank Martti Esala, Ansa Palojärvi and Mirva Céder, Agrifood Research Finland (MTT) for the ^{15}N analysis.

References

van Bochove, E., Prévost, D. and Pelletier, F. (2000). Effects of freeze-thaw and soil structure on nitrous oxide produced in a clay soil. Soil Science Society of American Journal, 64, 1638-1643.

Brooks, P.D., Stark, J.M., McInteer, B.B. and Preston, T. (1989). A diffusion method to prepare soil extracts for automated nitrogen-15-N analysis. Soil Science Society of American Journal, 53, 1707-1711.

Christensen, S. and Tiedje, J.M. (1990). Brief and vigorous N_2O production by soil at spring thaw. Journal of Soil Science, 41, 1-4.

Esala, M. (1991). Split Application of nitrogen: effects on the protein in spring wheat and fate of 15N-labelled nitrogen in the soil-plant system. Annales Agriculturae Fenniae, 30, 219-309.

Isermann, K. (1994). Agriculture's share in the emission of trace gases affecting the climate and some cause-orientec proposals for sufficiently reducing this share. Environmental Pollution, 83, 95-111.

Nykänen, H., Alm, J., Lång, K., Silvola, J. and Martikainen, P.J. (1995). Emissions of CH_4, N_2O and CO_2 from a virgin fen and a fen drained for grassland in Finland. Journal of Biogeography, 22, 351- 357.

Röver, M., Heinemeyer, O. and Kaiser, E.-A. (1998). Microbial induced nitrous oxide emissions from an arable soil during winter. Soil Biology and Biochemistry, 30, 1859-1865.

N$_2$O emissions from a water-saving rice production system (GCRPS) in North China

C. Kreye[1], K. Dittert[1], X. Zheng[2], X. Zhang[3], S. Lin[3], H.Tao[1,3] and B. Sattelmacher[1]
[1]*Institute of Plant Nutrition and Soil Science, Kiel University, Kiel, Germany*
[2]*Institute of Atmospheric Physics, Chinese Academy of Science, Beijing, China*
[3]*Department of Plant Nutrition, China Agricultural University, Beijing, China*

Introduction

In lowland rice production systems water needs are high, but at the same time water is becoming increasingly scarce (Li, 2001). Therefore, the Ground Cover Rice Production System (GCRPS) has been developed. In GCRPS, the soil is no longer submerged, but soil moisture is maintained at a level between saturation and field capacity. Additionally, the soil is covered by mulching materials such as plastic film or straw. In GCRPS, soil moisture conditions change from anaerobic to more aerobic conditions, which are likely to favour N$_2$O emissions (Bronson et al., 1997; Zheng et al., 2000). The objective of our study was to determine the impact of the new GCRPS on N$_2$O emissions, as compared with the traditional paddy rice system.

Materials and methods

A field experiment was carried out in 2002 near Beijing on a light loam soil with a deep ground water level (>10m). Gas emissions were studied in paddy rice, GCRPS plastic film and GCRPS straw mulch. Soil moisture in GCRPS was maintained at around 10kPa, irrigation management of the paddy was adapted to local practice with about 1 week drainage in July and alternate wetting and drying irrigation during the second half of August. Nitrogen fertiliser was applied as 180 kg ha^{-1} urea N, split into 3 applications. Gas samples were taken once per week during the vegetative phase for rice, using the closed chamber technique (Hutchinson & Mosier, 1981). Sampling frequency was increased after fertilisation or drainage. For cumulative flux calculations (mg m^{-2}), fluxes for intervals between measurements were interpolated linearly.

Results and discussion

Cumulative N$_2$O flux from GCRPS and the paddy were similar in magnitude with an insignificant tendency of increased flux from GCRPS in 2002 (Table 1). Nevertheless, N$_2$O fluxes were provoked by different events in GCRPS and the paddy. From GCRPS, highest emissions occurred after N fertilisation, while in the paddy (where N was applied to the flooded soil), fertilisation had a less pronounced effect. The paddy treatment was heavily affected by the event of soil drying during mid-season drainage, when most N$_2$O was emitted from these plots. The effects of drainage and fertilisation were less clear after the third fertilisation, as at this time, the paddy plots dried out occasionally.

Table 1. N_2O emissions from 3 treatments during the rice growing period 2002 in Beijing; (n=3).

Treatment	Total N_2O (mg m^{-2})	sd	% N_2O fertilisation[1]	% N_2O drainage[2]
Paddy	395.5 a	221.3	21 a	37
GCRPS plastic film	594.3 a	110.3	63 b	-
GCRPS straw mulch	471.4 a	87.1	54 b	-

Figures followed by the same letter are not significantly different at the probability level of 0.05 (Tukey test)

[1]sum of emissions over 14 days after each fertiliser application (% of total N_2O emission)

[2]sum of emissions over drainage period (% of total N_2O emission)

Conclusion

The major difference in N_2O emissions from the paddy and GCRPS can be seen in the clearly changed temporal pattern of fluxes. Nevertheless, the cumulative N_2O flux was similar from GCRPS and the paddy systems. Thus, from the environmental point of view, a change in management from the traditional paddy to GCRPS had no significantly negative effect on N_2O emissions.

Acknowledgements

This work was supported by the German Research Council, DFG, and the Natural Science Foundation of China, NSFC.

References

Bronson, K.F., Neue, H.-U., Singh, U. and Abao, E.B. (1997). Automated chamber measurements of methane and nitrous oxide flux in a flooded rice soil: 1. Residue, nitrogen, and water management. Soil Science Society of America Journal, 61, 981-987.

Hutchinson, G.L. and Mosier, A.R. (1981). Improved soil cover method for field measurement of nitrous oxide fluxes. Soil Science Society of America Journal, 45, 311-316.

Li, Y.H. (2001). Research and practise of water-saving irrigation for rice in China. In: R. Barker, R. Loeve, Y.H. Li and T.P. Tuong (eds). Water-saving irrigation for rice. Proceedings of an international workshop held in Wuhan, China, 23-25 March 2001. Colombo (Sri Lanka): International Water Management Institute. p 1-10.

Zheng, X., Wang, M., Wang, Y., Shen, R., Gou, J., Li, J., Jin, J. and Li, L. (2000). Impacts of soil moisture on nitrous oxide emissions from croplands: a case study on the rice-based agro-ecosystem in Southeast China. Chemophere - Global Change Science, 2, 207-224.

Controlling nitrous oxide emissions from agriculture: experience from The Netherlands

P.J. Kuikman, G.L. Velthof and O. Oenema
Alterra, Wageningen UR, P.O. Box 47, NL-6700 AA Wageningen, The Netherlands

Introduction

Agriculture in the Netherlands contributes approximately 20 M ton CO_2 equivalents or 10% of the country's total greenhouse gas emissions. Soils are the dominant sites of N_2O production (7.8 M ton CO_2 eq or 25.3 Gg N_2O in 1997) and most studies focus on measuring and understanding the spatial and temporal variations of N_2O emissions from soils. Between 2000 and 2003, a project (ROB - agro) was carried out to identify options for decreasing N_2O and CH_4 emissions from agriculture in The Netherlands. Implementation of these measures would help to satisfy the target of decreasing the total emission of greenhouse gases. This paper reports on the strategy, approach and results of the N_2O projects from the ROB - agro project.

Strategy and approach

A farm systems approach was used to identify potential mitigation options in coherent sub-projects with a focus on soil and grassland management, fertiliser and manure management, clover, grazing management, crop residues, and water management. We focussed on the main agricultural land users in The Netherlands, i.e. dairy farming (60%) with grassland and silage maize as fodder crops, and arable farming (30%) with potatoes, sugar beet and wheat as main crops. Each of the projects commenced with a systems analysis in which controls, options and the pros and cons of the various possible mitigation measures were identified. Special emphasis was given to trade-offs (unwanted effects, e.g., CO_2 and CH_4 emissions, NO_3^- leaching, NH_3 volatilisation, energy use) and to the economic cost of the measures (Oenema et al., 2002). Next, measures with high potential were tested in laboratory and field experiments (Burczyk et al., 2001; Velthof et al., 2002; van Groenigen et al., 2004). The results of the systems analyses and experiments were then integrated into a decision support system (MITERRA) to calculate effectivity and (side) effects, acceptability and economics of various policies and measures on a national basis. The measures were categorised into (i) management measures, (ii) technological measures and (iii) structural measures.

Results and discussion

Many of the tested measures showed a response in terms of changes in the emissions of N_2O, but few of the measures yielded consistent and uniform patterns for all sites tested, without unwanted side-effects. The very best and most simple measures appeared to be improving the N use efficiency and thereby decreasing the total N input into the system. As agriculture in The Netherlands is highly intensive and N is not used efficiently, there is much scope for decreasing N input and decreasing N_2O emissions. On average, N use efficiency at the farm level has increased by a factor of 2 between 1998 and 2003. As a result, N_2O emissions decreased by about 30%. Improving N use efficiency on animal farms means improving the N use efficiency in the whole chain: from soil to animal feed, to animal and animal waste and back to soil again. We found that the current manure policy is effective for decreasing N_2O emissions from agriculture.

Various additional measures have potential (Table 1), but these require much site-specific information and farmers' skill to be able to implement them successfully in practice. For example, anaerobically digested animal manure applied to land decreased emissions on sandy soil, and decreased emission on clay soil, relative to undigested manure. Manipulation of crop residues had an effect on sandy soil, but not on clay soil. Split applications of fertilisers tended to decrease emissions, while injections of animal manure and fertilisers at 5 or 10 cm depth in the soil tended to increase N_2O emissions, relative to broadcast applications. We conclude that site- and farm-specific information is needed to be able to implement these additional measures on all sites successfully

Table 1. The 10 most effective measures for mitigation of non - CO_2 greenhouse gases from agriculture in The Netherlands: values are based on 1990 data.

Measure	Effectivity (M ton CO_2 equivalent)
Replace NO_3^- by NH_4^+ on grassland	- 0.25
Split application of N fertilisation on grassland	- 0.10
Grassland renovation: no ploughing and improved grassland management	- 0.8 - 1.3
Replace mineral N by biological fixation by clover	- 0.1 - 1.0
Precision fertilisation (follow advice more strictly)	- 0.2 - 0.8
Feeding cattle: more maize in ration	- 0.2
Optimal manure management	- 0.5 - 0.8
Fertilisation: mineral N on arable. Manure on grass	- 0.5 - 1.0
Add nitrification inhibitors to manure	- 0.3
Anaerobic digestion of manure	- 0.7 - 1.5
Total	- 3.5 - 6.0

Conclusions

Management measures directed towards improving N use efficiency are the most promising and cheapest measures for decreasing N_2O emissions. Technological measures (fertiliser technology, application techniques, crop residue manipulation, manure treatment, irrigation) appear expensive, site-specific and often have unwanted side effects. Structural measures (i.e. decreasing the volume of production and number of animals via quotas) are also effective, but are very expensive. Recent structural measures by the government (buy-out of animal rights, and lowering the milk quota) have decreased N_2O emissions by some 5%. Summarising, the potential of cost-effective measures to decrease N_2O emissions is in the order of minus 40%.

References

Burczyk, P., Oenema, O., Kuikman, P.J. and Velthof, G.L. (2001). Nitrous oxide emission from grassland in different management conditions (fertilization regimes). Journal of Water and Land Development, 5, 57-67

Groenigen, J.W. van, Kasper, G.J., Velthof, G.L., Dasselaar, A. van den Pol - van and Kuikman, P.J. (2004). Nitrous oxide emission factors from silage corn fields under different mineral nitrogen fertilizer and slurry applications. Plant and Soil (accepted for publication).

Oenema, O., Velthof, G.L. and Kuikman, P.J. (2001). Technical and policy aspects of strategies to decrease greenhouse gas emissions from agriculture. Nutrient Cycling in Agroecosystems, 60, 301 - 315.

Velthof, G.L., P.J. Kuikman and O. Oenema (2003). Nitrous oxide emission from animal manures applied to soil under controlled conditions. Soil Biology and Fertility, 37, 221-230.

Fluxes of N_2O from permanent grassland with different levels of nitrogen supply

C. Lampe[1], F. Taube[1], M. Wachendorf[1], B. Sattelmacher [2] and K. Dittert[2]
[1]*Institute of Crop Science and Plant Breeding, Christian-Albrechts-University, 24098 Kiel, Germany*
[2]*Institute of Plant Nutrition and Soil Science, Christian-Albrechts-University, 24098 Kiel, Germany*

Introduction

N_2O emissions from agricultural soils are of concern since they contribute to global warming by destruction of stratospheric O_3 and they represent a loss of N from crop systems. This study was part of an integrated research project at the University of Kiel (Taube & Wachendorf, 2000) aimed at general improvement of N efficiency and reduction of N losses in dairy farming systems. N_2O fluxes were quantified over one year under four regionally relevant fertiliser combinations. The objective of this study was to obtain general information on rates of N_2O emissions for a specific region over a full year and to evaluate the effects of cattle grazing and of the amount of N fertiliser and its form (i.e. mineral fertiliser or slurry).

Materials and methods

A field experiment with 5 fertiliser treatments (Table 1) was conducted on the experimental farm 'Karkendamm' near Kiel, Germany, from April 2001 to March 2002. 15 micro-plots of 2.25 m^2 each were managed as a mixed system of permanent grassland sown with grass/clover (two cuttings and two successive grazings). The sandy soil was rich in humus (3.95% C) and had a pH of 5 to 5.5. ^{15}N labelled slurry was produced by feeding two steers with ^{15}N labelled hay and maize silage. After each N application, gas samples were taken daily for a period of two weeks, then at intervals of 2-3 days, and once per week during the winter period using the closed chamber method. The N_2O concentration of the gas samples from 75 sampling dates and the isotopic composition of N_2O were measured on a continuous flow mass spectrometer (Finnigan Delta Plus - Precon). The ratio of ^{15}N to ^{14}N in N_2O was used to differentiate N_2O originating from soil and fertiliser.

Table 1. N input (kg ha⁻¹) of the 5 experimental treatments.

Treatment	Mineral N[1] (kg ha⁻¹)	Slurry N[2] (kg ha⁻¹)	N_2 fixation[3] (kg N ha⁻¹)	Total N (kg ha⁻¹)
Control (C)	0	0	89	89
^{15}N slurry (S)	0	74	41	115
^{15}N 100N (M)	70+30	0	34	134
Slurry + ^{15}N 100N (MS)	70+30	74	59	233
^{15}N slurry + 100N (SM)	70+30	74	88	262

[1] CAN, [2] cattle slurry, [3] estimated by difference method (Haystead, 1981), n = 3.

Results and discussion

Total annual N_2O emissions ranged from 1.7 kg N_2O-N ha^{-1} yr^{-1} (M) to 4.9 kg N_2O-N ha^{-1} yr^{-1} (SM). They were not significantly different (Figure 1A). The measured rates are comparable to results of Velthof & Oenema (1995), who calculated annual N_2O-N releases from a sandy soil of 2.7 and 7.1 kg ha^{-1} for mown and grazed grassland, respectively. Both of their plots were fertilised with 313 kg mineral N ha^{-1} yr^{-1}. The lack of differentiation among our treatments was partly caused by the heterogeneous distribution of excrement from grazing cattle and by N_2 fixation of white clover equilibrating the N status in all treatments to some extent.

In the period before grazing (April 2 - July 31, 2001: 44 measuring dates), mean N_2O emission rates generally increased with increasing N inputs (Figure 1B). Mean N_2O-N fluxes were 15, 19, 25, 31 and 37 µg m^{-2} h^{-1} for the C, S, M, MS and SM treatments, respectively, and the flux from the highest N input treatment (SM) was significantly different from the Control (p = 0.05). ^{15}N ratios of emitted N_2O in mineral N and slurry plots revealed that 39% N_2O-N was emitted from ^{15}N-labelled mineral N fertiliser, which was significantly more than the 9% share from ^{15}N-labelled cattle slurry. This phenomenon was probably caused by different amounts of N being rapidly available for microbial processes from mineral fertiliser and organic manure. Overall, the soil N pool turned out as a large source of emitted N_2O under our experimental conditions.

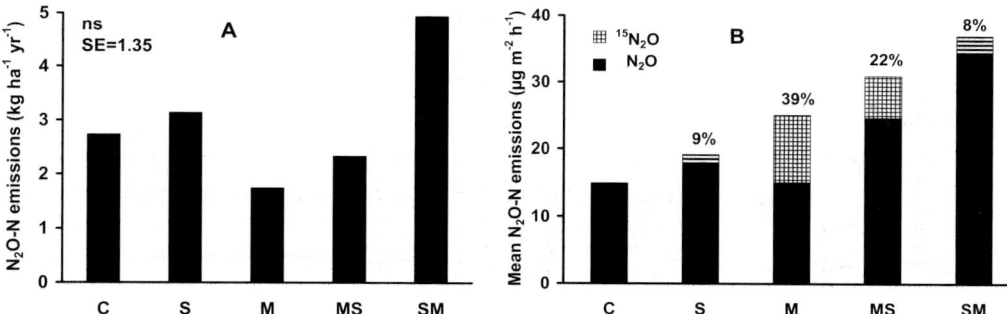

Figure 1. Total N_2O-N emissions (kg ha^{-1} yr^{-1}) from April 2, 2001 to March 4, 2002 (A) and mean N_2O-N emission rates (µg m^{-2} h^{-1}) for the cutting period (April 2, - July 31, 2001) and fertiliser derived N_2O (%) (B).

References

Haystead, A. (1981). Nitrogen Fixation and Transfer. In: J. Hodgson, R.D. Baker, A. Davis, A.S. Laidlaw and J.D. Leaver (eds). Sward Measurement Handbook, 229-242.

Taube, F. and Wachendorf, M. (2000). The Karkendamm Project: a system approach to optimize nitrogen use efficiency on the dairy farm. In: K. Søegaard, C. Ohlsson, J. Sehested, N.J. Hutchings and T. Kristensen (eds). Grassland farming. Balancing environmental and economic demands. Grassland Science in Europe, 5, 449-451.

Velthof, G.L. and Oenema, O. (1995). Nitrous oxide fluxes from grassland in the Netherlands: 2. Effects of soil type, nitrogen fertilizer application and grazing. European Journal of Soil Science, 46, 541-549.

Gaseous nitrogen emissions from effluent irrigated soils

Y. Master[1], R.J. Stevens[2], R.J. Laughlin[2], U. Shavit[1] and A. Shaviv[1]

[1]*Technion-IIT, The Faculty of Civil and Environmental Engineering, The Division of Agricultural Engineering, Haifa 32000, Israel*

[2]*Department of Agriculture and Rural Development, Agricultural and Environmental Science Division, Newforge Lane, Belfast BT9 5PX, Northern Ireland, UK*

Introduction

Irrigation with reclaimed effluent (RE) is essential in arid and semi-arid regions. One of the potential hazards of RE application is the stimulation of gaseous N losses and the possibility of affecting soil N processes. To our knowledge, no direct measurements of the N_2 and N_2O emissions from Mediterranean soils have been conducted so far. The ^{15}N gas-flux method was applied in a series of field and laboratory experiments to study the effect of RE irrigation on the emissions of N_2O, N_2, and NH_3 from a Grumosol (*Chromoxerert*) soil. Additional variables (e.g. moisture content) that influence the emissions were tested.

Materials and methods

The fluxes of N_2, N_2O, and NH_3 were measured from six Grumosol lysimeters during 50 hours following application of either freshwater (FW) or RE using a closed chamber (Master et al., 2003). The N fertiliser was applied either as $^{15}NH_4^+$ or $^{15}NO_3^-$. The gaseous N emissions and N transformation rates were also quantified in a set of short- (~50 hrs.) and long- (10-14 days) term laboratory incubations in which the soil and climatic conditions prevalent in the field were simulated.

Results and discussion

Short-term studies indicated that the effluent enhanced the emissions of NH_3 and N_2 emissions, but had no significant effect on those of N_2O (Table 1).

Table 1. Cumulative amounts of gaseous N (expressed in g N ha^{-1}), as N_2, N_2O and NH_3, lost from the Grumosol during the 48 (for N_2 and N_2O) and 72 (for NH_3) hours following fertiliser application. Numbers in parentheses are the standard deviations (n = 3 for N_2 and NH_3 and n = 6 for N_2O).

Treatment	N_2 loss	N_2O loss	NH_3 loss
FW	372 (91)	324 (31)	129 (43)
RE	493 (40)	320 (30)	245 (215)

† FW, fresh water; RE, reclaimed effluent.

Nitrification and denitrification were equally important to N_2O production under field conditions, however, the significance of the latter increased with higher moisture content (up to 75% of the N_2O amounts emitted from soils incubated under saturation). Saturated conditions significantly increased the amounts of N_2 and N_2O emitted to the environment (up to 3.1% of the applied N fertiliser).

Long-term studies showed that RE application more than doubled the emissions of N_2O from 0.7% of the applied N fertiliser in FW treatments to 1.5% in RE-treated Grumosol (Figure 1). No N_2 was detected under the experimental conditions, unless the soil was completely saturated.

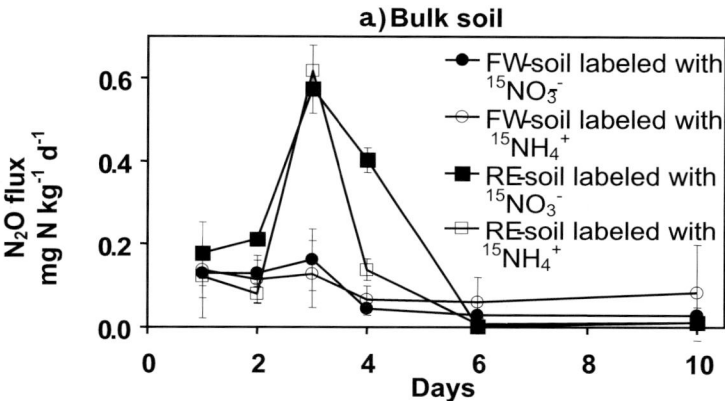

Figure 1. Nitrous oxide fluxes in a 10-day experiment from a Grumosol soil irrigated with either FW or RE and labelled either with $^{15}NH_4^+$ or $^{15}NO_3^-$. Error bars are the standard errors of the means (n = 3), or are smaller than symbols.

Conclusions

Nitrous oxide emissions from the Grumosol are stimulated by RE application, unless the soil is saturated, in which case the soil water content becomes the major factor influencing N_2O emissions. Additional observations of the RE effect included stimulation of nitrification (by 40%) and enhanced NO_2^- formation in the longer-term studies. These findings, combined with other known disadvantages of the secondary RE (salinity, sodicity, high content of pathogens and organic and inorganic pollutants, etc.) dictate a much more cautious approach in its application.

Acknowledgements

This project was partially funded by the Technion Grand Water Research Institute and the Israeli Ministry of Agriculture.

References

Master, Y., Laughlin, R.J., Shavit, U., Stevens, R.J. and Shaviv, A. (2003). Gaseous nitrogen emissions and mineral nitrogen transformations as affected by reclaimed effluent application. Journal of Environmental Quality, 32, 1204-1211.

Reducing losses of nitrous oxide from cattle slurry and mineral fertiliser applied to grassland by the use of DMPP

P. Merino[1], A. del Prado[2], S. Menéndez[3], L. Careaga[3], M. Pinto[1], J.M. Estavillo[3] and C. González-Murua[3]
[1]NEIKER, Bº Berrega, 1, 48160 Derio, Bizkaia, Spain
[2]IGER, North Wyke Research Station, Okehampton, EX20 2SB, United Kingdom
[3]Dpto Biología Vegetal y Ecología, Facultad de Ciencias, UPV/EHU, Apdo, 644, 48080 Bilbao, Spain

Introduction

Agricultural soils are major sources of N_2O. From an ecological point of view, using a nitrification inhibitor with NH_4^+ based fertilisers may be a potential management strategy to lower the fluxes of trace gases, thus decreasing their undesirable effects. The nitrification inhibitor 3,4-dimethyl pyrazole phosphate (DMPP) has proved successful in reducing N_2O emissions from mineral fertilisers (Linzmeier et al., 2001) at low concentrations (Zerulla et al., 2001), but its mitigation potential with slurry and in grassland soil has not been studied.

Materials and methods

In autumn 2002, a randomised complete block factorial design was carried out on a grassland site (at Derio, Bizkaia, Spain) to study the effects of the nitrification inhibitor DMPP (1 kg ha^{-1}) in reducing emissions of N_2O. Two types of fertilisers were applied with and without DMPP: cattle slurry and ammonium sulphate nitrate (26%). Static chambers were used to measure daily N_2O emissions after fertiliser application for the first seven days and then once a week during 135 days. Nitrous oxide emissions were measured using a closed air circulation technique in conjunction with a photoacoustic infrared gas analyser (Brüel & Kjaer 1302 Multi-Gas Monitor) over 40 minutes (Merino et al., 2001). Fluxes were calculated from the linear concentration increase in the chamber headspace over time.

Results and discussion

Peak N_2O emissions occurred four days after fertiliser application, with 1392 g N_2O-N ha^{-1} day^{-1} in the mineral fertiliser treatment. N_2O emissions were inhibited by the use of DMPP. Up to 165 days, following application of DMPP, N_2O emissions were reduced by 26.1% when applied with cattle slurry and 49.7% when applied with mineral fertiliser (Figure 1). These decreases agreed with the averages of 32% and 49% reported by Dittert et al. (2001) for grassland fertilised with slurry and Weiske et al. (2001) for fertilised cultivated soil, respectively. DMPP maintained soil NH_4^+ at higher levels between 7 and 22 days. Zerulla et al. (2001) found that nitrification was inhibited for 6 weeks, but, Weiske et al. (2001) found that the concentrations of NH_4^+ remained unaffected by DMPP treatment. The results indicate that DMPP decreases N_2O emissions both from cattle slurry and NH_4^+ sulphate nitrate fertiliser; its effect on soil NH_4^+ being effective up to 135 days after mineralised fertiliser treatment (Figure 2).

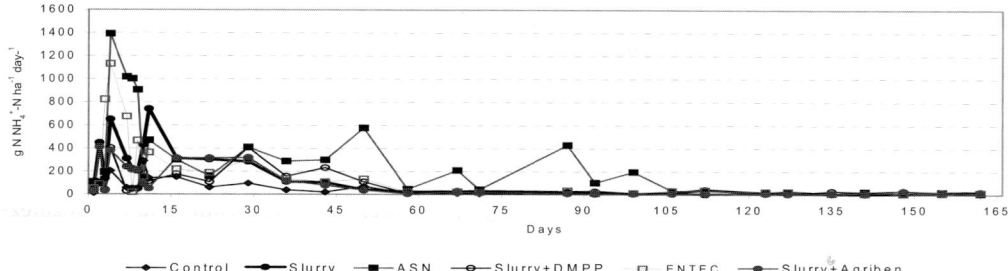

Figure 1. N$_2$O emission (g N-N$_2$O ha^{-1} day^{-1}) after application of 135 kg N ha^{-1}, soil temperature (°C) and soil water content (%WFPS); each value is the mean of four replicates.

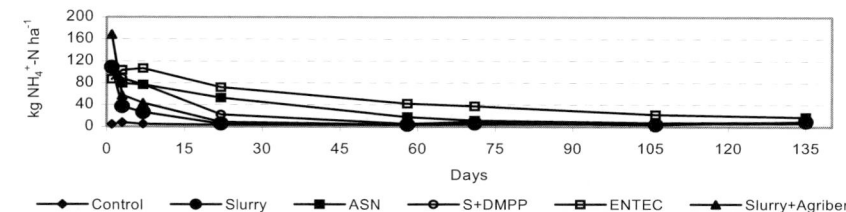

Figure 2. Typical course of the NH$_4^+$ content (kg N-NH$_4^+$ ha^{-1}) in the topsoil (10cm) after fertilisation with 135 kg N ha^{-1} as ASN and ENTEC.

Conclusions

Emissions of N$_2$O were reduced by the use of DMPP, either with mineral or slurry fertiliser up to 165 days. DMPP was more effective for soil NH$_4^+$, when applied with mineral fertiliser.

Acknowledgements

This project has been funded by the Spanish Commission of Science and Technology (CICYT project n° AGL2000-0543) and Universidad (UPV)-Empresa (COMPO) UE 02/A04

References

Dittert, K., Bol, R., King, R., Chadwick, D. and Hatch, D. (2001). Use of a novel nitrification inhibitor to reduce nitrous oxide emission from N-15-labelled dairy slurry injected into soil. Rapid Communications in Mass Spectrometry, 15, 1291-1296.

Linzmeier, W., Gutser, R. and Schmidhalter, U. (2001) Nitrous oxide emission from soil and from a nitrogen-15-labelled fertiliser with the new nitrification inhibitor 3,4-dimethylpyrazole phosphate (DMPP). Biology and Fertility of Soils, 34, 103-108.

Merino, P., Estavillo, J.M., Besga, G., Pinto, M. and González-Murua, C. (2001). Nitrification and denitrification derived N$_2$O production from grassland soil under application of DCD and Actilith F2. Nutrient Cycling in Agroecosystems, 60, 9-14.

Weiske, A., Benckiser, G., Herbert, T. and Ottow, J.C.G. (2001). Biology and Fertility of Soils, 34, 109-117.

Zerulla, W., Barth, T., Dressel, J., Erhardt, K., Horchler von Locquenghien, K., Pasda, G., Rädle, M. and Wissemeier, A.H. (2001). 3, 4-dimethylpyrazole phosphate (DMPP)-a new nitrification inhibitor for agriculture and horticulture. An introduction. Biology and Fertility of Soils (2001), 34, 79-84.

Influence of white clover on nitrous oxide fluxes in grassland

A. Mori[1,2], M. Hojito[1], H. Kondo[1], H. Matsunami[1] and D. Scholefield[2]
[1]*National Institute of Livestock and Grassland Science, Nishinasuno, Tochigi 329-2793, Japan*
[2]*Institute of Grassland and Environmental Research, North Wyke Research Station, Okehampton, Devon EX20 2SB, UK*

Introduction

Grassland soils develop rich microbial communities, because of continual additions of plant litter and dead roots to soil microorganisms. However, little is known about the effect of vegetation on the N_2O exchanges in grassland soil. Our study aimed to compare N_2O fluxes from grassland with or without white clover. White clover improves soil structure rapidly (Mytton et al., 1993), and its residue can supply both mineral N and organic C to soil microorganisms. N_2 fixation by white clover increases net N supply to the grassland and activates turnover of organic matter (Higashida, 1993). We hypothesised that white clover would change the physical, chemical, and biological properties of the soil, so that N_2O fluxes would be affected strongly in the grassland soil ecosystem.

Methods

A field study was carried out in unfertilised grassland plots at the National Institute of Livestock and Grassland Science, Japan (36°55′N, 139°55′E). The plots, laid out side by side, were planted with three different types of forage plant. The first plot (plot 1) was planted with cocksfoot (*Dactylis glomerata* L.), plot 2, with white clover (*Trifolium repens* L.), and plot 3, with cocksfoot and white clover mixture. The soil was classified as Entic Haplumbrepts, loamy, mixed, mesic. The 30-year averages of precipitation and temperature were 1561mm and 12.0°C, respectively. N_2O fluxes from the grasslands to the atmosphere were monitored weekly from April 2001 until March 2002 by a vented closed chamber method (Velthof & Oenema, 1995). Associated environmental variables (soil inorganic N, soil pH, soil moisture, soil temperature, grass yield and number of soil microorganisms) were also regularly determined.

Results and discussion

The N_2O fluxes from plots 1, 2 and 3 ranged from 1 to 10, 2 to 122 and 1 to 38 µg N m^{-2} h^{-1}, respectively (Figure 1). N_2O emissions were greatest from plot 2, smallest from plot 1 and intermediate from plot 3. The fluxes from each plot were significantly different from each other ($P < 0.01$). On an annual basis, plots 1, 2 and 3 were net N_2O sources of 0.39, 1.59 and 0.67 kg N ha^{-1} yr^{-1}, respectively. NH_4^+ and NO_3^- contents in the soil of plot 1 were maintained consistently, while those in plot 2 were greater than those in plot 1 from April to August. In plot 3, soil NH_4^+ contents were greater than those in plot 1 from April to August, however, NO_3^- contents were similar to those in plot 1. In June, November and March, the population of denitrifiers in plot 2 soil was greater than those in other plots. Large N_2O emissions from plots 2 and 3 resulted from the greater inorganic N contents in the soil. Additionally, the high temperature and soil moisture during summer, and the easily decomposed organic matter supply to the soil, might have stimulated the growth and activity of denitrifiers and increased the emission. However, nitrification could also have contributed to the large emission. No large emission was observed from plot 1, even in summer. A limited supply of inorganic N constrained the emission from plot 1.

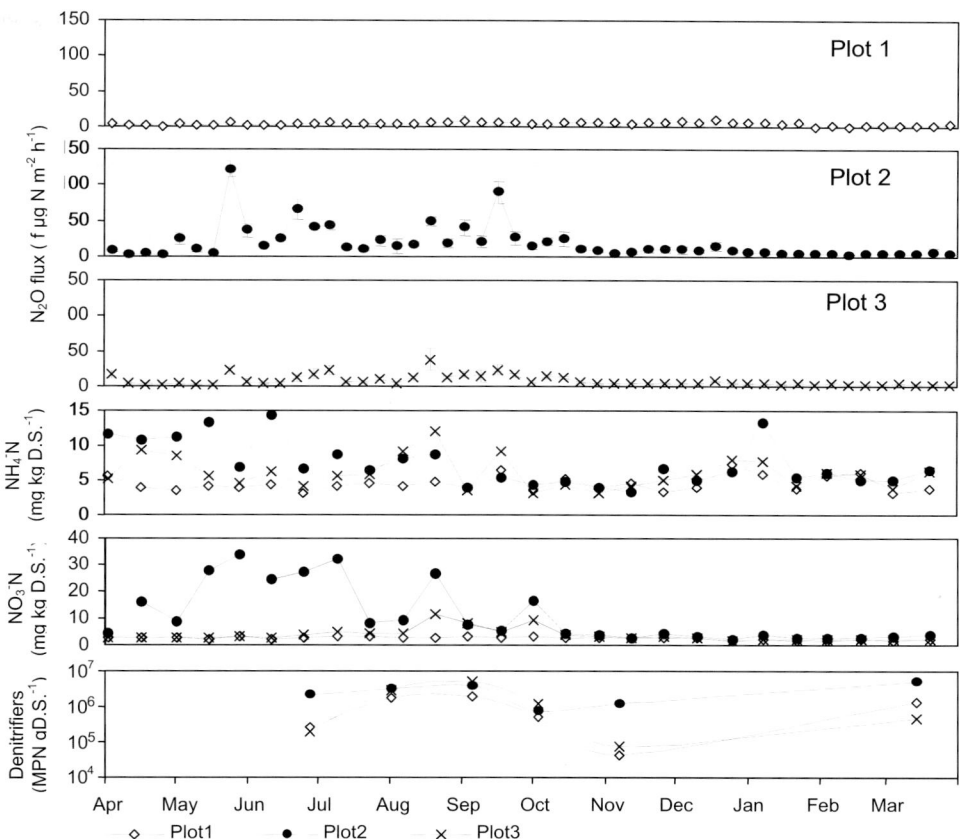

Figure 1. Seasonal changes of N_2O fluxes and soil environmental factors.

Conclusions

White clover increased the N_2O emission by increasing inorganic N contents in the soil. In summer, high soil temperature increased the release of inorganic N from the soil, facilitated the growth of denitrifiers and increased N_2O emission.

Acknowledgements

This research was supported by the National Institute of Livestock and Grassland Science and the Institute of Grassland and Environmental Research.

References

Mytton, L.R., Cresswell, A. and Colbourn, P. (1993). Improvement in soil structure associated with white clover. Grass and Forage Science, 48, 84-90.

Higashida, S. (1993). Soil microbial activities in the Tenpoku district and its contribution to grassland productivities. Report of Hokkaido Prefectural Agricultural Experiment Station, 80, 1-123.

Velthof, G.L. and Oenema, O. (1995). Nitrous oxide fluxes from grassland in the Netherlands: I. Statistical analysis of flux chamber measurements. European Journal of Soil Science, 46, 533-540.

The effect of agricultural ammonia deposition on nitrous oxide production by soils under coniferous and deciduous woodland cover.

T. Morrissey[1,2], P. Ineson[1] and D.R. Chadwick[2]
[1]*Department of Biology, PO Box 373, University of York, Heslington, YO10 5YW, UK*
[2]*IGER, North Wyke Research Station, Okehampton, Devon, EX20 2SB, UK*

Introduction

Agricultural soils are a major contributor to global N_2O emissions and re-emission of N by non-agricultural soils has been identified as an indirect source of N_2O with a re-emission level set by the IPCC at 1%, irrespective of soil type or land cover (IPCC, 1995). Non-conservation woodlands adjacent to farmlands have been suggested as possible NH_3 sinks to limit local agricultural pollution (Theobald et al., 2001). However, net N_2O production of woodland soils increases in areas of high N deposition in long term experimental studies (Brumme & Beese, 1992) and it is likely, therefore, that any reduction in NH_3 pollution caused by increased woodland cover around farms could result in increased N_2O emissions from the same areas. This laboratory study quantifies the effects of increased deposition of agricultural NH_4^+ on N_2O flux in woodland soils and investigates the following hypotheses:
1. Soils under N deposition will have a significantly higher N_2O flux than control soils.
2. There will be a significant difference in N_2O flux in cores from under oak (*Quercus robur*) and Scots pine (*Pinus sylvestris*) stands.

Materials and methods

Twenty four soil cores (30 cm deep, 15 cm diameter) were taken from a replicated woodland species comparison experiment, twelve each from both oak and Scots pine stands. All cores were maintained in the laboratory at ca.18°C. Production of N_2O was measured for each core using a gas chromatograph fitted with an electron capture detector. Prior to treatment, N_2O flux was measured from all cores and cores were paired according to flux, within blocks and cover. One core in each pair was treated with weekly applications of $(NH_4)_2SO_4$, equivalent to 50 kg N ha^{-1} y^{-1} in simulated rainfall; the other core received the same volume of de-ionised water.

Results and discussion

Cores from Scots pine stands showed a significantly higher (p<0.001) mean N_2O flux from $(NH_4)_2SO_4$ treated soils (307 ± 31.0 µg N_2O-N m^{-2} ha^{-1}) than from the control soil (119 ± 19.0 µg N_2O-N m^{-2} h^{-1}) after the start of amendment (Figure 1a). Cores from oak stands also showed a significant (p<0.001) increase in mean N_2O flux from $(NH_4)_2SO_4$ treated cores (337 ± 41.3 µg N_2O-N m^{-2} h^{-1}) compared with control cores (162 ± 25.3 µg N_2O-N m^{-2} h^{-1}; Figure 1b). There was no significant difference between total N_2O fluxes from oak and Scots pine cores, following the start of treatment with $(NH_4)_2SO_4$, in either treated (p=0.83), or control cores

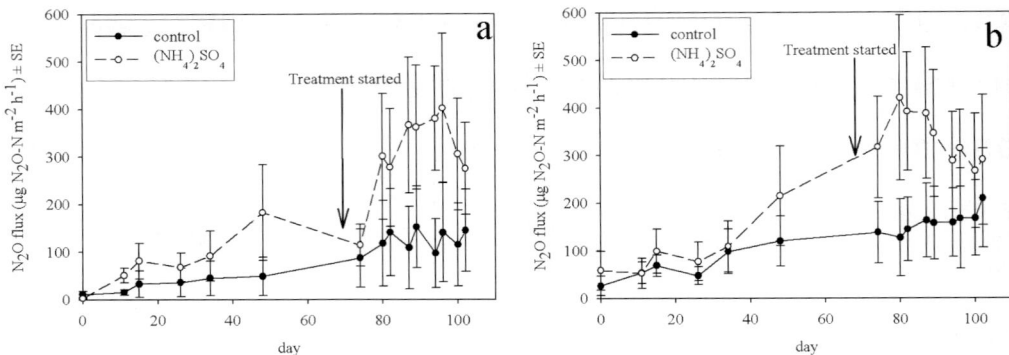

Figure 1. Mean weekly N_2O fluxes from a) Scots pine and b) oak covered soils during study period.

(p=0.44). The mean re-emission of applied N was considerably higher than the IPCC re-emission factor of 1%.

Conclusion

While the addition of $(NH_4)_2SO_4$ under controlled conditions caused a significant increase in total N_2O emission by both oak and Scots pine covered soil, there was no significant effect of tree cover observed in this study. It is possible that the discrepancy between the observed re-emission of N and that used by the IPCC, results from the absence of plant uptake of NH_4^+, which was suggested by Bowden et al. (1991) as a major limiter of nitrification in woodland soil, or perhaps, the short term nature of this study (Brumme & Beese, 1992). The positive response in N_2O flux observed here suggests that the use of woodland belts to limit agricultural N pollution may lead to increased emissions of this important greenhouse gas.

Acknowledgments

This work was funded by the BBSRC, studentship number 01/A2/D/07497.

References

IPCC (1995). Climate Change (1995). The Science of Climate Change. Contribution of Working Group I to the Second Assessment Report of the Intergovernmental Panel on Climate Change Cambridge University Press, Cambridge.

Bowden, R.D., Melillo, J.M., Steudler, P.A. and Aber, J.D. (1991). Effects of Nitrogen Additions on Annual Nitrous-Oxide Fluxes from Temperate Forest Soils in the Northeastern United-States. Journal of Geophysical Research-Atmospheres, 96, 9321 - 9328.

Brumme, R. and Beese, F. (1992) Effects of Liming and Nitrogen-Fertilization on Emissions of CO_2 and N_2O from a Temperate Forest. Journal of Geophysical Research-Atmospheres, 97, 12851-12858.

Theobald, M.R., Milford, C., Hargreaves, K.J., Sheppard, L.J., Nemitz, E., Tang, Y.S., Phillips, V.R., Sneath, R., McCartney, L., Harvey, F.J., Leith, I.D., Cape, J.N., Fowler, D. and Sutton, M.A. (2001). Potential for ammonia recapture by farm woodlands: design and application of a new experimental facility. The Scientific World, 1, 791 - 801.

Determination of N source in denitrification studies using stable isotope techniques

P.J. Murray[1], D.J. Hatch[1], E.R. Dixon[1], R.J. Stevens[2], R.J. Laughlin[3], K. O'Prey[2] and S.C. Jarvis[1]
[1]*Institute of Grassland and Environmental Research, North Wyke Research Station, Okehampton, Devon, EX20 2SB UK*
[2]*The Queen's University, Department of Agricultural and Environmental Science, Newforge Lane, Belfast, BT9 5PX UK*
[3]*The Department of Agriculture and Rural Development, Agriculture and Environmental Science Division, Newforge Lane, Belfast BT9 5PX, UK*

Introduction

A large proportion of atmospheric N_2O is derived from the process of denitrification in soils. Denitrification is known to be both temporally and spatially variable and can be promoted by N and C additions. The present work utilises a ^{15}N labelled fertiliser source to determine the interaction between C inputs and denitrification potential in topsoil and subsoil of a grassland soil under optimal conditions in a laboratory-based incubation experiment.

Methods

Soil cores (Crediton series, free draining sandy loam) of 90 cm length were divided into topsoil (0-20 cm) and subsoil (70-90 cm) fractions and sieved to <5mm. The soil was weighed into Kilner jars (1 litre volume) so that each jar contained the equivalent of 150g dry soil which was calculated from gravimetric water contents. Each jar was supplied with either artificial urine (Bristow et al., 1992), urea at typical field concentrations, urea with a five-fold increase in concentration, or a water control. There were four treatments for each soil and four replicates of each. Each jar also received labelled potassium nitrate (approximately 50 atom%) at the rate of 60 kg N ha^{-1}. Water was added to the soil to achieve 70% water filled pore space. Incubation took place at a constant temperature of 15°C. Aeration was restricted by inserting a needle into the septum of the sealed jar during the incubation period. Four hours before each sampling occasion, the jar was opened, the needle removed and fresh air allowed to circulate for approximately five minutes before re-sealing the jar. Headspace gas was collected regularly over a period of 311h using a 25 ml gas-tight syringe. Samples were stored in 12 ml evacuated glass tubes (Laughlin & Stevens, 2003) before being analysed for concentration and ^{15}N enrichment of N_2O and N_2 gas by isotope ratio mass spectrometry (Stevens et al., 1993) and CO_2 by gas chromatography.

Results

There was rapid denitrification in the topsoil of the added (labelled) mineral fertiliser, with the urine treatment being more active in stimulating the process (Figure 1a). In contrast, the native (unlabelled) N was utilised more slowly, but ultimately contributed over 50% of the total N_2O production (Figure 1b). In the subsoil (data not shown) no detectable denitrification

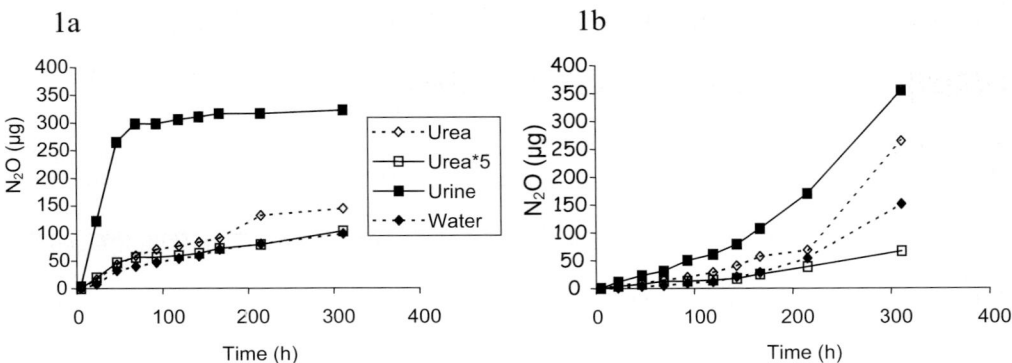

Figure 1. Cumulative evolution of N_2O from the topsoil derived from added $K^{15}NO_3$ (Figure 1a) or derived from native soil as indicated by natural abundance N_2O (Fig 1b) over a 311h incubation period.

occurred, but CO_2 measurements suggested that the urine treatment had stimulated some soil microbial activity.

Discussion

These results suggest that fertilised soils, that are also grazed, may exhibit enhanced denitrification when both urine and fertiliser N are present at the same time. Added fertiliser may then be lost as N_2O, rather than being utilised more fully by plant uptake (especially since the added N was denitrified more rapidly than native N sources). In this study, denitrification in the subsoil was found to be insignificant. However, the increase that we observed in CO_2 production in the subsoil is indicative that the amendments stimulated some extra microbial activity. Either denitrifying bacteria were not present in this subsoil, or the experimental conditions did not favour the synthesis of the appropriate enzymes within the period of our study. Other work on a similar soil (Clough et al., 1999) found that the subsoil, when supplied with both N and C, was capable of significant denitrification.

Acknowledgements

This work was conducted as part of the Natural Environment Research Council GANE Thematic Programme. The Institute of Grassland and Environmental Research is supported by the Biotechnology and Biological Sciences Research Council.

References

Bristow, A.W., Whitehead, D. C. and Cockburn, J. E. (1992). Nitrogenous constituents in the urine of cattle, sheep and goats. Journal of the Science of Food and Agriculture, 59, 387-394

Clough, T.J., Jarvis, S.C., Dixon, E.R., Stevens, R.J., Laughlin, R.J. and Hatch, D.J. (1999). Carbon induced subsoil denitrification of N-15 labelled nitrate in 1 m deep soil columns. Soil Biology and Biochemistry, 31, 31-41

Laughlin, R.J. and Stevens, R.J. (2003). Changes in composition of nitrogen-15-labeled gases during storage in septum-capped vials. Soil Science Society of America Journal, 67, 540-543.

Stevens, R.J., Laughlin, R.J., Atkins, G.J. and Prosser, G.J. (1993). Automated determination of nitrogen-15-labelled dinitrogen and nitrous oxide by mass spectrometry. Soil Science Society of America Journal, 57, 981-988.

Using a system of undisturbed, in situ soil lysimeters to determine nitrogen transformations in a sub-surface clay interface

P.J. Murray[1], E.R. Dixon[1], S.J. Granger[1], D.J. Hatch[1], S.C. Jarvis[1], R.J. Laughlin[2] and R.J. Stevens[2]
[1]*Institute of Grassland and Environmental Research, North Wyke Research Station, Okehampton, Devon, EX20 2SB, UK*
[2]*Queen's University, Belfast, Newforge Lane, Belfast, BT9 5PX, UK*

Introduction

Over 35% of lowland grassland systems in the UK are based on poorly drained soils. In this type of soil, transfer of materials mainly occurs in lateral flow. The region of soil between the permeable and impermeable layers could be a key area for transformations of N. A soil block experiment was designed to investigate the fate of N introduced at the clay interface with the soil undergoing wetting and drying. There were four treatments with three replicates of each.

Materials and methods

The soil consists of slightly stony clay loam topsoil over mottled stony clay. This is seasonally waterlogged with most excess water moving laterally through the topsoil. Two parallel trenches were excavated across a gentle slope and the remaining soil was held in place with boards and posts and then divided into 12 blocks (1 m^2) by inserting metal plates to a depth of 60 cm. Two different irrigation systems were set up, so that half the blocks received a constant amount of water every day (4 mm d^{-1}), based on the average winter rainfall pattern experienced at the site. The remaining blocks received the same total amount, but in a different pattern, so that they received their weekly allocation in the first three days and were then allowed to dry out in the remaining four days. This was to simulate the wetting and drying that would occur under more natural conditions and which may open up alternative pathways of aeration and drainage through the soil. Water content was monitored at two depths in one block from each irrigation system, using water content reflectometers attached to a data logger. Either an organic, or an inorganic form of N was added (urea or potassium nitrate) through stainless steel tubes positioned at the top end of each block and pushed down to just above the clay layer (ca. 30 cm depth). Both forms of N included a ^{15}N label, and KCl was added as a conservative tracer. Soil solution and gases were sampled from the clay interface layer, at three locations down the slope of each block, using *Macro Rhizon Samplers* (Meijboom & van Noordwijk, 1992) and silicone tubing gas samplers (Jacinthe & Dick, 1996).

Gases evolved from the surface were measured using a drainpipe ring and plug to form a closed headspace chamber. Drainage was collected from the front of the blocks (i.e. the downslope) by means of a steel plate and gutter inserted into the clay interface layer. Gases were analysed for concentration and ^{15}N enrichment of N$_2$O and N$_2$ gas. Soil water samples were analysed for Cl$^-$, urea, NO$_2^-$, NO$_3^-$ and NH$_4^+$. The ^{15}N component of NO$_2^-$ (Stevens & Laughlin, 1994) and

NH_4^+ (Saghir et al., 1993; Hauck, 1982) was analysed by methods which generate N_2O and analysed by stable isotope mass spectrometry.

Results

Nitrate treatment
Isotope ratio analysis showed that the added NO_3^- was rapidly converted to NO_2^-, N_2O and N_2 gas. There was little movement of NO_3^- away from the application point and no significant difference between watering regimes.

Urea treatment
Urea was rapidly hydrolysed to NH_4^+ and this moved down the slope. Ammonium concentrations were lower with daily watering at the application point, but higher with daily watering further down the slope

Conclusion

We have demonstrated that there is microbial activity at the clay interface region of this soil which is capable of rapidly producing NH_4^+ (hydrolysis) and NO_2^- and N_2O (denitrification) which are all potential environmental pollutants. There are implications for clay soils where cracks may open up, enabling nitrogenous compounds (e.g. urine from grazing animals or unused fertiliser), to reach the subsoil. Nitrate reaching the subsoil interface does not necessarily enter water systems, but may potentially add to atmospheric pollution through N_2O production. On the other hand, NH_4^+ reaching the base of the interflow layer either directly, or after mineralisation, has the potential to pollute watercourses.

Acknowledgements

This project was funded by the Natural Environment Research Council and is part of the Global Nitrogen Enrichment (GaNE) programme. IGER is supported by the Biotechnology and Biological Sciences Research Council.

References

Jacinthe, P.A. and Dick, W.A. (1996). Use of silicone tubing to sample nitrous oxide in the soil atmosphere. Soil Biology and Biochemistry, 28, 721-726.

Meijboom, F.W. and van Noordwijk, M. (1992). Rhizon Soil Solution Samplers as artificial roots. In: L Kutschera et al. (eds) Root Ecology and its practical application, Proceedings 3rd ISSR Symposium, 2-6 September 1991, Wien, Austria, 793-795.

Stevens, R.J. and Laughlin, R.J. (1994). Determining Nitrogen-15 in nitrite or nitrate by producing nitrous oxide. Soil Science Society American Journal, 58, 1108-1116.

Saghir, N.S., Mulvaney, R.L. and Azam, F. (1993). Determination of nitrogen by microdiffusion in mason jars. 1. Inorganic nitrogen in soil extracts. Communications in Soil Science and Plant Analysis, 24, 1745-1762.

Hauck, R.D. (1982). Nitrogen Isotope Ratio Analysis, 735-779. In: A.L. Page et al., (ed). Methods of Soil Analysis, 2nd edition part 2. Agronomy Monographs 9 ASA and SSSA, Madison, WI.

Urea concentration affects short-term N turnover and N$_2$O production in grassland soil

S.O. Petersen[1], S. Stamatiadis[2], C. Christofides[2], S. Yamulki[3] and R. Bol[3]
[1]Danish Institute of Agricultural Sciences, Dept. of Agroecology, Tjele, Denmark
[2]GAIA Centre, Ecology & Biotechnology Laboratory, Kifissia, Greece
[3]Institute of Grassland and Environmental Research, North Wyke Research Station, Okehampton, EX20 2SB, UK

Introduction

For Western Europe it is estimated that, on average, 8% of total N excreted by dairy cattle is deposited during grazing (IPCC, 1997). The intake and excretion of N is influenced by factors such as feed composition, lactation stage and pasture quality, and the excretion of excess N as urea in the urine can therefore vary considerably. It is well-known that plant roots may be scorched by urine deposition because of high levels of NH$_3$ in the soil following urea hydrolysis. We hypothesised that NH$_3$ could also be a stress factor for soil organisms, including nitrifying and denitrifying bacteria, and hence influence N$_2$O emissions. This laboratory study was conducted to investigate short-term effects of urea concentration on N$_2$O emissions and associated mechanisms.

Materials and methods

Solutions containing 0 (*CTL*), 5 (*LU*) and 10 g l^{-1} urea-N (*HU*) were added to sieved and repacked pasture soil (sandy loam, 2.7% C, 0.18% N, pH (CaCl$_2$) of 5.5 and CEC of 87 meq kg^{-1}) at a rate of 4 l m^{-2}. Also, 5 g l^{-1} urea-N was added to soil amended with 50 µg cm^{-3} NO$_3^-$-N in order to simulate N turnover in overlapping urine spots (*LUN*). The urea was labelled with 25 atom% ^{15}N. The final soil moisture was 60% WFPS. All treatments were incubated at 14°C. Carbon dioxide and N$_2$O evolution rates were determined after ca. 0.2, 0.5, 1, 3, 6 and 9 d. At the last four samplings, the replicates used for gas flux measurements were destructively sampled for determination of pH, electrical conductivity (*EC*), inorganic and total N, as well as dissolved organic C and phospholipid fatty acid composition. On Day 3, soil was also sub-sampled for determination of potential NH$_4^+$ oxidation (PAO) and denitrifying enzyme activity (DEA). The amount and isotopic composition of soil total N, NO$_3^-$, N$_2$O and N$_2$ were determined by IRMS; labelling of N$_2$ was insignificant.

Results and discussion

Total recovery of urea-N during the experiment was 84±1%. Soil NO$_3^-$ accumulated exponentially to concentrations of 90, 60 and 100 mg N kg^{-1} in *LU*, *HU* and *LUN* after 9 d. Of this, 47, 40 and 58 mg N kg^{-1} were derived from urea. Nitrification was thus delayed in the *HU* treatment. Between 33 and 52% of the NO$_3^-$ produced was derived from soil N, although initial soil NH$_4^+$ was <5 mg N kg^{-1}. This suggests a significant initial turnover of the NH$_4^+$ pool. Total concentrations of NH$_4^+$ after 1 d, corresponded to 51-61% of urea-N added, and after 3 d, to 80-85%. The transient disappearance could result from microbial assimilation in response to a decrease in osmotic potential; the *EC* levels in *LU*, *HU* and *LUN* corresponded to osmotic potentials of -0.05 to -0.12 MPa after 1 d, decreasing to between -0.14 and -0.19 MPa after 9 d. A negative interaction between osmotic stress and high NH_4^+ concentrations has been observed for, particularly, NO_2^- oxidation (Harada & Kai, 1968). The concentration of $NH_3(aq)$ calculated, suggested that nitrification rates could be significantly reduced in the *HU* treatment (Monaghan & Barraclough, 1992), as was also observed in this study. The PAO was not, however, reduced in *HU*, compared with the other urea treatments three days after urea deposition, indicating that inhibition of NH_4^+ oxidation in the soil was reversible. DEA was clearly affected by the urea amendment, probably as a result of the change in pH (Simek et al., 2002).

Emissions of N_2O during 0-9 d decreased in the order *LU>HU>LUN>CTL* and corresponded to only 0.1-0.2% of urea-N added. The emission rates for N_2O derived from soil were relatively constant in *LU*, *HU* and *LUN*. In *HU*, the emission of N_2O derived from urea increased dramatically between day 6 and 9, concomitant with an accumulation of NO_2^- to 8 mg N kg^{-1}.

Accumulated CO_2 evolution, disregarding the calculated contribution from CO_2 added in urea, was twice as much from *HU* as from *LU*. The CO_2 emission from *LUN* was lower than from *LU* for unknown reasons. The level of microbial biomass, as reflected in concentrations of PLFA, was greater in *LUN* than in the other treatments, but the absence of higher respiration rates indicates that this may have been due to a difference in extractability. In the *HU* treatment, an initial decrease in biomass was followed by a phase (Day 3 to 9) with extensive growth. The ratios of cy17:0/16:1w7c also indicated that the high urea concentration resulted in an initial phase of microbial stress which was followed by rapid microbial turnover (Petersen et al., 2002).

The time course of N_2O emissions, the correspondence with NO_2^- accumulation in *HU*, and the indications of microbial stress provided evidence that NH_4^+ oxidation was the main source of N_2O in the system investigated. The N dynamics observed were consistent with nitrifier-denitrification (Wrage et al., 2001). This implies that management practices which reduce the level of excess N in cattle urine may also reduce N_2O emissions from grazed pastures.

Acknowledgement

This study was carried out as part of the FP5 project MIDAIR (EVK2 CT-2000-00096).

References

Harada, T. and Kai, H. (1968). Studies on the environmental conditions controlling nitrification in soil. I. Effects of ammonium and total salts in media on the rate of nitrification. Soil Science and Plant Nutrition, 14, 20-26.

IPCC (1997). IPCC Guidelines for National Greenhouse Gas Inventories. Volume 3. London: Intergovernmental Panel on Climate Change.

Monaghan, R.M. and Barraclough, D. (1992). Some chemical and physical factors affecting the rate and dynamics of nitrification in urine-affected soil. Plant and Soil, 143, 11-18.

Petersen, S.O., Frohne, P.S. and Kennedy, A.C. (2002). Dynamics of a soil microbial community under spring wheat. Soil Science Society America Journal, 66, 826-833.

Simek, M., Jisova, L. and Hopkins, D.W. (2002). What is the so-called optimum pH for denitrification in soil? Soil Biology and Biochemistry, 34, 1227-1234.

Stark, J.M. and Firestone, M.K. (1995). Mechanisms for soil moisture effects on activity of nitrifying bacteria. Applied Environmental Microbiology, 61, 218-221.

Wrage, N., Velthof, G.L., van Beusichem, M.L. and Oenema, O. (2001). Role of nitrifier denitrification in the production of nitrous oxide. Soil Biology and Biochemistry, 33, 1723-1732.

Leached N and the nitrous oxide emission factor

D.S. Reay[1], K.A. Smith[1] and A.C. Edwards[2]
[1]*School of GeoSciences, Ecology and Resource Management, University of Edinburgh, Darwin Building, Mayfield Road, Edinburgh EH9 3JU, UK*
[2]*Macaulay Institute, Craigiebuckler, Aberdeen, AB15 8QH, UK*

Introduction

A major proportion of global N_2O emissions come from anthropogenic sources, particularly from the agricultural N cycle (Mosier et al., 1998). A substantial part of agriculturally-related emissions is believed to derive from N lost from agricultural land after leaching and run-off into drainage waters (Dowdell et al., 1979). This 'indirect' source of N_2O is accounted for in greenhouse gas budgets using the IPCC emission factor (EF) 'EF_5-g', where for each kg of NO_3^--N in the drainage water, 15g of N_2O will be emitted (IPCC, 1996). However, this 'indirect' N_2O EF carries with it a large amount of uncertainty. Several recent studies having suggested that the current value of EF_5-g may be too high (e.g. Nevison, 2000, Hiscock et al., 2002). Here we examine the N_2O EF for NO_3^- present in the drainage waters of two intensively farmed areas of Eastern Scotland, UK, with the aim of establishing how representative the current IPCC value for EF_5-g is for such high-N areas.

Methods

Study sites were on the Bush Estate in Midlothian, UK, and at the top of the Ythan catchment in Aberdeenshire, UK. Both catchments are dominated by intensive arable agriculture, with fields commonly receiving N applications >150 kg N ha^{-1} yr^{-1}. Water samples were taken both at field drain outfalls and at intervals along open drainage ditches. Water sampling and analysis of dissolved NO_3^- and N_2O in drainage waters at the two sites were as described in Reay et al. (2003). EFs were calculated as the direct proportion between NO_3^- and N_2O concentration in each water sample.

Results

At both study sites, N_2O concentrations many times (up to 250 times) greater than air equilibrium were identified. In samples taken from open drainage ditches, rapid outgassing of dissolved N_2O was evident, with N_2O loads rapidly (within ~ 100m) decreasing towards air equilibration. Conversely, NO_3^- was conserved in the open drainage ditches, leading to an ever-increasing divergence between dissolved N_2O and dissolved NO_3^- as distance from drainage outfall increased. The N_2O EF for dissolved NO_3^- in drainage waters at the Midlothian site was only 0.07%, compared with the 1.5% IPCC default value. The EF for this site rose to 0.15% for drainage waters emerging from a field drain and before rapid outgassing of N_2O in the open drainage ditch could occur. An intensive and long-term sampling programme at the Aberdeenshire study site produced an identical trend to that identified at the Midlothian site. Again, water samples from open drainage ditches gave EFs around twenty times lower than that of the IPCC default value (Figure 1). Water samples from a range of field drains over a full calendar year at the Aberdeenshire site produced a mean N_2O emission factor for dissolved

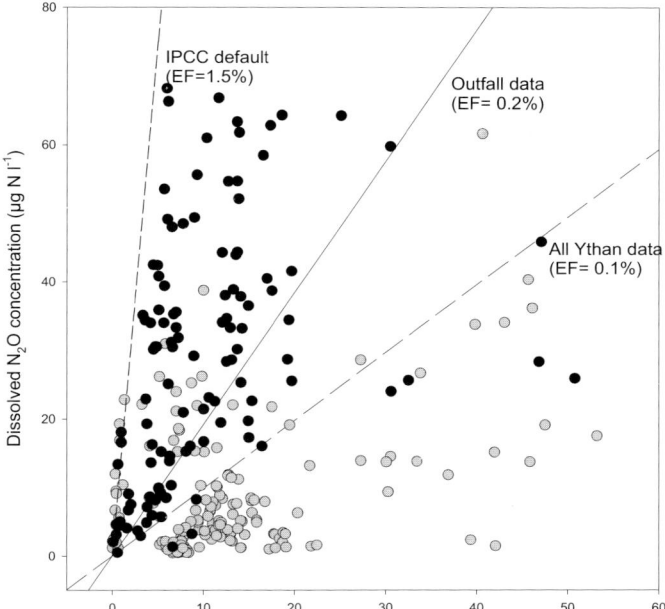

Figure 1. Dissolved N$_2$O verus nitrate concentrations for drainage waters in Aberdeenshire (March 2002-July 2003). Black symbols denote outfall samples.

NO$_3^-$ of 0.2%, double that obtained when open ditch samples were included, but much less than the IPCC default.

Discussion

The combination of rapid degassing of dissolved N$_2$O in open drainage ditches, with the conservation of dissolved NO$_3^-$, partly explains the low EFs reported here and in other studies for surface waters. However, such divergence between dissolved NO$_3^-$ and N$_2$O because of outgassing does not explain the low EF for field drain outfall waters at the two study sites. For the drainage systems investigated in our study, an N$_2$O EF for dissolved NO$_3^-$ of ca. 0.2% appears to be more appropriate than the current IPCC value.

Conclusions

Rapid outgassing of dissolved N$_2$O may lead to a large degree of error when calculating an EF for dissolved NO$_3^-$. An increasing body of work, including our own, suggests that the current default EF for N$_2$O arising from leached N in drainage waters is too high. We recommend either a lowering of the current EF 'EF$_5$-g', or the introduction of a new EF specifically for surface waters draining agricultural land, 'EF$_5$-d'.

Acknowledgements

This work was funded as part of the Natural Environment Research Council's thematic programme 'Global Nitrogen Enrichment' (GANE). The authors thank Yvonne Cook and Peter Glenister, of the Macaulay Institute, for their assistance with sample collection.

References

Dowdell R.J., Burford, J.R. and Crees, R. (1979). Losses of nitrous oxide dissolved in drainage water from agricultural land. Nature, 278, 342-343.

Hiscock, K.M., Bateman, A.S., Fukada, T. and Dennis, P.F. (2002). The concentration and distribution of groundwater N_2O in the Chalk aquifer of eastern England. In: Non-CO_2 Greenhouse Gases: Scientific Understanding, Control Options and Policy Aspects (J. Van Ham, A.P.M. Baede, R. Guicherit and J.G.F.M. Williams-Jacobse, eds). Proc. Third Int. Symp. Non-CO_2 Greenhouse Gases, Maastricht, The Netherlands, Millpress, 179-184.

IPCC (1996) Houghton, J.T., Meira, L., Filho, G., Lim, B., Treanton, K., Mamaty, I., Bonduki, Y., Griggs, D.J. and Callender, B.A. (eds). Revised 1996 IPCC Guidelines for National Greenhouse Gas Inventories. IPCC/OECD/IEA. UK Meteorological Office, Bracknell, UK.

Mosier, A.R., Kroeze, C., Nevison, C., Oenema, O., Seitzinger, S. and van Kleemput, O. (1998).

Closing the global atmospheric N_2O budget: Nitrous oxide emissions through the agricultural nitrogen cycle. Nutrient Cycling Agroecosystems, 52, 225-248.

Nevison, C. (2000) Review of the IPCC methodology for estimating nitrous oxide emissions associated with agricultural leaching and runoff. Chemosphere - Global Change Science, 2, 493-500.

Reay, D.S., Smith, K.A. and Edwards, A.C. (2003). Nitrous oxide in agricultural drainage waters. Global Change Biology, 9, 195-203.

Seasonal subsoil denitrification of leached ^{15}N-labelled nitrate

S.M. Thomas[1], T.J. Clough[2], G.S. Francis[1], D.I. Hedderley[1] R.R. Sherlock[2] and M.H. Beare[1]
[1]Crop & Food Research, Private Bag 4704, Christchurch, New Zealand
[2]Centre for Soil and Environmental Research, Lincoln University, PO Box 84, New Zealand

Introduction

Reducing the leaching of NO_3^- from agricultural soils to groundwater is an important environmental goal for both maintaining/improving water quality and reducing indirect N_2O emissions. However, few direct studies have examined the fate of leached NO_3^-, and in particular, the role of denitrification in removing NO_3^- from subsoils. There is also very limited information to quantify the amount of N_2O that is emitted as a result of subsoil denitrification, either directly from the soil surface or indirectly after NO_3^- and N_2O are leached. Indirect emissions of N_2O from leached N are a significant component of national greenhouse gas inventories, although the estimate is highly uncertain due to lack of data. In New Zealand these emissions account for approximately 23% of the N_2O inventory. We conducted an experiment using ^{15}N-labelled NO_3^- to examine seasonal changes in the production and losses of N_2 and N_2O from denitrification of leached NO_3^- in subsoils.

Materials and methods

The experiment was conducted using twelve 1 m deep intact cores of a poorly draining soil (Aqui Haplustepts) used for cropping or pasture. Cores were extracted in 25 cm diameter PVC pipes, replaced in the field, and subjected to field temperatures. ^{15}N-labelled NO_3^- (as KNO_3) was added to the top of the subsoils (at 20 cm depth) at 300 kg Nha^{-1} with an enrichment of approximately 40 atom%. Simulated rainfall was applied during four events of 40 mm within 16 days of the $^{15}NO_3^-$ application. An artificial shallow water table was established at a depth of 95 cm, 28 days after the amendment. Surface emissions, changes in subsoil concentrations and leaching losses of N_2 and N_2O were estimated from GC and MS analysis of gas samples collected from surface headspace chambers, five depths within the soil profile, and from leachate collection bags. Changes in solution mineral N and water soluble C in the soil profile and leachate were also determined. Temperature and moisture were measured in the soil profile using thermocouples and TDR probes. Measurements were made for 350 days.

Results and discussion

Greatest surface emissions of N_2O occurred soon after the application of $^{15}NO_3^-$ and simulated rainfall. The N_2O:(N_2O+N_2) ratio ranged from 0.03 to 0.23 for the surface losses. After 350 d, approximately 10 to 30% of the ^{15}N-labelled NO_3^- had been lost as N_2 and N_2O from the soil surface (Figure 1). In the subsoil, the greatest production of N_2O and N_2 occurred soon after the added NO_3^- was leached down the soil profile (data not shown). Concentrations of N_2O stayed relatively constant between 50 and 130 d, but then steadily declined to 350 d. In comparison, the decline in subsoil N_2 concentrations was much lower after 50 d. Consequently there was a marked, steady reduction in the $N_2O:(N_2O + N_2)$ ratio from day 50 onward.

The amount of NO_3^- leached was also highly variable between cores and ranged from approximately 2 to 43% of the NO_3^- applied. However, leaching could not explain the long, steady reduction of NO_3^- in the subsoil (Figure 2). A small amount of the applied N was leached as N_2O, ranging between 0 and 0.27%. This is much lower than the IPCC default value that assumes 0.45 % of the applied N is leached as N_2O. The amounts of C leached from the cores ranged from 1 to 2.3 g C m^{-2}. Carbon concentrations in leachate had declined to low values by the end of the study, presumably through its consumption by denitrifiers.

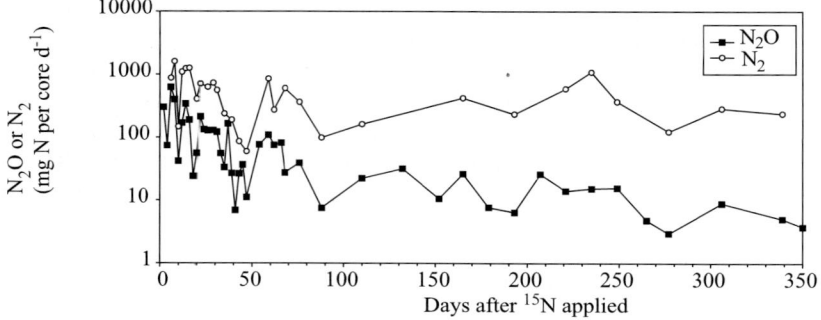

Figure 1. Mean N_2O and N_2 surface losses produced by subsoil denitrification (NB log scale).

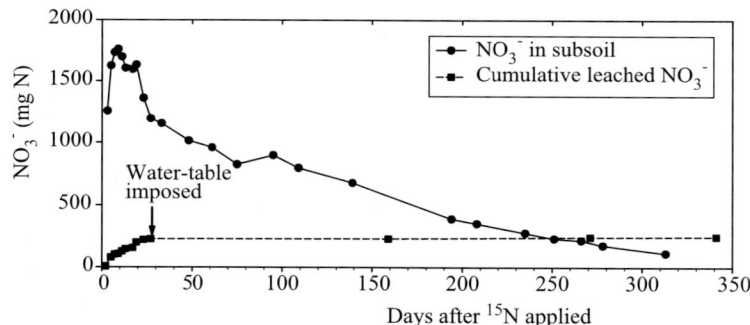

Figure 2. Changes in the amount of NO_3^- in the subsoil and NO_3^- leached from the subsoil.

Conclusions

Denitrification was an important process for removing leached NO_3^- from these subsoils. High moisture content, high NO_3^- availability, slow movement of gas and solution stimulated denitrification, with supply of C the main limitation on subsoil denitrification rate.

Acknowledgement

AGMARDT funded the study through a postdoctoral research fellowship.

N$_2$O emissions from intensive vegetable production systems

S.M. Thomas, H.E. Barlow, G.S. Francis and D.I. Hedderley
Crop & Food Research, Private Bag 4704, Christchurch, New Zealand

Introduction

Application of high rates of N fertiliser to vegetable crops poses a significant environmental risk from N leaching and runoff to water bodies and emissions of N$_2$O. In New Zealand, N fertiliser rates for intensive potato production are typically 300 to 450 kg N ha^{-1}. High rates of NO$_3^-$ leaching have been measured from these crops, but N$_2$O emissions have not been measured. In addition to high N application rates, potato crops are often irrigated, potentially further enhancing N$_2$O production. Other management effects such as soil compaction by tractor traffic in row crops, also greatly enhance N$_2$O emissions (Flessa et al., 2002). This field experiment examined the influence of N fertiliser rate, tractor compaction in furrows and water-filled pore space (WFPS) on N$_2$O emissions from an irrigated potato crop.

Materials and methods

The experiment was established in late spring (Nov 2002) on a well-drained soil (Typic Haplustepts). After cultivation, fertiliser (as calcium ammonium nitrate) was applied to each plot at rates of 0, 225 or 450 kg N ha^{-1}. Potatoes were planted at 55 cm spacings in rows 78 cm apart and then all plots were machine ridged such that each plot contained both compacted (by tractors) and uncompacted furrows. In each plot, a PVC cylinder (30 cm i.d.) was pushed into the soil to a depth of 5 cm at each of three sampling positions: the ridge (between tubers), the compacted furrow and the uncompacted furrow. Soil mineral N was measured at each sampling position in each plot on seven occasions during December to February. Headspace gas samples were taken (after covering cylinders for 20 and 40 min) on 24 occasions from November to March. Gas samples were analysed for N$_2$O concentration by gas chromatography. The experiment was a split-plot, randomised block design with five replicates. Soil temperature (at 5 cm depth) and moisture (at 0-20 cm depth) were recorded hourly using thermocouples and TDR. WFPS was calculated from measured bulk density and water content and an assumed particle density (2.65 g cm^{-3}). Daily rainfall and spray irrigation (applied five times) were measured throughout the experiment.

Results and discussion

In addition to 65 mm of rainfall, 285 mm of irrigation was applied to the crop. Mineral N contents before fertiliser application were 70 to 90 kg N ha^{-1}. After fertiliser application, most of the fertiliser N was in the ridges and least in the furrows (Figure 1a). The mineral N contents in the potato ridges were approximately 280 and 740 kg N ha^{-1} for the 225 and 450 kg N ha^{-1} treatments, respectively. Fertiliser and soil mineral N had small, but significant effects on N$_2$O emissions. More N$_2$O tended to be emitted from the fertilised plots, however emissions did not differ between the 225 and 450 kg N ha^{-1} treatments. Sampling position had a much larger effect on N_2O emissions (p<0.001), in the order: compacted furrow > fertilised ridge > uncompacted furrow (Figure 1b). N_2O emissions were strongly related to changes in WFPS (Figure 1c). Highest N_2O emissions occurred when WFPS was high, following rain or irrigation. Even after large irrigation applications (depth = 85 mm), ridge WFPS remained relatively low (<50%). WFPS varied with sampling position in the order: compacted furrow > uncompacted furrow > ridge. The lowest N_2O emissions occurred from the uncompacted furrows that had lower WFPS than the compacted furrows and relatively low mineral N concentrations (Figure 1b).

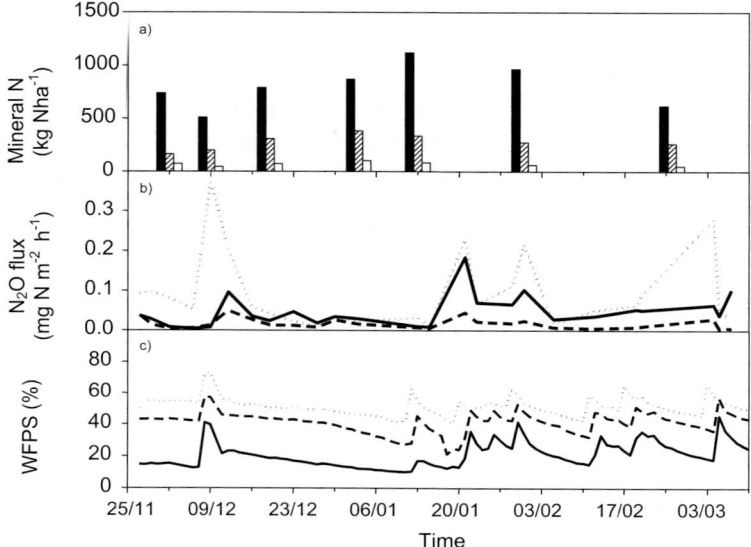

Figure 1. a) Mineral N concentrations in potato ridges (solid), compacted furrow (hatched) and uncompacted furrow (open); b) N_2O surface flux from potato ridges (solid line), compacted furrow (dotted line) and uncompacted furrow (dashed line); and c) changes in WFPS for ridges (solid line), compacted furrows (dotted line) and uncompacted furrows (dashed line). Data are for the 450 kg N ha^{-1} treatment (n=5).

Summary and conclusions

The influence of tractor compaction on N_2O emissions was more important than the rate of applied N fertiliser. Both tractor compaction and irrigation affected N_2O emissions by increasing WFPS, leading to enhanced denitrification. However, the effect of irrigation was less important in the well aerated, uncompacted fertilised ridges and furrows. Consequently, management that limits compaction in row crops will help reduce N_2O emissions.

Acknowledgements

This research was funded by the New Zealand Foundation for Research Science and Technology.

References

Flessa, H., Ruser, R., Schilling, R., Loftfield, N., Munch, J.C., Kaiser, E.A. and Beese, F. (2002). N_2O and CH_4 fluxes in potato fields: automated measurement, management effects and temporal variation. Geoderma, 105, 307-325.

Improving New Zealand predictions of N leaching for estimating indirect N_2O emissions

S.M. Thomas[1], S.F. Ledgard[2] and G.S. Francis[1]
[1]Crop & Food Research, Private Bag 4704, Christchurch, New Zealand
[2]AgResearch, Ruakura Research Centre, Private Bag 3123, Hamilton, New Zealand

Introduction

Indirect N_2O emissions from N that is leached or in runoff are considered to make a major contribution (approximately 23% or 5.7 Gg N_2O-N) to the New Zealand (NZ) N_2O emission inventory. This value is currently calculated using the IPCC methodology, i.e. by applying an emission factor (EF_5) to the amount of N that is leached or in runoff (N_{LEACH}). N_{LEACH} is calculated as a proportion ($Frac_{LEACH}$) of the N inputs from fertiliser and animal excreta. The value of $Frac_{LEACH(NZ)}$ is currently set at 0.15 for NZ inventory reporting. These estimates are uncertain, particularly for the value for EF_5. In addition, it is unclear whether $Frac_{LEACH}$ is appropriate for NZ's agricultural system that is dominated by year-round grazing of clover/grass pastures. This study examined the magnitude and uncertainty of NZ estimates of indirect N_2O emissions. One objective of the study was to determine whether the value of $Frac_{LEACH(NZ)}$ is appropriate, by comparing IPCC estimates using this value with predictions from a field-scale N leaching model for a range of typical farming systems.

Approach and methods

N_{LEACH} estimated with the IPCC calculation were compared with predictions using the OVERSEER® nutrient budget model developed for NZ conditions. Estimates were compared for 'typical' dairy, sheep and beef, arable cropping and intensive vegetable farming systems. Comparisons between years were also made for the livestock systems using data for 1990, 2000 and projections for 2010. Recommendations were then made on the appropriateness of (i) using the IPCC calculations and (ii) the value for $Frac_{LEACH(NZ)}$.

Results and discussion

The IPCC estimates for N_{LEACH} for livestock systems (Tables 1 and 2) were much greater than those predicted by OVERSEER®, while IPCC estimates for arable (Table 3) and intensive vegetable systems (data not shown) were lower than the OVERSEER® estimates. NZ has approximately 10.4 M ha of pastoral farms, and approximately 188,000 ha of arable and 53,000 ha of intensive vegetable cropping. Based on the model predictions, we recommend that the value of $Frac_{LEACH(NZ)}$ is reduced from 0.15 to 0.07 for pastoral systems.

The IPCC calculations do not take account of differences in soils or climate, which are important in determining the amount of N that is leached. These calculations are also inappropriate for estimating leaching from NZ arable soils as most of the leached N is derived from net N mineralisation of soil organic matter. In NZ, N leaching from arable soils is largely affected by management and crop rotations, and not fertiliser inputs. Therefore, we recommended that well calibrated, field-scale N leaching models should be used to estimate N_{LEACH} instead of

Table 1. N leaching estimates for NZ dairy farms using IPCC-based calculations and the OVERSEER® nutrient budget model.

	1990			2000			2010		
	Low	Av.	High	Low	Av.	High	Low	Av.	High
Cows per ha	2	2.4	3.2	2.5	2.7	3.5	2.4	3	4
N Input (kg N ha^{-1}):									
N_{FERT}	0	47	150	0	100	200	40	129	300
N_{EX}	167	266	250	213	360	500	220	418	347
N_{LEACH} (kg N ha^{-1}):									
IPCC	25	47	60	32	69	105	39	82	97
OVERSEER	16	26	49	20	40	64	25	48	92

Table 2. N leaching estimates for NZ sheep and beef farms using IPCC-based calculations and the OVERSEER® nutrient budget model.

	1990			2000			2010		
	SI High[1]	Av.	NI Int[2]	SI High	Av.	NI Int	SI High	Av.	NI Int
Stock units per ha:	0.9	6.6	12	1.1	7.3	13.2	1.3	8	14.5
N Input (kg N ha^{-1}):									
N_{FERT}	1	2	4	2	7	8	3	12	15
N_{EX}	12	85	176	11	93	192	10	101	212
N_{LEACH} (kg N ha^{-1}):									
IPCC	2	13	27	2	15	30	2	17	34
OVERSEER	2	4	14	2	5	16	2	6	18

[1]South Island high country farms, [2]North Island intensive finishing farms

Table 3. N leaching estimates using IPCC-based calculations and the OVERSEER® nutrient budget model for a 'typical' cropping rotation (recently out of clover/grass pasture).

	Crop in 'typical' rotation							
	Ryegrass seed	Clover seed	Winter Wheat	Spring Peas	Winter wheat[1]	Spring barley	Spring barley	Rotation average
N input (kg N ha^{-1}):								
N_{FERT}	180	0	250	0	250	150	150	140
N_{LEACH} (kg N ha^{-1} yr^{-1}):								
IPCC Frac$_{LEACH(NZ)}$	27	0[2]	37.5	0[2]	37.5	22.5	22.5	21[2]
OVERSEER	34	36	66	61	33	50	53	48

[1]Winter wheat following a cover crop.

[2]The N input for calculating Frac$_{LEACH}$ using IPCC formula does not include crop N fixation.

using the IPCC calculations. While we can improve the certainty of N_{LEACH}, EF_5 remains highly uncertain. EF_5 is the highest of the default emission factors for calculating national inventories using the IPCC method (2.5% of N_{LEACH}) and is based on a very small data set. Therefore, it is important that this value is better quantified in the future.

Acknowledgements

This study was funded by the New Zealand Ministry of Agriculture and Forestry.

Nitrogen losses during storage and following the land spreading of poultry manure

R.E. Thorman[1], B.J. Chambers[2], R. Harrison[1], D.R. Chadwick[3], R. Matthews[3] and R.J. Nicholson[1]
[1]ADAS Boxworth, Battlegate Road, Boxworth, Cambridge, CB3 8NN, UK
[2]ADAS Gleadthorpe, Meden Vale, Mansfield, Notts., NG20 9PF, UK
[3]Institute of Grassland Environmental Research, North Wyke Research Station, Okehampton, Devon, EX20 2SB, UK

Introduction

Thirty-six percent of UK agricultural NH_3 emissions (82 kt NH_3-N) are estimated to arise from the management of solid manures, with storage and land application responsible for ca. 50% of these. As a result of the EC Directive 96/1 on Integrated Pollution Prevention and Control (IPPC), farmers require practical guidelines on the best methods of storing and spreading solid poultry and pig manures to limit NH_3 emissions. The objectives of this study were to quantify the effects of contrasting manure storage and cultivation practices, following land spreading, on emissions of NH_3 and the greenhouse gas N_2O and to develop an integrated set of practical management guidelines to minimise N emissions.

Materials and methods

Replicated (3 per treatment) poultry manure heaps (ca. 15 m^3) were stored in concrete bunkers constructed at both ADAS, Gleadthorpe and IGER, North Wyke. Conventional storage was compared with turned, 'A'-shaped, roofed and sheeted heaps. The broiler litter at both sites was obtained from the same commercial farm. Emissions of both NH_3 (all treatments) and N_2O (selected treatments) were measured using portable, fan-ventilated enclosures at appropriate intervals over a 6-month storage period. Leachate quantities were also measured and samples taken for nutrient analysis. Following storage, the manures were applied to land and were either left on the soil surface or incorporated by ploughing or discing after 4 or 24 hours. Emissions of NH_3 were measured using passive diffusion samplers, with two dynamic chambers on each plot, for a period of 30 days. Nitrous oxide emissions were measured from selected treatments over a 3 month period using static chambers (2 per plot) and either gas chromatography or photo-acoustic infra-red analysis.

Results

At both sites, NH_3 emissions were greatest from the heaps stored under a roof (12.4 -19.0% of total N inputs to store) where the broiler litter surface remained porous and open. The sheeted heap, as expected, gave the lowest ($P<0.05$) NH_3 losses (1.3-2.3% of total N inputs to store), with the sheet acting as a barrier to NH_3 emission. Conventional, turned and A-shaped heaps, which were stored outdoors and developed a surface crust, gave intermediate NH_3 loss results. In contrast, N_2O losses from the heaps were small (< 0.5% of total N inputs to store), with emissions increasing over the storage period to give maximum fluxes of ca. 2 g N_2O-N t d^{-1}, probably due to development of anaerobic sites within the heap. There were no significant differences (*P*>0.05) in cumulative N_2O emissions between the conventionally stored and sheeted heaps. N losses in leachate varied from nil for the roofed treatment to 4.4% of total N inputs to the store from the sheeted heap.

Following land application of the conventionally stored and sheeted heaps at ADAS Gleadthorpe, NH_3 emissions were higher (*P*<0.01) from the sheeted (26% of total N into store) than the conventionally stored material (11% of total N into store). This indicates that some of the NH_3 retained by sheeting the heaps had subsequently been lost after surface application (Figure 1). To realise the full NH_3 reduction benefits of sheeting the heaps, the spread manure had to be rapidly incorporated into the soil following land spreading. Ploughing within 24 hours and discing within 4 hours of land spreading reduced (*P*<0.05) NH_3 emissions compared with surface spreading. Preliminary results show that larger N_2O losses were consistently measured from the ploughed (ca. 0.7 kg N_2O-N ha^{-1}) than the surface applied treatments (ca. 0.3 kg N_2O-N ha^{-1}).

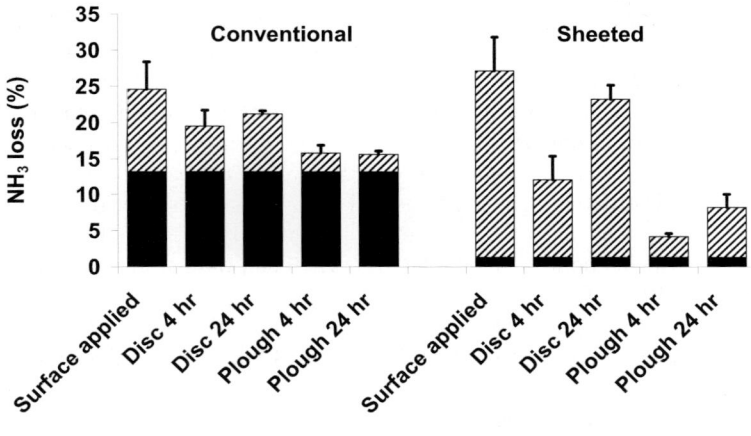

Figure 1. Mean NH_3 losses during storage and following land spreading as a percentage of total N into the store (ADAS Gleadthorpe).

Conclusions

Ammonia emissions from stored broiler litter were highest from heaps stored under a roof and lowest from sheeted heaps. However, NH_3 emissions following land application were higher from sheeted than conventionally stored manures. Ploughing within 24 hours and discing within 4 hours of manure application reduced NH_3 emissions. Ploughing manure into the soil increased N_2O emissions. These conflicts need to be considered, as ultimately, a decision on 'pollution swapping' trade-offs has to be made.

Acknowledgement

Funding of this work by the Department for Environment, Food and Rural Affairs (Defra) is gratefully acknowledged.

From N_2 fixation to N_2O emission in a grass-clover mixture

M. Thyme and P. Ambus
Risø National Laboratory, Plant Research Department, P.O. Box 49, 4000 Roskilde, Denmark

Introduction

In organic farming, biological N_2 fixation in grass-legume swards provides a major N input to the system, but knowledge is sparse regarding the amount of fixed N_2 lost from grasslands as N_2O. Nitrifying and denitrifying bacteria are the main contributors to N_2O production in soils. Currently, no contribution from biological N_2 fixation in grass-legume swards is included in the national N_2O inventories, partly because of uncertainties in quantifying N_2 fixation in grasslands (Mosier et al., 1998). According to the guidelines issued by The Intergovernmental Panel on Climate Change (IPCC), inventories for N_2O emissions from agricultural soils should be based on the assumption that 1.25% of the added N is emitted as N_2O (IPCC, 1997). The standard N_2O emission factor of 1.25% could be considerably unrepresentative for biologically fixed N_2 for two reasons. First, only a part of the fixed N is mineralised during the lifetime of the crop, and second, the release of inorganic N into the soil occurs slowly following degradation of organic residues. Therefore, a $^{15}N_2$-tracer-experiment was initiated on grass-clover to assess the contribution of recently fixed N_2 as a source of N_2O and the translocation of N from clover to companion grass.

Materials and methods

A mixture of white clover (*Trifolium repens* L. cv. Klondike) and perennial ryegrass (*Lolium perenne* L. cv. Fanda) was sown in pots using topsoil from an organic crop rotation. The ^{15}N-labelling approach consisted of introducing $^{15}N_2$ into both the above- and below-ground atmosphere to trace the biological N_2 fixation. A minimum-volume gas tight growth cabinet was developed, which could host 12 pots of 15 cm $\times$ 15 cm size. In this growth cabinet, three incubations were conducted with grass-clover mixtures at 4, 6 and 8 months of age. At each incubation event, the pots were situated in the growth cabinet for 14 days during which period the atmosphere was enriched in $^{15}N_2$ to 0.4 atom% excess. After the labelling period, half of the grass-clover pots were sampled. The N_2 fixation during the labelling period was established by relating the excess ^{15}N content of the plant material to the ^{15}N enrichment of the atmospheric N_2. During the following seven days, emission of $^{15}N_2O$ was measured from the remaining half of the pots using a static chamber method.

Results and discussion

At 4 months, N_2 fixation measured in grass-clover shoots and roots constituted 339 mg N m^{-2} d^{-1} (Figure 1). This is three to 13 times larger than the daily means of N_2 fixation determined in harvested shoot material in the field (Høgh-Jensen & Schjoerring, 1997; Vinther & Jensen, 2000), probably because of optimal growth conditions at this stage of the experiment. Following a severe aphid attack, N_2 fixation dropped dramatically at 6 months. Translocation of fixed N

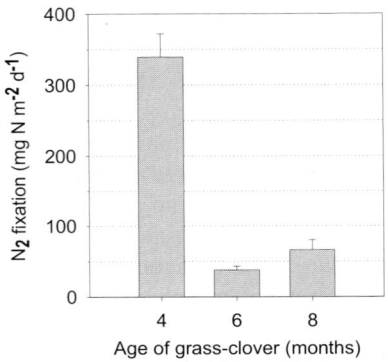

Figure 1. Biological N_2 fixation measured in grass-clover shoots and roots; n = 4, means ± SE.

from clover to grass shoots was observed at 6 and 8 months and represented 1 mg N m^{-2} d^{-1}. The fraction of fixed $^{15}N_2$, which was emitted as $^{15}N_2O$ at 4 months accounted for 3 (±0.5ppm) of the fixed N_2. These numbers only include $^{15}N_2O$ emission, which was not equalised by $^{14}N_2O$ uptake. The results are preliminary; since fixed N present in the soil has not yet been estimated.

Conclusions

Biological N_2 fixation plays an important role as N input to the grass-clover system. The aphid attack on the clover component led to translocation of fixed N to companion grass, which agrees with the view that N transfer is indirect, *i.e.* caused by turnover of organic clover residues. Emission of N_2O-N derived from recently fixed N_2 was not detected at 6 and 8 months, probably because the ^{15}N enrichment of the clover rhizodeposition was too low. In conclusion, the results at 4 months indicate that only a small proportion of the fixed N_2 is lost as N_2O. Even if a longer turnover time for clover N is considered, the N_2O emission factor for biologically fixed N_2 in grass-clover mixtures might be lower than the standard emission factor of 1.25% suggested by IPCC.

Acknowledgements

This work was funded by The Danish Research Centre for Organic Farming (DARCOF).

References

Høgh-Jensen, H. and Schjoerring, J.K. (1997). Interactions between white clover and ryegrass under contrasting nitrogen availability: N_2 fixation, N fertilizer recovery, N transfer and water use efficiency. Plant and Soil, 197, 187-199.

IPCC (1997). Greenhouse gas inventory. Reference manual. Volume 3, Intergovernmental Panel on Climate Change, Bracknell, UK.

Mosier, A., Kroeze, C., Nevison, C., Oenema, O., Seitzinger, S.and van Cleemput, O. (1998). Closing the global atmospheric N_2O budget: Nitrous oxide emissions through the agricultural nitrogen cycle. Nutrient Cycling in Agroecosystems, 52, 225-248.

Vinther, F.P. and Jensen, E.S. (2000). Estimating legume N_2 fixation in grass-clover mixtures of a grazed organic cropping system using ^{15}N methods. Agriculture, Ecosystems and Environment, 78, 139-147.

Nitrous oxide emission from an irrigated soil fertilised with pig slurry in Central Spain

A. Vallejo[1], J.A. Díez[2], L. García-Torres[1], P. Hernáiz[2] and S. López-Fernández[1]

[1]*Escuela Técnica Superior de Ingenieros Agrónomos, Universidad Politécnica de Madrid, Ciudad Universitaria. 28040 Madrid, Spain*
[2]*Centro de Ciencias Medioambientales CSIC, Serrano 115, 28006 Madrid, Spain*

Introduction

Nitrous oxide emissions and denitrification losses have been rarely measured in irrigated crops in Southern European countries, despite the fact that the surface areas used for such crops are important (Sánchez et al., 2001; Teira-Estmages et al., 1998).). In Spain alone, $2.3 \ 10^6$ ha of soil are used for irrigated crops. Conditions in these soils favour denitrification as high moisture contents (from irrigation) coincide with high soil temperatures, thus affecting denitrification rate (Maag & Vinther, 1999). By applying slurry to the soil, the activity of denitrifying bacteria is stimulated by the extra easily degradable organic C which is applied (Rochette et al., 2000). The aim of this study was to quantify the N_2O emission and the denitrification losses that occur in irrigated crops in Central Spain and, at the same time, evaluate the effect of the pig slurry application.

Materials and methods

The field experiment was carried out at La Poveda Field Station (30 km south-east of Madrid, Spain) on maize crops in 2001 and 2002. The crop was irrigated 10 times at rates varying from 32 to 52 mm per session. The soil is a *Calcaric Fluvisol* and has a sandy-loam texture (13% clay) and pH=8.1. Fertiliser treatments were: surface-applied pig slurry (SPS), immediately incorporated pig slurry (IPS), urea (U) and a control treatment (control) without any fertiliser. All the fertilisers were applied on 4th April using the same rate (200 kg N ha^{-1}). The N_2O was sampled by means of the closed chamber method. Denitrification was also estimated in the field with a core incubation method in the presence of acetylene (C_2H_2). Nitrous oxide emission, via the nitrification and denitrification pathways, was measured with in-field incubation of soil cores (Müller et al., 1998).

Results and discussion

During the first twenty days after applying the fertilisers (4th to 24th April), an increase in N_2O occurred in all the fertilised treatments reaching a maximum emission of 12.7 and 13.0 mg N m^{-2} d^{-1} (13-16 days after applying the fertiliser) for the U treatment in 2001 and 2002, respectively. Also, during the whole irrigation period (July-August), the emission was increased considerably and reached maximum values for ISP treatment of 14.1 and 10.9 mg N m^{-2} d^{-1}, in 2001 and 2002, respectively. In this period, the emission of N_2O was basically produced by the denitrification process (82 to 95% in 2001, and 69 to 79% in 2002). In the period of May to June and in September to October, emission of N_2O was very low and took place via the nitrification process as the water filled pore space (WFPS) did not surpass the denitrification threshold activation value (which was 63% in this soil). The cumulative N_2O-N emission showed differences between the fertiliser treatments and the control. The incorporated pig slurry treatment increased emission slightly in comparison wih surface application (Table 1).

Table 1. Cumulative N_2O emission and denitrification losses from the field plots during the whole experimental period in 2001 and 2002.

Fertiliser treatments	N_2O emission (g N m^{-2})		Denitrification losses (g N m^{-2})	
	2001	2002	2001	2002
Control	0.30	0.29	0.97	0.71
U	0.37	0.59	2.01	2.19
SPS	0.35	0.46	2.53	1.63
IPS	0.43	0.51	2.85	2.41

The losses through denitrification occurred mainly during the irrigation period and the type of fertiliser had a considerable effect. The highest losses in both years corresponded to the IPS treatment (Table 1), a fact that can be justified by the higher content of soluble organic C in soil activating the denitrification rate (Rochette et al., 2000). The plots with SPS had smaller denitrification losses than those of ISP treatment because it is probable that part of the NH_4^+-N was lost through volatilisation in this period. The denitrification rate differences between samples of each treatment could be explained by changes in WFPS and the soil temperature during the maize growing season.

Conclusions

Nitrous oxide emission and denitrification losses were important in an irrigated soil in a Mediterranean climate, especially in the irrigation period (July - August). It was further observed that the emission was activated when the soil was fertilised with urea or pig slurry.

Acknowledgements

This study was funded by the Spanish Commission of Science and Technology (CICYT, project AGL-1554-C02.)

References

Maag, M. and Vinther, F.P. (1999). Effect of temperature and water on gaseous emissions from soils treated with animal slurry. Soil Science Society of America Journal, 63, 858-865.

Müller, C., Sherlock, R.R. and Williams, P.H. (1998). Field method to determine N_2O emission from nitrification and denitrification. Biology Fertility of Soils, 28, 51-55.

Rochette, P.,Van Bochove, E., Prévost, D., Angers, D.A., Côte, D. and Bertrand, N. (2000). Soil carbon and nitrogen dynamics following application of pig slurry for the 19th consecutive year. I. Nitrous oxide fluxes and mineral nitrogen. Soil Science Society of America Journal, 64, 1396-1403.

Sánchez, L., Díez, J.A., Vallejo, A. and Cartagena, M.C. (2001). Denitrification losses from irrigated crops in central Spain. Soil Biology and Biochemistry, 33, 1201-1209.

Teira- Esmatges, M.R., Van-Cleemput, O. and Porta-Casanellas, J. (1998). Fluxes of nitrous oxide and molecular nitrogen from irrigated soils of Catalonia (Spain). Journal Environmental Quality, 27, 687-697.

Can tillage practice affect the contribution of nitrous oxide to the total greenhouse gas production from arable agriculture?

C.P. Webster, T.S. Scott and K.W.T. Goulding
Rothamsted Research, Harpenden, Herts, AL5 2JQ, UK

Introduction

Cultivation intensity will have far reaching effects on soil conditions, modifying factors such as bulk density, moisture, mineral N, distribution of plant residues and pore continuity. These conditions influence many soil processes, including those which produce and consume greenhouse gases. Therefore, use of particular tillage practices may be an option to mitigate global warming. We measured the flux of carbon dioxide (CO_2), N_2O and methane (CH_4) from adjacent fields of winter wheat under differing cultivation treatments. This contribution will concentrate on N_2O.

Experimental

Measurements were made at Wood Farm, a commercial arable farm near Hemel Hempstead, Hertfordshire, UK (51°46.51'N, 0°28.26'W). The soil is a silty-clay loam, rather poorly draining and subject to poor trafficability. We used adjacent fields which received contrasting autumn cultivations of minimum tillage (MT) and ploughing (P) (see Table 1). In November, eight flux chambers (area 0.07 m^{-2}) were placed in each field to measure N_2O and CH_4 flux as described by Smith et al., (1995). Carbon dioxide flux was determined using an eddy covariance method (Moncrieff et al., 1998). Soil temperature, moisture, bulk density and mineral N content were also measured.

Table 1. Comparison of the husbandry used on the two fields of this study.

	Minimum Tillage		Plough
Prior crop	W. Wheat		W. Wheat
Straw	Baled & removed		Chopped & incorporated
Cultivation	Shakerator, disc & top tilth		Plough, pressed & top tilth
Variety	W. Wheat, var. Claire		W. Wheat, var. Consort
Sowing date	27.9.02		25.9.02
N applications	28.2.03	Ammonium sulphate	38 kg N ha^{-1}
(both fields)	23.4.03	Ammonium nitrate	108 kg N ha^{-1}
	27.5.03	Ammonium nitrate	40 kg N ha^{-1}

Results

The 2002-03 winter was wet (and mild) but the following spring and summer were drier and warmer than the long-term average. A consistent trend was the greater moisture content of

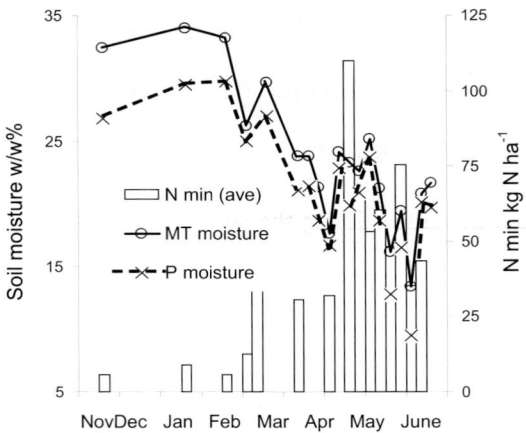

Figure 1. Soil moisture and mineral N status at Wood Farm.

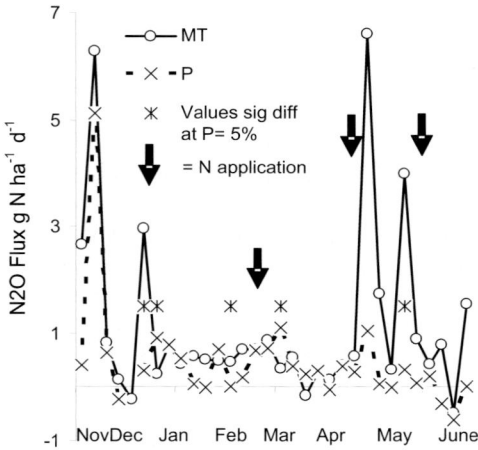

Figure 2. Nitrous oxide flux at Wood Farm.

the MT field (Figure 1). The top soil mineral N status (Figure 1, average of the two fields) was low during the winter, presumably heavy rainfall contributed to this. High levels in summer after N application indicate that dry weather inhibited crop N uptake and therefore, possibly also denitrification.

Rates of N_2O flux were low (Figure 2) and show large variability. Over 8 months (Nov '02 to June '03) we estimate that the cumulative losses of N_2O-N from the MT and P fields were 0.28 and 0.12 kg N ha^{-1}, 0.15% and 0.06% of the N applied, respectively; very small compared with the IPCC, EF of 1.25%. Maximum rates of N_2O loss were not associated with particular soil moisture and temperature combinations. CH_4 fluxes were small and close to our detection limit. There was no significant or even consistent difference between treatments.

Conclusions

- Cultivation intensity altered soil conditions. Soil moisture was greater on the MT treatment.
- Levels of N_2O flux were low. Generally losses were greater from the MT field, but cumulative losses from the two treatments were not significantly different. Peaks in production were not associated with particular soil conditions. Possibly denitrification dominated production during winter and nitrification during the spring. N fertiliser application had little impact on N_2O production.
- CH_4 flux rates were also small and with no significant difference between treatments.
- During this season any difference in greenhouse gas emission between treatments will be dominated by CO_2.

Acknowledgements

This is a collaborative project with the Institute of Atmospheric & Environmental Science, University of Edinburgh and the RAC, Cirencester. This work is funded by the BBSRC (grant no. 206/D16053).

References

Smith, K.A., Clayton, H., McTaggart, L.P., Thompson, P.E., Arah, J.R.M. and Scott, A. (1995). The measurement of nitrous oxide emissions from soil by using chambers. Philosophical Transactions of the Royal Society of London, Series A, 351, 327-338.

Moncrieff, J.B., Beverland, I.J., O'Neill, D.H. and Cropley, F.D. (1998). Controls on trace gas exchange observed by a conditional sampling method. Atmospheric Environment, 32, 3265-3274.

Soils as sources of N-trace gases in Germany - results from calculations with biogeochemical models

C. Werner[1], M. Kesik[1], H. Papen[1], C. Li[1,2] and K. Butterbach-Bahl[1]

[1]*Institute for Meteorology and Climate Research, Karlsruhe Research Centre, Garmisch-Partenkirchen, Germany*
[2]*Institute for the Study of Earth, Oceans, and Space, University of New Hampshire, USA*

Introduction

With regard to global climate change, soils are of significant importance as sources of atmospheric trace gases such as N_2O and NO. Total emissions of the active greenhouse gas N_2O from all soils are estimated to be in the range of 5-15 Tg N yr^{-1}, and therefore contribute approximately 55-65% to the total global atmospheric N_2O budget (IPCC, 1997). Considering the sources of the secondarily active greenhouse gas (NO), soils are, with approximately 21 Tg NO-N, of similar importance for the global atmospheric NO budget as NO-emissions associated with fossil fuel combustion.

The rates of N-trace gas production and consumption in soils by the microbial processes of nitrification and denitrification depend on a multitude of factors such as the diffusivity of the soil, texture, soluble organic C content, soil pH, N availability, soil moisture and human management. All these factors vary substantially on spatial as well as temporal scales, thus estimates of the source strength of soils for N-trace gases are still highly uncertain at regional and/or global scales. The IPCC approach presently used, relies on emission factors which specify the fraction of N_2O emitted to the atmosphere e.g. from N-fertilisers applied to agricultural soils. Substantial improvements of current estimates can only be achieved if mechanistic models (able to simulate N-trace gas emissions based on the processes involved in N-trace gas production, consumption and emission), are employed for inventory studies.

Materials and methods

In this study, the biogeochemical models DNDC (agricultural soils) and PnET-N-DNDC (forest soils) were linked to a GIS-database, which holds all the spatially and temporally differentiated input information and model drivers to calculate national inventories of N_2O- and NO-emissions from soils in Germany from 1990 to 1999. These mechanistic models use general ecological drivers such as climate, soil properties, vegetation and human management impact to derive soil environmental forces, which govern substrate fluxes, microbial activity and associated N-trace gas fluxes in different soil layers, which may be eventually released to the atmosphere (e.g. Li et al., 1992, 2000). Because of the multitude of input factors, an extensive GIS database was set up from a number of different geographical datasets, statistics and census information. These datasets included detailed information about soil properties, geographic location and extent of landuse types. Phenological observations were used to estimate the timing of crop management, statistical data and various datasets were imported to describe the utilisation and management of agricultural land and forests. Climatic model drivers such as daily temperature and precipitation were obtained for 1452 climate stations throughout Germany.

Results

Within the 10 year period, total N_2O-emissions from agricultural soils in Germany ranged from 94 to 172 kt N_2O-N yr^{-1} (average, 110 kt N_2O-N yr^{-1}). A pronounced inter-annual variation, only driven by meteorological conditions was observed. Compared with the agricultural soils, N_2O-emissions from forest soils in Germany were much lower and amounted to an average of 139 kt N_2O-N yr^{-1} (minimum, 128; maximum, 154). NO emissions from agricultural soils were calculated to be in the range of 42 to 49 kt NO-N yr^{-1}, with an average emission of 45 kt NO-N yr^{-1}. The figures for forest soils were lower and showed only small inter-annual fluctuations (11 - 12 kt NO-N yr^{-1}; average 11 kt NO-N yr^{-1}). A complex emission pattern was also apparrent and the regional source strength varied greatly.

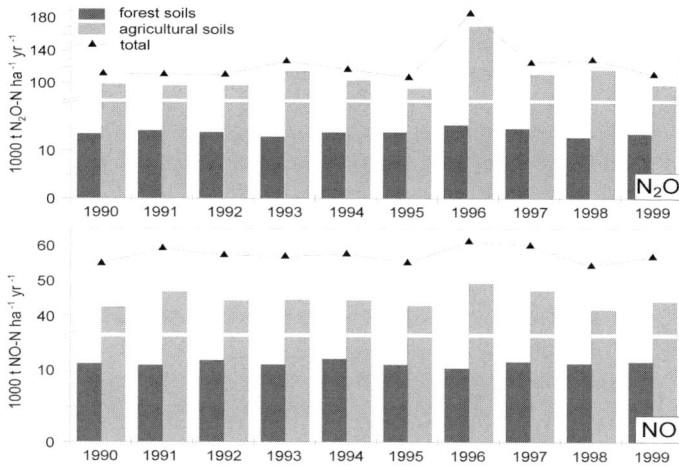

Figure 1. Total source strength of agricultural and forested soils for N-trace gas emissions.

Discussion

Our estimates show that soils contribute to approximately 50 - 60% of the total national N_2O source strength of Germany. Simulated emissions of NO from agricultural and forest soils were approximately 60% lower than for N_2O. However, our estimates indicate that agricultural and forest soils in Germany are a significant source of atmospheric NO_x within a range of approximately 8% (annual average) up to 30% (individual summer days). The results of the multi-year simulation (utilising observed meteorological data from 1990-1999 and various other detailed datasets) show that the modeled inter-annual variations of N-gas emissions, which were obviously induced by different climatic conditions and the timing of farming events, were as high as 73%. The high inter-annual variations, as well as the heterogeneous emission pattern, underline the benefits provided by process-oriented biogeochemical models in estimating the source strength and regional distribution of N-gas emissions from soils and the advantage of this methodology over the approach used by the IPCC.

Acknowledgements

This study was funded by the Umweltbundesamt (UBA), Germany.

References

IPCC (1997). Greenhouse gas inventory. Reference manual. Intergovernmental Panel on Climate Change. Bracknell, UK.

Li ,C., Frolking, S. and Frolking, T.A. (1992). A model of nitrous oxide evolution from soil driven by rainfall events: 1. Model structure and sensitivity, Journal of Geophysical Research, 97, 9759-9776.

Li, C., Aber, J., Stange, F., Butterbach-Bahl, K. and Papen, H. (2000). A process-oriented model of N_2O and NO emissions from forest soils. 1. Model development. Journal of Geophysical Research, 105, 4369-4384.

Denitrification in top soil and sub soil, data and model results

K.B. Zwart

Alterra, Centre of Soil Science, P.O. Box 47, 6700 AA Wageningen, the Netherlands

Introduction

Grassland and arable land in the Netherlands have had a surplus of nutrients applied for a number of decades. As a result, ground water NO_3^- levels exceed EU levels in a large area of the country. The major explanation of this situation is two-fold: (i) the large national production of animal manure and (ii) a high input of chemical fertilisers, needed to maintain a highly productive cropping system. Dutch legislation with respect to nutrient application became more stringent with the introduction of EU legislation on ground water quality regarding NO_3^- levels and since then, the situation has improved. However, NO_3^- concentrations in the upper ground water still exceed EU levels. Surprisingly little is known about the quantity of NO_3^- that may be denitrified in deeper soil layers in Dutch grassland and arable land. It is assumed that approximately 50% of the N-surplus will be denitrified, but data for deeper soil layers are lacking. For that reason denitrification rates were measured in soil samples up to a depth of 2.0 m below the surface and compared with other soil chemical and biological properties.

Materials and methods

Soil was sampled in 10 layers of 0.2 m each from grassland from 8 different dairy farms and from 5 different arable farms. Soils were analysed for total C, total N, 0.01M $CaCl_2$ extractable organic C and N (SOC and SON) and mineral N. Potential denitrification was measured using the acetylene-inhibition-method (AIM) in the presence of excess NO_3^- and anaerobic conditions. In addition, denitrification was measured by following NO_3^- concentrations in soil incubated for 12 weeks under anaerobic conditions. Soil respiration and N mineralisation were measured in samples incubated for 12 weeks. Oxygen concentration was followed in undisturbed soil columns taken from part of the sampling sites, and incubated under various moisture conditions. All incubations were at 20°C. Annual denitrification over 2.0 m depth was estimated for three soil typical profiles, differing in soil organic matter (SOM) in the sub soil. Total amounts of biodegradable C, the probability of anaerobic situations and an average NO_3^--N content of 22 mg kg^{-1} soil and an average soil temperature of 10°C were used as an input in a model based on that of Hénault & Germon (2000).

Results and discussion

The average potential denitrification and respiration rates were higher in grassland soils than in soils from arable land. Potential denitrification decreased sharply below 0.6 and 0.4 m for grassland and arable land, respectively (Figure 1). Both methods used to establish potential denitrification gave similar results (Figure 2). In addition, undisturbed soil columns with low SOM contents remained well oxygenated, even after prolonged incubation at low pF-values and 20°C (Figure 3). Potential denitrification was linearly correlated with respiration

(R^2=0.755) and N-mineralisation (R^2=0.501) (Figure 4). The results indicated that denitrification in the sub soil is of little significance, unless in addition to NO_3^-, biodegradable SOM is present. The results from the model calculations also indicated that in many sandy soils low in SOM, denitrification is a minor quantitative component in the soil.

N-balance. The low denitrification rates in the sub soil may at least partly explain why the EU NO_3^- levels are exceeded in ground water in large areas in the Netherlands.

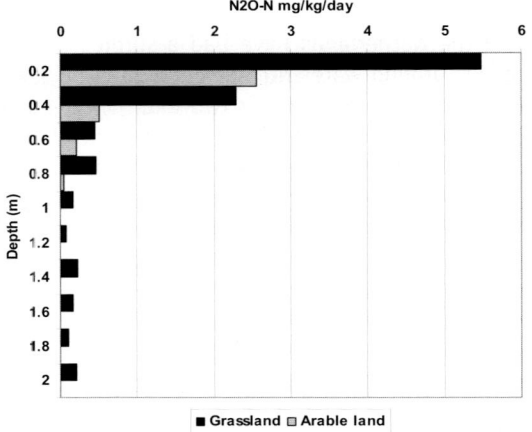

Figure 1. Potential denitrification rates in grassland and arable land up to 2 m below the surface.

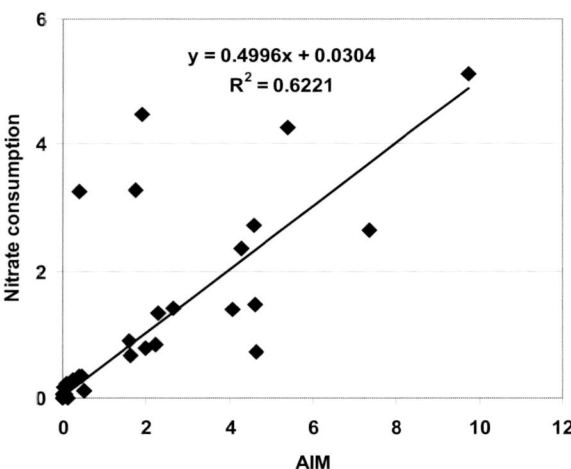

Figure 2. Relation between AIM and anaerobic NO_3^- consumption (mg NO_3^--N kg^{-1} soil d^{-1}).

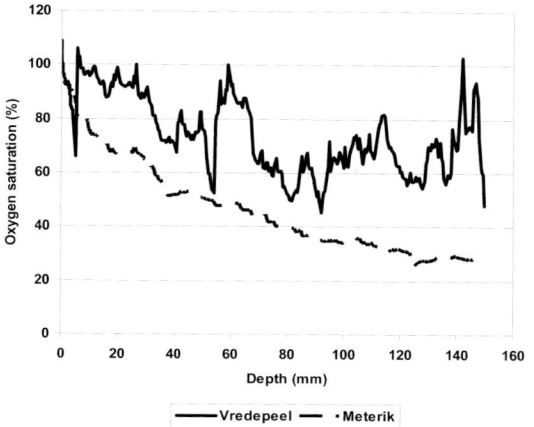

Figure 3. Oxygen saturation at different depths in undisturbed soil cores from two different arable fields.

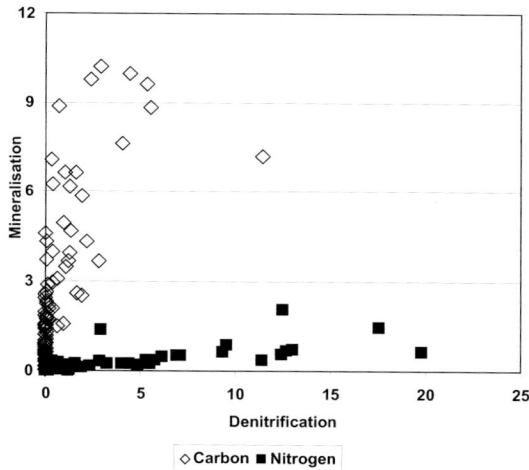

Figure 4. Relation between potential denitrification and C or N mineralisation (mg N or C kg^{-1} soil d^{-1}).

Acknowledgements

This project was financed by the Dutch Ministry of Agriculture, Nature and Food Quality, Programme 398-II

References

Hénault, C. and Germon, J.C. (2000). NEMIS, a predictive model of denitrification on the field scale. European Journal of Soil Science, 51, 257-270.

SECTION 4
CONTROLLING LOSSES TO WATER

Controlling losses to water

M.A. Shepherd[1] and E.I. Lord[2]
[1]ADAS, Gleadthorpe Research Centre, Mansfield, Nottinghamshire, NG20 9PF, UK
[2]ADAS, Woodthorne, Wergs Road, Wolverhampton, WV6 8TQ, UK

Abstract

A large proportion of NO_3^- in ground and surface waters derives from agriculture, especially in Western European countries under intensive agriculture. The Nitrates Directive (Anon., 1991) and, latterly, the Water Framework Directive (Anon., 2002a) aim to address diffuse N pollution. This paper assesses the types of measures that are appropriate for limiting NO_3^- loss and views some of the evidence for the effects of measures on improvements in water quality. To date, evidence is limited and it is our assertion that action programme measures under the Nitrates Directive will have minimal effect on NO_3^- loss. More drastic measures (e.g. land use change) may be necessary in areas of intensive agriculture.

Keywords: groundwater, Nitrates Directive, NO_3^- leaching, surface water

Introduction

Concerns over NO_3^- in water were initially associated with supposed health risks though, perhaps, the evidence is now less compelling (Leifert and Golden, 2000). Nevertheless, there are statutory limits on NO_3^- levels in water, and excessive levels impact on other aspects of water use (Table 1). Ammonium and NO_2^- are toxic to fish.

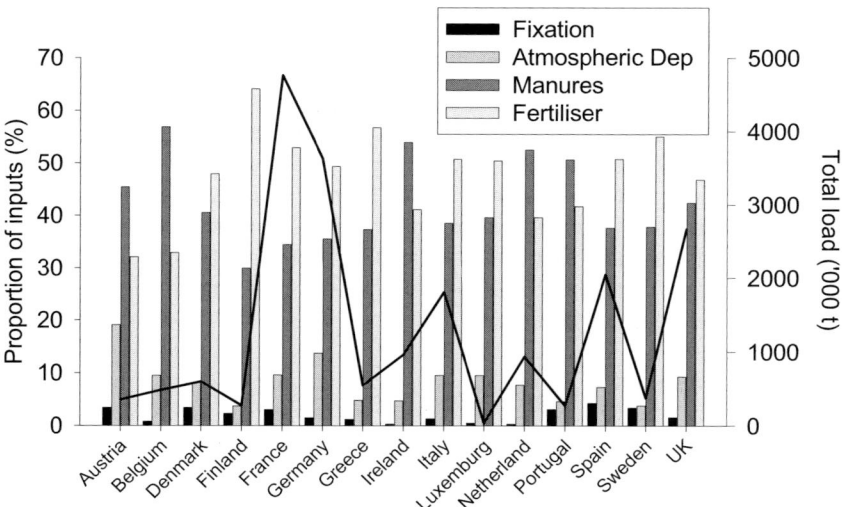

Figure 1. Nitrogen loadings and sources to agriculture in European Union States (histograms show proportions of inputs, solid line shows total load) (from Anon., 2002a).

Agriculture plays a dominant role in many catchments, with the majority of N inputs to fields coming from fertiliser and manure inputs (Figure 1). The greatest problems of NO_3^- contamination are in the intensively farmed Western European States (Figure 2). (Anon., 2002a) reported that over 20% of groundwater in the European Union (EU) and between 30 and 40% of lakes and rivers were showing excessive NO_3^- concentrations.

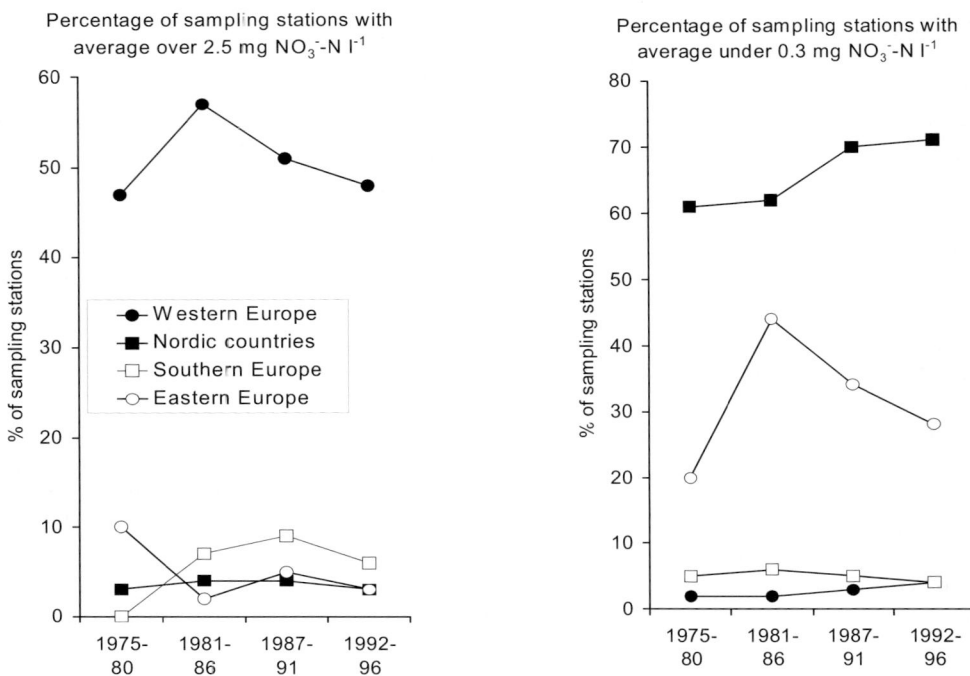

Figure 2. Summary of average NO_3^- -N concentrations in surface freshwaters across Europe (Crouzet et al., 1999).

Nitrogen from agricultural sources accounts for between 50 and 80% of the NO_3^- entering Europe's water. In England, ca. 70% of NO_3^- in rivers is thought to derive from agriculture (Anon., 2002b). There is also evidence of non-compliance with the Freshwater Fish Directive (FWFD) threshold limit for NH_4^+-N concentrations in some surface waters. It is clear that farming practices that decrease N loss need to be adopted. The aims of this paper, therefore, are to assess the types of measures that are appropriate for limiting N loss from agriculture and to assess some of the evidence to date for the effects of measures on improvements in water quality. We focus primarily on NO_3^-.

Mitigation measures

Much is already known about N loss from agriculture (e.g. Anon., 2000b). There is a huge scientific literature on mitigation methods to decrease N losses to water, e.g. Dampney et al., (2002) reviewed methods across Europe (Table 1). Factors that encourage NO_3^- loss include: over-fertilisation of crops; significant periods of bare soil during the winter drainage period; excessive numbers of livestock and consequential overloading with manure; inappropriate use

Table 1. Summary of the key methods for decreasing NO$_3^-$ loss from fields.

Source management:	Transport management:
• Avoid autumn fertiliser N applications unless there is a definite crop need.	• Maintain green cover over winter (including use of cover crops).
• Avoid autumn applications of slurry, poultry manure and liquid digested sludge.	• Split spring N fertiliser applications on soils prone to leaching.
• Reduce stocking rates to reduce manure loading per ha.	• Restrict manure application rates and timings to safe time windows, also avoiding periods of high rainfall
• Restrict livestock access to watercourses.	when soils are excessively wet.
• Reduce N inputs through animal feedstuffs.	• Introduce riparian buffer strips (but also need
• Use a reliable N recommendation system that takes account of all N sources.	complementary in-field control practices to control runoff).
• Irrigate drought-prone crops to maximise N use efficiency.	• Avoid liquid manure application on drained cracking clay soils, especially grassland.

of manure, including not accounting for N content when applying fertiliser and applying slurries/poultry manure in the autumn.

In essence, the aim should be to minimise the soil NO$_3^-$ pool present in the autumn (and, therefore, available for subsequent leaching during winter) by carefully managing manure and fertiliser N inputs to the preceding crop. Autumn applications of additional N, either as manure or fertiliser, should also be avoided.

Choice of crop will also influence the autumn soil NO$_3^-$ pool, depending on that crop's N balance (i.e. N applied as manure and fertiliser minus N removed from the field in produce), and the rate at which the residue N mineralises.
Consequently, some crops can be considered a higher leaching risk than others (Figure 3).

The Nitrates Directive requires the development of a (voluntary) Code of Good Agricultural Practice to offer general protection of all waters. Additionally, farmers within Nitrate Vulnerable Zones (NVZs, i.e. areas where concentrations of NO$_3^-$ in water exceed, or are likely to exceed in the future, 50 mg l^{-1}) are required to follow 'action programmes' to decrease NO$_3^-$ loss. In short, the main responsibilities for farmers are to:
• make better use of fertilisers and manure (i.e. fertilise crops according to best practice to avoid over-fertilisation)
• have closed periods for applications of some manure/sewage sludge (on sandy soils)
• increase manure storage on some farms as a result of closed application windows
• restrictions on maximum N loadings per farm and per field
• maintain better records of fertiliser and manure management

Assessing the impacts of control measures

Several previous UK studies have tried to assess the impacts of changes in agricultural practice on NO$_3^-$ loss.

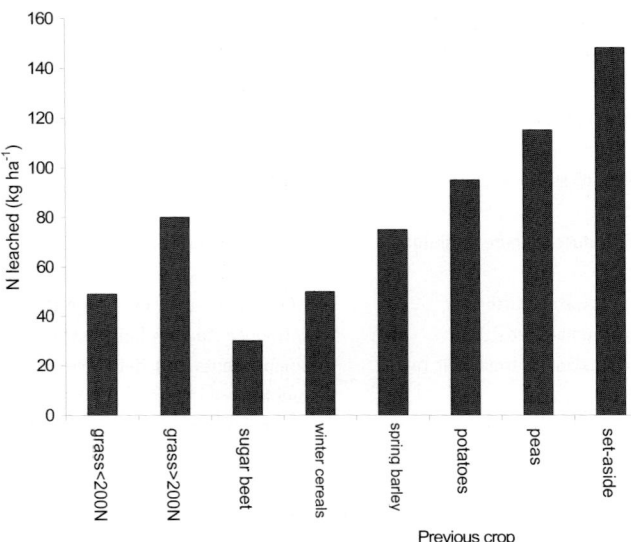

Figure 3. Nitrate leaching risk according to crop type: values are averages for a range of sites on drained soils with 'good nitrate practice' (from Goulding, 2000).

1. Nitrate Sensitive Areas

The Nitrate Sensitive Area (NSA) Scheme was an agri-environment scheme operated in 32 areas of England where it had been assessed that groundwater was highly vulnerable to continued NO_3^- leaching from agricultural land (MAFF, 1998). The NSA scheme benefited from a generally good uptake by farmers in the areas, and it provided an excellent opportunity to measure the effects of changed farming practices on NO_3^- loss. A hugely valuable (unique) database has been developed of (a) detailed data on cropping and husbandry practices for each field in the NSAs and (b) measurements of actual NO_3^- leaching from the soil zone from a representative sample of fields (using porous ceramic pots). A review of these data estimated that the NSA Scheme would reduce NO_3^- losses from these areas by around 20-25%, with a wide range. The greatest reductions were on areas with large intensive pig and poultry holdings. The NSA scheme was considerably more stringent than the NVZ Action Programme, and included the option to convert arable land to extensive, unfertilised, grassland. This, and the export of manures from pig and poultry holdings (which is much less likely to occur on large NVZs) were the major contributors to the large estimated impact.

2. Potential impact of NVZs on sandy/shallow soils

An early study of the potential impact of NVZs (Lord, 1995), made prior to the detailed development of the action plan, used the Pilot Nitrate Sensitive Areas (summarised in Lord et al., 1999) as input data. It made the deliberately optimistic assumption that all rules would be adhered to, and it assumed that restrictions on manure timing would apply to all arable land, and to all the soils within the NSAs (the soils in NSAs were commonly, but not entirely, sandy or shallow). The present NVZ rules on manure timing are less stringent before winter-sown arable crops; and would not apply at all to a proportion of the land in the study areas.

The study also assumed that export of pig and poultry manures was a viable option, because the study areas were small. As NVZ areas increase, this option becomes less practicable. The study concluded that NVZ rules in these areas could reduce NO_3^- leaching relative to prior conditions by ca. 22% if manure export were possible (Table 2).

Table 2. Calculated reductions in NO_3^- leaching as a result in changed practices.

Reduction of N leached in the winter of application:	
Pig and poultry manures: mainly export	13%
Cattle manures	< 1%
Reduction in fertiliser inputs:	
Better allowance for manures	4%
Non-manured land (mainly grass)	5%

The reductions from pig and poultry manure were mainly from the assumed export of more than half of these manures outside the NSAs, since the NSA area was small relative to the land area required for manure disposal on those NSAs with large pig or poultry holdings. This would not be relevant in large NVZs. The current NVZ Action Programme rules on manure timing are less stringent. The reduction in N fertiliser input assumed for grassland relies on a detailed interpretation of recommendations that would need to be more actively promoted to dairy farmers if it were to succeed. Using the data in the report, the overall reduction in NO_3^- leaching under the current NVZ Action Programme for these study areas (on sandy or shallow soils and assuming no export of manures), was estimated provisionally as ca. 10%. The estimate for grassland is based on observed reductions in fertiliser use within NSAs, and may be optimistic for NVZs where advisory input is much less and farmer enthusiasm is also less because there are no payments.

The impact of the NVZ Scheme will depend on the agricultural activity in the area - especially livestock numbers - and on soil type. Using this study as a (very rough) guide, the expected impact of the current NVZ scheme for different farming systems and soils is summarised in Table 3. These estimates assumed compliance with the principles of the scheme, which in some instances is difficult to ensure without detailed follow-up. For example, in principle, some reduction (in theory up to 10%) could be expected in areas of intensive dairy farming, but this would probably require an explicit advisory campaign because the fertiliser recommendations are complex to interpret. The greatest potential reductions, rising perhaps to 20%, would be expected on sandy/shallow soils in areas with intensive pig and poultry holdings.

Table 3. Estimation of the potential impact of changed management practices under the Nitrates Directive for arable and grassland production systems with or without additional pig or poultry manure availability.

Livestock	Land Use	Soil	Potential impact
Few pigs/poultry	Arable	All	<5%
	Grass - non-intensive	All	<5%
	Grass-intensive dairy	All	5-10%
Many pigs/poultry	All	Clay, medium	5-10%
	All	Sandy, shallow	10-20%

3. Study on implications of the Water Framework Directive

This study attempted to estimate the scope of the potential impact of the provisions of the Water Framework Directive. Three contrasting agricultural areas of the UK were chosen. For each, the MAGPIE/NEAP-N model system (Lord & Anthony, 2000) was used to estimate current NO_3^- concentrations as a long-term mean. Simple assumptions were made about N contributions from non-agricultural land. A target was set of a mean of not more than 50 mg l^{-1} NO_3^- for each mapping unit of 100 km^2.

In order to achieve this, measures were introduced within each square, in increasing order of severity. First, the NVZ rules for sandy or shallow soils were applied to the whole area (assuming zero application of high-N manures in autumn on arable land). Then, as much arable land as necessary was entered into the 'NSA Basic Scheme', which included the use of cover crops and some restriction on N fertiliser inputs. Then, pig and poultry manures were exported (or numbers reduced), then arable land was entered into the 'NSA Premium Scheme' (including conversion to extensive grassland, and then grazing livestock numbers were reduced (dairy, then other cattle and sheep). The introduction of the NSA scheme was assumed to improve the efficiency of use of pig and poultry manures but not to eliminate their effect on leaching. The calculations are theoretical, in that it would not be possible to introduce a scheme in this way, but they give a feel for the scale of change necessary in different parts of the country.

This study shows that compliance with the 50 mg l^{-1} NO_3^- limit on average, i.e. for groundwaters, would require measures substantially in excess of the current NVZ rules in much of eastern England (Table 4). However, compliance could be achieved with more limited disturbance to agricultural practice in western areas, because of both greater rainfall and the greater area of relatively extensively farmed grassland. The study implies that even after compliance with NSA/NVZ rules on manure timing and management, there are many areas where concentrations of pigs and poultry are incompatible with compliance with the EU limit.

If the target of 50 mg l^{-1} was applied to aggregated areas greater than 10 km^2, the required restrictions would be somewhat less stringent because small losses in one area could be offset against large losses in another. Thus, even in the Dorset area, where current mean concentrations averaged across the whole area were below the limit, the study found the need to restrict practices within some of the area.

Conclusions

Current assessments of European water quality suggest that diffuse water pollution from agriculture is likely to remain a priority issue for many years. Improvement will be driven by the implementation of both the Nitrates Directive and the Water Framework Directive. Table 5 shows that some countries will have a greater challenge than others, related to the general intensity of the agriculture in that country.

We believe that the main challenges that lie ahead in order to meet the environmental targets for NO_3^- loss that are being set are (a) implementing changes on farms and (b) facing the fact that, in some circumstances, current NVZ actions will be insufficient to meet the targets.

Table 4. Theoretical changes in farm management in three areas of England required to reduce average NO_3^- loss < 50 mg l^{-1}.

Location	Norfolk	Salop	Dorset
Land area (ha)	161,600	208,500	190,700
% Arable	65	36	23
% Grass	10	37	47
Annual Rainfall mm	671	725	929
Soil Water Capacity mm	260	278	279
Change necessary to comply with 50 mg l^{-1} in every 100 km²:			
Arable: proportion under NSA rules[1]	0.96	0.76	0.11
Housed livestock: proportion removed	0.79	0.65	0.02
Proportion of arable converted to extensive grassland	0.15	0.24	0.00
Dairy: proportion removed	0.07	0.02	0.00
Beef/sheep: proportion removed	0.02	0.00	0.00
Breakdown of NO_3^- concentration contributions (mg l^{-1} NO_3^-):			
Current	105.6	77.3	40.9
After NVZ rules applied to whole area	91.1	67.6	36.1
After NSA rules applied to whole area	79.9	61.2	32.1
After removal of all poultry, pigs	59.6	50.0	27.3
After conversion of arable land	23.4	31.7	19.9
After removal of all dairy cattle	14.3	11.1	6.0
Non-agricultural whole area	4.3	3.8	2.0

[1]NSA = Nitrate Sensitive Area, a precursor to NVZs in England, as described in the text.

Table 5. EC assessment of performance of Member States in implementing the Nitrates Directive (Anon, 2002a).

	Starting position	Results
Denmark	Difficult (high NO_3^-)	Good
The Netherlands	Difficult (high NO_3^-)	Bad
Belgium	Difficult (high NO_3^-)	Very bad
France, Germany, Ireland, Luxembourg, UK	Medium	Mediocre
Austria, Finland, Sweden	Easy (low NO_3^-)	Good
Portugal	Easy (low NO_3^-)	Mediocre
Greece, Italy, Spain	Easy (low NO_3^-)	Bad

On this latter point, monitoring water quality (a requirement of both the Nitrates and Water Framework Directives) will assess progress, but it may well be that current regulations will be insufficient to reverse the trend of increasing NO_3^- concentrations in some waters. The desk studies reported in this paper suggest in the dry east of England, for example, that more stringent measures may be required. If so, there is a balance between science and policy that needs to be struck, in bringing about what could be quite substantial changes in land management with minimal adverse effects on the rural economy, i.e. managing change sustainably.

However, success in decreasing diffuse water pollution also depends not only on the development of techniques, but also on their uptake by land managers. Table 1 suggests that there has been sufficient research to broadly identify the main approaches that can be adopted to decrease N loss. However, there is still much to be done, not least in incorporating some of these approaches into farming systems. This requires the agricultural community to adopt such practices, either voluntarily, or with compensation (e.g. agri-environment schemes) or through regulation. Assessing the cost benefit of mitigation methods is necessary to move implementation forward. This information is lacking and has been identified as a priority by the European Commission (Anon., 2002a).

Acknowledgements

This paper is based on work funded by the Department for Environment, Food and Rural Affairs (Defra), which we gratefully acknowledge.

References

Anon (1991) Council Directive 91/676/EEC, concerning the protection of waters against pollution caused by nitrates from agricultural sources. Official Journal of the European Communities, L375/1.

Anon (2000a) Council Directive 2000/60/EC establishing a framework for Community action in the field of water policy. Official Journal of the European Communities L327/1.

Anon (2000b). Soil Use and Management Special Edition 16.

Anon (2002a). Commission report of 17 July 2002 on implementation of Council Directive 91/676/EEC concerning the protection of waters against pollution caused by nitrates from agricultural sources. Synthesis from year 2000 Member States reports. Report - COM(2002) 407 final, Brussels.

Anon (2002b). The Government's Strategic Review of diffuse water pollution from agriculture in England. Defra, June, 2002.

Crouzet, P, Leonard, J., Nixon, S. Rees, Y., Parr, W., Laffon, L., Bogestrand, J., Kristensen, P., Lallana, C., Izzo, G., Bokn, T., Bak, J. and Lack, T.J. (1999). Nutrients in European Ecosystems. Environmental Assessment Report No. 4, European Environment Agency, Copenhagen, Denmark.

Dampney, P., Mason, P., Goodlass, G. and Hillman, J. (2002). Methods and measures to minimise diffuse pollution of water from agriculture - a critical appraisal. Report for Defra, Project NT2507.

Goulding, K. (2000). Nitrate leaching from arable and horticultural land. Soil Use and Management, 16, 142-152.

Leiffert, C. and Golden, M.H. (2000). Dietary nitrate: a re-evaluation of the beneficial and other effects. The International Fertiliser Society, London. Proceedings 456.

Lord, E.I. (1995). Expected impact of proposed Nitrate Vulnerable Zones on nitrate leaching. Study areas: Pilot Nitrate Sensitive Areas. Internal report prepared for MAFF-RMED. June 1995.

Lord, E.I., Johnson, P.A. and Archer, J.R. (1999). Nitrate Sensitive Areas - a study of large-scale control of nitrate loss in England. Soil Use and Management, 15, 1-7.

Lord, E.I and Anthony, S.G. (2000). MAGPIE: A modelling framework for evaluating nitrate losses at national and catchment scales. Soil Use and Management, 16, 167-174.

MAFF (1998). Report on the review of the Pilot Nitrate Sensitive Areas Scheme.

Nitrate leaching from arable crop rotations in organic farming

J.E. Olesen, M. Askegaard and J. Berntsen
Department of Agroecology, Danish Institute of Agricultural Sciences, Box 50, DK-8830 Tjele, Denmark

Abstract

Nitrate leaching from crop rotations for organic grain production were investigated in a field experiment on different soil types in Denmark from 1997 to 2002. Three experimental factors were included in the experiment in a factorial design: 1) proportion of grass-clover and pulses in the rotation, 2) cover crop (with and without), and 3) manure (with and without). Two four-course rotations were compared. They had one year of grass-clover as a green manure crop, either followed by spring wheat or by winter wheat. The nitrate leaching was measured using ceramic suction cells. The nitrate leaching did not differ between the rotations, as a change in leaching following the grass-clover was compensated by a reverse effect in the grain crops. Use of cover crops reduced N leaching by 23 to 38% at crop rotation level with the highest reduction on the coarse sandy soil. Simulation of N leaching using the FASSET model showed that a practice of using part of the summer period in the grass-clover as a bare fallow to control couch grass could increase leaching substantially, in particular on the sandy soil.

Keywords: organic farming, crop rotation, cover crops, manure, nitrate leaching

Introduction

The proportion of organic farming in Denmark has increased considerably over the past decade, and the organic farmed area constituted 6.4 % of the agricultural area in 2001 (Anonymous, 2002). This organic farming was during the 1990's mainly based on dairy farms with a high percentage of grass-clover and fodder crops in the crop rotation in combination with a stock of ruminant animals. However, during the last few years, many Danish arable farms have converted to organic farming.

The increase in organic farming is partly explained by financial support from the Danish government and the EU through agri-environmental schemes (Stockdale et al., 2001). In Denmark many of these schemes are directed towards protecting groundwater and surface waters from pollution with nitrates and pesticides. In the Aquatic Action Plan II agreed by the Danish parliament in 1998 organic farming has been included as one of the measures to reduce nitrate leaching from Danish agriculture (Grant et al., 2000). However, large uncertainties remain in the estimates of nitrate leaching from organic farming (Hansen et al., 2000).

The aim of the study presented here was to quantify the nitrate leaching from arable organic farming as affected by use of cover crops and manure application. Additionally the effect of crop rotation design and the interaction with soil and climate was included.

Materials and methods

The crop rotation experiment was designed as a factorial experiment with three factors (Olesen et al., 2000). The experimental factors were (1) fraction of grass-clover and pulses in the rotation (crop rotation), (2) cover crop (with and without cover crop) and (3) manure (with and without animal manure applied as slurry).

Results are presented for three sites representing different soil types and climate regions in Denmark. Jyndevad is located in Southern Jutland and represents a coarse sandy soil with an normal annual rainfall of 964 mm. Foulum is located in Central Jutland on a loamy sand with an annual rainfall of 704 mm. Flakkebjerg is located in Western Zealand on a sandy loam with an annual rainfall of 626 mm. The soil properties of the ploughing layer are shown for all sites in Table 1 and the climate during the experimental period 1998 to 2001 is shown in Table 2.

Table 1. Soil texture, content of organic C and total N, and pH in the top 25 cm of the soil at the three experimental sites in autumn 1996 prior to the onset of the experiment. pH is taken as pH(CaCl$_2$)+0.5. Soil minerals, organic C and total N are measured in per cent of dry soil (Olesen et al., 2000).

Location	Clay	Silt	Fine sand	Coarse sand	Organic C	Total N	pH
	< 2 µm	2-20 µm	20-200 µm	200-2000 µm			
Jyndevad	4.5	2.4	18.0	73.1	1.17	0.085	6.1
Foulum	8.8	13.3	47.0	27.2	2.29	0.175	6.5
Flakkebjerg	15.5	12.4	47.4	22.9	1.01	0.107	7.4

Tabel 2. Average annual temperature, precipitation and simulated drainage in rotation 2 during the period 1998 to 2001.

Location	Temperature (°C)	Precipitation (mm)	Drainage (mm)
Jyndevad	9.0	962	743
Foulum	8.1	716	415
Flakkebjerg	8.7	654	304

Four four-year crop rotations were compared (Olesen et al., 2000). However, results from two of the rotations are presented here (Table 3). These rotations differ with respect to the cereal following the grass-clover green manure crop. The grass-clover crop was followed by spring wheat in rotation 1, and by winter wheat in rotation 2. Crop rotation 1 was only represented at Jyndevad. Minor changes in the crop choice were made in 2001 (Table 3). All fields in all rotations were represented every year in two replicates. The plot size was 378, 216 and 169 m^2 at Jyndevad, Foulum and Flakkebjerg, respectively.

The undersown cover crops were either a pure stand of perennial ryegrass (*Lolium perenne*) or a mixture of perennial ryegrass and four clover species (hop medic *Medicago lupulina*, trefoil *Lotus corniculatus*, serradella *Ornithopus sativus* and subterranean clover *Trifolium subterraneum*). In 2001 chicory (*Chicorium intybius*) was also included with ryegrass in all cover crop mixtures. These cover crops were undersown in the cereal or pulse crop in spring. In the spring cereals the sowing took place on the same day as the cereal or pulse crop was sown, except for Jyndevad

Table 3. Structure of the two different four-course crop rotations with and without cover crops. The sign ':' indicates that a grass-clover ley, a clover or a ryegrass/clover cover crop is established in a crop of cereals or pulses. The sign '/' indicates a mixture of peas and spring barley.

Cover crop	Entry point	Rotation 1	Rotation 2
Without	1	S. barley:ley	S. barley:ley
	2	Grass-clover	Grass-clover
	3	Spring wheat[1]	Winter wheat[3]
	4	Lupin[2]	Peas/barley[4]
With	1	S. barley:ley	S. barley:ley
	2	Grass-clover	Grass-clover
	3	S. wheat:Grass[1]	W. wheat:Grass[3]
	4	Lupin:Grass-clover[2]	Peas/barley:Grass-clover[4]

[1]S. oats in 2001, [2]Peas/barley in 2001, [3]Winter rye at Jyndevad in 2001, [4]Lupin in 2001.

where sowing was delayed in order to permit weed harrowing in the rotations with cover crops. In the winter cereals the cover crop was sown in April just after the first weed harrowing.

The plots receiving manure were supplied with anaerobically stored slurry at rates corresponding to 40% of the N demand of the specific rotation. The N demand was based on a Danish national standard (Plantedirektoratet, 1997). The N demands from grass-clover and from peas/barley were set to nil. Cereal crops received slurry corresponding to a target of 50 kg NH_4-N/ha.

The experimental treatments were started in 1997. In 1996 a spring barley crop undersown with grass-clover was grown at all sites. No pesticides were applied in 1996. All locations were previously under conventional cropping. The crops during the five years prior to initiation of the experiment included different arable crops at Jyndevad and Flakkebjerg, and grass-clover and cereal crops at Foulum (Djurhuus & Olesen, 2000).

The experiment was unirrigated at all sites except at Jyndevad. All straw and grass-clover production was incorporated or left on the soil in all treatments. In 2000 and 2001 the grass-clover at Jyndevad was ploughed in early June (rotation 1) or late June (rotation 2) followed by harrowing several times to control couch grass (*Elymus repens*). In rotation 1 this bare fallow was followed by sowing of a cover crop of winter rye, winter vetch and rapeseed in mid July. In rotation 2 the following winter rye was sown mid August.

Leaching of nitrogen was measured using porous ceramic cups in selected plots. Four suction cells are permanently installed in each of these plots at a depth of 80 cm at Jyndevad and 100 cm at the other sites. The leaching is measured at all sites in those plots that corresponded to entry point one in the rotations when the experiment was initiated in 1997 (Table 3). At Foulum and Flakkebjerg leaching was additionally measured in all plots in rotation 2 without cover crops and with fertiliser. At Jyndevad leaching was measured in all plots with the manure treatment. Every 1-4 weeks, depending on site and precipitation, a suction of approximately 80 kPa was applied 3 days prior to sampling. The samples were bulked with equal sample volume from each of the four replicates per plot before analysis of nitrate concentrations.

The water balance was calculated using the Evacrop model (Olesen & Heidmann, 1990) for which inputs were daily meteorological measurements (precipitation, temperature and potential evapotranspiration). The observed precipitation at 1.5 m height was corrected to ground level using the methodology of Plauborg et al. (2002). Nitrate leaching was estimated using the trapezoidal rule (Lord & Shepherd, 1993), assuming that nitrate concentrations in the extracted soil water represents flux concentrations on the observation dates, and that concentrations change between measurement points in proportion to drainage. The accumulated annual leaching was calculated from 1 April to 31 March. The mean leaching was calculated for the four years from April 1998 to March 2002.

The FASSET soil-plant-atmosphere model (Olesen et al., 2002) was used to simulate the nitrate leaching from crop rotations 1 and 2 in all combinations with and without cover crops and with and without manure application. The simulations were also performed for soil and climate data from Foulum and Flakkebjerg. The observed management at the Jyndevad site from 1996 to 2002 was used for all simulations. However, irrigation was omitted at these two sites. The annual leaching was calculated from 1 April to 31 March and results are presented for the four years from April 1998 to March 2002.

Results and discussion

The nitrate-N concentration measured in the suction cups showed considerably larger and quicker fluctuations over time at Jyndevad compared with both Foulum and Flakkebjerg (Figure 1). This reflects the low retention of water and nutrients in the coarse sandy soil. The nitrate-N concentration tended to decrease over time at Foulum and to increase at Flakkebjerg. This may be an effect of the site history prior to onset of the experiment, which included a large proportion of grass crops at Foulum and continuous cereal cropping at Flakkebjerg. This thus indicates that soil N fertility plays an important role for the general or background N leaching at a given site.

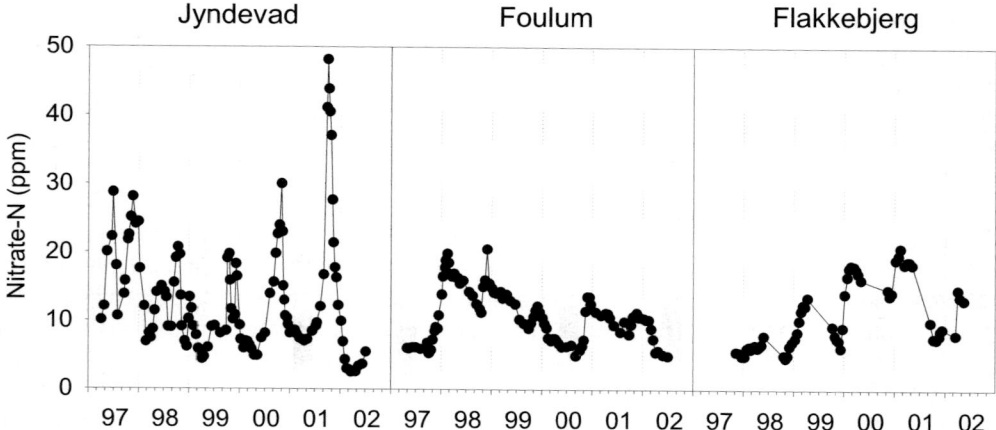

Figure 1. Mean measured nitrate-N concentration in soil solution from in rotation 2 without cover crops and with manure application.

There were only small effects of manure application on nitrate leaching (data not shown). On average manure application increased N leaching by 1 kg N/ha/yr. The results also showed no difference between crop rotations 1 and 2 in N leaching (Figure 2). There was a substantial effect of cover crops on N leaching. Cover crops thus reduced N leaching by 38% at Jyndevad and 23 to 25% at Foulum and Flakkebjerg (Figure 2). The reduction from cover crops was identical at Jyndevad in rotations 1 and 2 at rotation level (Figure 2). The magnitude of the leaching reduction is in line with other studies on cover crops in Denmark, which have shown the highest reductions on sandy soils and least reduction on loamy soils (Hansen & Djurhuus, 1997).

The pattern of the N-leaching was substantially different in the two rotations (Figure 3). In rotation 2 the highest leaching occurred following the ploughing of grass-clover in the autumn prior to winter wheat, whereas leaching peaked after the pulse crop in rotation 1. Other studies have also shown substantial N leaching following the cultivation of grass-clover leys, in particular with winter cereals (Watson et al., 1993; Djurhuus & Olsen, 1997).

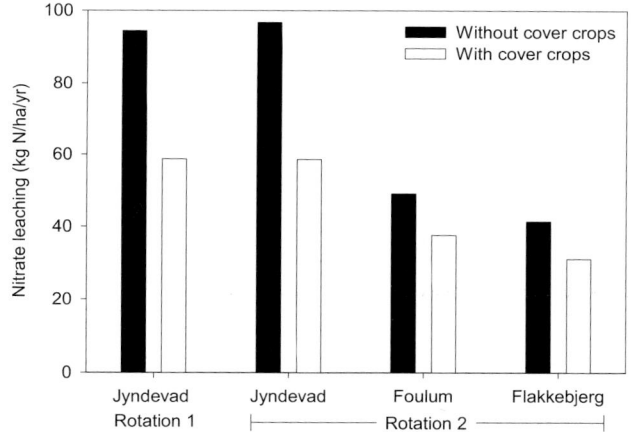

Figure 2. Mean annual measured nitrate leaching for rotations 1 and 2 at the three sites for the period 1998 to 2002.

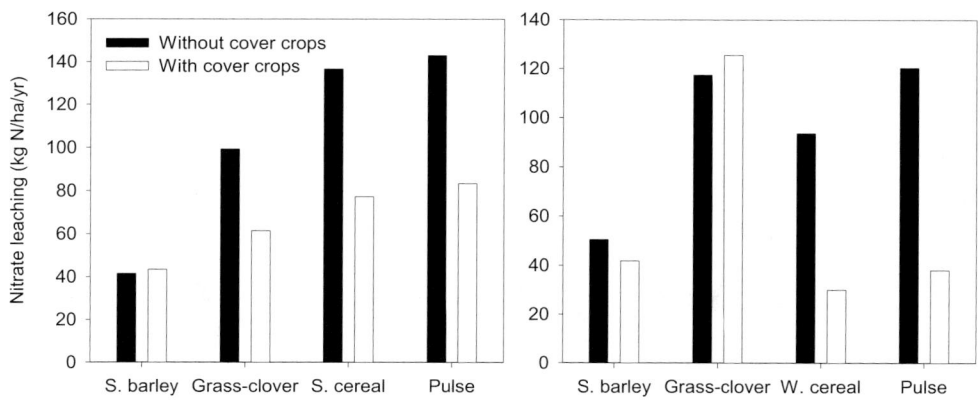

Figure 3. Mean annual measured nitrate leaching in rotations 1 (left) and 2 (right) at Jyndevad with manure application and with and without cover crops for the period 1998 to 2002.

The cover crops effectively reduced nitrate-N concentration during the winter period (Figure 4). However, the nitrate-N concentration at Jyndevad was high at Jyndevad in the autumn until the grass cover crop had grown to fully cover the ground. This did not happen until late October after which the nitrate-N concentrations were reduced markedly. This suggests a need for cover crops that are more effective in taking up soil nitrogen during the autumn period. Faster growing undersown grass crops may achieve this. However, such effective cover crops may also compete with the cereal crop thus reducing yields.

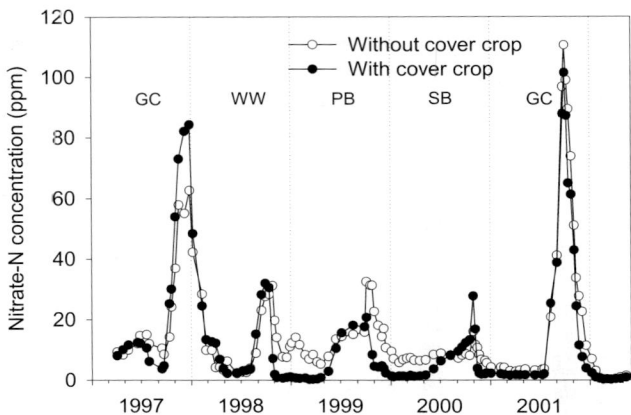

Figure 4. Mean measured nitrate-N concentration in soil solution at Jyndevad in rotation 2 with manure application in treatments with and without cover crop. The crop codes shown are GC: grass-clover, WW: winter wheat, PB: pea/barley, and SB: spring barley.

The simulated N leaching showed only small differences between the three sites using the observed management from Jyndevad (Figure 5). However, N leaching was substantially reduced following grass-clover at Foulum and Flakkebjerg. The simulations also showed a substantial effect of cover crops on N leaching. Surprisingly there was no difference between the two rotations in N leaching from the grass-clover crop, despite the fact that grass-clover was followed by spring wheat in rotation 1 and winter wheat in rotation 2. This may be explained by the fact that a bare summer fallow was applied to the grass-clover in 2000 and 2001. In the simulations this increased N leaching more in rotation 1 than in rotation 2, which may have been an effect of differences in the timing of the bare fallow between the two rotations. As the bare fallow and harrowing during autumn offers some of the few possibilities of controlling perennial weeds in organic farming, this indicates the dilemma between obtaining good weed control and reducing N leaching losses.

Conclusions

There was only a small effect of using either a winter cereal or a spring cereal on nitrate leaching at rotation level. Using a spring cereal reduced N leaching following the grass-clover, but this was compensated by a higher leaching in the following grain crops. Use of cover crops reduced N leaching by 23 to 38% at crop rotation level with the highest reduction on the coarse sandy soil. Simulation results showed that a practice of using part of the summer period in the grass-

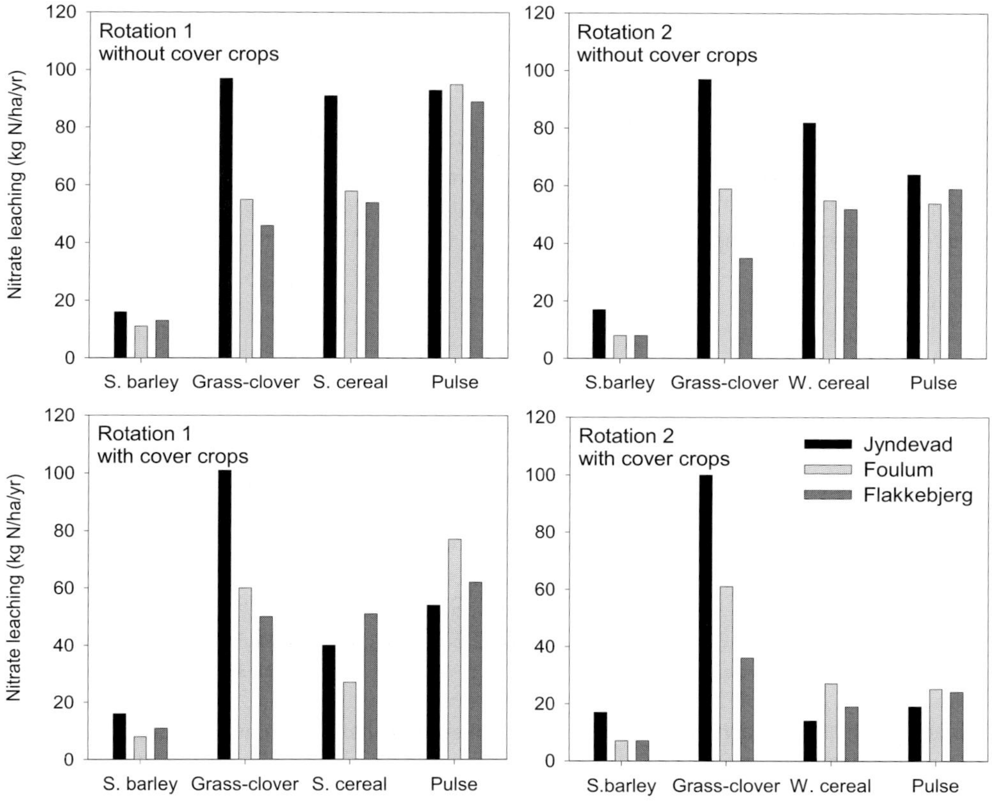

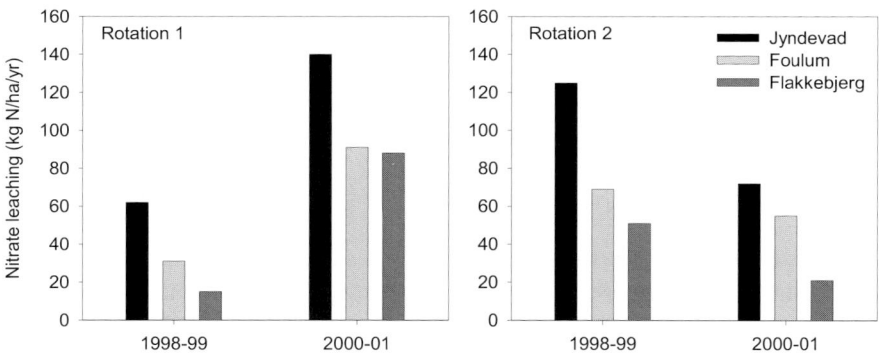

Figure 5. Simulated nitrate leaching for crop rotations 1 and 2 with and without cover crops and an average over four seasons from years 1998 to 2002.

Figure 6. Simulated nitrate leaching for grass-clover from the two periods (each two years) in rotations 1 and 2. The grass-clover during the first period was managed as a green manure crop, whereas the grass-clover during the second period was managed as a bare fallow during the summer period to control couch grass.

clover for a bare fallow to control couch grass could increase leaching substantially, in particular on the sandy soil.

Acknowledgements

The project was funded by the Danish Directorate for Development under the Ministry of Food, Agriculture and Fisheries. The project was an integral part of the activities under Danish Research Centre for Organic Farming.

References

Anonymous, 2002. Økologiske jordbrugsbedrifter 2001, autorisation, produktion. Plantedirektoratet, Copenhagen, Denmark.

Djurhuus, J. and P. Olsen (1997). Nitrate leaching after cut grass/clover leys as affected by time of ploughing. Soil Use and Management 13, 61-67.

Djurhuus, J. and J.E. Olesen (2000). Characterisation of four sites in Denmark for long-term experiments on crop rotations in organic farming. DIAS report No. 22. Danish Institute of Agricultural Sciences, Tjele, Denmark.

Grant, R., G. Blicher-Mathiesen, V. Jørgensen, A. Kyllingsbæk, H.D. Poulsen, C. Børsting, J.O. Jørgensen, J.S. Schou, E.S. Kristensen., J. Waagepetersen and H.E. Mikkelsen (2000). Vandmiljøplan II - midtvejsevaluering. Ministry of Environment and Energy, National Environmental Research Institute, Copenhagen.

Hansen, B., E.S. Kristensen, R. Grant, H. Høgh-Jensen, S.E. Simmelsgaard and J.E. Olesen (2000). Nitrogen leaching from conventional versus organic farming systems - a modelling approach. European Journal of Agronomy 13, 65-82.

Hansen, E.M. and J. Djurhuus (1997). Nitrate leaching as influenced by soil tillage and catch crop. Soil & Tillage Research 41, 203-219.

Lord, E.I. and M.A. Shepherd (1993). Developments in the use of porous ceramic cups for measuring nitrate leaching. Journal of Soil Science 44, 435-449.

Olesen, J.E., M. Askegaard and I.A. Rasmussen (2000). Design of an organic farming crop rotation experiment. Acta Agriculturae Scandinavica, Section B, Soil and Plant Science, 50, 13-21.

Olesen, J.E. and T. Heidmann (1990). EVACROP. A program for calculating actual evapotranspiration and drainage from the root zone. Version 1.01. Research note no. 9, Dept. of Agrometeorology, Danish Institute of Agricultural Sciences, Tjele, Denmark.

Olesen, J.E., B.M. Petersen, J. Berntsen, S. Hansen, P.D. Jamieson and A.G. Thomsen (2002). Comparison of methods for simulating effects of nitrogen on green area index and dry matter growth in winter wheat. Field Crops Research 74, 131-149.

Plantedirektoratet (1997). Vejledning og skemaer, mark- og gødningsplan, gødningsregnskab, grønne marker 1997/98. Plantedirektoratet, Copenhagen.

Plauborg, F., J.C. Refsgaard, H.J. Henriksen, G. Blicher-Mathiesen and C. Kern-Hansen (2002). Vandbalance på mark- og oplandsskala. DJF rapport - Markbrug 70. Danish Institute of Agricultural Sciences, Tjele, Denmark, 45 pp.

SAS Institute (1996). SAS/STAT Software: Changes and Enhancements through Release 6.11. Cary, NC, USA: SAS Institute Inc.

Stockdale, E.A., N.H. Lampkin, M. Hovi, R. Keatinge, E.K.M. Lennartsson, D.W. Macdonald, S. Padel, E.H. Tattersall M.S. Wolfe and C.A. Watson (2001). Agronomic and environmental implications of organic farming systems. Advances in Agronomy 70, 261-327.

Watson, C.A., S.M. Fowler and D. Wilman (1993). Soil inorganic N and nitrate leaching on organic farms. Journal of Agricultural Science, Cambridge 120, 361-369.

Nitrogen rate, surplus or residue? Performance of selected indicators for nitrate leaching

H.F.M. ten Berge[1], S.L.G.E. Burgers[1], M.J.D. Hack-ten Broeke[2], A. Smit[2], J.J. de Gruijter[2], G.L. Velthof[2], J.J. Schröder[1], J. Oenema[1], F.J. de Ruijter[1], S. Radersma[3], I.E. Hoving[4] and D. Boels[2]
[1]Plant Research International, P.O.Box 16, 6700 AA Wageningen, The Netherlands
[2]Alterra , P.O. Box 47, 6700 AA Wageningen, The Netherlands
[3]Applied Plant Research, P.O.Box 430, 8200 AK Lelystad, The Netherlands
[4]Animal Sciences Group, P.O.Box 2176, 8203 AD Lelystad, The Netherlands

Abstract

Results from three monitoring programmes were pooled and analysed to assess the suitability of three different indicators for NO_3^- leaching on sandy soils: N application rate, N surplus, and residual mineral soil N in autumn (Nmin). One large data set refers to 'Focus Nitrate', a land use based monitoring scheme including 478 observation spots each of approximately 20 m^2, distributed among 60 combinations of crop, soil and groundwater regimes (LUU's, land use units) on 34 farms. The other two data sets are from whole-farm studies, including 11 dairy and 19 arable farms, respectively, in which observations refer to farm-averages. The three data sets were used to develop regression models for NO_3^- concentration in the upper groundwater system. For arable systems, Nmin was the best indicator (lowest prediction error in farm-average NO_3^-) and variation in NO_3^- concentration could not be related to N-balance components (N-rate and N-surplus). For dairy systems, all three indicators performed approximately equally well, and NO_3^- concentration was clearly related to N-balance components. This contrast between dairy and arable systems remains unexplained. The prediction error of farm-average NO_3^- for a new case (farm and year) remained above ±38 mg l^{-1} (± 2 x SE of prediction) with any of the derived relations, and was much higher for some of the models.

Keywords: arable, dairy, indicator, monitoring, nitrate, spot observation, whole-farm

Introduction

Leaching of NO_3^- from agricultural lands to groundwater is a problem in The Netherlands, as well as in other regions of Europe, where farming has intensified over the past decades. While action programmes are being implemented to reduce NO_3^- emissions, there is a need to monitor progress at regional and national scales. Monitoring requires quantitative yardsticks or indicators. Such indicators may be utilised in several ways, or serve different purposes. The first is to signal trends in the environmental performance of the farming sector at regional, national, or subsector scales. The second is to estimate NO_3^- losses to groundwater and surface waters, at any desired scale. The third is steering the environmental performance of individual farm holdings, where (monetary) penalties or premiums can be attached to the attainment of preset threshold values of the indicator concerned. The focus of this paper is to evaluate the suitability of selected 'candidate' indicators for the second purpose and is limited to the quantitative estimation of NO_3^- losses to groundwater; for this we use the results obtained from three ongoing monitoring schemes, conducted in The Netherlands on sandy soils.

Methods

Selected indicators
'Nitrate' refers in this paper to the concentration of NO_3^- (mg NO_3^- l^{-1}) observed by sampling the upper groundwater once per year (see Table 1). The values are considered to represent the depth-averaged concentration in the first 1m of groundwater at the time of sampling. No dynamic behaviour within years can be inferred from the collected data. Neither did we attempt to normalise NO_3^- values for the time of sampling. The following indicators are evaluated with reference to the observed NO_3^- concentrations. They all refer to the season preceding the year when NO_3^- was measured: (a) the N surplus on the soil balance, defined as the difference between the annual N-input and the N-offtake in crop and animal products, (b) the rate of applied N and (c) the amount of residual mineral N (N_{min}) in the soil profile at the start of the 'leaching season'. These indicator variables, as well as the NO_3^- concentrations, were observed and aggregated at different scales, depending on the source project (Table 1).

Source projects and data sets
Data were collected in three monitoring schemes (source projects) that are still ongoing. The 'Focus Nitrate' project aims to establish empirical indicators for NO_3^- concentration in groundwater in dairy and arable farms, (Hack et al., 2003; Burgers (pers. comm.); Smit et al., 2003). The 'whole-farm dairy project' involves intensive farms, with annual production often exceeding 12,000 kg milk ha^{-1}. The main purpose here is to reduce the N-surplus at the farm gate balance by improved management, and so to reduce NO_3^- emissions (Oenema et al., 2001). The 'whole-farm arable project' also aims to reduce farm N-surplus and NO_3^- loss by improved management (De Ruijter & Smit, 2003). Details of methodology are summarised in Table 1 for all three projects.

In contrast to the whole-farm studies, monitoring in the 'Focus Nitrate' study was based on a procedure to acquire an even representation of soil types, groundwater regimes, and crops prevailing in the sand districts of The Netherlands. About 60 particular combinations of soil-crop-groundwater regime were identified, and are referred to as 'land use units' (LUU's). All observations refer to 20m^2 'spots' distributed among these LUU's. Crops were grass, maize and 'other field crops'. Within the latter, subclasses were recognised depending upon the respective N_{min} values expected for each crop on the basis of literature (Van Enckevort, 2002). These were 'low' (N_{min}<60 kg ha^{-1}; e.g. carrots, chicory, white cabbage), 'medium' (N_{min} 60 to 120 kg ha^{-1}) , and 'high' (N_{min} > 120 kg ha^{-1}; e.g. potato, lettuce, leek, and chinese cabbage). Groundwater regimes were classified according to the multi-annual mean peak level of the groundwater table: deeper than 80 cm below soil surface ('dry'), between 40 and 80 cm (medium) and shallower than 40 cm ('wet').

Regression analyses
Data from all monitoring seasons were pooled per source project, and regarded as independent observations for the purpose of linear regression analysis. For 'Focus Nitrate', the data were subdivided into sets for grass, maize, and 'other field crops', respectively. Regression models were developed for each of these subsets, with the overall purpose of explaining the variance in (groundwater) NO_3^- concentration at the single spot level, using LUU, additional soil properties, weather characteristics, and indicator variables as potential regressors. The search

Table 1. Details of three NO$_3^-$ monitoring schemes conducted in the Netherlands.

Monitoring scheme	'Focus Nitrate'	Dairy whole-farm	Arable-whole-farm
• No. of farms[1]	34	11	19
• No. of objects	478 'spots' each 20 m^2 in 60 LLU's	11 farms	19 farms
• Crops	grass; maize; field crops: arable, vegetables, tree crops	grass (mostly pasture) maize	field crops: arable, vegetables, tree crops
• Soils	sand; silt loam (loess)	sand; silt loam (loess)	sand
• Years[2]	2000-2001	1998-2001	2001
• Nitrate sampling	late March-early June top 1 m groundwater 1 borehole/spot	April-September top 1 m groundwater 48 bore holes/farm	April-September top 1 m groundwater 48 bore holes/farm
• Nmin sampling	Nmin 0-30, 30-60, 60-90 cm 1 Oct-1 Dec. 3 per spot mixed	none	Nmin 0-30, 30-60, 60-90 cm 1 Oct-1 Dec. 20 per field mixed
• Other observations	N-rates and N-surpluses at field and whole farm; soil properties[3] at 'spots'; groundwater regime daily precipitation (at farm)	N-rates and N-surpluses at field and whole farm	N-rates and N-surpluses at field and whole farm
• Data aggregation before regression	spot	farm	Farm
• Most successful indicator	N$_{minNO3}$	total N-rate N-surplus soil balance	N$_{min}$

[1]as included in the data analysis reported here.

[2]refers to crop season; NO$_3^-$ was measured in subsequent year.

[3]including profiles of soil texture, organic matter content, total organic C and total N, potential mineralisation, and potential denitrification

included neural network analysis to verify that the variance explained by the model was near to the maximum fraction attainable with the available data. In both whole-farm studies, farm mean NO$_3^-$ values per season were first assessed, and single borehole observations were not used in the analysis. Similarly, all regressors considered (N-rate, N-surplus on soil balance, N$_{min}$), refer to the whole-farm level. Regression models were then determined for the dairy farms and the arable farms separately.

Results

'Focus Nitrate' project

The simplest model performed almost as well as the more complex models that included such variables as the fraction of soil organic matter, C:N ratio of soil organic matter, potential mineralisation, and details on precipitation and groundwater level for specific periods within the year. Among the LUU-determinants (soil, crop, groundwater), groundwater regime was always dominant and significant; 'soil' and 'crop' were sometimes significant. The combined effects of particular LUU-determinants were expressed in a LUU-specific constant C_i (for LUU=i) (mg l^{-1}).

Next, the most important regressor in all cases was the amount of mineral N found in the soil profile in autumn (N_{min}). The best performance was obtained if we used the sum of N_{min} over all three depth intervals, but accounted for only the NO_3^- part of N_{min}, denoted as $N_{min,NO3}$ (kg ha^{-1}). Additionally, for the 'other field crops', the cumulative precipitation during April 1st - October 1st (P_{summer}, mm) was a significant and relevant regressor. In all three crop categories (grass, maize, 'other field crops'), the presence of peat horizons (exceeding 5 cm thickness) in the soil profile had a significant and relevant effect. The general model for the NO_3^- concentration (mg l^{-1}) is thus given by:

$$NO_3 = C_i + a \cdot N_{min,NO3}^{0-90} + b \cdot Switch_{peat} + d \cdot P_{summer} \qquad (1)$$

with $Switch_{peat}$= 1, if peat was present and $Switch_{peat}$= 0, if absent. The superscript 0-90 refers to the total sampling depth interval of 0 to 90 cm. P_{summer} is valid for 'other field crops' only and is ignored for grass and maize. The regression constant C_i has a different value for each combination of crop category (and for subclasses), soil type and groundwater regime. Regression coefficients a and b have different values for grass, maize and 'other field crops', but are constant within categories and thus, independent of soil type, groundwater regime, and crop subclass. Figure 1 shows the response of NO_3^- to $N_{min,NO3}$ based on Eq.1, for different crop categories and groundwater regimes, and all soil types pooled. This model explained 17% of the variance in NO_3^- for grass at the spot level, 21% for maize and 46% for 'other field crops'. Adoption of other regressors had little additional effect. The performance of this regression model is regarded as poor. This is partly due to single ('spot') observations being associated with high spatial (within-field) variation. The resulting regression models are thus, associated with large prediction errors when used to predict NO_3^- for a new 'spot'. Upscaling to farm level brings much improvement. The average of all $N_{min,NO3}$ observations in each LUU is then used to predict the mean NO_3^- value for each LUU through Eq.1.

Predictions of NO_3^- concentration at the whole-farm level are composed from these individual LUU_i- NO_3^- values, given a specified farm composition (the respective areas of LUU_i). These whole-farm level predictions (for dairy and arable farms) are associated with prediction errors (plus and minus twice the s.e. of the prediction) of ± 38 to 56 mg l^{-1} (95% confidence interval),

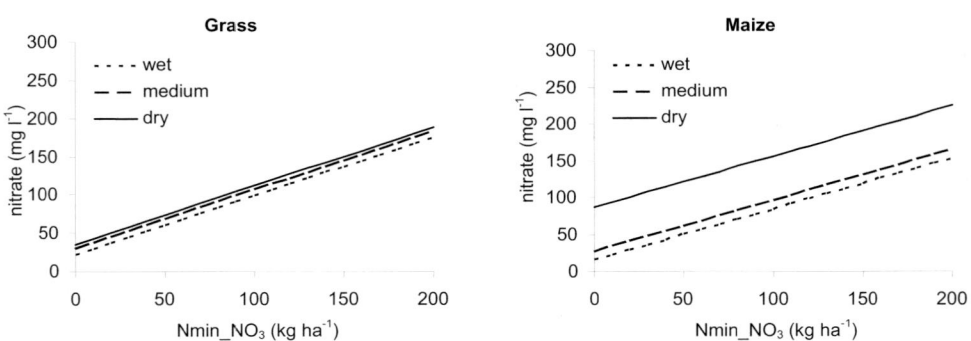

Figure 1. Response of NO_3^- concentration in groundwater to residual mineral soil N (Nmin-NO_3) according to the regression model obtained from the 'Focus Nitrate' data (Eq.1) for grass and maize, in the absence of peat layers in the soil profile. Relations for 'other field crops' are comparable. Based on pooled 2000/2001 and 2001/2002 data.

depending on the precise farm composition (Hack et al., 2003; Burgers et al., 2003). The model derives a considerable part of its 'predictive capacity' from the LUU-characteristic C_i. If Eq. 1 were reduced by ignoring all other RHS terms, thus retaining only C_i., the s.e. increases by approximately 5-15 mg l^{-1}, and hence the prediction error by ± 10-30 mg l^{-1}, depending on the accuracy by which the true mean $N_{min,NO3}$ for LUU_i is estimated (i.e. on sample numbers).

Dairy whole-farm study

The improvement in overall nutrient use efficiency, realised during 1999-2002, resulted in gradually decreasing N-rates and N-surpluses at the whole-farm soil surface N-balance, as well as decreasing groundwater NO_3^- levels at most farms. All four years (1999-2002) were favourable for crop growth, yet no indications were found that there was a trend in precipitation, or other growth conditions, which might have caused the observed decrease in NO_3^- values. Hence, we conclude that the pattern of falling NO_3^- concentrations is directly related to reduced N-inputs and N-surpluses. Figure 2 shows the relation between N-surplus on the whole-farm soil balance, and farm mean NO_3^- concentration. Table 2 shows the correlation coefficients for the relation between several N-balance components and farm-average NO_3^- concentration. Except in the first year (1999), total-N rate to the soil balance gave the highest correlation with NO_3^-. Effective N-rate (viz. 50% of N in animal manures and 100% of N in fertilisers considered 'effective' for crop uptake, within the same season) performed slightly less well, as did the N-surplus on the soil balance, and the N-surplus on the farm gate balance. Linear regression models for farm-average NO_3^- values based on one of these management variables (in Table 2) only, are associated with prediction errors that are comparable to those cited for the 'Focus Nitrate' study.

If, for example, the NO_3^- concentration for a new case (farm in year) were to be estimated with a linear regression model containing only N-surplus on the whole-farm soil balance as regressor, and if this surplus were known exactly, the prediction error would still be ± 40 to 50 mg l^{-1} (95% confidence interval), according to the data available from this project.

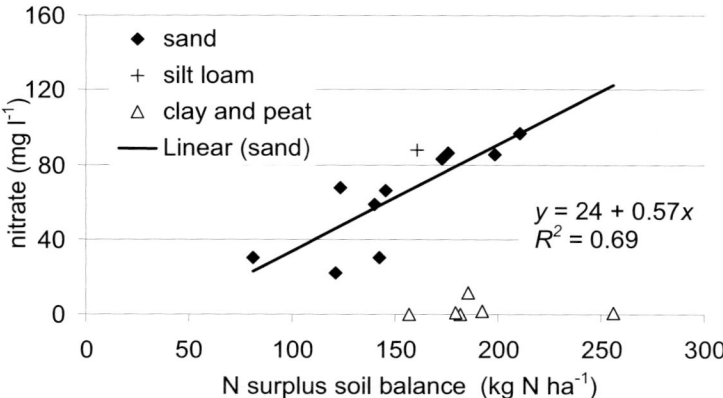

Figure 2. Three-year mean of the annual farm mean NO_3^- concentration versus three-year mean N-surplus on the annual farm soil balance. The regression line relates to all farms on sandy soils and one on silt loam in the 'whole-farm dairy study'. Observations for dairy farms on clay and peat are included for comparison only (not referred to in text).

Table 2. Correlation coefficient (R^2) between various selected indicators at whole-farm level in Year x and farm mean NO_3^- concentration in groundwater in Year x+1, for three consecutive seasons. Whole-farm study for 11 intensive dairy farms in The Netherlands.

N-balance year (x)	Farm gate N-surplus kg ha^{-1}	N-surplus[b] farm soil balance kg ha^{-1}	N-rate total-N kg ha^{-1}	N-rate effective-N kg ha^{-1}
1999	0.50	0.66[a]	0.50[a]	0.45[a]
2000	0.41	0.57	0.70	0.53
2001	0.45	0.55	0.72	0.61
1999-2001 pooled	0.46	0.50[d]	0.64[e]	0.52
using 3-year mean	0.53	0.69	0.72	0.58

[a]available for five farms only, [b]exclusive of N lost by volatilisation manures and grazed pastures; including atmospheric NH_4^+ deposition, [c]inclusive of N lost by volatilisation from manures and grazed pastures; including atmospheric NH_4^+ deposition, [d]NO_3 (mg l^{-1}) = -3 + 0.44*Nsurplus and [e]NO_3 (mg l^{-1}) = -57 + 0.29*Nrate

Arable whole-farm study

As in the 'Focus Nitrate' study, groundwater depth had a strong effect on NO_3^- concentration. Groundwater depth alone explained 42% of the total variance in farm-average NO_3^- concentration. This fraction increased to 60% when total N_{min} (NO_3^- and NH_4^{+}; 0-90 cm depth) was adopted as AN additional regressor. Replacing N_{min} by the N-surplus on the soil balance, resulted in virtually no improvement; neither was total N-rate a useful regressor, nor effective-N rate or N-surplus derived when accounting only for effective-N rate as input. The associated prediction errors are ±105 mg l^{-1} for the model with groundwater depth and N_{min} as regressor, and ±115 mg l^{-1} if N_{min} was not included in the model (at 95% confidence level). These errors are approximately twice as large as the maximum errors obtained in 'Focus nitrate' (with the different procedure applied there; the data, however, differed too). Apparently, N-rate and N-surplus had no obvious effect on NO_3^- concentration in groundwater under arable systems, and it is not fully clear why this is so, when significant correlations were found for dairy systems. Similarly, no significant effect of N management on N_{min} remained in this data set, after removal of one high-leverage point. Preliminary results for the 2002/2003 season seem to confirm this 'absence of pattern'.

Discussion of indicators: their pros and cons

The indicators discussed are: N-rate, N-surplus, and Nmin. In listing the disadvantages of each, for estimating NO_3^- leaching, we ignore the process of denitrification because any of the indicators can be flawed, indiscriminately. However, it is acknowledged to be relevant in many soils. N-rate (amount of N annually applied per ha) ignores variation in crop N-offtake, which in reality modulates NO_3^- loss. It also fails to account for NO_3^- losses derived from mineralised organic reserves in the soil. Total-N rate, in our case, includes atmospheric deposition, while NH_3 losses from applied manure are not subtracted and are hence, part of the applied N. Effective-N rate was assessed assuming that 50% of N in cattle slurry is as effective for crop uptake (within the same season) as fertiliser-N (which is assumed to be 100% effective). In the whole-farm dairy study, total-N rate was the best indicator for NO_3^- (Table 2). The better performance of total-N rate, as compared with effective-N rate confirms, in our view, that most of the organic-N in cattle manure does becomes available, albeit beyond the scope of a single year. Successful as this indicator may be for dairy systems, the results from both the 'Focus Nitrate' spot study (dairy as well as arable) and the whole-farm arable study, showed that neither N-rate, nor any other of the measured N-balance components contributed significantly to the explanation of the variance in NO_3^-. We have no good explanation for this contrast.

N-surplus does account for variation in crop N-offtake. However N-surplus is flawed as an indicator by mineralisation from organic-N reserves (soil-N supply), except when soil-N pools are in steady state equilibrium with the input regime. N-surplus is more sensitive to this drawback than N-rate, because of 'the surplus paradox'. At any given N-rate, high mineralisation rates tend to augment N-losses but reduce N-surplus (the latter because of higher N-offtake). So, lower N-surpluses are associated with higher N-losses, if the source of variation in N-surplus is a variable soil-N supply. This is perhaps why total-N rate performed better than N-surplus in the whole-farm dairy study. N-surplus performed poorly in the arable farm study, as well as in the 'Focus Nitrate' spot study. As with N-rate, we cannot explain the contrasting performance of N-surplus in the dairy versus arable farm studies. Neither is it clear why N-surplus worked well in the whole-farm dairy study, but not at the 'spot' level for grass and maize in the 'Focus Nitrate' set, nor when scaled up to whole-farm level starting from 'spot' data. One cause may be the low accuracy in estimations of N-surplus at the 'spot' level.

Residual mineral N (N_{min}) is less than the previous indicators afflicted with disturbance by mineralisation, varying crop offtake, and varying soil-N supply. It is, on the other hand, more subject to weather conditions and fails to account for summer leaching below the root zone. N_{min} is a much better predictor than N-balance components (rate, surplus) in arable systems. This is true of both the 'spot' study and the whole-farm arable study. In the whole-farm study, we had access to only 'total N_{min}' data ($NO_3^- + NH_4^+$ N). The 'spot' study revealed that N_{minNO3} worked better in that data set. It is possible, therefore, that NO_3^- predictions by the regression model obtained in the whole-farm study can be improved too, by using N_{minNO3} instead of total N_{min}. However, it is unlikely that this would reduce the prediction error from the current 105 mg l^{-1} (above) to the level of 38-56 mg l^{-1}, obtained with the LUU-based approach in 'Focus Nitrate'. As the whole-farm dairy study did not include monitoring of N_{min}, no direct comparison can be made between the indicator performance of N-rate and N-surplus on the one hand, and N_{min} on the other, at least not within the same data source. We can compare, however, the N_{min}-based predictions for dairy systems (as obtained from upscaled regression models in the 'Focus Nitrate' project) with N-rate (or N-surplus) based predictions derived from the whole-farm dairy study. The ranges of the prediction error in farm-average NO_3^- concentration are then roughly identical.

It would seem that N_{min}, or even better N_{minNO3}, is the indicator to be preferred overall, performing much better in arable systems than the alternatives studied, and equally well in dairy systems. Yet, one peculiarity remains to be addressed: the NO_3^- value expected at $N_{min}=0$ (Figure 1). This intercept is, admittedly, the result of the applied statistical procedure, but is firmly justified by the data: many observations of N_{minNO3} were sufficiently low to confirm that the intercept is no artifact arising from extrapolation beyond the reach of actual data. This suggests that the relations obtained may be used to estimate groundwater NO_3^- from observed N_{minNO3} in monitoring schemes where data collection follows the protocols maintained in 'Focus

Nitrate'. The question remains on how to interpret intercept values in Figure 1, in order to use N_{min} as a guide towards N-saving farming practices? According to Figure 1, even N_{min}-values as low as zero would not suffice to approach the NO_3^- threshold of 50 mg l^{-1} on soils with groundwater regime 'dry', except under grass. This intercept may be the result of NO_3^- leaching during the growing season (from applied N and/or mineralised N from organic reserves). Alternatively, it may result from the two months sampling period for N_{min}. Weather conditions during October-November may have reduced an existing relation between N_{min} and NO_3^-, by lowering N_{min} values through rainfall events, or increasing them via continued mineralisation. Groundwater NO_3^- is less variable, and variations are unlikely to be directly associated with those in N_{min}, six months earlier. This would reduce the N_{min}-nitrate relation to a translation of the entire line (Figure 1) towards lower N_{min}-ranges, and perhaps also to a decreasing slope. This renders the relationship in Figure 1 unreliable for adoption in models that aim to evaluate the impact of nutrient management on NO_3^- leaching, as well as for the assessment of 'safe' threshold values to be used in steering N-management. We are not aware of convincing experimental proof of such high intercepts in single-location trials, with contrasts in N_{min} generated via graded N-applications (N-rates), and with simultaneous N_{min} sampling in all treatments. Another puzzling aspect of N_{min}, this time in the whole-farm arable data set, is the absence of a response in N_{min} to N-rate and N-surplus (both at whole-farm level). This contradicts the results usually found in single-location N-response trials in arable crops (Van Enckevort et al., 2002) as well as in grass and maize (ten Berge et al., 2002), and points at the dominant effect that local conditions (soil, weather and crop properties) may have on N_{min} in real-farm multi-location data sets.

Conclusions

In this study, we attempted to assess the performance of potential indicators for NO_3^- concentration in groundwater using the results from three NO_3^- monitoring projects, each with its own particular aim and methodology. Although the composite nature of the data sets does not allow final conclusions, we found sufficient evidence to support the following: (1) N_{minNO3} combined with basic information on LUU: (intersection of soil, crop, and groundwater), provides the best indicator for monitoring NO_3^- in groundwater under arable systems; N-rate and N-surplus were poor indicators in arable systems. Depending on the desired monitoring scale, application of the N_{min} indicator requires a sampling scheme designed to assess mean N_{minNO3} for each LUU within a farm or region. Sampling should most probably be confined to a short period at the end of the growing season. It should be debated whether the precision gained by monitoring N_{min} is worthwhile, given the predictive capacity already attained with LUU characteristics only. (2) Farm-average NO_3^- under dairy systems can be estimated via LUU-based N_{min} monitoring with the restrictions just given, or equally well from recorded information on N-rate or N-surplus in the whole-farm soil balance. N-surplus at the farm gate balance performed less well, even with the fairly homogeneous set of dairy farms studied. Among the options to express N-rate, total-N rate is to be preferred over effective-N rate as an indicator for NO_3^- in dairy systems. (3) On soils where groundwater remains deeper than 80 cm, relatively high values are found to be associated with very low Nmin values, based on our regression analysis. Until this is better understood, the relations provided must not be used to model leaching in response to N-inputs at particular sites. (4) The error (95% confidence) associated in our approaches with predicting the farm-average NO_3^- concentration, with the help of farm-level or upscaled 'spot-level' or regression relations, is roughly equal to the target concentration (50 mg l^{-1}), as defined by the EU-Nitrates Directive.

References

Berge H.F.M. ten, Burgers, S.L.G.E., Schröder, J.J. and Hofstad, E.J. (2002). 'Partial balance'-regression models for Nmin. In: H.F.M. ten Berge (ed), A review of potential indicators for nitrate loss from cropping and farming systems in The Netherlands. Plant Research International Report 31, Wageningen, The Netherlands. pp 25-60.

Enckevort, P.L.A. van, van der schoot, J.R. and van den Berg, W. (2002). Estimation of residual mineral soil nitrogen in arable crops and field vegetables at standard recommended N-rates. In: H.F.M. ten Berge (ed), A review of potential indicators for nitrate loss from cropping and farming systems in The Netherlands. Plant Research International, Wageningen, The Netherlands. pp 77-90.

Hack-ten Broeke, M.J.D., Burgers, S.L.G.E., ten Berge, H.F.M., van Enckevort, P.L.A., de Gruijter, J.J., Hoving, I.E., Smit, A. and Velthof, G.L. (2003). Ontwikkeling van een indicator om te Sturen op nitraat. Gegevens en regressie-analyse voor het eerste meetseizoen (2000-2001). Reeks Sturen op Nitraat 4, Alterra-rapport 772, Wageningen. (65 pp). In Dutch.

Oenema, J., Koskamp, G.J. and Galama, P.J. (2001). Guiding commercial farms to bridge the gap between experimental and commercial dairy farms: the project 'Cows and Opportunities'. Netherlands Journal of Agricultural Science, 49, 277-296.

Ruijter, F.J. de, and Smit, A.L. (2003). Relaties tussen nitraat in het grondwater en potentiële indicatoren voor nitraatverlies op de voorloperbedrijven van Telen met Toekomst. Plant Research International, Wageningen, The Netherlands. 28 pp.

Smit, A., Hack-ten Broeke, M.J.D., ten Berge, H.F.M., Burgers, S.L.G.E., Chardon, W., van Enckevort, P.L.A., de Gruijter, J.J., Hoving, I.E. and Velthof, G.L. (2003). Gegevensverzameling Sturen op Nitraat. Reeks Sturen op Nitraat 3, Alterra-rapport 658, Wageningen. (48 pp). In Dutch.

Safety-nets and filter functions of tropical agroforestry systems

G. Cadisch[1], E. Rowe[2], D. Suprayogo[3] and M. van Noordwijk[4]

[1]*Dept. Agricultural Science, Imperial College London, TN25 5AH, Wye, Kent, UK*
[2]*Dept. Plant Sciences, Wageningen University, 6700 AK Wageningen, The Netherlands*
[3]*Fakultas Pertanian, Jurusan Tanah, Universitas Brawijaya, Malang, Indonesia*
[4]*ICRAF - SE ASIA, PO Box 161, 16001 Bogor, Indonesia*

Abstract

The design of productive and efficient intercropping agroforestry systems, which minimise environmental pollution, depends on achieving complementarity between component species' resource capture niches. This concept was tested by assessing spatiotemporal patterns of vertical and lateral water and N capture and use by trees and intercrops in agroforestry systems in Indonesia. Using trees to provide a biological root safety-net in intercropping systems, will only be successful if trees do not adversely affect crop growth. Sensitivity analyses using the dynamic agroforestry model WaNuLCAS suggested that N interception efficiency increases with tree root length density and depth of soil in the safety-net layer. However, N interception efficiency is limited by tree N demand, and it decreases as the proportion of tree N derived from atmospheric fixation increases. Hence non-N_2 fixing trees may provide a more effective safety-net for N. Thus the current perception of the ideotype, fast growing, N_2 fixing, agroforestry tree has to be revised towards a slower growing, but drought resistant species, that is deep rooting with few active topsoil roots and which produces slowly decomposing pruning residues. Scaling up these concepts offers opportunities for enhancing the filter efficiency of farms by appropriate selection and positioning of agroforestry options within the landscape.

Keywords: competition, complementarity, ideotype, N leaching, root length density

Introduction

Trends in resource management strategies differ substantially between farmers in developed and developing countries. In affluent northern countries, resource excess is a major concern and reduction in spatial variability and precision agriculture are aimed at maximising the use of nutrient resources, thus avoiding leakage and maximising profits. In contrast, resource poor, small-scale farmers in the tropics try to exploit spatial variability (strategic placement) and control routes for nutrient loss (safety-nets and filters), generally with the aim of maximising the use of scarce nutrient resources to minimise risks and improve their livelihoods, rather than avoiding pollution. The design of productive and efficient intercropping agroforestry systems, which minimise environmental pollution, depends on achieving complementarity between component species' resource capture niches. This challenges the commonly held perception that fast growing N_2 fixing trees are the best intercropping partners on poor soils. Such trees may be highly competitive, and by producing high quality prunings are likely to increase the turnover and potentially, the losses of soil N. Furthermore, the ability of tree based systems to act as a safety-net against vertical N losses through the soil and lateral N losses through the landscape is not only determined by the distribution of their roots in the soil profile, but also by their N sink strength and the residence time of N in the root zone (rainfall intensity, soil properties).

There are several mechanisms by which filter and safety-net features in an ecosystem can be achieved:
- Biological (root safety-net)
- Biochemical (plant residue quality)
- Chemical (soil retardation mechanisms)
- Physical (preferential flow through soil macro-pores)

These mechanisms can be exploited using spatial and temporal arrangements of filtering elements within farming systems. In this paper we will focus on the impact and design of the biological safety-net feature of tree roots in simultaneous (hedgerow) agroforestry systems (Figure 1).

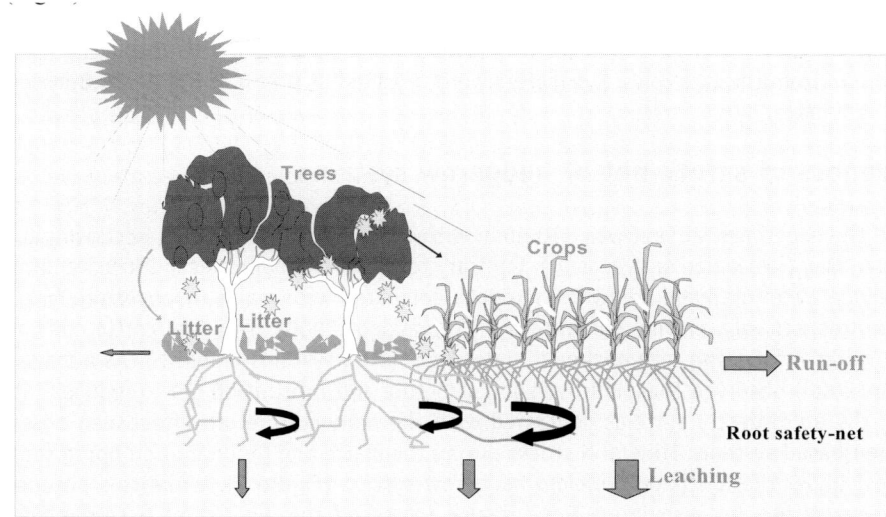

Figure 1. Tree root safety-net in a simultaneous agroforestry system.

The potential modes of action of an effective biological safety-net include:
- Tree nutrient sourcing below the main crop rooting zone, resulting in the interception of nutrients that would otherwise be leached
- Reduced water drainage (e.g. increased water use)
- Reduced concentration of nutrients in the soil solution (e.g. high tree+crop N uptake)
- Increased by-pass flow of soil solution with low nutrient concentration through macropores, due to improved soil structure

Roots of perennials can often exploit deeper soil nutrients better than annuals, since their deep roots are present throughout the year, whereas annual crop roots are only active in deeper soil layers during the later stages of crop growth.

Case study: intercrop hedgerow systems in Sumatra

An experiment was carried out at the Biological Management of Soil Fertility (BMSF) project site of Universitas Brawijaya/ICRAF/Wye College/PT Bunga Mayang, near Karta, North Lampung, Sumatera, Indonesia ($4^o31'$ S, $104^o55'$ E), on a field with a gentle slope (4%). The soil was a Plinthitic Kandiudult with a sandy clay topsoil and clay-plinthite accumulation at depth (>0.8 m, with high aluminium saturation). Cropping systems were established in 1985-1986; one with no hedgerows ('monocrop'), and hedgerow intercropping systems including either the non-fixing leguminous tree *Peltophorum dasyrrhachis* (P-P) or the N_2-fixing *Gliricidia sepium* (G-G). Hedgerows were 4 m apart and trees were 50 cm apart within the row. Two crops were generally grown per year, e.g. maize (December-March) followed by groundnut. Maize received 90 kg N ha^{-1} at planting and both crops received adequate P+K. Mean maize yields (t ha^{-1}) over four cropping seasons in 1991-1992 were 2.30 (monocrop), 2.33 (G-G) and 3.32 (P-P) (van Noordwijk et al., 1995). Spatiotemporal patterns of vertical and lateral water and N capture and use by trees and intercrops in these agroforestry systems were elucidated by monitoring solute movement using vacuum lysimeters (Suprayogo, 2000) and isotopic tracing of ^{15}N labelled urea fertiliser and tree prunings (Rowe, 1999). In the maize-groundnut systems, the vacuum lysimeters were placed at 125 cm from the crop rows: in the hedgerow system they were installed 0, 0.7 and 2 m from the tree rows. Lysimeters were installed at 0.2, 0.4, 0.6 and 0.8 m depths with four replicates.

Effectiveness of safety-net of hedgerow systems

Rainfall in the experimental year was 3102 mm with the main rainfall events occurring during December to May (i.e. maize crop season). Contrary to the initial hypothesis, inclusion of trees did not reduce water drainage below 0.8 m depth compared with maize monoculture (i.e. 933 vs 797 mm, respectively, during January-May) (Suprayogo, 2000). This was partly due to the increased water infiltration rate observed under tree-based systems (Table 1). As a trade off, Suprayogo (2000) observed that improved soil structure and infiltration led to reduced lateral run-off in the agroforestry systems compared with the monoculture, an effect also observed in other tree-based systems. Mineral N stocks during the maize and groundnut cropping seasons were largest under the N_2-fixing *gliricidia* hedgerow system (Figure 2). While this potentially provided more N for the associated crops it also increased the amount of N susceptible to leaching. The amount of mineral N under the non-fixing leguminous tree *peltophorum* was lower than under the maize monocrop, posing the smallest risk of leaching and gaseous losses.

The estimated N leaching losses during the two cropping seasons amounted to 65 kg N ha^{-1} in the maize-groundnut monoculture system (Table 1). The agroforestry systems varied in their ability to reduce N leaching. While the non-fixing tree *peltophorum* significantly reduced mineral N leaching compared with the monocrop system, the *gliricidia* hedgerow system did not. The difference in the efficiency of trees in reducing N losses was not attributable to a single property, but has to be viewed as a result of several interactions occurring in the intercropping system. The main reasons for the enhanced capability of the *peltophorum* hedgerow system to reduce N leaching were identified as: i) slow release of N from prunings of *peltophorum* (Table 1) from lower N quality (non-fixing) and high active polyphenol content (protein-binding capacity) of its leaves (Handayanto et al., 1994), ii) lower amounts of *peltophorum* prunings to be recycled

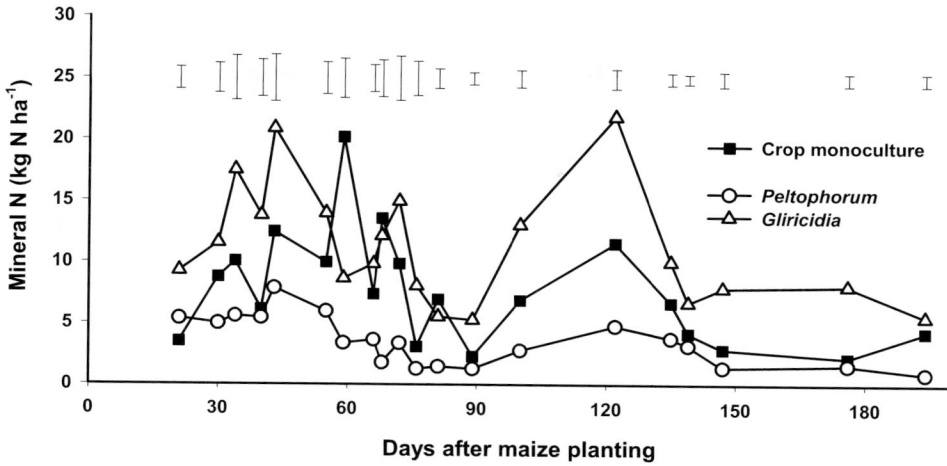

Figure 2. Soil mineral-N dynamics (0-0.8 m) under a maize-groundnut monocrop and Gliricidia sepium (N₂ fixing) or Pelthoporum dasyrrachis (non-fixing) hedgerow intercropping systems (adapted from Suprayogo (2000)). Bars indicate SED.

Table 1. Effect of maize-groundnut monoculture or hedgerow intercropping systems on infiltration, cumulative vertical mineral N movement (N leached), litter decomposition (N recovery in litter), crop urea-N recovery and the 'non-accounted for' part of the soil-plant isotopic ^{15}N balance (adapted from Suprayogo, 2000 and Rowe, 1999).

Cropping systems	Infiltration cm/day	N leached g N/ha (%[1])	^{15}N recovery in litter % of applied	^{15}N recovery in crop % of applied	^{15}N system N deficit % of applied
Peltophorum	36.6	20 (-73)	31	34	17
Gliricidia	35.2	64 (-3)	6	22	34
Monoculture	34.5	65 (0)	na	42	32
SED	0.8	6	10	4	

[1] Percentage of - decreased or + increased of total leached mineral-N in relation to control

during the wet season because of slower growth, iii) less competition with crop for nutrients (Table 1) and water during the main cropping season partly because of relatively low topsoil root length density (Figure 3), iv) less light competition from a compact canopy shape and slow growth, resulting in better crop growth and v) slightly higher infiltration rates (Table 1) from the increased occurrence of macropores with associated enhanced by-pass flow (Suprayogo, 2000).

Location of vertical safety-net

Soil nutrients below the active rooting depth of annual crops can be accessed by trees in two ways: (i) the retrieval of nutrients already present in layers below the effective extent of rooting of annual crops ('nutrient pumping', Mekonnen et al., 1997 and (ii) the interception of nutrients moving below the rooting zone of crops ('root safety net', Van Noordwijk et al., 1996, Figure 1). Conditions conducive to 'nutrient pumping' include soils without physical or chemical barriers

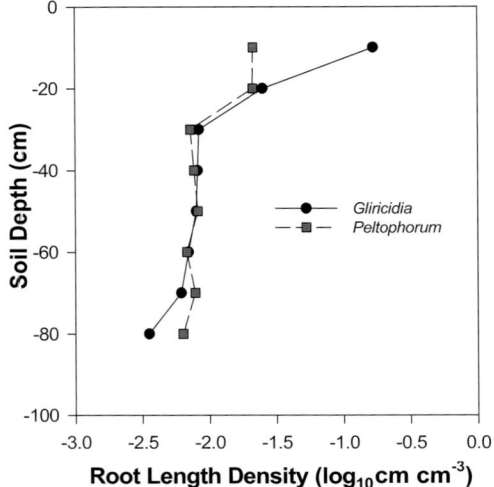

Figure 3. Tree root length densities.

to deep rooting and with large amounts of plant-available nutrients (leached or parent material derived) below the rooting zone of crops, and trees which have a high N demand and deep rooting system. Favourable conditions for the 'safety-net' role of perennials include soils with physical or chemical barriers to deep rooting (as in the case study, where rooting depth was restricted below 0.8 m because of high aluminium saturation and plinthic soil structure) and the accumulation of mobile soil nutrients. In both cases, the distribution and density of roots, the demand of plants for nutrients, and the distribution and concentration of available nutrients and water will influence the efficiency of extraction of nutrients by trees.

Assessing root length density (Lrv) of the two tree species under investigation revealed that *gliricidia* had a much higher Lrv in the topsoil than *peltophorum* (Figure 3) confirming the previous observations of a high competitiveness of *gliricidia* with the associated food crop. Thus, although root length densities per se were not strongly different between the two trees at lower soil depths (Figure 3, considering the log scale), *peltophorum* had a greater proportion of its roots in the deeper soil than *gliricidia* which may, in part, explain the higher safety-net efficiency of the *peltophorum* based system. Rowe et al., 2001, assessed the nutrient acquisition activity of tree roots by assessing plant recovery of ^{15}N injected at different soil depths and distances from these trees. They demonstrated that *peltophorum* derived a greater proportion of its N from deeper soil layers (>40 cm) than *gliricidia*, making it a more efficient safety-net option.

Efficiency of tree root safety-nets

Potential and limitations of agroforestry options for improving N recycling efficiency and reducing environmental losses were further examined using the dynamic agroforestry model WaNuLCAS (van Noordwijk & Luisiana, 1999). In particular, we assessed the role of the root length density (Lrv) and rainfall on the tree root safety-net efficiency. The safety-net zone (i.e. the zone below the crop root zone which is explored by tree roots) was, in the simulation, defined as the 40-60 cm soil layer, in order to reflect restricted root subsoil exploration in this soil because of an aluminium-rich plinthic subsoil. Safety-net efficiency (SNE) was defined as:

$$SNE = 100 \times \frac{TN_{upt}}{Leach_{out} + TN_{upt}}$$

where *SNE* is safety-net efficiency (%), TN_{upt} is tree N uptake from the safety-net layer and $Leach_{out}$ is N leached beneath safety-net layer.

The model simulated the December-March high rainfall period, without N fertiliser application. Simulations suggested that a maximum 22% of leached N would be intercepted by tree roots in the 40-60 cm soil layer at a density of 4 cm cm^{-3} (Figure 4). Measured tree root length densities at this depth however, were in the range 0.005-0.015 cm cm^{-3}, sufficient to intercept less than 5% of leaching N according to this simulation. The safety-net zone beneath crop roots may, however, be considerably thicker than the 40-50 cm soil layer, especially at early stages of crop growth, or in soils which allow deeper rooting, resulting in higher interception efficiencies due to enhanced mineral N residence times. Thus, in the humid tropics during the rainy season, even a relatively dense mat of roots would not intercept all N leached during periods of low tree demand, or during heavy rainfall events when residence time is short.

Additional simulations (not presented) demonstrated that the thickness of the safety-net layer, and hence the maximum rooting depth of trees increased the interception efficiency. With a safety-net layer 4 m thick and a reduced rainfall of 2200 mm, the interception efficiency was also less dependent on Lrv than in the above example in the range between 0.02 (15%) and 0.1 (18%) cm cm^{-3}. This indicates a saturation of the safety-net which might occur because of asynchrony between leaching and tree N demand. A reduction in tree N demand also reduces safety-net efficiency; for example, satisfying a proportion of demand by N$_2$-fixation results in an approximately linear reduction in safety-net efficiency (Cadisch et al., 2003). Using ^{15}N injections at depth and the ^{15}N natural abundance technique to estimate N$_2$ fixation, Rowe et

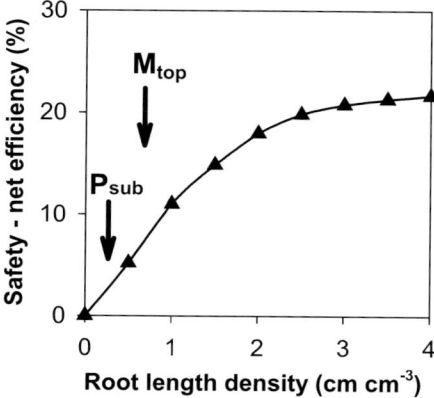

Figure 4. Relationship between tree root density (40-60 cm layer) and safety-net efficiency to intercept mineral N during the rainy season in Lampung, Indonesia. (adapted from Cadisch et al., 2003. P_{sub} = Peltophorum dasyrrachis Lrv in subsoil; M_{top} = maize Lrv in topsoil.

al. (1999) confirmed that the non-fixing legume *peltophorum* took up more N (42 kg N ha⁻¹) from beneath the main crop rooting zone than the N_2 fixing legume *gliricidia* (21 kg N ha⁻¹).

Smaller amounts of rainfall resulted in greater interception efficiencies (Figure 5). In intercropping systems in tropical dry land areas, tree and crop roots intercept all of the mineral N available in the soil profile, with subsoil N capture by trees of little importance due to restricted vertical N movements. As rainfall increases, simulated overall interception efficiencies decrease and subsoil capture becomes more important, the latter being highest in our case at ca.1200 mm rainfall. With greater rainfall, safety-net efficiencies decline because of the shortened nutrient residence time in the safety-net layer.

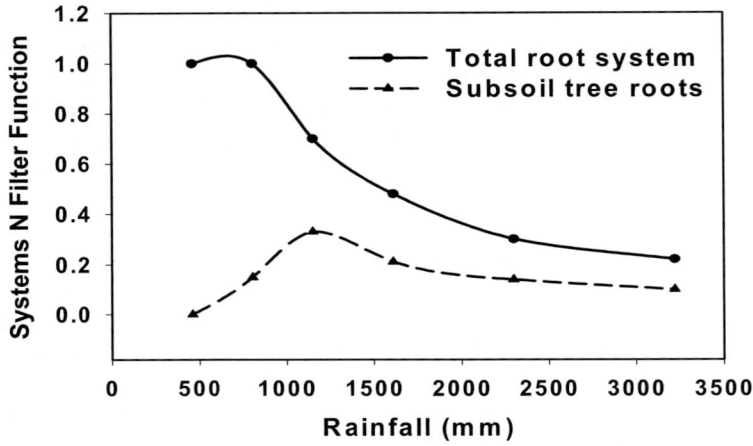

Figure 5. Simulated safety-net or filter efficiencies of combined crop and tree total root system or subsoil tree roots only (adapted from van Noordwijk & Cadisch, 2002).

Ideotype tree for simultaneous agroforesty systems

The need to design new agroforestry systems and hence the search for suitable tree species has led to formulations of an ideotype tree for soil fertility improvement and multiple other purposes. Our current results challenge the commonly held perception that the ideotype tree is a fast growing N_2 fixing multipurpose tree. While this is likely to be a good description for a fallow tree in a sequential agroforest, the ideotype tree for efficient safety-net functioning in simultaneous agroforestry systems has other attributes (Table 2). Based on our current experiences, any competition that trees impose on crops, severely affects crop growth as well as their N demand and hence their N retention capacity in the system. Fast growing trees are thus not optimal as they compete too strongly with crops for nutrients and water. The latter effect (water competition) is often underestimated in humid tropical zones, but is important in many cases because there are pronounced short dry spells even during the rainy season, which when occurring during flowering, severely affect crop yield and N sink strength. Although fast growing trees could be pruned more regularly, in reality farmers are reluctant to invest in pruning activities because of labour constraints. However, trees need to produce enough biomass to sustain soil organic matter levels and hence need to be drought resistant (e.g. as in the case of *peltophorum*) to compensate for slower growth during rainy season.

Table 2. *Current and proposed ideotype tree definitions for effective safety-net functioning and complementarity in resource use in simultaneous agroforestry systems.*

Current perception of ideotype tree	Proposed ideotype tree
Fast growing	Slow-medium growing but drought resistant
Deep rooting	Deep rooting + few active topsoil roots
N₂ fixing	Not necessarily N₂ fxing
High residue quality	Low-medium residue quality

In temperate and/or intensive systems some of the problems of competition with superficial rooting trees may be overcome by fertiliser additions and irrigation during dry spells, but these may add to the inefficiency of resource capture when total N sink strength is insufficient. Often a single tree may not fulfil all criteria. Mixing plant species for spatial and temporal complementarity is an option, which has proved useful in improving total resource capture (Gathumbi et al., 2002).

Spatial arrangements of agroforestry systems

Increasingly, it is recognised that the way agroforestry sytems are up-scaled also has a major impact on their filter efficiency at the landscape level. This requires new approaches, since lateral flows affect the behaviour of neighbouring plots. Suyamto et al. (2003), using a GIS based landscape-dynamic model (FALLOW), have assessed how the positioning of trees in the landscape affects the effectiveness of their watershed function in Sumatra. The results suggested that the maintenance of riparian forest had a more efficient filter function than trees allocated only to sloping zones, the ridge top zone, or positioned randomly in the landscape. The FALLOW approach to assessing the efficiency of agroforestry systems takes into account the interrelationship between management at different scales and brings into play the role of the associated community in identifying best options to improve nutrient recycling efficiency and minimise environmental pollution problems. Such exploratory model simulation approaches are useful for the design and management of new agroforestry systems in order to achieve multiple objectives.

Conclusions

Using trees to provide a biological root safety-net in intercropping systems will only be successful if the trees do not adversely affect crop growth. Thus slow growing, but drought resistant and deep rooting trees (with few active roots in the topsoil) are good at providing resource complementarity and root safety-net functions. Trees do not have to be N₂ fixers to provide an active safety-net but should provide slowly decomposing residues. Sensitivity analyses using the dynamic agroforestry model WaNuLCAS suggest that interception efficiency increases with tree root length density in the safety-net layer and with the thickness of this layer, but decreases as biological N₂ fixation increases.

References

Cadisch, G., de Willigen, P., Suprayogo, D., van Noordwijk, M. and Rowe, E. (2003). Catching and competing for mobile nutrients in soils. In: M. van Noordwijk, C. Ong and G. Cadisch (eds.) Belowground Interactions in Tropical Agroecosystems with Multiple Plant Components. CAB International, Wallingford, UK, (in press).

Gathumbi, S.M., Ndufa, J.K., Giller, K.E. and Cadisch, G. (2002). Do mixed species improved fallows increase above- and below-ground resources capture? Agronomy Journal, 94, 518-526.

Handayanto, E., Cadisch, G. and Giller, K.E. (1994). Nitrogen release from prunings of legume hedgerow trees in relation to quality of the prunings and incubation method. Plant and Soil, 160, 237-248.

Mekonnen K., Buresh, R.J. and Jama, B. (1997). Root and inorganic nitrogen distribution in sesbania fallow, natural fallow and maize fields. Plant and Soil, 188, 319-327.

Rowe, E. (1999). The safety-net role of tree roots in hedgerow intercropping systems. Department Biological Sciences. Wye College, University of London.

Rowe, E., Hairiah, K., Giller, K.E., van Noordwijk, M. and Cadisch, G. (1999). Testing the safety-net role of hedgerow tree roots by ^{15}N placement at different soil depths. Agroforestry Systems, 43, 81-93.

Rowe, E.C., van Noordwijk, M., Suprayogo, D., Hairiah, K., Giller, K.E. and Cadisch, G. (2001). Root distributions partially explain ^{15}N uptake patterns in *Gliricidia* and *Peltophorum* hedgerow intercropping systems. Plant and Soil, 235, 167-179.

Suprayogo, D. (2000). The effectiveness of the safety-net of hedgerow cropping systems in reducing mineral N-leaching in Ultisols. Department Biological Sciences. Wye College, University of London.

Suyamto, D.A., van Noordwijk, M. and Lusiana, B. (2003). FALLOW model: assessment tool for landscape level impact of farmer land use choices. In: MODSIM proceedings, Townsville, Australia, 6p.

van Noordwijk M. and Cadisch, G. (2002). Access and excess problems in plant nutrition. Plant and Soil, 247, 25-40.

van Noordwijk, M., Lawson, G., Soumare, A., Groot, J.J.R. and Hairiah, K. (1996). Root distribution of trees and crops: Competition and/or complementarity. In: C.K. Ong and P. Huxley (eds). Tree-Crop Interactions. CAB International, Wallingford, UK, pp 319-364.

van Noordwijk, M. and Luisiana, B. (1999). WaNuLCAS a model of water, nutrient and light capture in agroforestry systems. Agroforestry Systems, 43, 217-242.

van Noordwijk, M., Sitompul, S.M., Hairiah, K., Listyarini, E. and Syekhfani, M. (1995). Nitrogen supply from rotational or spatially zoned inclusion of Leguminosae for sustainable maize production on an acid soil in Indonesia. In: R.A. Date (ed). Plant Soil Interactions at Low pH. Kluwer Academic Publishers, Netherlands, pp 779-784.

Comparison of the efficiency of different catch crops on potentially leachable nitrate

P.Y. Bontemps, R. Lambert, C. Devillers and A. Peeters
Laboratoire d'Ecologie des Prairies, Université Catholique de Louvain, Belgium

Introduction

Catch crops have proved to be effective for maintaining low NO_3^- profiles after the harvest of certain risky crops. Traditionally, Belgian farmers grow mustard (*Sinapsis alba*) as a catch crop, in part because of the relatively low cost of its seeds. However, this crop has a drawback: in certain conditions, it encourages the proliferation of nematodes harmful to sugar beet. The aim of this trial was to test other catch crops (Monfort, 1985; Laurent et al., 1995; Pousset, 2000).

Materials and methods

The trial was established on a farm in Walloon Brabant, after a pea crop which had received no N fertiliser. The different species sown were *Brassica napus* (Bn), *Brassica rapa* (Br), *Lolium multiflorum* (Lm), *Phacelia tanacetifolia* (Pt), *Raphanus sativus* (Rs) and *Secale cereale* (Sc). A bare plot was used as control. The trial was organised in 4 replicates.

The catch crops were sown on 21 August 2001 with a cereal seeder. Nitrate profiles were sampled on 21 November that year. Sampling probes under vacuum were used to sample soil solution: three samples were taken at a depth of 150 cm in each plot in order to obtain an average sample for each treatment.

Results and discussion

Table 1 and Figure 1 show that the control soil contained much NO_3^-; the various crops are classified by decreasing order of effectiveness. There was a very wide range of results, from 17 to over 190 kg $N-NO_3$ ha^{-1} in the profile to 150cm depth. It appears that the catch crops ensure an average reduction of two-thirds of the N residues, the average residual amount being 57 kg $N-NO_3$ ha^{-1} under catch crops. In addition to being the two most efficient N traps, *Raphanus sativus* and *Brassica rapa* are also the crops with the most consistent results, with a standard deviation of < 5 kg $N-NO_3^-$ ha^{-1}.

Table 1. Results of the NO_3^- profiles (kg $N-NO_3^-$ ha^{-1}).

	Control	Lm	Sc	Pt	Bn	Br	Rs
Mean	193.2	94.1	83.8	80.2	42.7	22.3	17.0
Standard deviation	32.7	20.5	20.5	17.5	10.5	3.1	2.7
Fisher test	a	b	bc	bc	bc	c	C

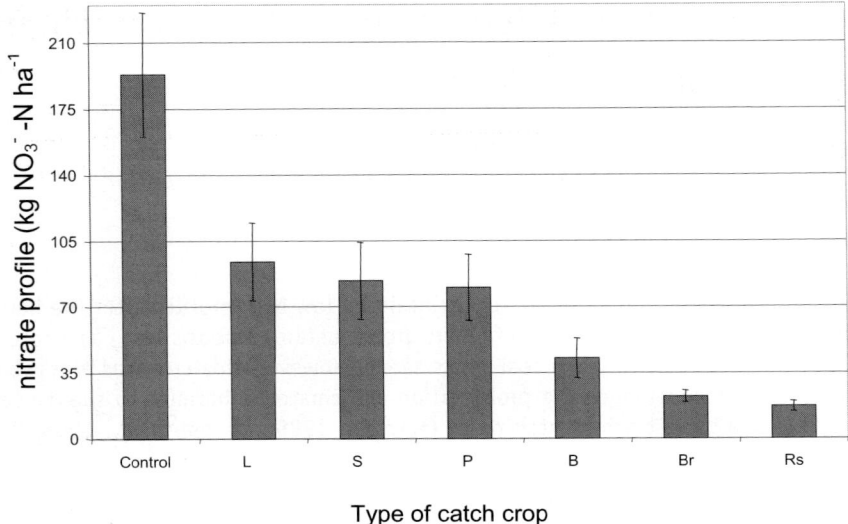

Figure 1. Results of the NO_3^- profiles for different catch crops.

Conclusion

In conclusion, the *Brassicaceae* were the most effective catch crops in this trial, followed by *Phacelia* and then by grasses. *Raphanus sativus* is especially interesting under Belgian conditions because it is one of the most efficient catch crops and anti-nematode varieties are available.

References

Laurent, F., Machet, J.M., Pellot, P. and Trochard, R. (1995). In: F. Laurent (ed). Azote et interculture Nitrogen and intercropping. Institut Technique des Céréales et des Fourrages, 38-49.

Monfort, B. (1985). La technique des engrais vert en Belgique, CARABE, 55 pp.

Pousset, J. (2000). Engrais verts et fertilité des sols, Editions Agridécisions, Paris, 287 pp.

Inverse stochastic modelling of water and N drainage from lysimeters

M. Decrem[1,2], K.C. Abbaspour[2], J. Nievergelt[1], F. Herzog[1] and W. Richner[1]
[1]*Federal Research Station for Agroecology and Agriculture (FAL Reckenholz), CH-8046 Zürich, Switzerland*
[2]*Federal Institute for Environmental Science and Technology (EAWAG), CH-8600 Dübendorf, Switzerland*

Introduction

Losses from agricultural diffuse sources are, in most cases, the dominant N source in the catchments of the Swiss Plateau. To estimate the effects of the Swiss policy measures aimed at reducing N leaching from arable land, a project has been launched to develop a computer-based system capable of estimating N leaching losses at the landscape scale. In this study, a process-based simulation model was calibrated against N discharge measurements from lysimeters, applying an inverse modelling method that allows for the estimation of the model's predictive uncertainty.

Materials and methods

A data set from a long-term, ongoing lysimeter experiment at FAL Reckenholz was used in this study (Nievergelt, 2002). The 12 lysimeters (2.0 m in diameter, 2.5 m deep) are used as free draining systems where the amount and quality of soil water discharge is monitored. For simplicity, this paper examines only the case of lysimeter 10. The soil type under investigation is classified as a Cambisol derived from moraine deposits, having a loamy soil texture. The crop rotation included cereals, potatoes, rape, sugar beet, temporary grassland, and phacelia used as a catch crop, and can be considered representative for normal cropping practices. Only mineral fertiliser was applied, in amounts according to the guidelines of the Swiss Federal Office for Agriculture. The soil was cultivated manually. Leachates were collected every 14 days and analysed by standard methods to determine the concentration of NO_3^-. To estimate N exports, plant samples were collected at harvest and analysed for N content. Standard daily weather data were obtained from a nearby meteorological station, for the simulation period of Jan. 01, 1990 to Dec. 31, 1999.

The process-orientated dynamic model LEACHM (Hutson, 2003) was applied to simulate the fate of N inputs in the soil across the period of interest. The 'tipping bucket' capacity-based technique was used to define water movement through the soil profile. Some input values were taken from the literature (e.g. the rooting depths of the crops), or were based directly on observations (e.g. the potential N uptakes). However, the bulk of the input values related to the water regime and the N turnover were derived indirectly from calibration against the time series of cumulative drainage outflow and N drainage losses.

The Latin Hypercube Inverse Modelling (LHIM) method was used for the calibration and uncertainty estimation of the model. In the application of LHIM, a large number of model runs are made, each parameterised with random sets of values from Latin Hypercube sampling on uniform distributions across the range of each one. Each run is assessed by some chosen objective function, calculated from a comparison of observed data and simulated responses. At each time step, the simulations are ranked to form the cumulative distribution of the output variable from which chosen percentiles can be selected to represent the model uncertainty. An approximation of the variance-covariance matrix is used to determine the 95% confidence interval of the optimising parameters. The parameter ranges are updated accordingly, in a way that they are always centred on the current optimal values. This procedure is iterated a few times until (i) the prediction uncertainty represented by the 2.5th and 97.5th percentiles encompasses (or respects) a substantial amount of the measured data (i.e. more than 90%), and (ii) the average prediction uncertainty is smaller than the standard deviation of the measured data.

Application and results

Water and N drainage were determined by a total of 37 model parameters, including the initial water and N contents of the layered soil profile. A range of plausible values was assigned to each of the parameters to represent the uncertainty of the *a priori* estimates. Each iteration comprised 1000 model runs which were assessed by a 2-variable objective function, based on the root mean square error (RMSE) involving both measured cumulative water and N discharge fluxes. The model structure, with its simple 'tipping-bucket' concept, was able to reproduce the measured cumulative drainage outflow and the derived uncertainty bounds encompassed most of the observed data (Figure 1a). However, the calculated percentiles indicated that large uncertainties were associated with the predicted cumulative N discharge flux (Figure 1b). The inability to capture an important N discharge peak (Figure 1b), which followed ploughing of temporary grassland, resulted in uncertainty bounds for the end of the simulation period amounting to 300 kg N ha^{-1}, of the same order as the measured value.

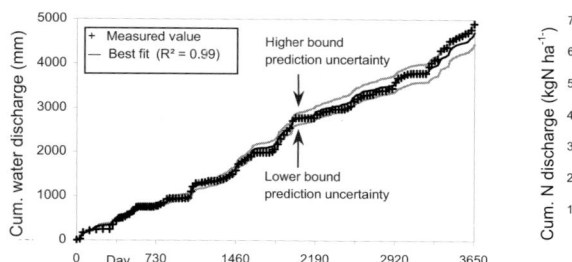 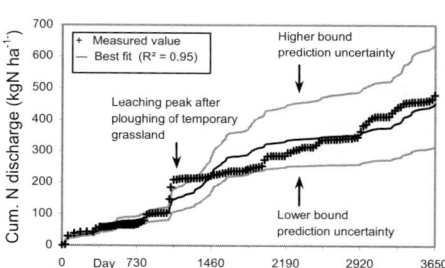

Figure 1. Prediction uncertainty and best fit for cumulative water and N discharge fluxes.

Conclusions

LEACHM is not intended to simulate N turnover in soil after incorporation of grassland. To fully account for the increased N leaching losses that arise from the ploughing of temporary grassland, modifications in both the model structure (e.g. addition of a more labile soil mineralisable N pool) and the application of LHIM (e.g. introduction of weight coefficients in the objective function and calibration against rates instead of cumulative values) are necessary. These modifications will be implemented in the next round of our modelling study.

References

Hutson, J.L. (2003) LEACHM - Model description and user's guide. School of Chemistry, Physics & Earth Sci., The Flinders University of South Australia, Adelaide SA 5001.

Nievergelt, J. (2002). Nitrat und Fruchtfolgen 20 Jahre lang beobachtet (Nitrate leaching in a crop rotation from 1981 to 2000). Agrarforschung, 9, 28-33.

Nitrogen concentrations in an intensively farmed livestock catchment

O. del Hierro[1], M. Pinto[1], A. Artetxe[1] and A. del Prado[2]
[1]*NEIKER A.B., Berreaga, 1, 48160 Derio, Bizkaia, Spain*
[2]*IGER, North Wyke Research Station, EX20 2SB Okehampton, UK*

Introduction

In intensively farmed areas, the high input of nutrients can result in large N and P losses from land and have adverse effects on groundwater, surface water and the atmosphere (Neeteson, 2000). The aim of this study was to evaluate the influence of an intensive dairy farm on the water quality of the Iñola River.

Materials and methods

The study was conducted from 2000 to 2002 on a 9.7 km^2 subcatchment (Figure 1) of Urrunaga watershed (121 km^2), which is one of the main sources of freshwater to the Basque population. The average rainfall is 1200 mm yr^{-1}. A dairy farm (200 ha of grassland, 900 cows, 10,000 l milk cow^{-1}yr^{-1}, clay soils) was selected in order to evaluate its influence on the water quality of the nearby rivers. Water sample points were divided into forestry (3, 5), forestry-livestock (6, 10), livestock (1, 4, 8, 9, 13, 15, 21) and point sources (7, 12, 14) (Figure 1). The results will be focused on SP 3 (forestry before entering the farm), SP 1 and SP 4 (as the river goes through the farm) and SP 9 (at the end of the farm and previously diluted by the forestry, SP 5).

Results and discussions

The high rate of N applications (370 kg N ha^{-1} yr^{-1}) affected the river N concentrations (Table 1). The manure application management is by far the most important factor controlling river runoff and drainage water quality (Eghball & Gilley, 1999). The study showed a N spatial pattern

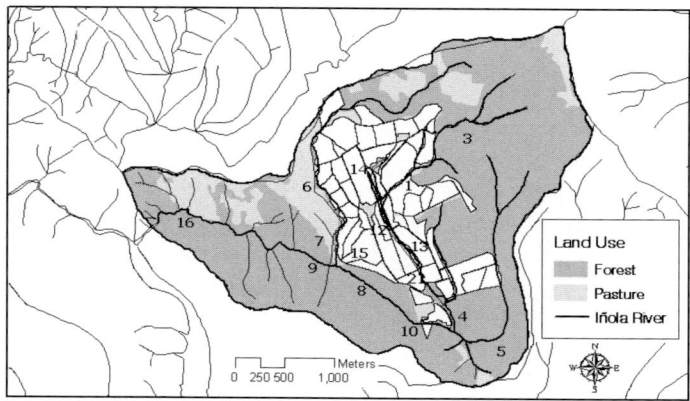

Figure 1. Land uses and location of water sample sites in the Iñola River subcatchment.

with increasing N concentrations as the distance to the farm decreased, recording the lowest N concentrations in the forestry sampling points (3 and 5). The results showed a seasonal pattern in stream N concentration and a spring peak was observed due to mineral fertilisation. Furthermore, an autumn peak (after first heavy autumn rains), following high summer manure applications, was observed during the study period (Figs. 2 and 3).

Table 1. Flow-weighted mean concentrations in the river sampling sites during the study period and comparison with the EC maximum admissible level for drinking water (European Community, 1980).

SP	NO_3^--N (mg l^{-1})			NH_4^+-N (mg l^{-1})			NO_2 -N (mg^{-1})			TKN (mg l^{-1})		
	Min	Mean	Max	Min	Mean	Max	Min	Mean	Max	Min	Mean	Max
3	0.24	0.94	1.51	0.02	0.14	5.58	<0.01	0.02	0.44	0.04	0.96	3.26
1	0.25	2.91	8.40	0.01	0.23	5.67	0.01	0.03	0.12	0.08	1.21	3.14
4	0.59	8.64	21.61	0.02	0.36	5.17	0.01	0.09	0.30	0.18	1.85	4.26
9	1.02	4.08	11.54	0.02	0.55	4.94	0.01	0.11	0.67	0.10	1.66	4.38
EC limit for drinking water (EC 80/778):												
	11.3 mg NO_3^--N l^{-1}			0.38 mg NH_4^+-N,			0.03 mg $NO_2^--N^{-1}$,			1.0 mg TKN l^{-1}		

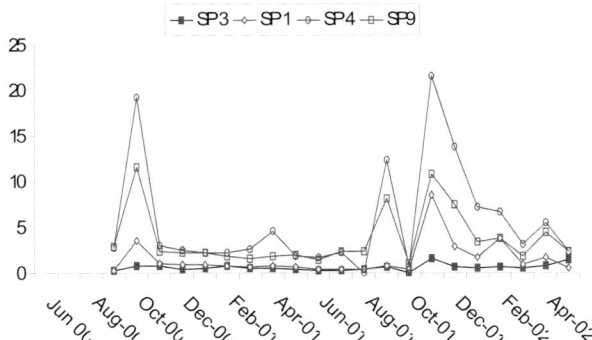

Figure 2. Seasonal pattern in concentration of NO_3^--N.

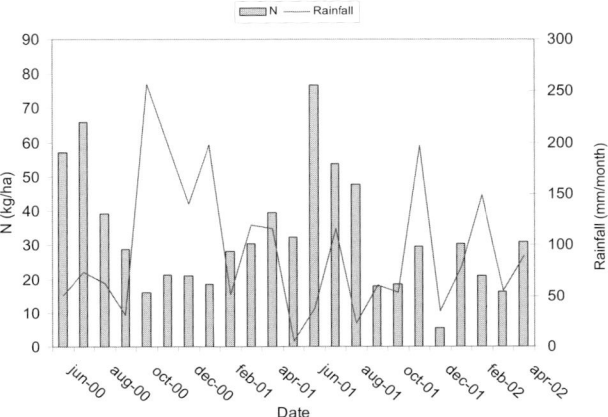

Figure 3. Monthly Rainfall and N fertilisation.

There was a strong association between stream N concentrations and farm management, with increasing N concentrations as the river goes through the farm, indicating a need for correct manure applications management.

References

Eghball, B. and J.E. Gilley. 1999. Phosphorus and nitrogen in runoff following beef cattle manure or compost application. Journal of Environmental Quality, 28 p-1201-1210.

European Community. 1980. Council directive relating to the quality of water for human consumption. EC 80/778. Official Journal of European Community, L229,11-29.

Neeteson, J.J (2000). N and P management on Dutch dairy farms: legislation and strategies employed to meet the regulations. Biology and Fertility of Soils, 30, 566-572.

NGAUGE DSS as a tool to assist UK dairy farmers to comply with EU nitrate legislation

A. del Prado, L. Brown and D. Scholefield

Institute of Grassland and Environmental Research, North Wyke Research Station, Okehampton, EX20 2SB, UK

Introduction

NGAUGE DSS (Brown et al., 2003) was utilised as a tool to study economic and environmental implications of changing N inputs to dairy farms within UK NVZs from current levels to those giving compliance with 170, 230 and 250 kg N ha^{-1} limits on 'manure-N'. Abatement strategies and their impact on possible 'pollution-swapping' effects were fully investigated.

Scenarios and parameters

A typical UK dairy farm (Jarvis, 1993) simulation was run for 6 types of soil/drainage/management and for 5 different UK regions (Devon, Cheshire, 'Dry East Midlands', 'Wet East Midlands' and West Wales). Slurry application and mineral fertiliser were simulated, following typical timetable patterns (Smith et al., 2001; MAFF, 2000). The DSS was run over a N fertiliser range of 0-420 kg N ha^{-1}, resulting in fertiliser response graphs for N leaching and plant production. Results from current UK fertiliser uses (low, average and high) were compared with those obtained under NVZ compliance rules (230 and 170 kg N ha^{-1} manure-N).

Strategic timing of manure application and abatement measures, such as mineral fertiliser optimisation (NGAUGE optimisation procedure), were also simulated. NGAUGE is capable of optimising fertiliser input, according to several criteria and in relation to herbage, fertiliser or N-loss targets. The model predicts the optimum distribution of monthly fertiliser input to cut or grazed fields on an annual basis, according to the criterion and target set by the user. The criterion used in the study was 'highest herbage with best N-use efficiency ratio'. This ratio is defined as 'total plant N/total N losses'. Exporting slurry to beef farming systems, possible benefits in using FYM rather than slurry and surface extensification were also studied.

Possible 'pollution-swapping' was investigated through optimisation of fertiliser inputs: N losses from denitrification, leaching and NH$_3$ volatilisation were compared for a limited number of scenarios. NGAUGE was also utilised in order to study the potential to optimise N use at a catchment scale.

Results

Use of NGAUGE to simulate current practice and transition to NVZ rules indicated that for compliance at NVZ$_{170}$, the annual fertiliser rate would need to be reduced to ca. 133 kg N ha^{-1} (averaged over the farm) and this would result in an average reduction in DM yield of 15% and in the amount of N leached of 35%. Current average fertiliser rates are approximately equivalent to NVZ$_{220}$. Soil conditions, weather and previous management history had major influences on

N leaching. The model predicts that it would not be possible to comply strictly with NVZ_{170} (peak N concentration <11.3 mg l^{-1}) on sandy soils in any grassland region of the UK under 'normal management' and average weather conditions, but it would be possible to comply if mean N concentration was considered. For clay soils with poor drainage, peak N concentrations were predicted to be <11.3 mg l^{-1} with NVZ $_{170,\ 230}$ and $_{250}$ kg N ha^{-1}. Strategic timing of manure application only improved N leaching results when manure was applied in March, instead of January (at NVZ_{170}, sandy and clay soils showed a reduction of 16 and 31% in N leached with a small gain in dry matter yield). The optimisation procedure showed some scope for reducing N leaching through a different mineral fertiliser distribution in sandy soils (20-25% in N leaching with no loss in production) with some associated, but acceptable 'pollution-swapping' with it. On the other hand, on clay loam soils the reductions could only be achieved by reducing denitrification.

Exporting slurry from dairy farms, following current N fertiliser level use to beef grassland outside the NVZ zones, resulted in a reduction of N leached up to compliance levels, particularly when combined with optimisation of fertiliser and best manure application distribution. The model also predicted that it would be possible to reduce leaching by 13 and 23% of that from the same system operating at NVZ_{170}, under conventional management, by exporting 30 or 50% of slurry, respectively, coupled with mineral fertiliser optimisation. This would have little effect on herbage production levels. If the manure was exported as FYM rather than slurry, NGAUGE predicted that a small reduction in N leaching (up to 5%) would be achievable without yield penalties. Two example strategy -scenario combinations of beef systems for manure importing were considered, both with sandy soils. Both strategies enabled compliance with the EC Nitrates Directive on the dairy and beef farms, but increased leaching, as expected, on the beef farms. NGAUGE was also used to optimise N use at greater scales between farms within the catchment. The effects of fertiliser distribution pattern between farms was investigated, based on N inputs to and outputs from 10 'model' dairy farms (each of 50 ha), set in a sandy catchment (1000 ha) in Devon. Nitrogen use at the catchment scale was more efficient with equal fertiliser inputs to all farms.

Acknowledgements

The authors would like to thank to DEFRA for funding this project.

References

Brown, L., Scholefield, D., Jewkes, E.C., del Prado, A. and Lockyer, D.R. (2003). NGAUGE: A decision support system to optimise N fertilisation of UK grassland for economic and/or environmental goals. (This volume).
Jarvis, S.C. (1993) Nitrogen cycling and losses from dairy farms. Soil Use and Management, 9, 99-105.
MAFF (2000). Fertiliser recommendations for agricultural and horticultural crops (RB209). HMSO, London.
Smith, K.A., Brewer, A.J., Crabb, J. and Dauven, A. (2001) A survey of the production and use of animal manures in England and Wales. 111. Cattle manures. Soil Use and Management, 17, 77-87.

Comparison between the risk of nitrogen leaching from temporary cut grassland and maize

B. Deprez, D. Knoden, H. de Blander, R. Lambert, C. Decamps and A. Peeters
Laboratoire d'Ecologie des Prairies, Université catholique de Louvain, Place Croix du Sud, 5 bte 1, 1348 Louvain-la-Neuve, Belgium

Introduction

Maize cropping has become increasingly popular in Belgium and now occupies some 9% of the Agricultural Area in Wallonia. Various factors can explain this increase: e.g. ease of cultivation, high yields, easy harvesting and conservation and good nutritive value. However, maize cultivation also has considerable drawbacks, especially in NO_3^- leaching during intercropping (Le Gall et al., 1990). The increase in maize cropping took place to the detriment of temporary cut grassland which, in contrast, provides an excellent protection against NO_3^- leaching while the cover is established. The purpose of this study was to compare the impact of a maize crop with that of a temporary grassland on the potential leaching of NO_3^-, while respecting the good agricultural practices for the two crops.

Methods

The trial lasted two successive years on a loamy soil. The land is divided into four complete random blocks. The treatments were a maize crop and two *Lolium multiflorum* (Lm) crops respectively, with one (Lm 1x) or two (Lm 2x) spreadings of slurry. The dressings used are shown in Table 1. The input from slurry was based on analysis, using an efficiency coefficient (% of total N available during the first year after spreading) of 50% for the cattle slurry and 60% for pig slurry. The N balance was calculated from the difference between inputs and outputs. The inputs included mineral fertilisation and the available fraction of the organic manure. The exports were calculated by multiplying the yields in dry matter by the N content of the forage crop. The NO_3^- profiles were measured in November at a depth of 1.5 m.

Table 1. N fertiliser rate, yields, N balances and NO_3^- profiles in November.

Crop	Year	Available N from slurry kg N ha^{-1} (a)	Mineral fertiliser kg N ha^{-1} (b)	N Input kg N ha^{-1} (c=a+b)	N Output kg N ha^{-1} (d)	N balance kg N ha^{-1} (c-d)	Nitrate kg NO_3^--N ha^{-1}	Yields t MS ha^{-1}
Maize	2000	162	90	252	230	22	92	18,4
	2001	183	40	223	226	-3	65	18,1
Lm 1x	2000	48	335	383	356	27	2	16,9
	2001	183	220	403	288	115	7	13,5
Lm 2x	2000	24+70	315	409	350	59	2	16,8
	2001	95+88	220	403	319	84	18	13,7

Lm 1x: single slurry addition; Lm 2x split slurry addition

Results and discussion

As shown in Table 1, the amount of NO_3^- in the soil at the end of the season was considerably greater in the maize plots than in the grassland. The grassland took up almost all the NO_3^- in the soil, even when the N application was much greater than the exports. This feature highlights the importance of N incorporation into the soil organic matter and uptake of N by grass in autumn. Having two applications of slurry did not affect the amount of NO_3^- in the profile (Figure 1).

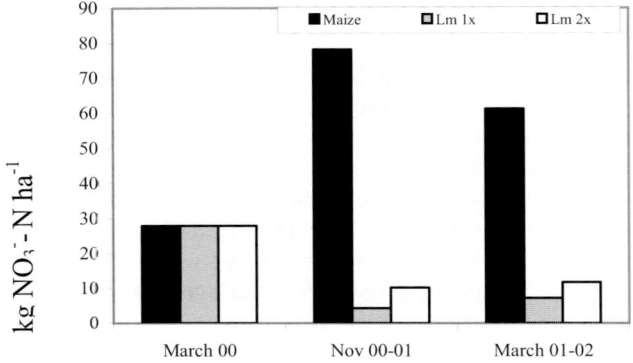

Figure 1. Total NO_3^- content at a soil depth of 1.5 m, Mean of the two years of culture.

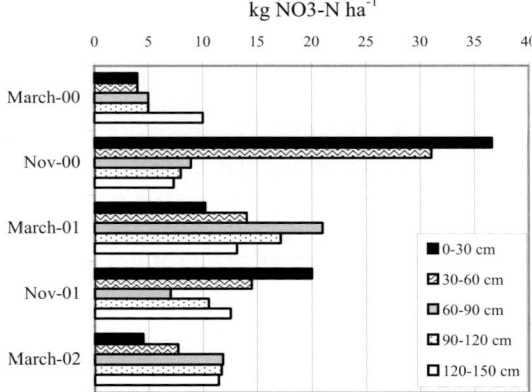

Figure 2. Nitrate profile under maize crop.

Because of continuous N mineralisation and low N uptake by maize at the end of the season, the soil profiles generally had a high NO_3^- content, even when fertilisation was well planned (Frankinet *et al.*, 2001, Lambert *et al.*, 2002). Moreover it is difficult to establish a catch crop after the maize harvest and thus, there is a high risk of NO_3^- leaching.

Acknowledgements

Financial support by the Ministère des Classes Moyennes et de L'agriculture is gratefully acknowledged (convention n° S-5978 section 2).

References

Frankinet, M., Renard, S., Dautrebande, S. and Casse, C. (2001). Gestion intégrée de l'azote en cultures arables et normes nitriques (Integrated management of nitrogen in arable cultures and nitric standards). Rapport d'activité final. Contrat de recherche S.S.T.C. NP/42/023. Plan d'appui scientifique à une politique de Développement durable. Appui scientifique à la recherche prénomative dans le secteur alimentaire dans un contexte de développement durable, 76p.

Lambert, R.,Van Bol, V., Maljean, J-F and Peeters, A. (2002). Projet pilote pour la protection des eaux de la nappe des sables bruxelliens. Rapport d'activité final (Project controls for the protection of groundwater of bruxellian's sands. Final report).

Le Gall, A., Legarto, J. and Pfimlin, A. (1997). Place du maïs et de la prairie dans les systèmes fourragers laitiers. III. Incidence sur l'environnement (Place of maize and grassland in the dairy forage systems.III. Impact on the environment). Fourrages, 150, 147-169.

Temporal and spatial denitrification patterns in nitrate retention by three riparian buffer zones

K. Dhondt[1,2], P. Boeckx[1], O. Van Cleemput[1] and G. Hofman[2]
[1]Ghent University, Laboratory of Applied Physical Chemistry, 9000 Ghent, Belgium
[2]Ghent University, Department of Soil Management and Soil Care, 9000 Ghent, Belgium

Introduction

The aim of this study was to investigate temporal and spatial denitrification patterns in NO_3^- retention along the topohydrosequences formed at the upland-wetland interface in three adjacent riparian zones with different vegetation cover (i.e. mixed vegetation, forest, grass).

Materials and methods

Within the three riparian zones, NO_3^--N and Cl^- analyses were performed on groundwater samples collected at different depths (0.5-1.0, 1.0-1.5 and 1.5-2.0 m) from a network of dipwells installed in several transects parallel to the slope of the topography. In each riparian site, soils were sampled ($n = 10$) at four depths (0-30, 60-90, 120-150, 180-210 cm). A composite sample was prepared for each depth to determine lateral and vertical stratification of denitrification enzyme activities (DEA) at three different temperatures (5, 10 and 15°C). The acetylene inhibition method with a laboratory incubation of soil slurries amended with 50 µg NO_3^--N g^{-1} soil was employed.

Results and discussion

During the growing season, the topohydrographic landscape setting resulted in very effective groundwater NO_3^- retention by the three riparian zones, irrespective of the vegetation cover (Figure 1 and Figure 3a-b). At soil temperatures of 10-15°C, DEA in each soil layer was high and clearly related to the amount of organic C (Figure 2 and Figure 4a-b). In winter time, the riparian zones lost some of their buffering capacity towards groundwater NO_3^-. There was no plant uptake and sub-surface DEA was highly suppressed at 5°C.

Conclusion

This study revealed that the topohydrographic landscape setting, together with subsurface denitrification in C enriched soil horizons, control groundwater NO_3^- fluxes in riparian zones.

Acknowledgements

This research was funded by the Flemish Impulse Program for Nature Development of the Flemish Community (Research Project AMINAL/NATUUR/VLINA/ 9904) and by the Special Research Fund (BOF) of Ghent University.

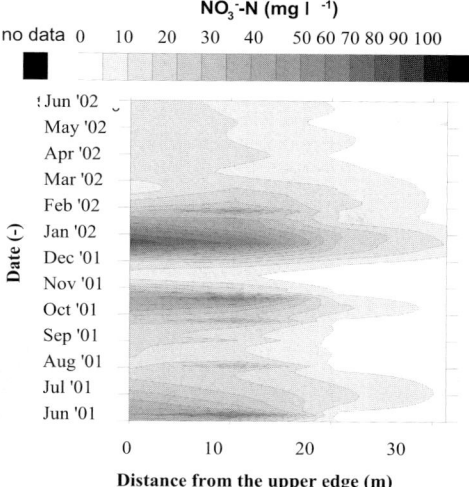

Figure 1. Contour map of groundwater NO₃⁻-N concentrations at a depth of 1.0-1.5 m in the mixed vegetation riparian zone.

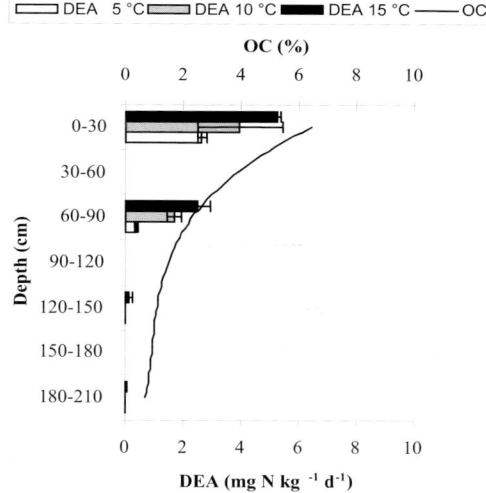

Figure 2. Variation of organic C and temperature dependent DEA in the mixed vegetation riparian zone.

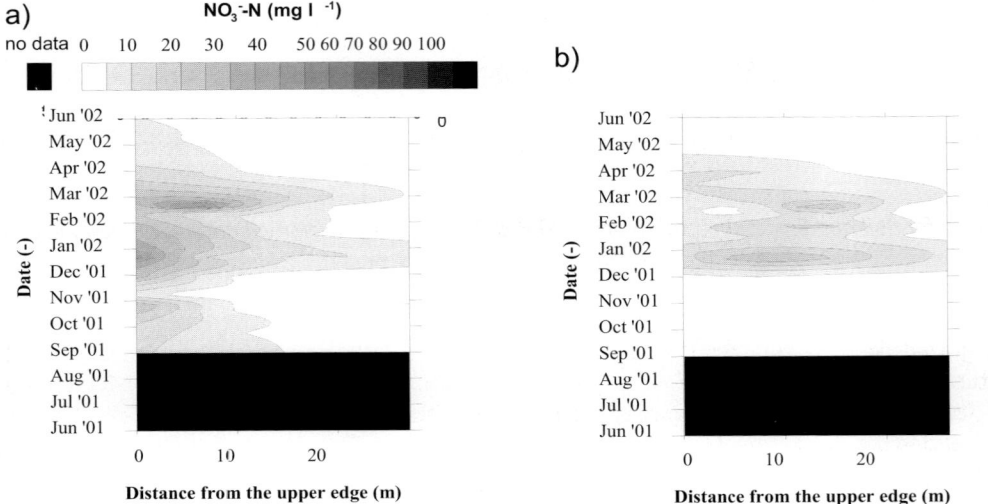

Figure 3. Contour map of groundwater NO_3^--N concentrations at a depth of 1.0-1.5 m in (a) the forest, and (b) the grass riparian zone.

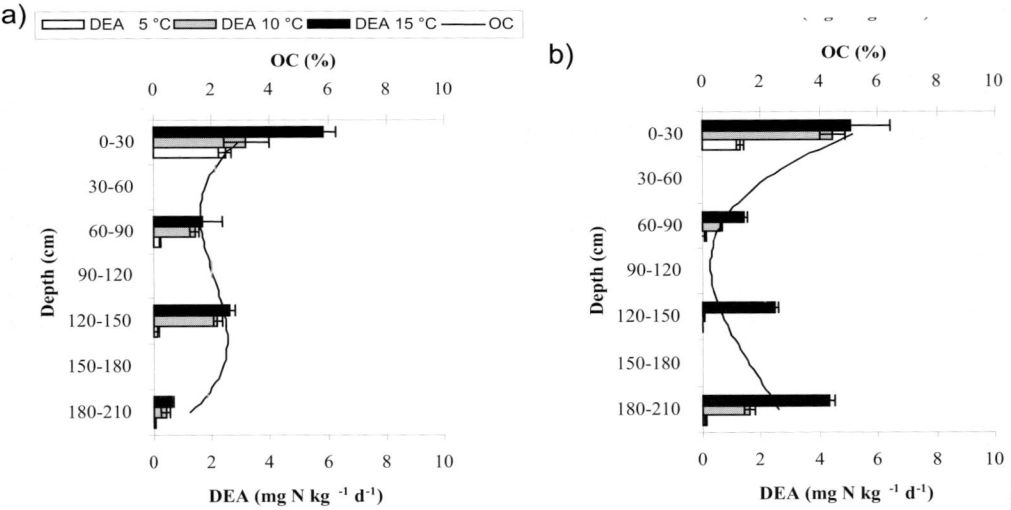

Figure 4. Variation of organic C and temperature dependent DEA in (a) the forest, and (b) the grass riparian zone.

Effective reductions in nitrate leaching and nitrous oxide emissions by the use of a nitrification inhibitor, dicyandiamide (DCD), in a grazed and irrigated grassland

H.J. Di and K.C. Cameron
Centre for Soil and Environmental Quality, PO Box 84, Lincoln University, Canterbury, New Zealand

Introduction

In grazed dairy pasture systems, a major source of NO_3^- leached and N_2O emitted is the N returned in the urine from the grazing animal (Di & Cameron, 2000; 2002a, b, c; Cameron et al., 2002). The objective of this study was to use lysimeters to measure directly the effectiveness of treating soil with a nitrification inhibitor, dicyandiamide (DCD), in decreasing NO_3^- leaching and N_2O emissions in a grazed dairy pasture under irrigation.

Materials and methods

The soil used was a free-draining Lismore stony silt loam (Pallic orthic brown soil; Udic Haplustept loamy skeletal) from an existing dairy farm in mid-Canterbury, in the South Island of New Zealand (171⁰43'E, 43⁰45'S). The pasture was a mixture of perennial ryegrass (*Lolium perenne* L.) and white clover (*Trifolium repens* L.). Undistorbed soil monolith lysimeters, 50 cm diameter and 70 cm deep, were collected and installed in a field-lysimeter facility for the study. Treatments on the lysimeters included applications of urea at the rate of 200 kg N ha^{-1} yr^{-1}, split into 4-8 applications, and dairy cow urine at 1000 kg N ha^{-1} yr^{-1} in single applications. The urine rate was to simulate the N loading rate under a dairy cow urine patch. To assess the effectiveness of DCD in decreasing NO_3^- leaching and N_2O emissions, DCD was either applied or withheld in paired urine treatments. DCD was applied in solution immediately after each urine application at the rate of 15 kg ha^{-1}. Leachates were collected and analysed for NO_3^-, NH_4^+ and NO_2^-. Nitrous oxide emissions were determined using a closed chamber method and analysed on a GC. Herbage was cut periodically and dry matter yield and N offtake were determined

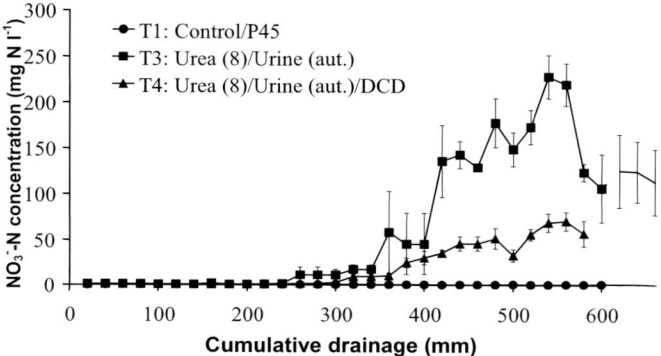

Figure 1. Nitrate N concentration in the drainage water from lysimeters.

Results

Nitrate N concentration in the drainage water from the lysimeters was significantly decreased by the application of DCD (Figure 1). The use of DCD decreased NO_3^--N leaching by 76% for the urine N applied in the autumn, and by 42% for urine N applied in the spring, giving an average reduction of 59%. This would reduce the NO_3^--N leaching loss in a grazed field from 118 to 46 kg N ha^{-1} yr^{-1}. The NO_3^--N concentration in the drainage water would be reduced accordingly from 19.7 to 7.7 mg N l^{-1}, with the latter being below the drinking water guideline of 11.3 mg N l^{-1}. Emissions of N_2O were decreased by 76% after urine application in the autumn (from 27 kg N_2O-N ha^{-1} without DCD to an average of 6 kg N_2O-N ha^{-1} with DCD over the 6 months experimental period). Nitrous oxide flux was decreased by 78%, following urine application in the spring (from 18 without DCD to 4 kg N_2O-N ha^{-1} with DCD over the 3 month period) (Figure 2). In addition to the environmental benefits, the use of DCD also increased herbage production by more than 15%.

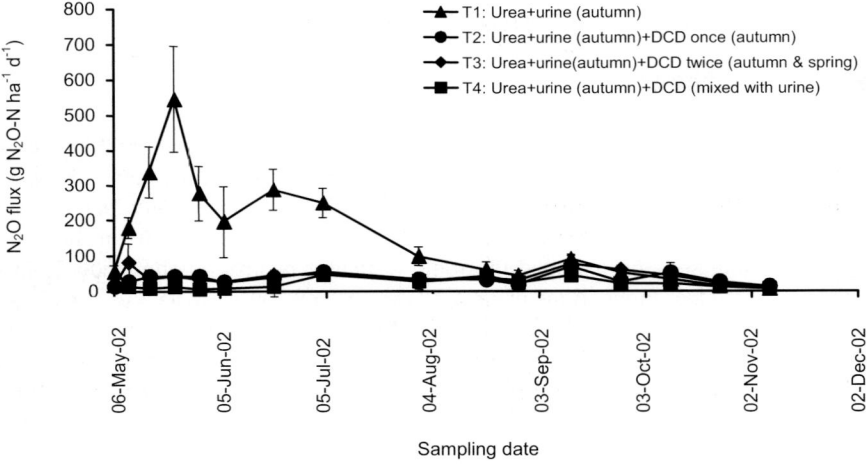

Figure 2. Nitrous oxide emissions as affected by the treatment of DCD.

Conclusions

The use of DCD to treat the soil is extremely effective in reducing NO_3^- leaching and N_2O emissions from urine patches in a grazed dairy grassland system. The use of DCD therefore has the potential to make dairy farming more environmentally sustainable..

References

Cameron, K.C., Di, H.J. and Condron, L.M. (2002). Nutrient and Pesticide Transfers from Agricultural Soils to Water in New Zealand. In: Agriculture, Hydrology and Water Quality. P. Haygarth and S.C. Jarvis (eds). CAB International, Wallingford, Oxon, U.K. 373-393.

Di, H.J. and Cameron, K.C. (2000). Calculating nitrogen leaching losses and critical nitrogen application rates in dairy pasture systems using a semi-empirical model. New Zealand Journal of Agricultural Research, 43, 139-147.

Di, H.J. and Cameron, K.C. (2002a). Nitrate leaching in temperate agroecosystems: sources, factors and mitigating strategies. Nutrient Cycling in Agroecosystems, 64, 237-256.

Di, H.J. and Cameron, K.C. (2002b). Nitrate leaching and pasture production from different nitrogen sources on a shallow stony soil under flood irrigated dairy pasture. Australian Journal of Soil Research, 40, 317-334.

Di, H.J. and Cameron, K.C. (2002c). The use of a nitrification inhibitor, dicyandiamide (DCD), to reduce nitrate leaching from cow urine patches in a grazed dairy pasture under irrigation. Soil Use and Management, 18, 395-403.

Nitrate leaching and N_2-fixation in grasslands of different composition, age and management

J. Eriksen and F.P. Vinther
Danish Institute of Agricultural Sciences, Department of Agroecology, P.O. Box 50, 8830 Tjele, Denmark

Introduction

Cut grassland systems usually have high N efficiency and consequently low NO_3^- leaching, whereas the introduction of grazing animals increases the loss potential dramatically (Jarvis, 2000). It has been demonstrated that NO_3^- leaching from unfertilised grass-clover swards is lower than from mineral-N fertilised swards grazed by cattle. The possible explanation for this is that the N_2 fixation by pasture legumes is regulated by a natural feedback mechanism driven by soil inorganic N levels. The feedback mechanism acts as a limit to N inputs from legumes and consequently regulates the potential for N losses (Ledgard, 2001). Here we report 5 years of NO_3^- leaching data from four cropping sequences with different grassland frequency and management for both unfertilised grass-clover and fertilised grass. Effects of one year of N_2 fixation in 1, 2 and 8-year-old grass-clover pastures are reported.

Materials and methods

Four cropping sequences (long-term grazed, long-term cut, cereals followed by 1 and 2 year grazed leys) were established for both unfertilised grass-clover (perennial ryegrass, *Lolium perenne* L./white clover, *Trifolium repens* L.) and fertilised perennial ryegrass (300 kg N ha^{-1}). Each plot was equipped with ceramic suction cups and NO_3^- leaching was estimated during 1997-2002, as described by Eriksen (2001). In 2001, N_2 fixation was estimated in 1, 2 and 8-year-old grass-clover pastures using an enriched ^{15}N-dilution, method as described by Vinther & Jensen (2000). In the cropping sequence with cut grass, annual herbage production was determined during 1994-2001.

Results and discussion

For better comparison of years with different drainage, the annual flow-weighted mean N concentration (NO_3^- leaching per volume of drainage) was calculated (Figure 1). In cropping sequences with cereals, followed by re-seeded grass, parallel patterns were observed. Nitrate leaching decreased with increasing time from grassland ploughing and without any difference between systems with grass-clover and ryegrass. After reseeding of grass-clover, leaching continued, but decreased after the 1st year of grazing, followed by a small increase in the 2nd year, whereas reseeding of grazed ryegrass caused an increase in NO_3^- leaching in both the 1st and 2nd years compared with the last year of cereals. In grazed grassland, in production years 4 to 8, there was a distinct difference between the two sward types. In grass-clover, leaching per volume drainage decreased from a moderate level in year 4 to a very low level in years 5 to 8. This indicates a lowering of N inputs in grazed grass-clover swards with increasing age, since many studies have suggested that NO_3^- leaching from grass systems are

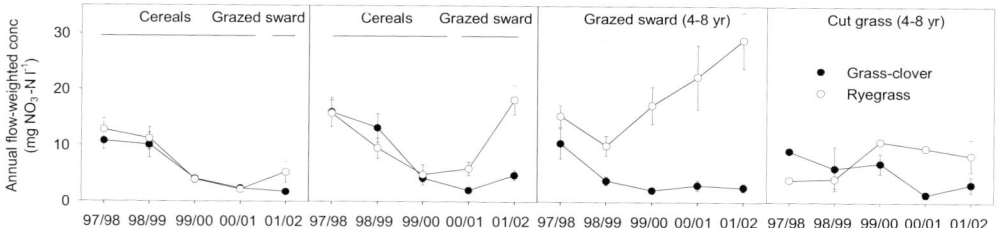

Figure 1. Flow-weighted NO₃⁻ concentration (NO₃⁻ leaching per volume drainage) in four cropping sequences with unfertilised grass-clover or fertilised ryegrass. Error bars: ±SE.

similar, at similar total N inputs, regardless of the source (Ledgard, 2001). In contrast to this, losses from grazed ryegrass increased considerably and almost linearly from year 5 to 8. In cut grassland in production years 4 to 8 NO_3^-, leaching was at a low level for both sward types, but with grass-clover, was greatest in year 4 and lowest in year 6 to 8. N_2 fixation was lower in 8-year-old swards compared with fully established 2-year-old swards as a consequence of lower dry matter production, lower clover content and a lower proportion of clover fixed N_2. During the lifetime of the old grass-clover swards, N_2 fixation dropped from 260 kg N ha^{-1} in the first three years (Eriksen, 2001) to 115 kg N in production year 8, in contrast to the fertilised grass sward that received 300 kg N ha^{-1} in all years. Thus, over the years, a considerable difference between sward types has built up which may, at least partly, explain the differences in NO_3^- leaching.

The deposition of the dung and urine by the cows during grazing is unknown, as for practical and economic reasons, one herd grazed all plots. However, some relationship is expected between grass production and manure deposition, simply because cows graze more in highly productive plots which inevitably leads to higher deposition. As dry matter production was maintained in older cut ryegrass swards and decreased by almost 50% in cut grass-clover swards, deposition of excreta is expected to be proportionally greater in ryegrass swards, assuming some similarity between cut and grazed grass production. This was confirmed by higher and considerably more variable NO_3^- concentrations in drainage from old grazed ryegrass swards compared with grass-clover, probably due to urination by the grazing cattle. These observations suggest that the high NO_3^- leaching losses in ryegrass were an indirect effect of fertiliser inputs on grass production leading to increased grazing and thus increased recycling of N in dung and urine to the sward.

Conclusions

In young swards, NO_3^- leaching from grass-clover and ryegrass systems was similar, but with increasing age leaching from the fertilised ryegrass increased dramatically, compared with a constant low level from the unfertilised grass-clover. Decreasing N_2 fixation in older swards may partly explain smaller losses in grass-clover than in fertilised ryegrass. The clover component appears to equalise differences in soil N availability in swards of different age.

References

Eriksen, J. (2001). Nitrate leaching and growth of cereal crops following cultivation of contrasting temporary grasslands. Journal of Agricultural Science, Cambridge, 136, 271-281.

Jarvis, S.C. (2000). Progress in studies of nitrate leaching from grassland soils. Soil Use and Management, 16, 152-156.

Ledgard, S.F. (2001). Nitrogen cycling in low input legume-based agriculture, with emphasis on legume/grass pasture. Plant and Soil, 228, 43-59.

Vinther, F.P. and Jensen, E.S. (2000). Estimating legume N_2 fixation in grass-clover mixtures of a grazed organic cropping system using two ^{15}N methods. Agriculture, Ecosystems and Environment, 78, 139-147.

Action programme and practices for reduction of nitrate from agricultural sources in Central Greece

T. Georgiou, T. Karyotis, I. Katsilouli, A. Haroulis, M. Toulios and G. Argyropoulos
National Agricultural Research Foundation, Institute for Soil Mapping and Classification, 1, Theophrastou Str., 41335, Larissa, Greece

Introduction

Thessaly is located in central Greece, occupies an area of about 14,000 km^2, and the cultivated land covers 36% of this. The eastern part of Thessaly is dry and climatic data show a water deficit in the period between March and October. Irrigation demands are predominantly covered (75.3%) by groundwater abstraction via a dense network of shallow and/or deep boreholes and (24.7%) by surface water. The aquifers in the pilot area have been classified as low to moderately vulnerable to N pollution and signs of pollution are shown in more than 65% of the monitored ground water boreholes where NO_3^- concentration was > 25 mg l^{-1}. Moreover, NO_3^- originating from fertilisers has caused eutrophication in the River Pinios delta. In the pilot project concerning NO_3^- pollution, more than 3,200 farmers were involved between 1996 and 2002. The objective of this study was to draw up an applicable action plan, in order to mitigate the NO_3^- pollution of ground water by means of a set of proper legislative measures and agricultural practices.

Methods

Thessaly has been designated as a 'vulnerable to nitrates pollution zone' and incentives were launched by the European Union in the frame of Directive 91/676 (EEC, 1991) and Regulation 1257/99 (EEC, 1999). The project was applied to the main irrigated crops such as: cotton, sugar-beet, wheat and maize. Soil samples were taken from the fields during the growth period, each year from 1996 to 2000. Water boreholes were also monitored during the irrigation period and both NO_3^- and NH_4^+ were determined. In accordance with the Nitrates Directive (EEC, 1991) and with the regulations in support of rural development (EEC, 1999), the following specific actions and practices have been adopted by the Greek Ministry of Agriculture: rotation schemes, land cover during the rainy winter period, use of fertilisers in balance with crop demands, N input reduction, frequent soil analysis, mandatory fertilisation plans, and restrictions of cultivation on steeply sloping soils and irrigation.

Results and conclusions

Analysis of experimental data suggested that fertilisation efficiency was increased and can be achieved by splitting the recommended amount of applied N fertiliser and/or by using fertigation. Additionally, further decrease seems to be feasible if water-soluble fertilisers are used, because of the higher N efficiency index.

The availability of labour is the key constraint in explaining farmers' reluctance to change practices (e.g. for splitting N applications or maintaining buffer strips). Research and studies are needed to investigate the socio-economic dimensions of Action Programmes and to provide the robust socio-economic assessment required for implementation of the regional policy. A cost-effectiveness analysis needs to combine expertise for assessing the impact of selected measures with those for assessing the costs of these measures. Results of the project have indicated that farmers have started altering their viewpoint on crop fertilization, towards a rational and more scientific consideration. It seems they have been convinced that N fertilisation reduction does not necessarily imply an equivalent reduction in yield, especially for cotton, which is the main crop. It is understood by farmers that N crop requirements are lower, especially in the initial stages of growth. A considerable reduction in N fertilisers has been recorded in the pilot project area since the initiation of the programme. The reduction in the applied N fertilisers was estimated at ca. 15,000 tonnes for the pilot area. Decreased N fertilisation in Thessaly has lowered the expenses for buying N fertilisers as presented in Table 1. The application of proper rotation schemes by using crops with less water demands in relation to other characteristics (soil conditions, slope, climate, etc.) may have an impact on the reduction of NO_3^- in groundwater. In practice, the overall reduction is higher considering that participating farmers have also reduced the amounts of applied fertilisers on the land they own outside the project area.

Several practices and restrictions are dictated in the action plan and comprise mainly the reduction of the pollutant input and saving irrigation water. Targeted measures may be undertaken that correspond to regional needs, farmers incomes, flexibility and social acceptance. Financial support for a more environmentally-friendly agriculture and dissemination of knowledge is necessary for the implementation of the Nitrates Directive 91/676/EEC.

Table1. Decreased N fertilisation and total saving value of fertilisers.

Fertiliser type	Decreased N application (t)	Saving in m
Pre-sowing application		
16-20-0	46,875	10.2
11-15-15	68,175	18.0
20-10-0	37,500	7.7
20-10-10	37,500	9.3
Fertigation		
46-0-0 (urea)	16,305	2.2
48-0-14	57,690	42.3
33-0-0	22,388	3.6
20-20-20	37,500	66.0

References

EEC, Council Directive 91/676/EEC concerning the protection of the water against pollution caused by nitrates from Agricultural Sources. (1991). Official Journal of European Communities L, 375, 1-5.

EEC, Council Regulation 1257/99 on support for rural development from the European Agricultural Guidance and Guarantee Fund. (1999) Official Journal of European Communities L, 160, 80pp. (in Greek).

A field study of nitrate leaching from tillage crops in Ireland

K.V. Hooker[1,2], K. Richards[1], C.E. Coxon[2]and R. Hackett[3]
[1]*Teagasc, Johnstown Castle, Wexford, Ireland*
[2]*Department of Geology, Trinity College Dublin, Dublin 2, Ireland*
[3]*Teagasc, Oak Park, Carlow, Ireland*

Introduction

Arable land, which accounts for 9.2% of the utilised agricultural area in Ireland, contributes NO_3^- to natural waters (Ryan et al., 2001). Neill (1989) found a direct relationship between river NO_3^- concentrations and the percentage of ploughed land in catchments in the south-east of the country. This is the region with the highest proportion of tillage land. However, apart from this there is little relevant Irish data, especially in terms of the potential of abatement strategies, to address the losses.

The present study aims to determine the contribution of arable farming practices to NO_3^- loads in this region, both at field and catchment scale. The ability of cover crops and alternative tillage methods to reduce NO_3^- leaching under Irish conditions will also be examined. Preliminary NO_3^- leaching results from spring barley and winter wheat, the two principal cereal crops in Ireland, are presented for the first drainage season.

Materials and methods

The study site was situated on a well drained clay loam soil overlying a gravel aquifer. There was four treatments, spring barley and winter wheat, each at high and low N input levels (Table 1). There were three replicate plots per treatment. Six ceramic suction cups were installed at a depth of 1.5 m in each plot (18 per treatment). Water samples were taken weekly from each cup between October - December 2002 and fortnightly thereafter. The samples were analysed for NO_3^--N and NH_4^+-N colorimetrically.

Results

Mean NO_3^--N concentrations were ≥ 10.0 mg l^{-1} for all treatments (Table 1). The variability of the data between replicate plots of the same treatment and within individual plots was high. To date, there are no apparent differences between treatments or temporal trends in NO_3^- concentrations.

Discussion and conclusion

It is hypothesised that the highly variable results may reflect localised denitrification and/or preferential flow in this clay loam soil. Monitoring will continue for a further two years, during which time the reasons for the spatial variability will be investigated further.

Table 1. Nitrate concentrations in soil solution under spring barley and winter wheat, treatments, October 2002 - Feb. 2003.

Treatment		Samples	NO_3^--N (mg l^{-1})			Standard
Crop	N input (kg ha^{-1} y^{-1})	(No.)	Max.	Min.	Mean.	Deviation
Spring barley	105	170	39.6	2.6	11.7	6.9
Spring barley	137.5	171	48.3	1.7	14.1	8.9
Winter wheat	187.5	192	50.2	1.8	14.4	9.3
Winter wheat	225	222	36.0	0.1	10.0	7.3

With the exception of the high N winter wheat treatment, mean concentrations breached the limit of 11.3 mg l^{-1} NO_3^--N set in the EU Nitrate Directive (European Community, 1991). Over 90% of all samples breached the 2.6 mg l^{-1} dissolved inorganic N limit used by the Irish Environmental Protection Agency (EPA) to classify eutrophic estuaries (EPA, 2001). It is clear from these first results that such NO_3^- concentrations at 1.5 m depth are a concern, particularly given the gravel aquifer and likely NO_3^- contribution to the adjacent River Barrow, whose estuary has been classified as eutrophic (EPA, 2001).

Acknowledgements

This project was funded by Teagasc, the Irish Agriculture and Food Development Authority.

References

European Community. (1991). Council directive concerning the protection of waters against pollution caused by nitrates from agricultural sources. EC 91/676. Official Journal of the European Community L375, 1-8.

EPA, Environmental Protection Agency. (2001). An assessment of the eutrophic status of estuaries and bays in Ireland. EPA, Wexford, Ireland.

Neill, M. (1989). Nitrate concentrations in river waters in the south-east of Ireland and their relationship with agricultural practice. Water Research, 23, 1339-1355.

Ryan, M., Sherwood, M., and Fanning, A. (2001). Leaching of Nitrate-N (NO_3-N) from cropped and fallow soil - a lysimeter study with ambient and imposed rainfall regimes. Irish Geography, 34, 34-49.

Dissolved organic nitrogen concentration in two grassland soils

D.L. Jones[1], J.F. Farrar[2] and V.B. Willett[1]
[1]*School of Agricultural and Forest Sciences, University of Wales, Bangor, LL57 2UW, UK*
[2]*School of Biological Sciences, University of Wales, Bangor, Gwynedd LL57 2UW, UK*

Introduction

It has been suggested that research investigating the dynamics of N in terrestrial environments may be fundamentally flawed because of the lack of consideration of a key cog in the N cycle, namely the uptake of organic N by plants (Chapin, 1995). In light of this statement, some authors have suggested that we need to thoroughly re-evaluate all the literature on ecosystem nutrient flows (van Bremmen, 2002). Despite these recent reports, however, the phenomenon of dissolved organic N (DON) uptake by plants is not new and has been considered many times during the last century (Jones & Darrah, 1994). The possibility that plants can circumnavigate the need to take up inorganic N by taking up low molecular weight DON appears plausible, however, direct evidence to support the significance of this process is clearly lacking. The aim of this study was to determine the concentration of dissolved N species in two contrastingly managed grassland ecosystems and to discuss these results in the context of plant N acquisition.

Materials and methods

Soil was collected from two contrastingly managed UK grasslands located at the University of Wales, Henfaes Experimental Station (4°01'W 53°14'N). The first grassland site was a high production sheep grazed Eutric cambisol with *Lolium perenne* and *Trifolium repens* vegetation which receives regular fertilisation (120 kg N ha^{-1} yr^{-1}; pH 6.32; 35 g C kg^{-1}; C-to-N ratio = 14). The second grassland site is a low production, sheep grazed Leptic podzol with *Festuca ovina* and *Cynosurus cristatus* vegetation and which receives no fertiliser inputs (pH 4.7; 53 g C kg^{-1}; C:N = 12). At each location, nine intact soil cores were taken with a 5 cm diameter stainless steel corer from a depth of 0 to 15 cm and stored separately in gas-permeable plastic bags at 4°C. Soil solution was then extracted within 24 h by the centrifugation-drainage technique of Giesler & Lundström (1993). Typically, 10 to 20 ml of soil solution was extracted from each soil sample (i.e. soil solution extraction efficiency of 30 to 60%). The soil solutions were stored frozen in polyethylene vials at -20°C to await analysis. Soil solutions were obtained on a monthly basis over a period of 1 year.

Soil solutions were analysed for dissolved inorganic N (DIN; NO$_3^-$ and NH$_4^+$) with a San-plus System segmented flow autoanalyser (Skalar Ltd., York, UK) while free amino acids were determined by the fluorometric procedure of Jones et al. (2002). Free amino sugars were determined by HPLC using the method of Jones (1986). Total dissolved N (TDN) in solution was determined with a Shimadzu TOC-V-TN analyser (Shimadzu Corp., Kyoto, Japan). DON was calculated as TDN minus DIN.

Results

As expected, the concentration of soluble N was approximately 3 fold higher in the high production grassland in comparison with the low production grassland (Table 1). The major statistical significant difference in soluble N between the two soils was the level of NO_3^- which was 10 fold greater in the fertilised grassland ($P < 0.05$). In contrast, the levels of NH_4^+ and DON were similar in both grassland types. At both sites, low molecular weight DON in the form of free amino acids comprised only a small proportion of the total DON pool (5-10% of the total DON) while free amino sugars (e.g. glucosamine, galactosamine) were below the detection limits of the HPLC technique (0.001 mg N l^{-1}) at practically all sampling dates.

Table 1. Soluble N in soil solution from two contrasting managed grassland soils (mg N l^{-1}). Values represent annual means ±SEM from a monthly sampling regime (n = 12) except amino sugars (n = 6).

	High production grassland	Low production grassland
NH_4^+	0.7 ± 0.1	0.8 ± 0.1
NO_3^-	15.2 ± 8.8	1.5 ± 0.4
DON	6.7 ± 0.5	5.1 ± 0.5
Free amino acids	0.5 ± 0.1	0.5 ± 0.2
Free amino sugars	<0.1	<0.1

Discussion

The findings from this study suggest that DON constitutes a major form of soluble N in grassland soils. Taken together with measurements of DON in freshwaters from UK catchments with similar soil types, it would suggest that leaching of DON from soil is a major N loss pathway particularly in low input agricultural systems (Willett et al., 2004). Our results suggest that while DON may be present in high concentrations, only a small proportion of this is present as free amino acids which are known to be plant available. Based on other evidence, we hypothesise that the majority of the unidentified DON is of high molecular weight and relatively unavailable to plants.

Conclusions

This study demonstrates the need to re-evaluate the potential role of DON in studies measuring soluble N losses and cycling in agricultural land. Because of its potential importance in plant nutrient uptake, further investigations of the types of DON forms which are available to plants are warranted. Furthermore, N cycling models should be adapted to explicitly include DON as a major model component, once the factors determining its regulation in soil are better understood.

References

Chapin, F.S. (1995). New cog in the nitrogen-cycle. Nature, 377, 199-200.

Giesler, R. and Lundström, U. (1993). Soil solution chemistry - effects of bulking soil samples. Soil Science Society of America Journal, 57, 1283-1288.

Jones, B.N. (1986). Amino acid analysis by *o*-phthaldialdehyde precolumn derivitization and reverse-phase HPLC. p 121-151. In: J.E. Shively (ed). Methods of Protein Microcharacterization: A Practical Handbook, Humana Press.

Jones, D.L., Owen, A.G. and Farrar, J.F. (2002). Simple method to enable the high resolution determination of total free amino acids in soil solutions and soil extracts. Soil Biology and Biochemistry, 34, 1893-1902.

Jones, D.L. and Darrah, P.R. (1994). Amino-acid influx at the soil-root interface of *Zea mays* L. and its implications in the rhizosphere. Plant and Soil, 163, 1-12.

van Breemen, N. (2002). Nitrogen cycle - Natural organic tendency. Nature, 415, 381-382.

Willett, V.B., Reynolds, B.A., Stevens, P.A., Ormerod, S.J. and Jones, D.L. (2004). Dissolved organic nitrogen regulation in freshwaters. Journal of Environmental Quality (in press).

Simulation with STICS soil-crop model of catch crop effects on nitrate leaching during the fallow period and on N released for the succeeding main crop

E. Justes[1], F. Dorsainvil[2], M. Alexandre[1] and P. Thiébeau[3]
[1]*INRA, UMR ARCHE, Auzeville, BP27, 31326 Castanet-Tolosan, France*
[2]*Université d'Etat, Faculté d'Agronomie et de Médecine, BP 1441, Port-au-Prince, Haïti*
[3]*INRA, Unité d'Agronomie Laon-Reims-Mons, BP 224, 51686 Reims cedex, France*

Introduction

The efficiency of catch crops to prevent NO_3^- pollution has been widely demonstrated (e.g. Meisinger et al., 1991). However, additional work is needed to optimise their management in order to simultaneously minimise water transpiration, maximise N removal from the soil profile and N release for the next crop. Using a crop model seems to be the most efficient solution. STICS has been built as a generic soil-crop model (Brisson et al., 1998); it was initially validated for wheat and maize and the lastest 5.0 version (Brisson et al., 2003) has been parameterised and validated for catch crops, such as white mustard and Italian ryegrass (Dorsainvil, 2002; Alexandre, 2002). The aim of this work was to analyse the impact of catch crops on NO_3^- leaching during the fallow period and on N release for the succeeding main crop using the STICS model.

Materials and methods

Various scenarios of management practices were simulated with and without a catch crop (straw incorporated at wheat harvest). Different French pedoclimatic conditions were carried out using climatic data of the last thirty years, in particular those of Brittany (Quimper: 48°N, 356°W; sandy-silt soil) and Champagne (Chalons: 48.6°N, 4. 2°E, chalky soil) regions; their climatic characteristics are shown in Table 1. Different input variables were tested and in particular: (i) date of catch crop emergence, (ii) date of incorporation, (iii) initial amount and distribution of mineral-N in soil, and (iv) soil depth (60, 90, 120 cm). Various output variables were analysed: (i) N leaching and NO_3^- concentration in water at the end of the drainage period (25/04), (ii) N release after catch crop incorporation, and (iii) mineral-N in soil at the end of winter (N_{min}15/02), at the sowing date of a spring crop (N_{min}25/04) and at the end of crop establishment (N_{min}15/07).

Results and discussion

Only results concerning scenarios for the 90 cm depth soil, the low level of initial mineral-N (40 kg N ha^{-1}) and for the optimal date of incorporation (01/12 for mustard in Chalons and 15/02 for Italian ryegrass in Quimper) are shown in Table 2, for median, first and ninth decile. Results of the simulations were in agreement with available experimental data (e.g. Meisinger et al., 1991). Thus, catch crops were found to be efficient in reducing N leaching in the two situations tested. Of course, the later the incorporation date the lower the NO_3^- concentration

Table 1. Annual climatic conditions for two French stations tested (ET = EvapoTranspiration).

Station (department)	Years	Mean Temperature °C	Rainfall Mm	Potential ET mm	(Rainfall - P.ET) mm
Chalons (51)	1975 to 1999	10.3	628	664	-37
Quimper (29)	1970 to 2000	11.4	1230	675	555

in drained water. However, the decrease in NO_3^- concentration by maintaining a catch crop is minor when destruction occurred after mid-December (results not shown).

A pre-emptive competition effect for N was simulated for a moderate rainy climate such as Chalons, i.e., less N was available for the next main crop after catch crop in comparison with bare soil, as shown by Thorup-Kristensen (1994). In fact, N_{min}15/04 and also N_{min}15/07 were lower after mustard than after bare soil, indicating that the sum of N released from residues, plus the reduction in N leaching by a catch crop, was less than the depletion of soil N_{min}15/02 because of catch crop N uptake. However, this negative effect was not obtained in the rainier conditions of Quimper, where N release from residues increased soil N_{min}15/07 because of high N leaching in bare soil, leading to very low amounts of NO_3^- in soil after winter.

Table 2. Simulated results with STICS. Numbers represent successively: first decile-median-ninth decile. * no drainage was simulated, thus no NO_3^- concentration could be calculated.

Pedo-climate (department)	Scenario tested	N leached (kg N ha⁻¹)	NO_3 in water (mg NO_3^- L⁻¹)	Residue N released (kg N ha⁻¹)	N_{min}15/02 (kg N ha⁻¹)	N_{min}25/04	N_{min}15/07
Chalons (51)	Mustard	0-**2**-5	*-**7**-15	18-**27**-32	28-**36**-43	50-**55**-63	84-**95**-109
	Bare soil	2-**21**-42	54-**72**-87		51-**70**-78	57-**74**-89	93-**113**-129
Quimper (29)	Ryegrass	30-**62**-91	32-**37**-43	15-**26**-34	14-**19**-25	53-**61**-76	101-**122**-144
	Bare soil	84-**122**-141	62-**82**-131		19-**31**-52	36-**44**-70	86-**102**-123

Conclusions

The efficiency of catch crops to prevent NO_3^- leaching was confirmed in a larger range of tested conditions, indicating that catch crops are always useful, even when they are destroyed in early November (Alexandre, 2002). Moreover, in order to avoid, or strongly reduce the N pre-emptive competition effect, the catch crop should be incorporated before winter in a moderately rainy climate, which is earlier than is generally practised.

References

Alexandre, M. (2002). Evaluation par simulation avec le modèle STICS des effets des cultures intermédiaires piège à nitrates. Master Thesis, INP-ENSAT, 69pp.

Brisson, N., Mary, B., Ripoche, D., Jeuffroy, M-H., and Ruget, F. (1998). STICS: a generic model for the simulation of crops and their water and nitrogen balances. I. Theory and parameterization applied to wheat and corn. Agronomie, 18, 311-346.

Brisson, N., Gary, C., Justes, E., Roche, R., Mary, B., and Ripoche, D. (2003). An overview of the crop model STICS. European Journal of Agronomy, 18, 309-332.

Dorsainvil, F. (2002). Evaluation, par modélisation, de l'impact environnemental des cultures intermédiaires sur les bilans d'eau et d'azote dans les systèmes de culture. PhD Thesis, INA P-G, Paris, 124pp.

Meisinger, J.J., Hargrove, W.L., Mikkelsen, R.L., Williams, J.R. and Beson, V.W. (1991). Effects of cover crops on groundwater. In: W.L. Hargrove (ed), Cover crops for clean water. Soil and Water Conservation Society, Jackson, Tennessee, USA, 57-68.

Thorup-Kristensen, K. (1994). The effect of nitrogen catch crop species on the nitrogen nutrition of succeeding crops. Fertilizer Research, 37, 227-234.

Nitrate in soils and water originated from agricultural sources: a case study in Thessaly, Central Greece

Ir. Katsilouli[1], Th. Karyotis[1], Th. Georgiou[1], Th. Mitsimponas[1], A. Panagopoulos[2], A. Panoras[2], D. Pateras[1], A. Haroulis[1], G. Argyropoulos[1] and M. Toulios[1]
[1]*Institute for Soil Mapping and Classification, 1, Theophrastou Str., 41335, Larissa, Greece*
[2]*Land Reclamation Institute, 57400, Sindos, Greece*

Introduction

The aim of this study was to determine the concentration of inorganic forms ($NO_3^--N + NH_4^+-N$) of N in arable soils and in ground water, implementing the rules of the Nitrates Directive 91/676 (EEC, 1991) and Council Regulation 1257/99 (EEC, 1999) in the Thessaly district. Vulnerable areas were designated according to the results obtained from boreholes distributed over the arable land. A map has been compiled (Figure 1) showing the level of NO_3^- in the ground water samples. This map was based on criteria, such as: vulnerability of geological formations to NO_3^- pollution, NO_3^- concentration and level of exploitation of the aquifer. It can be used by scientists as a tool to develop guidelines leading to reduced leaching and establishing codes and practices for effective fertiliser N applications.

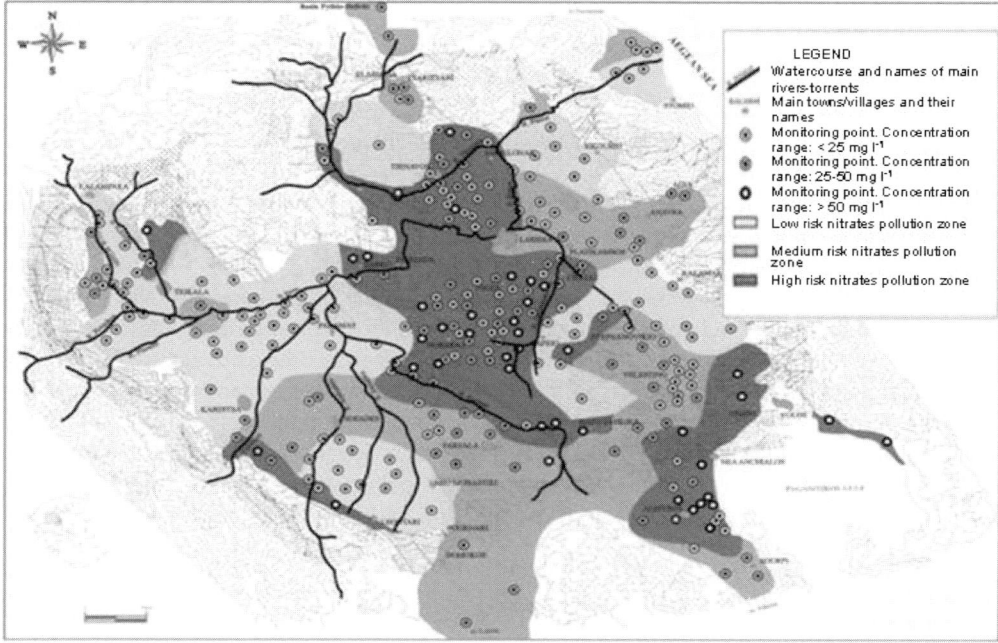

Figure 1. Nitrate levels in the pilot area of Thessaly.

Methods

The examined area of Thessaly has been divided into municipal districts. Soil and water (collected from boreholes during the irrigation period) samples were collected from 1996 to 2001. Determination of soil N was conducted before sowing, after basic fertilisation and a few days before harvesting. Soil sampling was performed before integration of the crop cycle, hence an amount of N was expected to be taken up by crops. A model was applied to calculate the required N fertilisation levels for the main crops of the pilot area (cotton, wheat, sugarbeet and cotton). Factors such as geomorphological, soil characteristics, irrigation practices and characteristics of various N fertilisers were taken into account.

Results and conclusions

Increased NO_3^- content in the soils was observed in the period 1996-2000 at the end of August, in comparison with the mean content before sowing. In contrast, in 2001 a considerable decline in NO_3^- was found (Table 1) and this may be due to heavy rainfall in summer, which raised the rate of leaching.

Table 1. Average inorganic nitrogen at different sampling periods.

Period	Number of examined fields	N (mg kg^{-1} soil) pre-sowing	N (mg kg^{-1} soil) after basic	N (mg kg^{-1} soil) Harvest
1996	148	5.7	27.3	11.9
1997	365	13.7	20.9	17.1
1998	306	8.5	14.5	15.5
1999	385	9.5	17.4	9.9
2000	219	8.4	16.3	14.7
2001	150	24.9	16.8	16.5

At the end of the growth period, the lowest concentration was 9.9 mg kg^{-1} soil (1999), whilst the maximum N content was 17.1 mg kg^{-1} soil (in 1997). This distribution may be attributed to over fertilisation, climatic factors and soil properties (structure, slope etc.). Results from 305 boreholes indicated that the mean NO_3^- concentration in 19% of the boreholes was higher than 50 mg l^{-1}. and an increased content was found in the drier parts of eastern Thessaly. Reduced N inputs without any considerable yield decline can be ascribed to the positive impact of improved agricultural practices (e.g. splitting fertilisation). This underlines the demands for a more balanced fertiliser pattern, in which the specific soil conditions, the rainfall pattern, and agricultural practices are taken into account. Utilisation of rational irrigation and financial incentives, are substantial for protection of ground water from NO_3^- pollution. Finally, NO_3^- leaching can be reduced by extension of drip irrigation in combination with proper drainage networks.

References

EEC. Council Directive (91/676/EEC) concerning the protection of the water against pollution caused by nitrates from agricultural sources (1991). Official Journal of European Communities, L375, 1-5.

EEC. Council regulation 1257/99 on support for rural development from the European Agricultural Guidance and Guarantee Fund (1999). Official Journal of European Communities, L160, 80pp. (in Greek).

Monitoring and mathematical modelling of nitrate leaching in an experimental field treated with pig slurry

P. Mantovi[1], L. Fumagalli[1] and G.P. Beretta[2]
[1]*Research Centre on Animal Production (CRPA SpA), Corso Garibaldi 42, 42100 Reggio Emilia, Italy*
[2]*University of Milan, Department of Earth Sciences "Ardito Desio", Via Mangiagalli 34, 20133 Milano, Italy*

Introduction

Situations where NO_3^- concentration in groundwater have increased, sometimes have been ascribed to the overuse of N from manure, exceeding the crop removal. Nitrate leaching through the unsaturated zone was studied by characterisation and five year monitoring of an experimental site in a vulnerable area of the Emilia-Romagna region (Reggio Emilia, Northern Italy). The trial was part of the GeTraMiN project 'Control of the Genesis, Transformation and Migration of Nitrates from the Soil to Surface Water and Groundwater', coordinated by CRPA.

Materials and methods

The experimental site (4000 m^2) was located along the foothills of the Apennine mountains (alluvial fan of the Enza river). The soil was fine textured (Udertic Ustochrepts fine, mixed, mesic), very deep, calcareous, well to moderately well-drained. The unsaturated zone had low permeability (up to 10^{-7} m s^{-1}), while the groundwater depth varied between 10-17 m. Three plots were each equipped with tensiometers (from 30 to 180 cm deep) and ceramic cup samplers (from 30 to 600 cm deep). Near the field perimeter, piezometers were installed. Water samples, collected every 2 weeks, were analysed for NO_3^- and NH_4^+ content.

During the monitoring period, cereals were cropped and pig slurry applications were made twice a year (just before the sowing and after harvesting of each crop) with a slurry tank with radar control for continuous regulation of distributed quantity. The total N in manure supplied to crops was about 400 kg N ha^{-1} yr^{-1}: the common dose for spring-summer cereals in this region. Two test areas were excluded from fertilisation. Ammonia air losses were measured at spreading with the wind-tunnel method. MACRO and SOILN mathematical field scale models (Larsson & Jarvis, 1999) were applied to simulate respectively, water flow and N dynamics, in the layered soil for the first three years of the work. Their results were verified with data derived from monitoring.

Results and discussion

Over the first 1.5 years of the trial, with extremely dry conditions, water flow did not go beyond the first 2 m of soil depth. As a consequence, NO_3^- accumulation was verified in the surface layer of the soil, and only partially attenuated by maize uptake. From 1999, as a consequence of some very rainy seasons, NO_3^- was transported through the first meters of the unsaturated zone to below the root depth (Figure 1). Nitrate leaching depth was associated with water

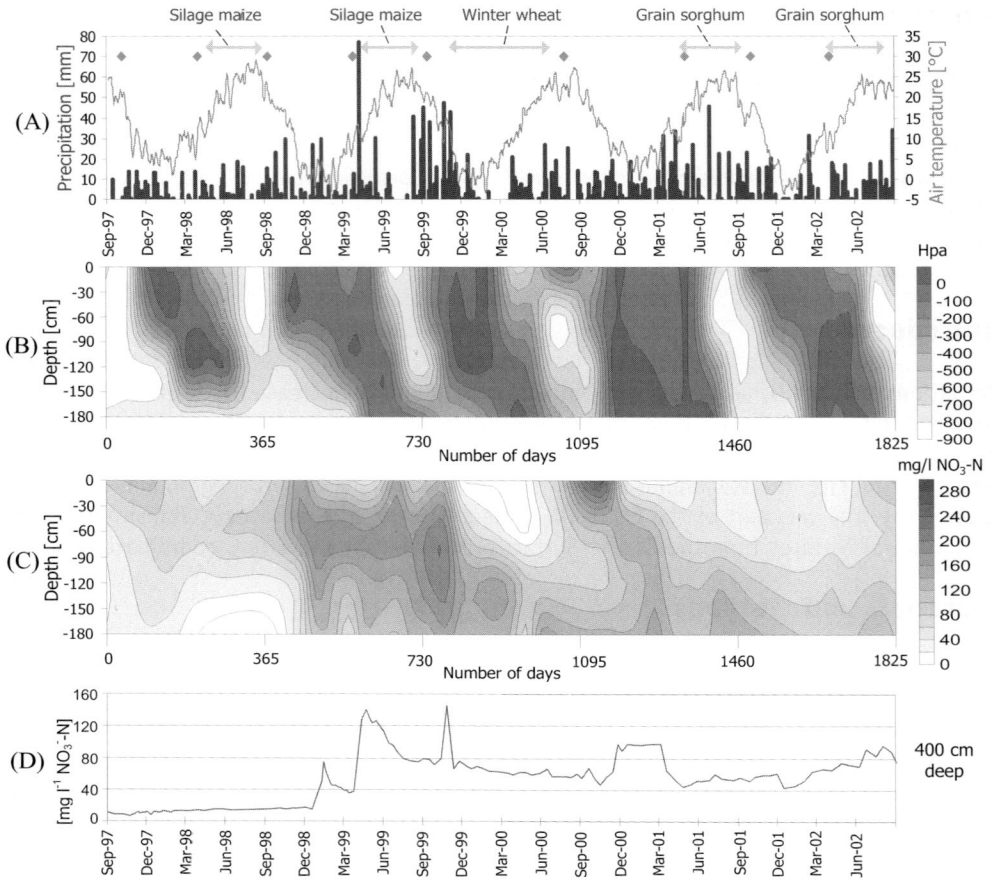

Figure 1. Meteorological conditions (A), soil matric potential up to 180 cm deep (B) and NO₃⁻ -N concentration in soil water up to 180 cm (C) and 400 cm deep (D).

Note: arrows represent the cultivation period, rhombi represents pig slurry application times.

infiltration in the soil, with migration of NO_3^- at least as far as 4 m deep. The maximum NO_3^- concentration in the soil solution of the topsoil, of about 300 mg NO_3^--N l⁻¹, was registered in early autumn 2000, after summer landspreading on wheat stubble. The autumn rainy conditions, associated with bare soil, were the most critical ones with respect to NO_3^- leaching through the unsaturated zone.

Results predicted by mathematical models were in line with the experimental data; hence these models seems to be suitable tools, even though they require some further refinements. For the first three years of the experiment, SOILN estimated that 310 and 9.6 kg NO_3^--N ha⁻¹, respectively, passed through the depths of 180 cm and 600 cm. These values were 23.5% and 0.7% of total N input on the field for the same period.

Acknowledgements

This project was funded by the Agricultural Council of the Emilia-Romagna Region

References

Larsson, M.H. and Jarvis, N.J. (1999). A dual porosity model to quantify macropore flow effect on nitrate leaching. Journal of Environmental Quality, 28, 1298-1307.

Nitrate leaching in arable cropland: effects of N-management

A. Smit and K.B. Zwart
Alterra, Soil Science Centre, PO box 47, 6700 AA Wageningen, The Netherlands

Introduction

Because of high fertiliser application rates on arable crops in the Netherlands, NO_3^- concentrations in the ground water often exceed 50 mg l^{-1}. In order to decrease these concentrations, the project 'Farming with a future' was started in 2000. The aim of this project is to find combinations of management measures which will reduce the N-surplus and subsequently the ground water NO_3^- concentrations. The measures may include substitution of organic manure by artificial fertilisers and precision in timing and placement of N applications. At the experimental arable farm, Vredepeel, a 1:4 rotation, with potatoes and sugar beet as major crops and including industrial vegetable crops, was designed in two systems (high N and low N) and implemented in 2000. N-surplus with the high N system is comparable to the common agricultural practice and that of low N, is much lower. Alterra participates in the project and is responsible for measuring and modelling the annual NO_3^- leaching for 2001 and 2002. In this paper, the results of two years' measurements are presented.

Materials and methods

Nitrate concentrations were measured in the upper ground water down to a depth of 2 m below the soil surface. In addition, the soil water was sampled at 50 cm below the soil surface by means of ceramic cups. The timing of sampling depended on the water flux. A new sample was taken every time the precipitation surplus (since the last sampling date), exceeded approximately 50 mm. The results shown in Figs. 1 and 2 are until 13 January 2003. Water samples were analysed for NO_3^- and NH_4^+ using standard techniques. Additional measurements for modelling the water flux, such as precipitation surplus, soil water content and the pressure head at several depths were made at the experimental farm.

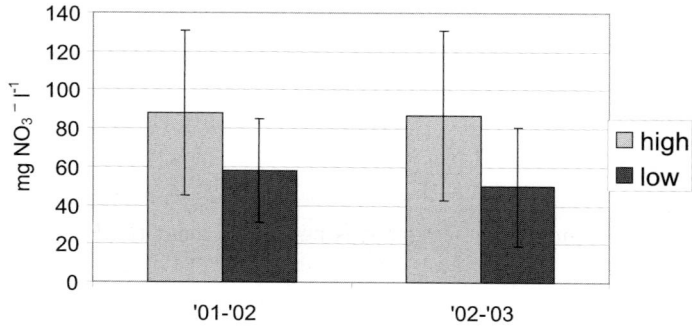

Figure 1. Average NO_3^- concentrations in the upper groundwater in the high-N and low-N systems in two years (2001-2002 and 2002-2003).

Results

The average ground water NO_3^- concentrations were lower in the low N system in both years (Figure 1).

However, within both systems, N leaching differed, largely depending on the crop, as shown for two fields (A and B, Figure 2 and Table 1). Cultivation of two leguminous crops within one year, resulted in the highest N leaching, probably as a result of rapid mineralisation of N-rich crop residues combined with a high water permeability of the soil. Substitution of beans by a catch crop, strongly reduced leaching. In maize, leaching was lower and a catch crop had less effect.

Table 1. Crops in two fields (A and B), split into high N and a low N systems.

Field	crop 2001	crop 2002
A-high N	Late silage maize	Pea + fresh beans
A-low N	Early silage maize + catch crop	Pea + tagetes or fodder radish
B-high N	Pea + fresh beans	Potato
B-low N	Pea + catch crop	Potato with straw ploughed in

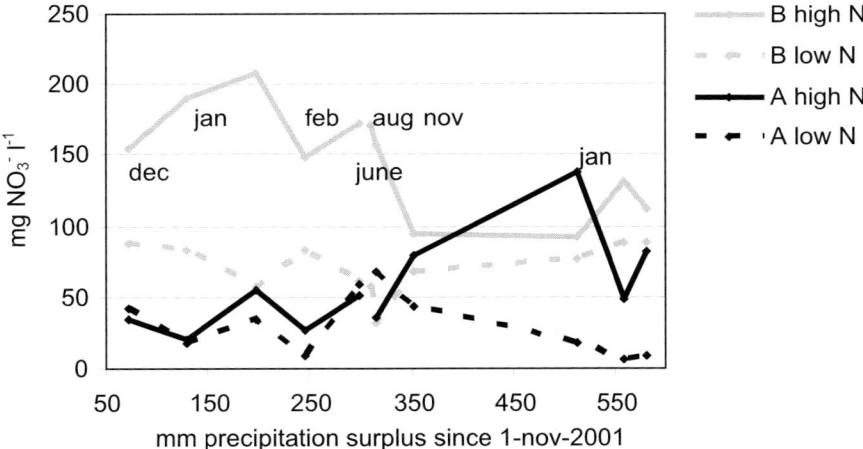

Figure 2. Nitrate concentrations in the upper ground water under the fields A and B (see also Table 1).

Conclusions

Reduced N-inputs resulted in a significant improvement in ground water quality without negative effects on crop yield. Further improvement is needed to meet the EU Nitrates Directive. This indicates that even more drastic measures are needed. Among these is removal of crop residues, including the catch crop, in order to decrease (soil) mineralisation.

Reducing nitrate contamination of groundwater from intensive greenhouse-based vegetable production in Almeria, Spain - management considerations

R.B. Thompson[1], M. Gallardo[1] and C. Gimenez[2]
[1]*Dpto. Producción Vegetal, Universidad de Almeria, 04120 La Cañada, Almeria, Spain*
[2]*Dpto. Agronomía, Universidad de Córdoba, 14080 Córdoba, Spain*

Introduction

Approximately 25,000 ha of plastic greenhouses are used for intensive horticultural production in the coastal region of Almeria, in south-eastern Spain. The major crops grown are: tomato, pepper, melon, watermelon, cucumber, aubergine, zucchini, and green beans. Eighty percent of cropping occurs in soil, the other 20% in open hydroponic systems. Most of the soil used is an artificial soil system (enarenado), in which 20-30 cm of imported loam soil is placed over the indigenous stony loam soil, and 8-12 cm of coarse sand mulch is placed over the loam soil. In some cases, the sand mulch is placed over the indigenous soil. Drip irrigation with fertigation is used. Every 2-3 years, manure is commonly applied under the sand mulch before cropping. Underlying aquifers have appreciable concentrations of NO_3^- (Jiménez et al., 1996), suggesting contamination from the horticultural industry. A study was conducted to assess the risk of NO_3^- leaching loss from this industry, and to identify contributing management practices.

Methods

Fifty three commercial greenhouses were studied, with ≥ 4 greenhouses for each of the 9 major vegetable species (see Introduction); the number for each species broadly reflected the relative acreage. After each crop, the sand mulch was cleared and the soil was sampled to measure residual NH_4^+-N and NO_3^--N. Thirty eight greenhouses were sampled to 40 cm depth, and the other 15, to 60 cm depth (the stone content of the soil prevented sampling to >40 cm depth in 38 greenhouses). Soil samples were immediately chilled, and were frozen within several hours of collection. Soil was extracted with 2M KCl (solution to soil ratio of 10:1), and analysis was by standard colorimetric methods. Crop N uptake was estimated for each crop. Farmers provided production data. On each farm, four representative plants were sampled for final biomass and N content. Pruning, and fruit N content data were obtained from four representative plants in 1-2 greenhouses for each of the nine species.

Participating farmers kept a record of all irrigations, which were used to calculate the total volume of irrigation water. These volumes were compared with estimated crop water requirements calculated with a mathematical model developed for greenhouse crops in Almeria, using solar radiation and temperature data (Cajamar, 2001). Participating farmers completed a questionnaire of nutrient and irrigation management practices.

Results

The frequency distribution of residual soil mineral N was positively skewed. Mean and median values of residual soil mineral N (combined NH_4^+-N and NO_3^--N) to 40 cm, were 318 and 257 kg N ha^{-1}, respectively. In 75% of greenhouses, there was >200 kg mineral N ha^{-1}; in 40%, >300 kg N ha^{-1}; and in 19%, >400 kg N ha^{-1}. On average, 57% of soil mineral N was in the form of NH_4^+-N. Ninety four percent of greenhouses had 100-199 kg NH_4^+-N ha^{-1}. The distribution of residual soil NO_3^--N was considerably skewed. Half had <100 kg NO_3^--N ha^{-1}, and a small number had much larger amounts, e.g. 15% had >200 kg NO_3^--N ha^{-1}, and 6% had >500 kg NO_3^--N ha^{-1}. In 15 greenhouses, the 40-60 cm depth was also sampled: the mean and median amounts of mineral N were 167 and 121 kg N ha^{-1}, respectively; the percentage distributions between NH_4^+-N and NO_3^--N, and within both NH_4^+-N and NO_3^--N, were similar to those of the 0-40 cm depth. For 8 of the 9 vegetable species examined, mean estimated crop N uptake was <350 kg N ha^{-1}, in 4 of 9 it was <200 kg N ha^{-1}.

The questionnaire data indicated that N fertiliser management was based on experience in 100% of farms. For the crop surveyed, soil testing was used on only 19% of farms, and none used foliar testing. Manure applications, within the previous 2 years, were made in 75% of greenhouses; yet in only 26%, was the manure N considered in N fertiliser planning. The N content of irrigation water was considered in only 43%. Irrigation management was based on experience in 87% of the greenhouses (15% farmer, 25% advisor, 47% both); the rest combined tensiometers with experience. Irrigation to leach salts was used in 98% of greenhouses (mostly before a crop); volumes exceeded 30 mm on 60% of farms. 57% of farms had disinfected the soil, of which 60% used volumes >20 mm. The comparison of total applied volumes of irrigation water to calculated crop water requirements indicated that one third of farms applied >150% of calculated crop water requirements.

Discussion

Soil mineral N (to 40 cm depth) after cropping was generally appreciable. Commonly, it was equivalent to a large proportion, or more, of crop N requirements. This suggested that, in many greenhouses during the most recent and/or preceding crops, the amount of N provided to crops was appreciably in excess of crop N uptake. The accumulation of large amounts of residual mineral N in soil is consistent with the nutrient management practices of a majority of farmers, which are based on experience and generally do not include soil testing, or consider N supplied by recent manure applications and in irrigation water. Considering both the soil mineral N data for the 0-40 cm and the more limited 40-60 cm data, it is clear that large amounts of mineral N were commonly present throughout the soil profile after cropping. Although an average of approximately 60% of mineral N was in the immobile form of NH_4^+-N, this would be at risk of leaching after nitrification, where appreciable drainage occurred. Using >150% of calculated crop water requirements as the criterion, approximately 30% of greenhouses applied excessive irrigation during cropping. The large additional irrigations applied for salinity management and soil disinfection, between crops, are likely to leach considerable amounts of mobile nutrients below the root zone.

This study demonstrated an appreciable risk of NO_3^- leaching in this horticultural industry, which is consistent with reported NO_3^- contamination of underlying aquifers (Jiménez et al., 1996). There is considerable scope for improving N and irrigation management practices.

References

Cajamar. (2001). ProHo v1.0. Programa de Riegos Para Cultivos Hortícolas Bajo Invernadero. Cajamar, Almeria, Spain

Jiménez Espinosa, R., Molina, L., Pulido Bosch, A. and Navarrete, F. (1996). Influencia de la agricultura intensiva en el contenido de nitratos de las aguas del Campo de Dalías: evolución temporal y espacial. Geogaceta 20, 1281-1284.

Nitrogen budgets at field, farm and polder scales in a polder used for dairy farming

C.L. van Beek, G.L. Velthof and O. Oenema
Alterra, PO Box 47, 6700 AA Wageningen, The Netherlands

Introduction

Nitrogen budgets can be derived at different spatial scales. Differences between such budgets may point to losses, supply, or retention of N. The present study reports on N budgets derived at field, farm and polder scales, in a small polder dominated by dairy farming. The aim of the study was to relate N budgets of fields and farms to N leaching losses to surface water at the polder scale. This insight is needed when N budgets are used as a tool to decrease the N leaching from agriculture to surface water.

Materials and methods

The study site was a hydrologically well defined peat land polder of 200 ha, in the western part of The Netherlands. The land inside this polder is used as grassland for intensive dairy farming (6 farms). Ditches and canals, covering ca.10% of the total surface area, are used to drain the fields and for division of land. The surface water level in the polder is regulated, i.e. during periods of water excess, surface water (with its constituents) is pumped out of the polder to regional surface waters and during periods of water shortage, regional surface water is let in. During 1999-2001, measurements were performed on discharge, denitrification and mineralisation at the field scale. Net mineralisation was based on N uptake by herbage on fields where N applications were withheld and losses via denitrification were estimated by monthly measurements using the acetylene inhibition technique. At the polder scale, N budgets consisted of inputs via inlet water and of outputs via the pumping station. At the farm scale, N budgets consisted of inputs via concentrates and fertilisers and output via milk. While at the field scale, budgets consisted of inputs via fertilisers, cattle excreta and manure applications and outputs via grazing and silage production.

Results and discussion

In early spring, denitrification was an important N removal process in the polder, while in winter, discharge at the pumping station was a major N output factor (Figure 1).

Annually, N discharge was about 40 kg N ha^{-1}, while 87 kg N ha^{-1} was lost via denitrification. Average N inputs and outputs are summarised in Table 1. Mean total N input at the field scale was large; dominant inputs were manure (250-350 kg N ha^{-1}yr^{-1}), fertiliser (approximately 150 kg N ha^{-1}yr^{-1}), and net mineralisation (263 kg N ha^{-1}yr^{-1}). Inputs at farm scales and polder scales were dominated by concentrates, fertiliser and net mineralisation. Outputs at the field scale occurred through harvested grass, via grazing and mowing (for silage), and through losses via leaching, denitrification and NH$_3$ volatilisation. Outputs at farm scales and polder scales were dominated by the export of agricultural products, and through N losses via leaching, denitrification and NH$_3$ volatilisation.

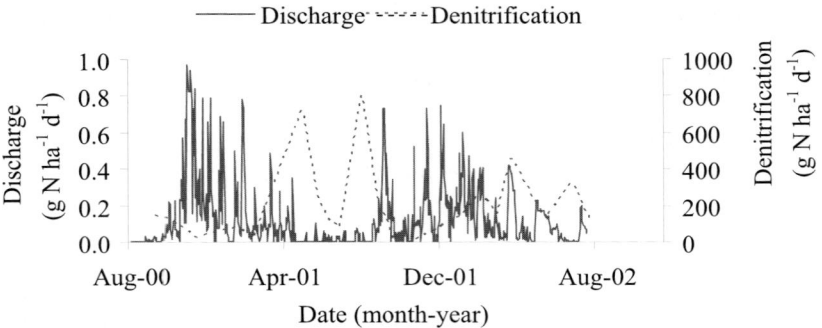

Figure 1. N discharge rates at the pumping station and denitrification rates in the polder.

Of the total N input at the polder scale, 14% was lost via denitrification, 6% was discharged from the polder and 3% volatilised as NH_3. At smaller time scales, the contribution of different loss pathways will alter, e.g. in winter, discharge is a major loss pathway, which is largely taken over by denitrification in summer (Figure 1). However, 62% of the N surplus at the polder scale was unaccounted for, which could be largely retrieved at the farm scale (Table 1). The results suggest losses and/or mismatches in stables and farm buildings (Van Beek et al, 2003).

Table 1. Average N flows (kg N ha⁻¹yr⁻¹) at field (land), farm (land) and polder scale (land and water) during 1999-2001.

Inputs	Field	Farm	Polder	Outputs	Field	Farm	Polder
Agricultural inputs	485	338	338	Agricultural outputs	387	96	96
Net mineralisation	263	263	263	Leaching	30	30	
Atm. deposition	31	31	31	Denitrification	87	87	88
Inlet station			1	Pumping station			40
				NH_3	12	18	18
Sum	779	632	633		516	231	242

N leaching from field to the polder ditches was estimated at 30 kg N ha⁻¹yr⁻¹, compared with 40 kg N ha⁻¹yr⁻¹ from the polder. This difference may originate from i) high variability between fields resulting in a few fields with much higher leaching losses compared with average fields (Van Beek et al, 2003), and/or ii) other nutrient pathways to the ditch, e.g. approximately 3 kg N ha⁻¹yr⁻¹ reached the surface water through atmospheric deposition.

Acknowledgements

This project was funded by the Dutch ministry of Agriculture, Nature and Food Quality (LNV research programme 398-II).

References

Van Beek, C.L., Brouwer, L. and Oenema, O. (2003). The use of farmgate balances and soil surface balances as estimator for nitrogen leaching to surface water. Nutrient Cycling in Agroecosystems (in press).

Nitrate leaching on loess soils as affected by N fertilisation and crop rotation

W. van Dijk, P. Dekker and J.R. van der Schoot
Applied Research Plant and Environment, Lelystad, The Netherlands

Introduction

On soils with a low ground water table, such as loess soils, the EC-level for NO_3^- in groundwater (50 mg NO_3^- l^{-1}), is often exceeded. Measurements of NO_3^- content on loess soils in the Netherlands showed that on 75 and 50% of the farms, levels were higher than 50 and 100 mg l^{-1}, respectively. Nitrogen management on farms must therefore be improved to meet environmental goals. In 1995, a long term field experiment was started to investigate the relationship between N fertilisation levels and NO_3^- leaching in different crop rotations.

Materials and methods

The experiment was carried out from 1995 to 2001 at a loess site in the Netherlands. This paper will focus on two rotations: ware potatoes-winter wheat-sugarbeets (R1) and continuous maize (R2). The R1 rotation represents the situation on arable farms, while the R2 rotation is often seen on dairy farms. Nitrogen was provided at recommended levels, either from mineral fertiliser (MF) or from a combination of cattle slurry and mineral fertiliser (CS+MF). Slurry was applied annually in the spring. It was assumed that ca. 65% of the total N applied with slurry was effective. Averaged N-rates are given in Table 1. Nitrate concentration in soil water was measured at a depth of 135-150 cm in November and March. The results of these two measurements were averaged to get the annual level of NO_3^- concentration.

Results and discussion

Nitrate concentration in soil water was lower for the R1 than for the R2 rotation because of differences between crops. Leaching was highest after potatoes and silage maize while low levels were observed after winter wheat and sugar beets. Variation between years was large. Comparable differences in N-losses between crops were mentioned by Van Enckevort et al. (2002). The results of this experiment show that in arable rotations, the EC level for NO_3^- can be achieved when crops are fertilised at recommended levels and when manure is used in a proper way. In the continuous maize system, leaching levels exceeded the EC level in the manure based treatment. Schröder et al. (1996) reported comparable results. However, they also showed that leaching levels can be decreased by about 50% when a winter catch crop was grown. In practice, much higher NO_3^- concentrations were measured compared with those observed in our experiment. This is probably due to a poor use of manure, i.e. high rates exceeding recommendations and application in autumn.

The differences in NO_3^- leaching between crops were not well related to the N-surplus. For example, N surplus for silage maize was much lower than for sugar beets, while the opposite was observed for NO_3^- concentration in soil water. However, for all crops, the higher N-surpluses in the manured plots did lead to increased NO_3^- concentrations, effects being strongest for maize and potatoes. The increased leaching is probably because of a higher N mineralisation level. However, organic matter content of the soil also decreased in the MF-treatments (results not shown), indicating that systems based on 100% mineral fertilisers are probably not sustainable in the long term.

Table 1. N-surplus and NO_3^- concentration in soil water as affected by crop rotation and fertilisation (average values of whole experimental period).

Rotation	Fertilisation	Crop	N-rate (kg ha[-1])		N-surplus[1,2]	Nitrate concentration
			Slurry	Mineral fertiliser	(kg ha[-1])	(mg l[-1])[2,3]
R1	MF	Potatoes	-	263	46	58 (44-73)
		Winter wheat	-	164	-2	20 (6-50)
		Sugarbeet	-	163	78	11 (8-16)
		Average	-	*197*	*41*	*30*
	CS+MF	Potatoes	223	123	128	72 (38-113)
		Winter wheat	117	80	31	26 (6-57)
		Sugarbeet	220	29	162	13 (11-17)
		Average	*187*	*77*	*107*	*37*
R2	MF	Silage maize	-	148	-24	40 (18-62)
	CS+MF	Silage maize	218	38	85	62 (37-95)

[1] N-surplus = (N_{slurry} + $N_{mineral fertiliser}$) - $N_{yield harvest product}$
[2] LDS crop x rotation (P<0,05) = 12 kg N ha[-1] and 11 mg l[-1] for N-surplus and NO_3^- concentration, respectively
[3] minimum and maximum levels in the experimental period are given in parentheses

Conclusion

The results indicate that in arable crop rotations on loess soils the EC level of 50 mg NO_3^- l[-1] in drainage water can be achieved under conditions of recommended fertiliser levels and a restricted manure use. In the continuous maize system, the EC level was only met when no manure was used.

References

Van Enckevort, P.L.A., van der Schoot, J.R. and van den Berg, W. (2002). Estimation of residual mineral soil nitrogen in arable crops and field vegetables at standard recommended rates. In: Ten Berge (ed). A review of potential indicators for nitrate loss from cropping and farming systems in the Netherlands. Reeks Sturen op Nitraat 2, Plant Research International report nr 31, 77-90.

Schröder, J.J., van Dijk, W. and de Groot, W.J.M. (1996). Effects of cover crops on the nitrogen fluxes in a silage maize production system. Netherlands Journal of Agricultural Science, 44, 293-315.

Loss of inorganic nitrogen in surface runoff from grazed grassland

C.J. Watson, R.V. Smith and E. Chisholm
Department of Agriculture and Rural Development, The Queen's University of Belfast Newforge Lane, Belfast, BT9 5PX, UK

Introduction

The main concerns about inorganic N in land drainage water from grassland are compliance with the EC Nitrates and Water Framework Directives (European Community, 1980; European Union, 2000), which define a maximum admissible concentration of 11.3 mg NO_3^--N l^{-1}, 0.38 mg NH_4^+-N l^{-1} and 30 µg NO_2^--N l^{-1} for potable use and the potential role of N in eutrophication of surface waters (European Community, 1991). Although there are no EC guideline concentrations for NO_3^- in waters capable of supporting freshwater fish, there are guidelines for NO_2^- and NH_4^+ in both salmonid and cyprinid waters (European Community, 1978). There is little information on N loss in surface runoff from grassland. The object of the current study was to measure the concentrations of inorganic N in surface runoff from grazed grassland and to assess the relative contributions made by land drainage and surface runoff to N loss.

Materials and methods

Four grassland plots (each 0.2 ha) received 250 kg N ha^{-1} yr^{-1} in six equal applications during the 2000, 2001 and 2002 growing seasons. Plots were grazed by beef steers from April to October to maintain a constant sward height of 7 cm. Plots were hydrologically isolated and artificially drained to v-notch weirs with weekly flow proportional monitoring of drainage water (Watson et al., 2000). Surface runoff collectors were installed across the width of the plots at the lowest point and were connected to portable water samplers (ISCO, Inc.). When surface runoff occurred, 24 x 200 ml water samples were taken at 20 min intervals and a composite sample was produced for the event. Concentrations of NO_3^-, NO_2^- and NH_4^+ in water samples were analysed using automated colorimetric procedures (Downes, 1978; Scheiner, 1976;) adapted for the Bran and Luebbe 'TRAACS 800' autoanalysis system.

Results and discussion

In the period June 2000 - February 2001, twelve surface runoff events were measured and these were normally associated with rainfall intensities greater than 4 mm hr^{-1}, which contributed only 2% to periods of rainfall. From March 2001 - February 2002, twenty-two surface runoff events were measured, while from March 2002 - February 2003, sixty two surface runoff events were observed. Annual mean NO_3^--N concentrations in both surface runoff and drainage water were well below the EC limit of 11.3 mg N l^{-1} (Table 1). The EC guideline concentrations of NH_4^+ for salmonid and cyprinid waters are 0.03 and 0.16 mg N l^{-1}, respectively, and the equivalent concentrations for NO_2^- are 3 µg N l^{-1} and 9 µg N l^{-1}, respectively. The mean concentrations in surface runoff from all plots from March 2002 to February 2003 were 0.69 mg NH_4^+-N l^{-1} and 38.7 µg NO_2^--N l^{-1}. These concentrations were well in excess of the guidelines set in the Freshwater Fish Directive. Very high concentrations of inorganic N were measured when surface runoff occurred immediately after fertiliser N application and these peaked at 43.9 mg NO_3^--N l^{-1}, 10.9 mg NH_4^+-N l^{-1} and 486 µg NO_2^--N l^{-1}. Elevated NH_4^+ and NO_2^- concentrations have been shown to adversely affect aquatic biota.

Table 1. Annual mean concentrations and loads of NO_3^--N, NH_4^+-N and NO_2^--N in drainage water and surface runoff (March 2002 -February 2003).

	NO_3^--N (mg l^{-1})		NH_4^+-N (mg l^{-1})		NO_2^--N (µg l^{-1})	
	Runoff	Drainage	Runoff	Drainage	Runoff	Drainage
Mean concentration	1.69	3.58	0.69	0.18	38.7	28.3
	NO_3^--N (kg N ha^{-1})		NH_4^+-N (kg N ha^{-1})		NO_2^--N (g N ha^{-1})	
	Runoff	Drainage	Runoff	Drainage	Runoff	Drainage
Annual load	1.90	25.5	0.99	1.06	46	143

The mean NH_4^+ and NO_2^- concentrations in drainage water entering the v-notch weirs from March 2002 to February 2003 were lower than in surface runoff and averaged 0.18 mg N l^{-1} and 28 µg N l^{-1}, respectively. Total NO_3^--N and NO_2^--N lost in runoff was substantially lower than that lost to drainage water, whereas the quantity of NH_4^+-N lost was comparable (Table 1).

Conclusions

Mean NH_4^+-N and NO_2^--N concentrations in surface runoff were well in excess of EC guidelines for waters supporting freshwater fish and are likely to cause stress to aquatic biota.

Acknowledgements

The technical support of staff within the Agricultural and Environmental Science Division is gratefully acknowledged.

References

Downes, M.T. (1978). An improved hydrazine reduction method for the automated determination of low nitrate levels in freshwater. Water Research, 12, 673-675.

European Community (1978). Council directive on the quality of fresh waters needing protection or improvement in order to support fish life. EC 78/659. Official Journal of the European Community L222, 1-9.

European Community (1980). Council directive relating to the quality of water for human consumption. EC 80/778. Official Journal of the European Community L229, 11-29.

European Community (1991). Council directive concerning the protection of waters against pollution caused by nitrates from agricultural sources. EC 91/676. Official Journal of the European Community L375, 1-8.

European Union (2000). Council directive establishing a framework for Community action in the field of water policy. 2000/60/EC. Official Journal of the European Community L327, 1-21.

Scheiner, D. (1976). Determination of ammonia and kjeldahl nitrogen by indophenol method. Water Research, 10, 31-36.

Watson, C.J., Jordan, C., Lennox, S.D., Smith, R.V. and Steen, R.W.J. (2000). Inorganic nitrogen in drainage water from grazed grassland in Northern Ireland. Journal of Environmental Quality, 29, 225-232.

Nitrogen losses in drainage water following pig slurry applications to an arable clay soil

J.R. Williams[1], B.J. Chambers[2] R.B. Cross[1] and R.A. Hodgkinson[2]
[1]*ADAS Boxworth, Battlegate Road, Boxworth, Cambridge, CB3 8NN, UK*
[2]*ADAS Gleadthorpe, Meden Vale, Mansfield, Notts., NG20 9PF, UK*

Introduction

Farm manure applications to agricultural land in the UK supply an estimated 450,000 tonnes of N each year (Williams et al., 2001). Typically, 60% of the total N content of pig slurry is present as readily available (NH_4^+) N, with the remainder present as organic N which will mineralise slowly over a period of months to years. On clay soils, manures are commonly applied in the autumn when soils are dry and can carry the weight of heavy application machinery, without causing damage to the soil structure. However, autumn applications (which account for *ca.* 50% of pig slurry applications to cereal cropped land) represent the biggest potential risk of diffuse nutrient pollution from manures, as crop nutrient uptakes are low and nutrients can be rapidly lost in drainflow and surface run-off. Medium loam and clay textured soils occupy *ca.* 60% of the area of Nitrate Vulnerable Zones (NVZs) in England which, under arable cropping, are likely to be drained to relieve surface water problems.

This paper presents results from the second year of a three year study to investigate the effects of a range of contrasting pig slurry application timings and cultivation practices on N concentrations and losses in sub-surface drainflow.

Materials and methods

Pig slurry (50 m^3 ha^{-1}, ca.180 kg ha^{-1} total N) was applied between September 2000 and March 2001 to hydrologically isolated plots (48 m x 12 m) on a clay textured soil of the Hanslope Association in cereal cropping at ADAS Boxworth (Cambridgeshire, UK). There were 3 replicates of nine treatments, viz. control (no slurry; autumn and spring plough), autumn plough (slurry applied September, December and March), autumn disc cultivation (slurry applied in September) and spring plough (slurry applied in September, December and March to uncultivated stubble), giving 27 plots in total. Each plot was drained with two sub-surface lateral perforated plastic pipes with gravel fill to within 30 cm of the surface, supplemented by mole drains at 2 m spacing and 50 cm depth. Drainflow volumes were measured continuously using V-notch weirs and drainage water samples taken automatically on a flow proportional basis. The samples were analysed for NO_3^--N, NH_4^+-N and soluble organic N. The N concentrations were combined with drainflow volumes to quantify total N losses kg ha^{-1}.

Results and discussions

The very wet autumn and winter, (when rainfall between 1 September and 30 April was 621 mm: double the long-term average), meant that soil conditions were too wet to establish a winter cereal crop. The mean over winter drainage volume was 298 mm. Nitrate-N losses from

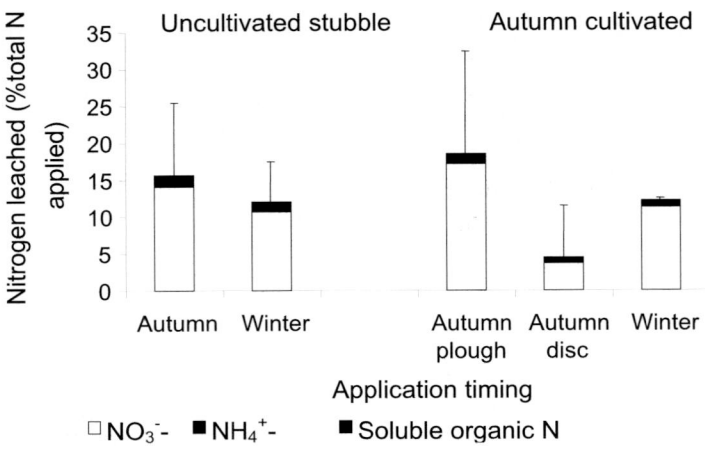

Figure 1. Nitrogen losses in drainage water 2000/01.

the autumn and winter application timings to uncultivated stubble (at 14% and 11% of the total N applied) were greater ($P<0.05$) than from the untreated control (32 kg ha^{-1} N). Nitrate-N losses from land that was ploughed, or disced following the autumn slurry application, were equivalent to 17% and 4% of the total N applied, respectively, and 11% of the total N applied from the December application to previously ploughed land, although these losses were not significantly different ($P>0.05$) from the untreated cultivated control (40 kg ha^{-1} N). The proportion of NO_3^-, NH_4^+ and soluble organic N lost from all treatments was similar at 90%, 10% and <1% of total N losses, respectively (Figure 1). The lower N losses from the winter application timing (compared with autumn) were largely a reflection of the lower drainage volume (114 mm) after application.

The second year results showed no differences in N losses between the autumn cultivated and uncultivated treatments. This was in contrast with the first year of the study (1999/2000), when slurry N losses from the autumn ploughed land (2% total N applied) were ca. 4 fold lower than from the uncultivated stubble (9% total N applied), where slurry remained on the soil surface (Williams et al., 2002). This may have been due to the greater amounts of rainfall in October and November 2000 (222 mm compared with 74 mm in 1999) which led to increased autumn drainage volumes (158 mm in 2000 compared with 20 mm in 1999), although the time interval between application and the start of drainage was similar in both years at 11 and 3 days in 2000 and 1999, respectively.

Acknowledgement

Funding of this work by the Department for Environment, Food and Rural Affairs is gratefully acknowledged.

References

Williams, J.R., Chambers, B.J., Smith, K.A. and Ellis, S. (2001). Farm manure land application strategies to conserve nitrogen within farming systems. Proceedings of the SAC/SEPA Conference. Agriculture and Waste Management for a Sustainable Future, 167-179.

Williams, J.R., Smith, K.A., Chambers, B.J. and Cross, R.B. (2002). Nitrogen and phosphorus losses in drainage water following pig slurry applications to a drained clay soil. Tenth International Conference of the FAO ESCORENA Network on Recycling of Agricultural, Municipal and Industrial Residues in Agriculture. Food and Agricultural Organisation of the United Nations, Rome, 109-114.

SECTION 5 – RECONCILING PRODUCTIVITY WITH ENVIRONMENTAL CONSIDERATIONS

Reconciling productivity with environmental considerations

D. Scholefield

Institute of Grassland and Environmental Research, North Wyke Research Station, Okehampton, Devon, EX20 2SB, UK

Abstract

This paper reviews briefly the problems of reconciling agricultural productivity with environmental considerations, mainly with reference to the N cycle and the need to comply with the EC Nitrates Directive and legislation affecting farming activities within Nitrate Vulnerable Zones. Whether current systems are environmentally sustainable is debated and strategies for improving the efficiency of N use are identified. The need to establish the scientific and technical scope of feasible options to achieve economic and environmental sustainability is highlighted. Future research priorities are identified and there is much scope for improvement, but there will be a limit to the stringency of environmental controls that can be accommodated. Beyond this limit, retention of a viable agricultural industry capable of 'feeding the nation' will not be feasible.

Keywords: environment, farming systems, nitrogen cycle, sustainability

Introduction

Pressures are growing to identify ways in which the economic viability of different agricultural systems can be reconciled with their environmental impacts. That is, agricultural systems must become both economically and environmentally sustainable. How to achieve this is a major question facing national policy makers, research scientists, farmers and their advisors throughout the developed world. While consumers demand cheap, healthy food of high quality, society in general, places increasing pressures on farmers to reduce emissions of pollutants to water and the atmosphere and at the same time to maintain or improve biological diversity and the aesthetic quality of the landscape.

Nitrogen is of pivotal importance to both sides of this complex equation, as it is usually the limiting element to plant growth and is a constituent of the major pollutants - NO_3^-, NH_3 and N_2O. The fundamental difficulty with reconciling higher productivity (the main agricultural driver) with lower environmental impacts is that the efficiency of incorporation of N inputs into saleable products generally decreases with increasing production per ha. Under what circumstances and to what degree it is possible to reconcile economic productivity with defined environmental impacts, may be revealed only by novel and ingenious applications of our knowledge of how N flows within and between the components of agricultural systems. However, although productivity and environment are linked intrinsically by N flows, we should remember that quantitative definition of what is an acceptable state for either, is both arbitrary and transient.

In this paper, the relationships between N flow to products and those to N loss pathways will be considered for conventional grassland and arable scenarios. The potential for improving overall efficiency is assessed with reference to implementation of novel strategies and to what is likely to be biologically, technically and politically feasible. The different approaches to the identification of the blueprints of sustainable agricultural systems are identified. The importance of using 'life cycle' and integrated approaches that can reveal synergistic, buffering and antagonistic interactions between effects of strategies (that may culminate in phenomena such as 'pollution swapping') is highlighted. Another important consideration is the scale at which economic and environmental sustainability is required to be manifested (e.g. the field, farm, river basin or political unit), as this can determine the likelihood of achieving success. Coupled with this is an emerging awareness of the overriding importance of site factors such as soil conditions, climate and topography in determining the ease with which environmental objectives can be met on productive farms and also in order to gauge the potential of a major reversion to mixed farming.

Sustainability criteria and efficiency of N use

As mentioned earlier, the criteria for both economic and environmental sustainability are arbitrarily defined and continuously changing, as the price of commodities change within local and global markets and as legislation limiting emissions of N compounds becomes ever more stringent. Thus, there is no *a priori* guarantee that economics and environment can be satisfactorily reconciled for any particular commodity at a given time and location, even by employing the most effective management strategies identified by research.

The problem may be condensed into one in which the overriding consideration, even for livestock production, is the efficiency of uptake of soil inorganic N by the plant. Plant uptake efficiency varies according to 3 types of factors. *Type one* includes inherent physiological traits that vary between crop species/varieties and for any given species, vary in time throughout the year. *Type two* factors are site conditions such as availability of soil water and other nutrients (e.g. P, K, and S), temperature and light intensity. *Type three* factors are farm managements including those of manures, fertilisers, crop harvesting and rotations. All of these factors and their interactions must be considered in relation to the total rate of supply of inorganic N to the soil from all sources, for which the plant is competing with the processes of N loss: denitrification, NH_3 volatilisation and leaching to watercourses. At low rates of supply, plant uptake competes very effectively with these processes and environmental impacts are normally small. As rate of supply increases however, there is increasing tendency for spatial and temporal mis-matches between supply and the plant's capacity for uptake, and then N losses begin to increase rapidly. Finally, at a certain rate of supply, the plant cannot increase uptake further and all inputs above this rate are susceptible to loss.

The N fertiliser response curve can be used as a means of tracking the course of this competition in the soil, but cannot be used to predict N losses, as it neither takes account of other inputs apart from fertiliser, nor of the proportion of N taken up that is partitioned to below-ground components. Nevertheless, the response curve has been used to base schemes of fertiliser application aimed at achieving economic optimum yields (e.g. Neetesen & Wadman, 1987). The 'economic optimum' rate is calculated simply as the maximum fertiliser rate above which any further yield response would not pay for the cost of additional fertiliser. Unfortunately, the economic optimum has been linked to a supposed discontinuity in the N input/N loss relationship at which the plant's 'demand' for soil N is thought to be uniquely satisfied. Thus credence has been given, particularly in arable farming, to the idea that it is environmentally safe to apply fertiliser up to the 'demand' level of the plant.

Figure 1 shows that, even for arable cropping, N losses are sensitive to fertiliser inputs over most of the range. Figure 2 shows that for some vegetable crops, the response to fertiliser of the harvestable component is very shallow, while the potential for N loss, with increasing fertiliser input, increases very steeply. The problem in common to all agricultural systems, to a greater or lesser extent, is that the small increments in yield following high fertiliser applications are often, nevertheless, essential for economic sustainability.

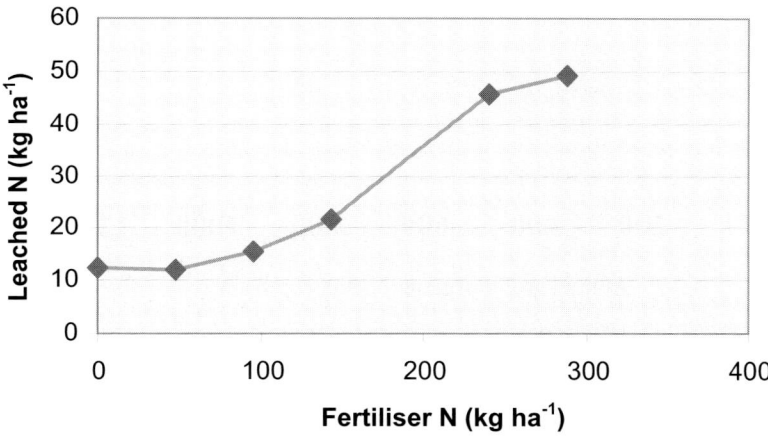

Figure 1. Relationship between fertiliser input and NO_3^- leaching for winter wheat at Broadbalk (from Goulding et al., 2000).

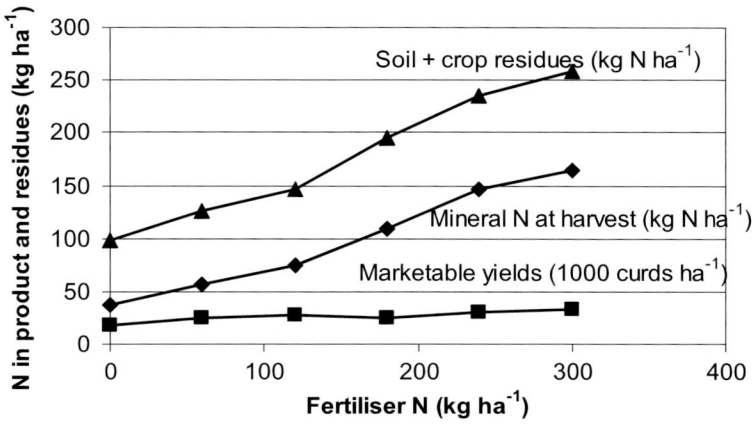

Figure 2. Relationship between N in product and N susceptible to be leached with increasing fertiliser rates applied to cauliflowers (unpublished data from C. Rahn, HRI Wellesborne, UK).

Controlling nitrogen flows and losses

Figure 3 shows that by plotting total soil inorganic N flux against the proportion of it taken up by the whole plant, a more useful model for comparing efficiency of N use between systems (in this case grazed grassland) may be obtained. It is important to note that the efficiency of N use, as measured by values on the Y axis in Figure 3, varies continuously with N input and that by merely reducing inputs (moving to the left along the regression line), efficiency can be increased substantially.

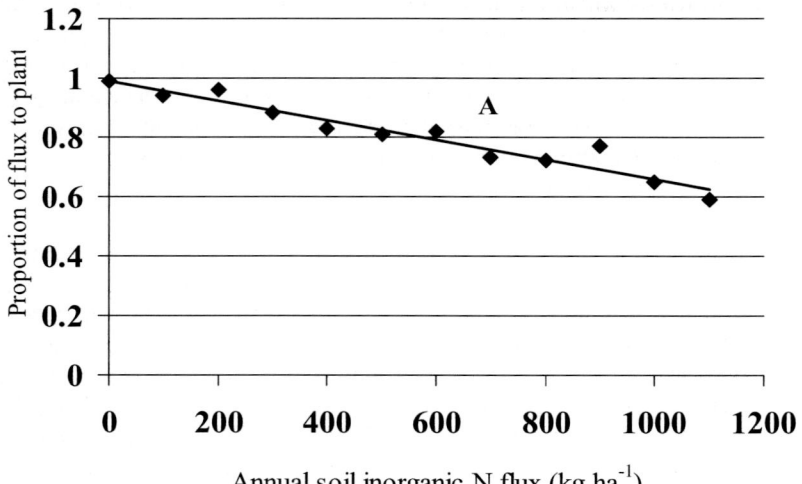

Figure 3. Relationship between the annual total flux of inorganic N through the soil and the proportion of it taken up by the plant for grassland grazed by beef cattle (from Scholefield et al., 1991).

The critical question is, however, does any single location on this line satisfy both economic and environmental criteria of sustainability for the system being evaluated? If so, then the solution to achieve sustainability may be one of simply reducing inputs. This solution may apply for situations such as over-fertilisation of cereal and vegetable crops as 'insurance' and for applications of large amounts of slurry to maize land which has already received inorganic fertiliser. If there is no point on the line at which economic and environmental criteria of sustainability can be satisfied then new systems must be designed with specifications lying above the line in Figure 3 (e.g. at point A on the graph). This may involve changes *to type 1, type 2* and/or *type 3* factors.

Are current commercial systems of agriculture environmentally sustainable?

To answer this question we must examine the N loss data obtained from systems studies and from applications of N cycling models and compare these data with relevant emission limits. Perhaps the most pertinent and quantifiable limit is the EC Nitrates Directive concentration limit of 11.3 mg NO_3^- -N l^{-1} for drinking water.

Survey data (Jarvis, 1999), measurements from grazing systems (Scholefield et al., 1993) and modelled outputs (Scholefield et al., 1991) show that for grassland, annual fertiliser inputs can range widely between 100 and 689 kg N ha^{-1}, and that NO_3^- leaching losses can range from <5 kg N ha^{-1} for low input extensively managed systems, to >150 kg N ha^{-1} for intensively managed dairy systems involving field grazing, and even as great as 338 kg N ha^{-1} for some 'double cropping' systems in Portugal (Trindade et al., 1997).

Nitrogen inputs to arable cropping systems are less variable than those to grassland and, consequently, the N leaching losses are also less variable, ranging typically, on average, between about 50 kg N ha^{-1} for winter wheat on cracking clay soils (see Figure 1, also Lord & Mitchell, 1998) and >90 kg N ha^{-1} for potatoes and sugar beet receiving typical fertiliser applications at the economic optimum. Losses from vegetable crops are potentially much greater, as N inputs tend to be large to ensure maximum yields of high value products (e.g. Figure 2 showing potential for large N losses from N rich residues of cauliflower and accumulation of soil inorganic N during autumn). Set against this potentially potent source of N loss, is the much smaller area of land typically devoted to vegetable growing.

Leaching losses are found to be sensitive to *type 2* site factors: for a given level of N input much greater losses occur from well-drained, sandy soils than from poorly drained clay soils (Scholefield et al., 1993; Neeteson et al., 1989). The previous history of management also has effects in both grassland and arable systems, with previously undisturbed, highly fertilised land contributing more potentially leachable N through mineralisation than impoverished long-term arable land. Weather patterns are also important, with greater leaching following hot, dry summers, than that following cool, wet ones. All of this variability due to site and management factors makes the establishment of robust relationships between economic output and environmental impacts for a given agricultural system very difficult. Additional complications are the lack of agreement on standard procedures for assessing compliance with the Nitrates Directive, in relation to scale of implementation and with regard to how NO_3^- concentrations are measured and/or calculated from N loads. For UK surface waters, simple empirical relationships between N load leached from the land and N concentrations in rivers can be derived that are applicable at the catchment scale (Scholefield et al., 1996). This coarse scale analysis reveals major effects by dilution of diffuse pollution with water draining from extensively managed uplands in the West of the UK, but not in the East. Thus, whereas in eastern UK, N loads averaging about 30 kg ha^{-1} are associated with river concentrations of <11.3 mg N l^{-1}, in western regions 70 kg N ha^{-1} would be allowable before the Directive is contravened. Since much of the arable agriculture is practised in the East and most of the grassland agriculture is situated in the West of the UK, these relationships and the available information on N leaching loads from arable (Lord & Mitchell, 1998; Lord et al., 1999) and grassland (Jarvis, 1993; Scholefield et al., 1993) systems indicate that currently, our agriculture is barely sustainable in relation to water quality.

Evaluation of sustainability at the field and farm scales will involve taking full account of all of the complexity in the N cycle that is expressed at these scales and for this, more detailed models are required. The decision support model NGAUGE (Brown et al., 2003, this volume) was used to evaluate the ease of compliance of UK dairy farms with Nitrate Vulnerable Zones legislation, which stipulates a limit on N in manure produced on the farm of 170 kg ha^{-1}. NGAUGE was used to simulate N cycling under current practice in several regions and with a range of site conditions and the implications for production and leaching of changing to achieve compliance with the legislation. The study revealed that for compliance, fertiliser inputs should be reduced to 133 kg N ha^{-1}, averaged over the farm, and this would result in average reductions in production and leaching of 15% and 35%, respectively (del Prado et al., 2003, this volume). However, the effect of soil conditions was indicated to be very important, such that on sandy soils, compliance with the NVZ limit for manure N would not necessarily ensure compliance with the Nitrates Directive limit on N concentration.

Gaseous losses of N through emissions of N_2O, N_2 and NH_3 can be substantial, particularly from systems of livestock production (Scholefield & Oenema, 1997). While at present there is no legislation applied at the farm scale to limit such emissions, national governments are bound by international agreements to expedite reductions in environmentally damaging gases. Unfortunately, where the emission of N by the aqueous route is small, due to a combination of site conditions, it is often the case that emission by a gaseous route is large, and vice versa. Table 1 shows N losses calculated using the NGAUGE model resulting from application of 214 Kg N ha^{-1} to grazed grassland fields with contrasting soil types. While NH_3 emissions are little affected, there is marked 'pollution swapping' between N losses through leaching and denitrification. These interactions between N loss pathways may also be brought into play with the implementation of mitigation strategies aimed at any single loss pathway. For example, injection of slurry in order to reduce NH_3 losses may increase N_2O emissions (Chadwick et al., 2000), while switching the temporal distribution of manure and inorganic N fertiliser application from late- to early-season to reduce leaching, will almost certainly increase N loss through denitrification.

Table 1. N losses (kg ha^{-1}) and peak drainage concentrations in parentheses from grazed grassland receiving 214 kg N ha^{-1} calculated using the NGAUGE model and showing the effects of contrasting soil conditions on 'pollution swapping'.

Soil conditions	Leaching	Denitrification	Ammonia
Sandy loam (mod drainage)	39 (15 mgl^{-1})	17	25
Clay loam (poor drainage)	9.2 (mgl^{-1})	37	23

Just as the spatial distribution of inputs to different fields on a farm will influence the ease of achieving sustainability (considered at the scale of the individual farm), due to interactions between site factors already described, there will be similar scaling effects at the catchment scale (due to the variability of N inputs to the farms within). To exemplify effects of between-farm variability at the catchment scale, the NGAUGE model was used to simulate two scenarios: *Scenario 1*: components of the N cycle were calculated for a catchment comprising 32 dairy farms, each of 50 ha (36 fields) and receiving 100 kg N ha^{-1} (160 t N in total).

Scenario 2: components of the N cycle were calculated for a similar catchment of 32 farms each of 50 ha, but with N inputs polarised so that only 8 of the farms receive fertiliser (each receive 400 kg N ha^{-1}), but the total applied to the catchment remains the same as scenario 1.

Table 2 shows that such polarisation of N input to livestock farming has two main effects. One is that considerably less milk can be produced overall, and the second is that the efficiency of N use is greatly reduced, such that on a sandy soil, leaching from the catchment is more than doubled and efficiency of N use (N in milk per unit of N leached) is reduced to a quarter of that achieved without polarisation.

Table 2. *Effects of polarisation of N input to a dairy catchment on N in product (milk) and N leached (kg ha⁻¹) for two soil types, simulated by the NGAUGE model.*

Soil type	Clay loam			Sandy loam		
N output	Milk	Leach	M/L	Milk	Leach	M/L
Scenario 1	59	5	11.8	63	18	3.5
Scenario 2	36	4.5	7.5	35	45	0.8
S1/S2	0.6	0.9		0.5	2.5	

Another aspect of environmental sustainability, that is assuming increasing importance with respect to agriculture, is biodiversity. Modern farming methods have resulted in much reduced botanical diversity due to increased fertility, intensive and frequent harvesting, soil disturbance and the replacement of indigenous species with a narrow range of crop species. Tallowin (2004) has found that, on a sample of Devon livestock farms, species numbers were inversely related to N inputs from fertiliser: while there were >30 species present with zero fertiliser N applied, this was reduced to <10 species with 300-400 kg N ha⁻¹ applied annually (Figure 4). There is no doubt that diverse populations of above-ground organisms rely on a diverse plant population for food and habitats and that the stability of populations and ecosystem dynamics depend on maintenance of evolved relationships between components (Vickery et al., 2001). If biodiversity is to be considered in a quantitative way, as part of the economics-environment equation in land use issues, there needs to be much better definition of biodiversity criteria, and ways to include these criteria in nutrient cycling models must be devised.

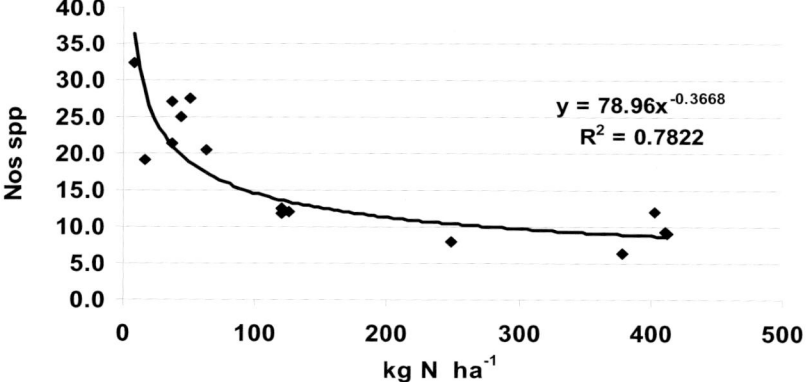

Figure 4. *Species number per field against N input for livestock farms in Devon.*

From the above brief consideration of criteria of sustainability and the efficiency of N use on some example systems of agriculture, it appears that in the UK at least, economically sustainable, intensive agriculture is some way from being environmentally sustainable. There are several solutions to this problem including:

i. subsidise uneconomic farming
ii. relax environmental constraints
iii. increase market value of products
iv. optimise present systems

v. design new systems from specially 'tuned' components and the most effective management strategies

Solution (i) is being implemented by way of the introduction of schemes of payment to farmers providing environmental goods. Solution (ii) is unlikely to be considered and government policy is more likely to introduce even more stringent controls on diffuse pollution from agriculture, e.g. as legislation to enforce the Water Framework Directive is developed. Solution (iii) would seem to be the most worthy of support if the public can be persuaded to pay more for better (healthier) food with guaranteed quality and indeed, if the farmer can actually produce such value added commodities. Organic farming can be classed as a type (iii) solution. To date, however, the demand for organically produced food has been limited and the claimed 'green' credentials of organic farming not entirely proven. In fact, most of our research effort for greater sustainability in agriculture has been channelled in pursuit of type (iv) solutions, but if these do not work, or cannot be implemented cost-effectively, more attention to identifying type (v) solutions may have to be given.

Type (iv) solutions evaluated in research studies include:
- Use of suitable indicators, such as farm nutrient balances and surpluses
- Use of technical diagnostics to assess levels of nutrients in plant, soil and animal components
- Better manure management including use of covered stores, injection of slurry and addition of nitrification inhibitors
- Better quantification of and accounting for nutrient supply through mineralization of organic residues
- Optimisation of rotations for efficient N use and pest control
- Reversion to mixed farming
- Tactical fertiliser application
- Precision fertiliser application
- Balance supply of all nutrients
- Use of more N-efficient plants (e.g. C4 plants such as maize in animal production)
- Optimise dietary protein and energy in ruminant feedstuffs
- Use a more efficient animal/feed combination
- Use N supplied through fixation by legumes rather than from inorganic fertiliser
- Use models and decision support tools to optimise nutrient inputs for specified economic and environmental targets
- Use cover crops in arable agriculture
- Switch to spring cereals from winter sown crops
- Strategic placement of nutrient buffers, farm trees and ponds

Information on the effectiveness of some example type (iv) solutions is given below.

Potential of some strategies for improving efficiency of N use

Use of indicators
Indicators of N use have been defined and used widely as an easily available means of evaluating and comparing efficiency of N use in and between different systems of livestock production

(e.g. Jarvis, 1999). By plotting economic and environmental indicators against each other on the same graph, it should be possible to identify farms that are buoyant economically, but have small environmental impacts. Figure 5 shows such an analysis for a sample of dairy farms located in south west England. While the general trend is one of increasing potential N loss with increasing profit, there are farms, such as the one highlighted in Figure 5, which appears to have a much greater efficiency of N use than the norm. On identification of these more efficient farms, the next step is to use detailed background information to ascertain whether their superior performance can be attributed to farm managements, or to any particularly advantageous economic factors exclusive to that farm.

Another use of indicators is as a means of implementation and control of nutrient management legislation. An important problem with their use in this context, however, is that, because of the great variability in indicator values, the likelihood of success in achieving their goal becomes less certain, the further the indicator is positioned away from that goal (Schroder et al., 2004).

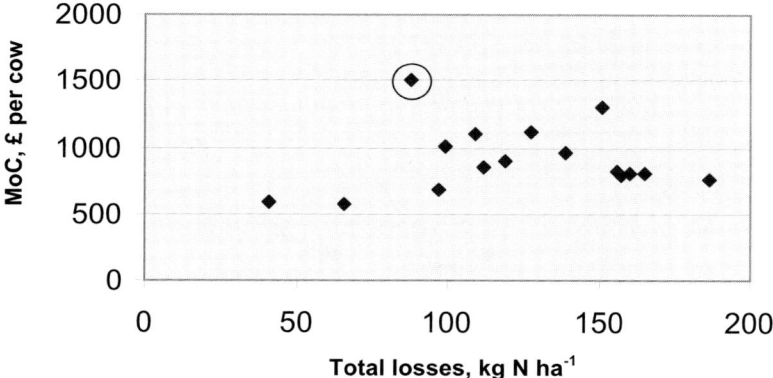

Figure 5. Plot of economic (margin over concentrate feed: MOC) against environmental (total N losses) indicators of performance of a sample of 16 dairy farms in SW England.

Use of technical diagnostics
Measurements of nutrient values of plant components, in soil horizons and of feedstuffs for animals, offers the means to 'diagnose' the efficiency of nutrient use at a given time and location. Values of such diagnostics can be used to trigger and moderate fertiliser applications as well as for inputting to sophisticated mathematical models of nutrient cycling. Each has advantages and disadvantages. The French concept of a Nitrogen Nutrition Index allows an evaluation of the degree of over- or under-fertilisation of a crop according to measurements made on the plant (Figure 6). The soil mineral N value gives an indication of the amount of N surplus to the plants requirements that is susceptible to leaching in the following drainage period (Figure 7). The former is biased to assessment of economic performance, while the latter better indicates environmental impact. Comparison and evaluation of the two have been made by Farruggia et al. (2004).

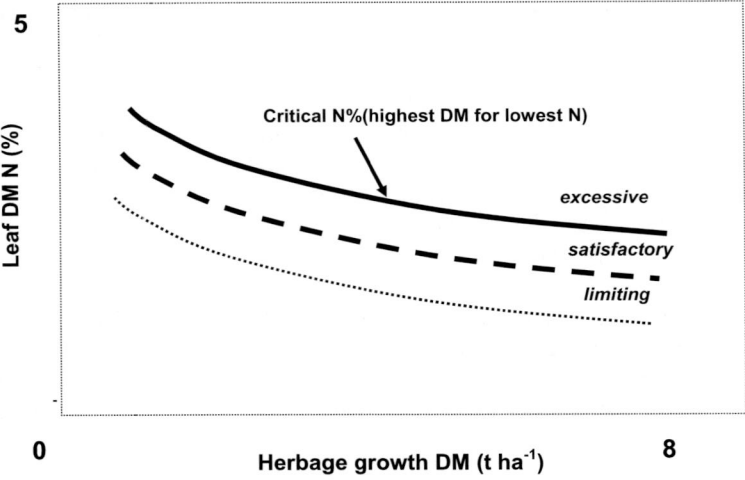

Figure 6. Plot of N content of leaf against dry matter yield of a grass sward increasing during the growing season, used as a diagnostic of crop N nutrition.

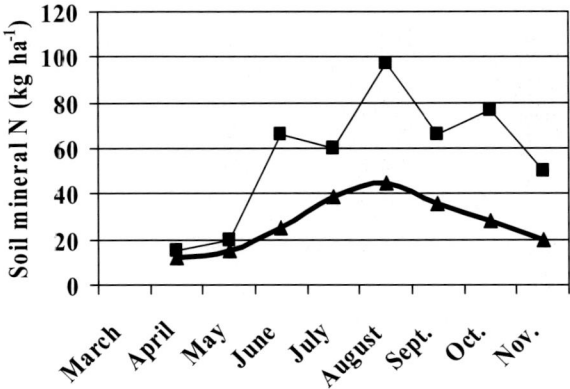

Figure 7. Soil mineral N content (0-30 cm depth) for a grass field receiving fertiliser at the recommended rates (square symbols) and for a field receiving 'tactical' applications of fertiliser according to results of frequent soil testing (triangular symbols).

Reversion to mixed farming

Two existing N cycle models were coupled together to simulate N flows through mixed ley-arable farming systems in the UK. The NCYCLE (Scholefield et al., 1991) derivative NFIXCYCLE and the arable SUNDIAL model (Smith et al., 1996) were joined to enable prediction of N (and C) accumulation under a grassland management and the subsequent use of that N for arable production in a series of rotations. Table 3 shows that for each of the defined efficiency indices, the value for the mixed farming system lies midway between those for the grassland and arable phases, and there was no overall advantage in a coupled system that was discernible by these particular models.

Table 3. Indices of efficiency of N use calculated for grass, arable and mixed ley-arable rotations, made using the coupled NFIXCYCLE and SUNDIAL models.

Efficiency index	Grass	Mixed	Arable
Product N/ N input	0.24	0.47	0.65
Product N + N losses/ N input	0.65	0.89	2.24
Product N / N losses	0.56	0.73	0.80

Balance supply of all nutrients

While it has been a widespread agronomic practice to balance N, P and K applications to agricultural land in Europe, according to crop requirements, S levels have been allowed to diminish as deposition from industrial sources have been attenuated. Figs. 8 and 9 show that redressing an S deficiency on a sandy soil increased the yield of cut herbage, and at high N input substantially reduced NO_3^- leaching. The effects of the same treatments applied to a clay soil with adequate S content were very small.

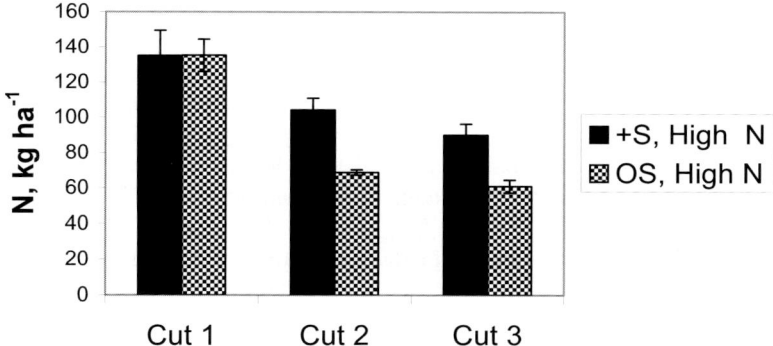

Figure 8. Effect of application of S to a cut grass sward on an S deficient, sandy soil (from Brown, 2000).

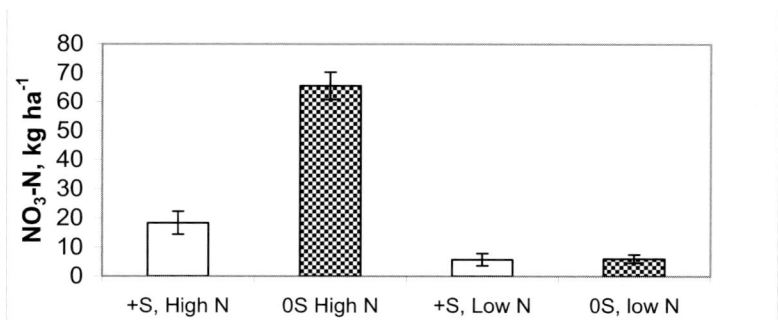

Figure 9. Effect of application of S to a cut grass sward on an S deficient sandy soil on NO_3^- leaching in response to high N application (clear bars) and low N application (hatched bars) (from Brown, 2000).

Use N from legumes rather than fertiliser

One of the most interesting comparisons between the fate of N from fixation and that from fertiliser has been the comparison of grass/white clover based systems and those based on grass alone receiving inorganic fertiliser. Early studies showed that N loss through NO_3^- leaching from grass/white clover is generally much smaller than from highly fertilised grass (e.g. Parsons et al., 1991) and this had suggested that legume-based pasture systems were environmentally benign. Nevertheless, it is now generally believed that the smaller N loss from grass/white clover is due to the lower level of production and that the two systems (grass/white clover receiving no N fertiliser and grass only receiving inorganic fertiliser) would release similar amounts of NO_3^- at equal levels of production per ha (Cuttle et al., 1992; Tyson et al., 1997; Houda et al., 1998). Moreover, NO_3^- leaching beneath clover leys in dairy rotations (Eriksen et al., 1999) and after ploughing (Djurhuus & Olsen, 1997) was found to be substantial, and not compatible with the objectives of environmental sustainability.

If this is so, we might expect NO_3^- leaching from beneath legume-based swards to increase with increasing legume content and the level of N fixation per ha. While annual amounts of N_2 fixed in legume-grass pastures range typically between 100-250 kg ha^{-1} (West & Mallarino, 1996; Thanopoulos & Ledgard, 2000), there is some recent evidence for this (Loiseau et al., 2001): these workers recorded leaching losses from lysimeters sown with pure white clover within the range 28-140 kg N ha^{-1} over a 6 year period, whereas the range for grass-white clover during the same period was 1-19 kg N ha^{-1}. Moreover, Halling & Scholefield (2001) reported a good correlation at two Swedish sites between the yield of legume in mixed swards cut for silage and the level of potentially leachable mineral N accumulated in autumn (Figure 10).

Although it now seems that high levels of clovers in grazed swards may give rise to substantial levels of NO_3^- leaching from beneath them, we have little information on the leaching potential from beneath alternative legumes. However, the small amount of information on lucerne-based swards in recent US studies (Toth & Fox, 1998; Russelle et al., 2001) indicates that, despite

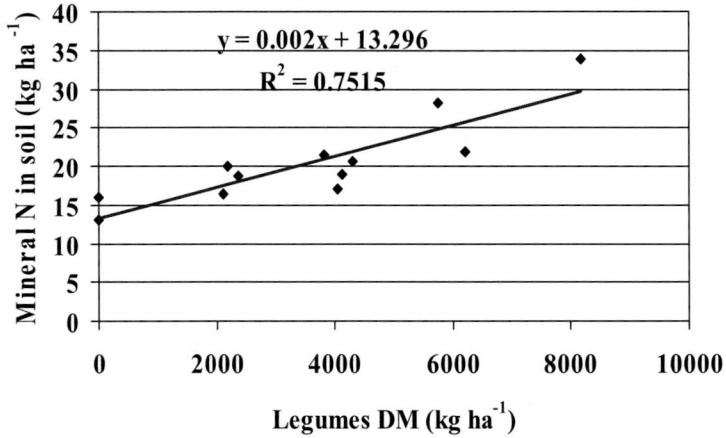

Figure 10. Relationship between the yield of legume in a mixed grass/legume sward and the level of mineral N accumulated in the soil during autumn at two Swedish sites.

giving dry matter yields as great as those from clover at many sites, NO_3^- leaching was much reduced and the ability of lucerne to actively sequester N from N-rich soil was demonstrated.

Conclusions and research needs

We have:
- Quantified knowledge of N cycles in current conventional systems
- Started to quantify states of current systems against sustainability criteria
- Identified sensitivity of both productivity and environmental impacts to changes in properties of system components
- Identified and implemented some strategies for improving sustainability

We now need to:
- Obtain better understanding of the links between the N, P and C cycles
- Adopt 'top-down' approaches to system definition
- Incorporate economic and biodiversity criteria more into research studies
- Develop ways to incorporate spatial and scaling effects
- Attempt to define whether, and by how much, improved sustainability is scientifically and technically feasible, before recommending changes to farming practice

Meeting the challenge of reconciling economic productivity with acceptable environmental impacts from agricultural systems demands that research effort is focused at the systems level, with major concerted activities in modelling, farm monitoring and conventional hypothesis-based scientific investigation. Progress will be accelerated by greater interactions between those concerned with N pollution and biodiversity issues and by a greater degree of consideration given to economic criteria in system studies. Clearly, the effective transfer to the practising farmer of the new knowledge obtained, will be critical to the implementation of novel strategies. Moreover, the greater the difference between the present and the newly identified sustainable systems, the greater the difficulties of implementation without concerted involvement of government policy makers and leaders of the food industry. Ultimately, retaining a productive agriculture capable of 'feeding the nation', but with acceptably small environmental impacts, may depend upon how willing we are to resist global market forces to ensure the supply of high quality, traceable food. Without clear policy objectives on this issue and the means and resources to achieve them, success will be limited.

References

Brown, L, Scholefield, D., Jewkes, E.C., Preedy, N., Wadge, K and Butler, M. (2000). The effect of sulphur application on the efficiency of nitrogen use in two contrasting grassland soils. *Journal of Agricultural Science* 135, 131-138.

Chadwick, D., Misselbrook, T. and Pain, B. (2000). Is Europe reducing its ammonia emissions at the expense pf the global environment? In: Proceedsings of the Second International Conference on Air Pollution from Agirucltural Operators, pp. 1-9. Americal Society of Agricultural Engineers Michigan, USA. ISBN 1-892769-12-3.

Cuttle, S.P., Hallard, M., Daniel, G. and Scurlock, R.V. (1992). Nitrate leaching from sheep-grazed grass/clover and fertilised grass pastures. Journal of Agricultural Science, 119, 335-343.

Djurhuus, J. and Olsen, P. (1997). Nitrate leaching after cut grass/clover leys as affected by time of ploughing. Soil Use and Management, 13, 61-67.

Eriksen, J., Askegaard, M. and Kristensen, K. (1999). Nitrate leaching in an organic dairy/crop rotation as affected by organic manure type, livestock density and crop. Soil Use and Management, 15, 176-182.

Farruggia, A., Gastal, F. and Scholefield, D. (2004). Assessment of the nitrogen status of grassland. Grass and Forage Science (in press).

Goulding, K.W.T., Poulton, P.R., Webster, C.P. and Howe, M.T. (2000). Nitrate leaching from the Broadbalk Wheat Experiment, Rothamsted, UK, as influenced by fertilizer and manure inputs and the weather. Soil Use and Management, 16, 244-250.

Jarvis, S.C. (1999). Accounting for nutrients in grassland: challenges and needs. In. A.J. Corrall (ed). Accounting for Nutrients. A challenge for grassland farmers in the 21st century. British Grassland Society Occasional Symposium No 33. pp. 3-13.

Jarvis, S.C. (1993). Nitrogen cycling and losses from dairy farms. Soil Use and Management, 9, 99-105.

Halling, M.A. and Scholefield, D. (2001). Correlation between yield of forage legumes in grass mixtures and accumulation of soil mineral nitrogen in Sweden. In. Proceedings of the 19th International Grassland Congress, Brazil, pp. 109-111.

Houda P.S., Moynagh, M., Svobada, I.F. and Andreson, H.A. (1998). A comparative study of nitrate leaching from intensively managed monoculture grass and grass-clover pastures. Journal of Agricultural Science, 131, 267-275.

Ledgard, S.F., Penno, J.W. and Sprosen, M.S. (1999). Nitrogen inputs and losses from clover/grass pastures grazed by dairy cows, as affected by fertiliser application. Journal of Agricultural Science, 132, 239-249.

Loiseau, P., Carrere, P., Lafarge, M., Delpy, R. and Dublanchet, J. (2001). Effect of Soil-N and urine-N on nitrate leaching under pure grass, pure clover and mixed grass/clover swards. European Journal of Agronomy, 14, 113-121.

Lord, E.I., Johnson, P.A. and Archer, J.R. (1999). Nitrate Sensitive Areas.

Lord, E.I. and Mitchell, R.D.J. (1998). Effects of nitrogen inputs to cereals on nitrate leaching from sandy soils. Soil Use and Management, 14, 78-83.

Neeteson, J.J., Greenwood, D.J. and Draycott, A. (1989). Model calculations of nitrate leaching during the growth period of potatoes. Netherlands Journal of Agricultural Science, 37, 237-256.

Neeteson, J.J. and Wadman, W.P. (1987). Assessment of economically optimum application rates of fertilizer N on the basis of response curves. Fertilizer Research, 12, 37-52.

Parsons, A.J., Orr, R.J., Penning, P.D. and Lockyer, D.R. (1991). Uptake, cycling and fate of nitrogen in grass/clover swards continuously grazed by sheep. Journal of Agricultural Science, 116, 47-61.

Russelle, M.P., Lamb, J.F.S., Montgomery, B.R., Elsenheimer, D.W., Bradley, S. and Vance, C.P. (2001). Alfalfa rapidly remediates excess inorganic nitrogen at a fertiliser spill site. Journal of Environmental Quality, 30, 30-36.

Scholefield, D. and Oenema, O. (1997). Nutrient cycling within temperate agricultural grasslands. Proceedings of the 18th International Grassland Congress, Saskatoon, Canada, 1997.

Scholefield, D., Lord, E.I., Rodda, H.J.E. and Webb, B.W. (1996). Estimating peak nitrate concentrations from annual nitrate loads. Journal of Hydrology, 186, 355-373.

Scholefield, D. and Smith, S.U. (1996). Nitrogen flows in ley-arable systems. In: D.Younie (ed). Legumes and Sustainable Farming Systems. University of Aberdeen, UK.

Scholefield, D., Tyson, K.C., Garwood, E.A., Armstrong, A.C., Hawkins, J. and Stone, A.C. (1993). Nitrate leaching from grazed grassland lysimeters: effects of fertiliser input, field drainage, age of sward and patterns of weather. Journal of Soil Science, 44, 601-613.

Scholefield, D., Lockyer, D.R., Whitehead, D.C. and Tyson, K.C. (1991). A model to predict transformations and losses of nitrogen in UK pastures grazed by beef cattle. Plant and Soil, 132, 165-177.

Schroder, J.J., Scholefield, D., Cabral, F. and Hofman, G. (2004). The effects of nutrient losses from agriculture or ground and surface water quality: the position of science in developing indicators for regulation. Environmental Science and Policy, (in press).

Smith J.U., Bradbury, N.J., Addiscott TM. (1996). SUNDIAL: A PC-based system for simulating nitrogen dynamics in arable land. Agronomy Journal, 88 (1): 38-43.

Tallowin, J.R.B., Smith, R.E.N., Goodyear, J. and Vickery, J.A. (2004). Spatial and structural uniformity of lowland agricultural grasslands in England. Agriculture, Ecosystems and the Environment (submitted).

Thanopalous, R. and Ledgard, S.F. (2000). How much N_2 is fixed by perennial clovers in Mediterranean sown pastures. Options Mediterraneennes, 45, 327-330.

Toth, J.D. and Fox, R.H. (1998). Nitrate losses from a corn-alfalfa rotation: lysimeter measurement of nitrate leaching. Journal of Environmental Quality, 27, 1027-1033.

Trindade, H., Coutinho, J., van Beusichem, M.L., Scholefield, D. and Moreira, N. (1997). Nitrate leaching from sandy loam soils under a double cropping forage system estimated fom suction probe measurements. Plant and Soil, 195, 247-256.

Tyson, K.C., Scholefield, D., Jarvis, S.C. and Stone, A.C. (1997). A comparison of animal output and nitrate leaching losses recorded from drained fertilized grass and grass'clover pasture. Journal of Agricultural Science, 129, 315-323.

Vickery, J.A., Tallowin, J.R., Feber, R.E., Asteraki, E.J., Atkinson, P.W., Fuller, R.J. and Brown, V.K. (2001). The management of lowland neutral grasslands in Britain: effects of agricultural practices on birds and their food resources. Journal of Applied Ecology, 38, 101-104.

West, C.P. and Mallarino, P. (1996). Nitrogen transfers from legumes to grass. In: (eds). R.E. Joost and C.A. Roberts. Nutrient Cycling in Forage Systems. Forage Systems Potash and Phosphate Institute, Norcross, GA. Pp 167-173.

Forage maize production as affected by tillage, N source and nitrification inhibitors

D. Báez, J. Coutinho, N. Moreira and H. Trindade
Department of Plant Science and Agricultural Engineering, UTAD, Ap. 1013, 5001-911 Vila Real, Portugal

Introduction

Double cropping systems based on maize (*Zea mays*) as a summer crop are an efficient way to increase forage production in areas with intensive land use. Organic fertilisers, such as cattle-slurry, are often considered by farmers as a waste product and their use may produce atmospheric pollution through gaseous N emissions. Several studies demonstrated that the use of nitrification inhibitors (NI) applied together with mineral and organic fertilisers reduce the environmental risk of N losses and also increase crop yields. In intensive cropping systems there is a considerable interest in no-tillage practices to reduce soil erosion and labour costs in crop production. The aim of this work was to evaluate forage maize production after the application of inorganic and cattle-slurry fertilisers containing two different NIs, 4-dimethylpyrazole phosphate (DMPP) and dicyandiamide (DCD) and to test the no-tillage effect.

Materials and methods

The experiment was carried out at Vila Real (Portugal) during 2002 on a poorly drained silty loam soil classified as Dystric Cambisols (FAO classification) with (0-30 cm) pH value of 6.4, OM content of 29 g kg^{-1} and 4.9 cmol(+) kg^{-1} of CEC. The trial was laid out as a two factor factorial with a split-block layout and three replications. One factor consisted of N source with 9 treatments, Ammonium Sulphate Nitrate (ASN, 19.5% NH_4^+-N 6.5% NO_3^--N), 2 NIs-stabilised N fertilisers Entec®26 (COMPO, 1% DMPP relative to NH_4^+-N in the ASN) and Nitrotop (ADP, 8% NH_4^+-N 16% Amide-N, 4% DCD-N relative to total-N) and 6 combinations of surface banded (SS) or injected (SI, 30 cm depth) cattle slurry, either with or without DMPP and DCD as NI. There was also a no N treatment control. The second factor consisted of two tillage systems, no-till (NT) and conventional tillage practice (T, mouldboard plough). In slurry treatments, DMPP (25% solution) was applied at a rate of 4 l ha^{-1} and DCD-N was applied at a rate of 5% relative to total N in the slurry. N application rate was targeted at 160 kg N ha^{-1} in all treatments. Silage maize, cv. Belmont (300 FAO, 95,000 seeds ha^{-1}), was sown on June 11 and harvested on September 23. At harvest, samples of maize yield components (cob and stover) were collected for dry matter yield and Kjeldhal N analyses.

Results and discussion

Compared with NT, tillage increased total and cob DM yield, respectively, by 1.98 and 1.22 t ha^{-1}, and improved cob DM content by 10% (from 22.8 to 33.2% DM). Stover DM yield was affected only by N source. The interaction between tillage and N source was significant for total, cob and stover yield. In the tillage treatments (Figure 1), mineral fertilisers Entec and

Nitrotop, showed the highest total yields (but not significantly different from ASN, SS+DCD and SI without DMPP treatments), whereas in NT systems no N source effect was found.

With respect to N uptake (Figure 2), tillage increased significantly stover and cob N uptake by 9.2 and 13.0 kg N ha^{-1}; in contrast, N uptake was not affected. In the tillage treatments, Entec showed the highest total N extraction, which was significantly different from slurries, whereas in the no till systems, similar to DM yield, no N source effect was found. Production differences between both sowing types, probably resulted from the soil texture and the lack of soil moisture at planting. No-tillage systems produce lower early season soil temperatures than conventional mouldboard ploughing, especially in fine-textured soils which influences emergence rates negatively. Reduced tillage systems also increase soil mechanical resistance, which can delay root development and increase water stress, if dry conditions occur early in the growing season. Comparing the two NIs (and although the differences were statistically not significant), the effects were not clear and depended on fertiliser form and tillage system. Our data suggest DMPP was advantageous for cob production and N uptake in cob and stover when added to mineral fertilisers (Entec 26) in the tillage treatments. However, DCD was more effective when applied together with slurry (SS in tillage system and with SI in no- till system).

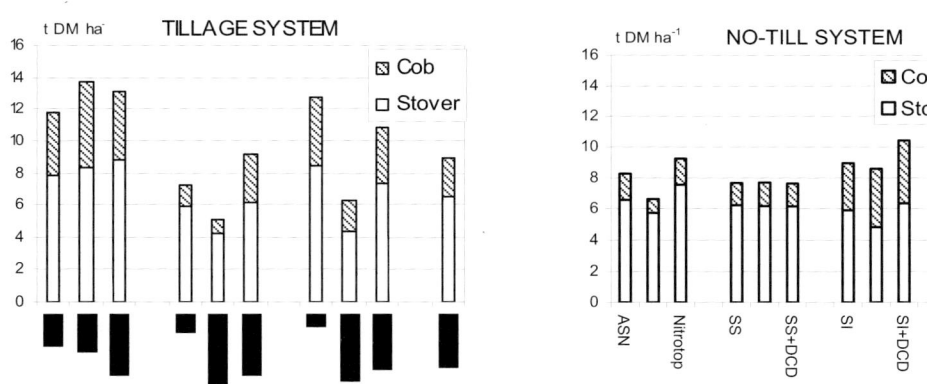

Figure 1. Average cob and stover dry matter (DM) yield (t ha^{-1}) of maize as affected by N source and tillage system.

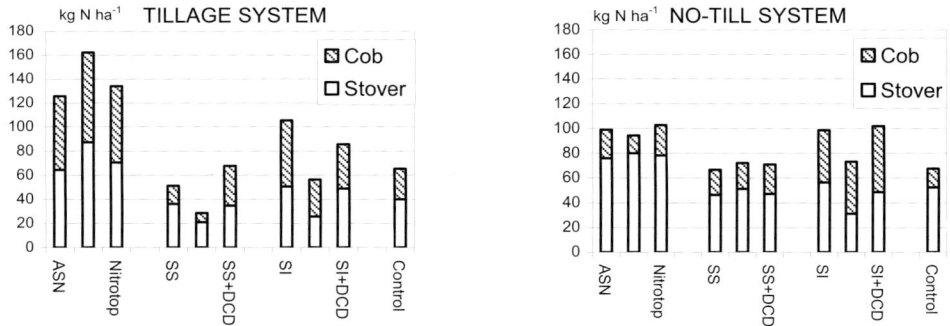

Figure 2. Average cob and stover N uptake (kg N ha^{-1}) of maize as affected by N source and tillage system.

Conclusions

Tillage increased total DM yield but not N uptake by maize. The effects of N source and NIs on DM yield and N uptake were especially found in the conventional tillage system in which DMPP had a positive effect in Entec and DCD with slurry when it was applied to the surface.

Acknowledgements

This study was funded by project N° 177-Acção 8.1 DE & D-Programa AGRO and D. Báez held a post-doc grant from Spanish government. We thank COMPO Agricultura and Adubos de Portugal for supplying DMPP and DCD.

Is the N balance a good indicator of nitrogen losses in arable systems?

N. Beaudoin[1], B. Mary[1], F. Laurent[2], G. Aubrion[2] and J.K. Saad[1]
[1]*INRA, Unité d'Agronomie, rue Fernand Christ, 02007 Laon Cedex, France*
[2]*Arvalis - Institut du végétal, Station expérimentale, 91720 Boigneville, France*

Introduction

A sustainable agriculture must avoid any excess of mineral N in order to save chemical fertiliser and reduce N losses. The N balance, i.e. the difference between N inputs and outputs in soil, is an indicator of this excess. It has been recently proposed in France as a measure for taxing bad agricultural practices and improving water quality. However, a low N balance may result in a reduction of grain quality and sometimes of yield. What is the environmental meaning of the N balance?

Materials and methods

The results come from two complementary mid term experiments (Thibie and Bruyères) established in 1990 in Northern France. The main crops are winter wheat, sugarbeet, winter barley and spring peas. All soils have a good internal drainage. The Thibie experiment compares the effect of 2 factors: the N rate (recommended, or 35% reduced rate) and the absence/presence of a catch crop (CC) after each main crop (Mary et al., 2002). Bruyères' catchment, spread over 200 ha, includes 21 fields. Agricultural practices consisted in optimised N fertilisation and CC sown before spring crops from 1990 to 1996; a further 20% reduction of N fertiliser was imposed from 1997 (Beaudoin et al., 1999). N fertilisation was exclusively added as mineral N and was optimised using the AZOBIL software (Machet et al., 1990). Leaching was calculated either from porous suction cups and drainage estimated at Thibie or by measuring soil mineral N at different dates and using the LIXIM model at Bruyères (Mary et al., 1999). Dry matter and N content of exported and returned products have been measured. The N balance was calculated as: $B = S + F - E$ where S = biological fixation (calculated), F = fertiliser, E = exported N.

Results

For 1991-1999, the mean fertilisation rates of the optimised and reduced treatments were 180 and 122 kg N ha^{-1} at Thibie and 152 and 118 kg N ha^{-1} at Bruyères (peas excluded), respectively. At Thibie, the mean balances were 34-39 kg N ha^{-1} yr^1 with optimised fertilisation and 15-18 kg N ha^{-1} yr^{-1} for the reduced one. The 32% fertiliser reduction reduced the N balance by 69% and leaching by 16%. In contrast, the establishment of CC did not change the N balance, but reduced leaching by 62% (Figure 1a). The drainage was decreased with CC (data not shown).

At Bruyères, the mean balance was +28 and -3 kg N ha^{-1} yr^{-1} with optimised fertilisation and reduced N, respectively (Figure 1b). From 1997 to 1999, reducing fertiliser rate by 20% gave a negative N balance, but increased leaching by 30%. This paradox is attributed to some recent failures in CC establishment. Figure 2 shows N balance versus N fertiliser rate for the winter

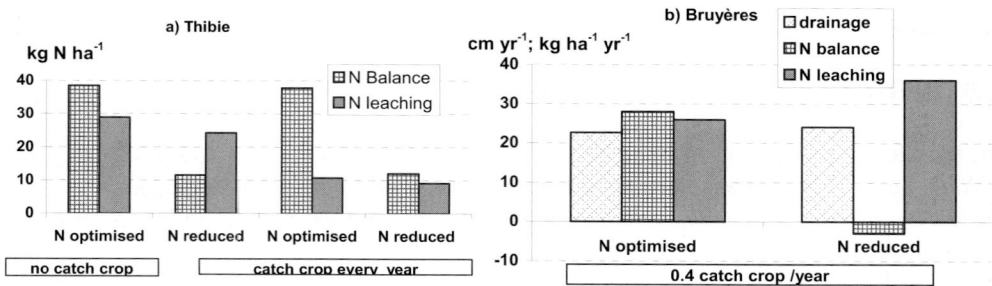

Figure 1. Mean values of N balance and N leaching at a) Thibie and b) Bruyères.

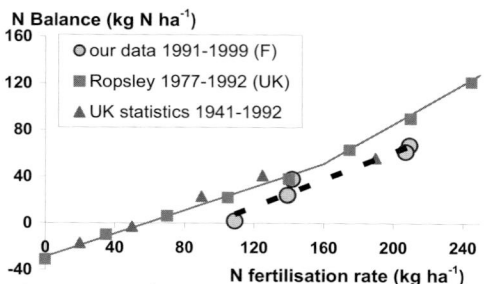

Figure 2. N balance vs N rate for w. wheat.

wheat crop. Our data fall below those reported by Davies & Sylvester-Bradley (1995), however the slope of the regression line (0.58) is similar.

Discussion

The results show that NO_3^- leaching is not correlated to N balance when fertilisation varies or when CC are introduced. A positive correlation can be found in systems which have a large excess of N, such as intensive cattle areas. Our results may be representative of arable systems with optimised fertilisation and moderate N balance: the N balance corresponds to the combination of 3 fluxes: gaseous emissions (NH_3, N_2, N_2O), soil immobilisation and leaching. The use of CC does not change the balance, whereas it strongly reduces leaching, resulting in N immobilisation and C-N storage in soil. Incorporating *vs.* exporting crop residues would have similar effects. Conversely, a reduction in mineral fertilisation markedly reduces the N balance, but has little effect on NO_3^- leaching. Its main effect probably results in decreasing gaseous N emissions. Lowering the N balance in arable systems would probably reduce more gaseous emissions than leaching. Therefore a nil N balance does not necessarily prevent losses and the optimum N balance to aim for, is slightly positive.

Acknowledgements

This work was funded by A.R.E.P. and the Water Agency of Seine Basin.

References

Beaudoin, N., Makowsky, D., Parnaudeau, V. and Mary, B. (1999). Impact of agricultural scenarios on nitrate pollution at the catchment scale . Proc. 10th Nitrogen Workshop, vol 2, IV-6.

Davies, D.B. and Sylvester-Bradley, R. (1995). The contribution of fertiliser nitrogen to leachable nitrogen in the UK : a review. Journal of Science, Food and Agriculture, 68, 399-406.

Machet, J.M., Dubrulle, P. and Louis, P. (1990). AZOBIL: a computer program for fertiliser N recommendations based on a predictive balance-sheet method. Proc. 1st ESA Congress, Paris.

Mary, B., Beaudoin, N., Justes, E. and Machet, J.M. (1999) Calculation of nitrogen mineralisation and leaching in fallow soils using a simple dynamic model. European Journal of Soil Science, 50, 549-566

Mary, B., Laurent, F. and Beaudoin, N. (2002) La gestion durable de la fertilisation azotée (Sustainable N. supply). Proc. 65th IIRB Congress, Brussels, 59-65.

A low disturbance technique for applying slurry on forage land

S. Bittman[1], L.J.P. van Vliet[1], C.G. Kowalenko[1], S. McGinn[2], A.K. Lau[3], N. Patni[1], T. Forge[1], N. McLaughlin[4], D.E. Hunt[1], F. Bounaix[1] and A. Friesen[1]
[1]*Agriculture and Agri-Food Canada (AAFC), PARC, Box 1000, Agassiz, BC, Canada, V0M 1A0*
[2]*AAFC, Lethbridge Research Centre, Lethbridge, AB, Canada*
[3]*University of British Columbia, Vancouver, BC, Canada*
[4]*AAFC, ECORC, Ottawa, ON, Canada*

Introduction

Manure nutrients are required for forage crops, but applying slurry manure onto land can lead to environmental concerns including water and air contamination. Some of these risks can be reduced by injecting manure into soil, but surface application by banding or broadcasting is more common on forages because injection can damage swards and is difficult to carry out on stony and sloping land. Techniques for applying manure with minimal soil disturbance and environmental impact are required. We developed an implement (SSD) that surface-bands slurry over vertical slots made by a soil aerator. The intent was to improve soil infiltration of the slurry with minimum soil disturbance and sward damage. This study compared effects of applying dairy slurry on a perennial grass sward by broadcasting, surface-banding and the SSD applicator, and compared implement draft required for the SSD and typical injectors.

Materials and methods

The SSD applicator (Holland Equipment, Norwich, ON) creates vertical aeration slots with 20 cm long ground-driven tines spaced 19 cm apart (10 slots m^{-2}); the slurry is banded directly over the slots so it soaks quickly into the soil. The size of slots is set by the depth and offset angle (0-10°) of the tines (15 cm depth and 2.5° offset in these trials). The study was conducted on sandy to silt loam soil at Agassiz, BC on established swards of tall fescue (*Festuca arundinacea* Schreb.) and orchardgrass (*Dactylis glomerata* L.) in 2000 and 2001. Dairy slurry was applied at 60-95 kg NH_4^+-N ha^{-1} (60-70 m^3 ha^{-1}, 5% dry matter). Ammonia emission was measured by micrometeorological flux gradient technique. Odour samples were assessed with a dynamic dilution olfactometer. Runoff was collected from a field with 3-5% slope. Draft was measured in the field with an instrumented drawbar.

Results and discussion

The SSD slots did not reduce grass yield, even when repeated 2-3 times per year (Table 1), unlike some injection systems. The SSD significantly increased the yield response of tall fescue (2000) and orchardgrass (2001) to manure (relative to broadcasting), on some cuts and over multi-cut totals. Surface banding was intermediate. Similar results were obtained for N uptake (not shown). About 50% of NH_4^+-N in broadcast-applied manure was lost as NH_3 within 14 d of application (Table 2). The SSD reduced NH_3 loss by 48%, relative to broadcasting. Most of the emission and emission reduction occurred in the first 24h after application. The SSD applicator reduced odour emission rate by as much as 36%, particularly in the 0.5h period after application (Lau et al., 2003). The SSD greatly reduced the amount of runoff and nutrients (Table 3), probably because of higher infiltration rates. This was particularly significant soon after the start of autumn rains; by late winter/spring, aeration slots were partially filled with soil and residues. The draft requirement for the SSD depended on depth of penetration and offset angle. At the 2.5° offset used in the trials, the SSD required less power than the Yeter (single offset disk) and the same power as the Husky (304-mm wide cultivator tooth) injector, despite 4 times as many emitters per unit applicator width.

Table 1. Yield (t ha⁻¹) of tall fescue (2000) and orchardgrass (2001) after receiving dairy manure with different applicators.

| | Manure | 2000 | | | | 2001 | | |
		Jul 7	Aug 30	Oct 19	Total	Jul 9	Sep 11	Total
Control	no	1.92c[1]	0.65b	0.40b	2.97c	1.03c	1.60b	2.63c
SSD	no	1.92c	0.71b	0.41b	3.04c	1.05c	1.58b	2.64c
Banding	yes	2.85ab	1.63a	1.34a	5.82ab	2.12a	2.87a	4.99ab
Broadcast	yes	2.56b	1.71a	1.44a	5.71b	1.79b	2.87a	4.66b
SSD	yes	2.98a	1.82a	1.44a	6.25a	2.13a	3.06a	5.19a

[1]values in columns not followed by same letter are significantly different at P< 0.05

Table 2. Emission of NH_3-N (kg NH_3-N ha⁻¹) from dairy slurry for 2 wk after broadcast or SSD application at 63 kg NH_4^+-N ha⁻¹ (mean of 4 trials).

Day	1	2	3	4-7	7-14	Total
Broadcast	25.4a[1]	2.1a	1.0a	1.6a	1.0a	31.1a
SSD	13.6b	1.1b	0.5a	1.1a	-0.3a	16.1b

[1]values in columns not followed by same letter are significantly different at P< 0.05

Table 3. Accumulated (Nov. 2001- April 2002) runoff and nutrient loadings after slurry application in late Oct. 2001.

	Runoff (mm)	NH_4^+-N (kg ha⁻¹)	Total N (kg ha⁻¹)	PO_4^{2-}-P (kg ha⁻¹)	Total P (kg ha⁻¹)
Broadcast	27.6a	1.2a	7.0a	0.12a	0.89a
SSD	15.0b	0.5b	1.3b	0.04b	0.28b

[1]values in columns not followed by same, letter are significantly different at P< 0.05

Conclusions

Results from two years of research in south coastal BC show that the SSD applicator improved crop yield and reduced NH_3 volatilisation, odour and surface runoff compared with conventional surface broadcast application. The SSD did not damage grass swards and required relatively low draft. Applicators 2.5 to 10 m wide have been in commercial production since Jan. 2001, and in extensive farm use in North America, proved to be durable, rapid and of relatively low cost.

Acknowledgements

This project was funded by Investment Agriculture, AAFC-MII, and Holland Equipment Ltd.

Reference

Lau, A.K., Bittman, S. and Lemus, G. (2003). Odour measurement for manure spreading using a subsurface deposition applicator. Journal of Environmental Science and Health,. 38, 233-240.

Effects of field history on the establishment of white clover in association with perennial ryegrass

L.M. Bommelé[1], D. Reheul[1], N. Van Eekeren[2] and F. Nevens[1]
[1]*Ghent University, Department of Plant Production, Coupure Links 653, Gent, Belgium*
[2]*Louis Bolk Institute, Driebergen, The Netherlands*

Introduction

Where ploughing is possible and the climate permits arable cropping, a ley-arable rotation is preferable to permanent grassland and permanent arable land (Younie & Hermansen, 2000). According to these authors, temporary grasslands provide clean grass (uncontaminated with larvae), provide an opportunity for control of perennial weeds and ensure a high clover content. According to Simon et al. (1997), the clover content in association with perennial ryegrass is strongly variable and unpredictable, as it is affected by environmental and sward-specific factors and some management practices. In an existing experiment in Belgium (Nevens & Reheul, 2003), grass/clover was sown in plots with a different history in order to study the establishment and changes in white clover content in newly sown grasslands.

Materials and methods

The experiment was conducted on a sandy loam soil at the experimental farm of Ghent University in Melle. During spring 2002, we sowed a mixture of 40 kg perennial ryegrass (*Lolium perenne* cvs. 'Plenty' + 'Roy') and 4 kg white clover (*Trifolium repens* L. cv. 'Huia'). The treatments were five backgrounds and four N application rates (0, 100, 300 and 400 kg N ha^{-1}). The backgrounds were: (1) grass/clover installed in ploughed down 35-year old grassland, (2) grass/clover installed in ploughed down 3-year old grassland, (3) grass/clover installed in ploughed 35-year old arable land, (4) grass/clover installed in ploughed 3-year old arable land, and (5) 35-year old permanent grassland as a reference. A part of the prepared seed bed was left uncultivated (fallow) to study the N-mineralisation of soil organic matter. The design was a split plot; the background being the main factor. At each silage cut, the grass was weighed in the field. A representative sample of ca. 700 g was dried (for 12 h at 75°C) to determine the total dry matter (DM) yield. Another fresh sample of about 300 g was taken in July, August, September and October 2002 to separate the clover from the grass ; and we calculated the clover share as a proportion in the total herbage yield of four successive cuts. We also measured the mineral N content of the soil under fallow and 0 N plots on a monthly basis from April to October 2002. The difference in soil N content between the fallow and 0 N plots gives an estimate of the available soil N for plant growth. At the end of the growing season, the NO_3^--N content of the soil profile (0 - 90 cm) was measured with a nitrate-specific electrode after a 1% $KAl(SO_4)_2$ extraction of the soil.

Results and discussion

The observed grass/clover DM yield obtained from the old plots and from the newly sown plots in 2002 ranged from 6100 (ploughed 3-year old grassland without fertilisation) to 15400 kg DM ha^{-1} yr^{-1} (permanent grassland, 400 N). None of the resown swards outyielded the permanent grassland, because of the spring sowing. The yield decreased from 2500 to 4000 kg DM ha^{-1} yr^{-1} as compared with the permanent grassland depending on the history and the mineral N application. The average clover establishment and its persistence were dependent on the mineral N application rate and on the background of the newly sown plots (Figure 1). Under low N fertilisation (0 and 100 kg N ha^{-1}), the clover content was significantly different between backgrounds. We found the highest clover content when the sward was established in arable plots. To provide information on the further development of the clover share, the experiment will be continued for at least one more year.

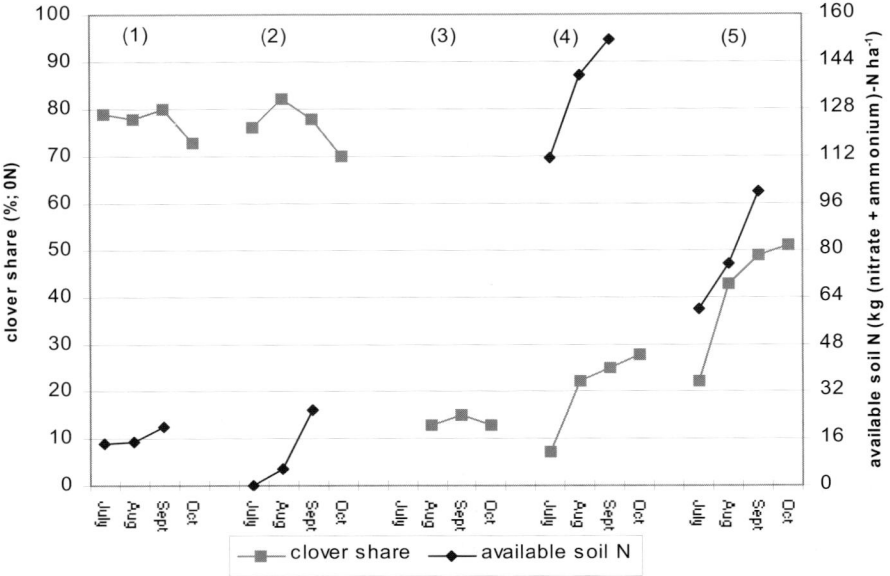

Figure 1. *White clover % in the total herbage yield and available soil N for plant growth (kg (NO$_3$+NH$_4^+$)-N ha^{-1}) in 0 N plots for four successive cuts from July to October 2002 in five backgrounds from left to right : (1) grass/clover installed in ploughed 35-year arable land and (2) 3-year arable land, (3) old permanent grassland as a reference, (4) grass/clover installed in turned down 3-year old grassland and (5) 35-year old grassland.*

The residual soil N did not exceed the limit of 90 kg NO$_3^-$-N ha^{-1} (data not shown), a legally defined threshold in Flanders. Probably, a large part of the mineralised N was taken up by roots and stubbles of the new grass swards or was fixed to soil particles.

Conclusion

The yield performance of permanent grassland was higher than the newly installed swards during spring 2002. During the year of installation, the establishment and the persistence of white clover was related to the mineralised soil N. We found the highest clover content when a new sward was grown in arable land.

References

Nevens, F. and Reheul, D. (2003). Permanent grassland and 3-year leys alternating with 3 years of arable land: 31 years of comparison. European Journal of Agronomy, 19, 77-90.

Simon, J.C., Leconte, D., Vertès, F. and Le Meur, D. (1997). Maîtrise de la pérennité du trèfle blanc dans les associations. Fourrages, 152, 483-498.

Younie, D. and Hermansen, J. (2000). The role of grassland in organic livestock farming. In: Soegaard *et al.* (eds), Grassland Farming, Balancing Envrionmental and Economic Demands, Proceedings of the 18th General Meeting of the European Grassland Federation, Aalborg, Denmark, 22-25 May 2000, 493-509.

Nitrogen losses in relation to rice varieties, growth stages, and nitrogen forms determined with the ^{15}N technique

N.C. Chen and S. Inanaga
Laboratory of Plant Nutrition, Faculty of Agriculture, Kagoshima University, Kagoshima 890-0065, Japan

Introduction

N losses in soil-plant ecosystems have long been a high-priority research area to maximise N use efficiency and to minimise environmental pollution caused by excess N application. Hydroponic studies (e.g. Ashraf, 1997) indicate that N can be lost from the plant itself. As a part of the JSPS project (P02214) 'Mechanisms of nitrogen volatilisation from plants', this work studied the relationship between N losses in relation to rice varieties and N forms of fertilisers applied at different growth stages using a ^{15}N labelling technique.

Materials and methods

One-month old Japonica or Indica seedlings were transplanted into 108 pots (6 l) with Kimura culture solution containing 20 mg kg^{-1} N as NH$_4$NO$_3$. The solution was renewed once a week and pH was adjusted to 5.5. At the tillering (TI), panicle initiation (PI) and heading stages

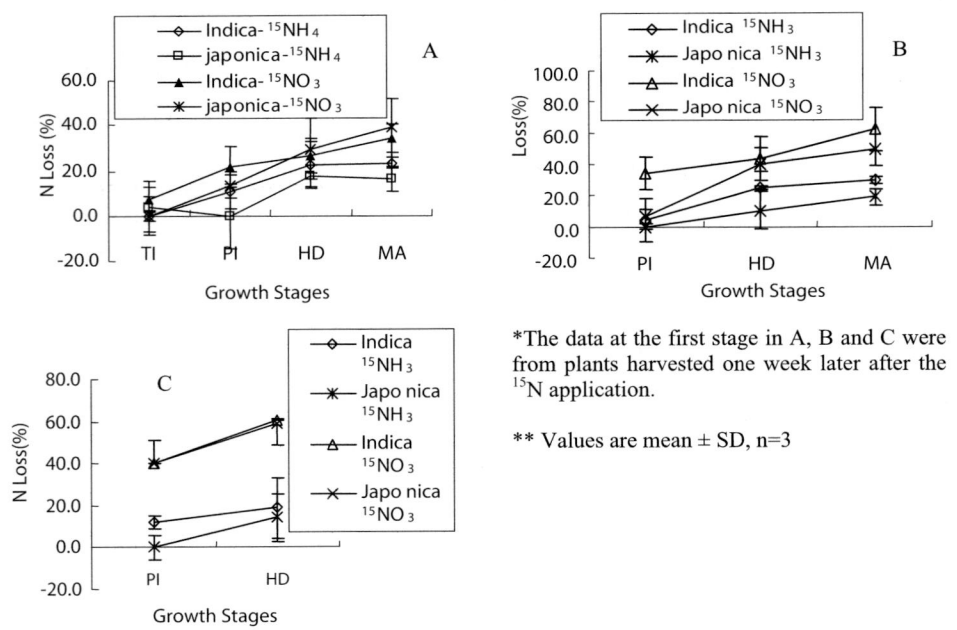

*The data at the first stage in A, B and C were from plants harvested one week later after the ^{15}N application.

** Values are mean ± SD, n=3

Figure 1. Plant N losses at different stages from N applied at TI, PI and HD stages.

(HD), NH_4NO_3 was replaced either with $^{15}NH_4NO_3$ or $NH_4{}^{15}NO_3$ for one week and then returned to the previous solution. Then they were harvested in triplicate either immediately after the feeding or at PI, or at HD or at maturity stage (MA). Nitrogen absorbed by rice plants was calculated by subtracting N left in the solution in the pots from N applied. Nitrogen losses were calculated as: N loss (%) = (N absorbed-N left in plants)/N absorbed x100.

Table 1. Significance of factors affecting Plant N Losses at MA Stage from N Applied at TI, PI and HD Stages.

Source	Sig.
Variety	0.195
Stages	0.009
N Forms	0.000
Variety * Stages	0.281
Variety * N Forms	0.378
Stages * N Forms	0.012
Variety * Stages * N Forms	0.673

Results and discussion

Nitrogen losses could be 20% to 60% from N applied at different stages increased with the growth stages (Figs. 1A - C). N losses from Indica were more than from Japonica; $^{15}NO_3{}^-$-N had a much higher N loss than $^{15}NH_4{}^+$-N, especially when N was applied at the heading stage (Figure 1C). SPSS analysis of N losses at MA stages from N applied at TI, PI and HD stages showed that N forms ($p<0.001$) and growth stages ($p<0.01$) had significant effects on N losses (Table 1). When N was applied at the TI stage, most of the $NH_4{}^+$ and $NO_3{}^-$ absorbed may be incorporated into organic N compounds; when it was applied at later stages, (especially at HD stage), there may be less incorporated into organic N compounds and might be released into air more easily after some physiological processes, resulting in higher N losses. The difference in N losses between $NH_4{}^+$ and $NO_3{}^-$ might be due to their different physiological processes after absorption into plant structures. The mechanisms are still to be elucidated.

Conclusion

A large amount of N can be lost from rice plants when N was applied at TI and especially at later stages. Nitrate absorbed into plants can result in more losses than $NH_4{}^+$-N, and Indica lost more N than Japonica.

References

Ashraf, M., Shamsi, S.R.A., Sajjad, M.I. and Azam, F. (1997). Nitrogen losses from tops of three rice varieties grown in nutrient culture solution. Pakistan Journal of Botany, 29, 319-322.

Effectiveness of alternative managements to reduce N losses from dairy farms

S.P. Cuttle[1] and M.M. Turner[2]

[1]*Institute of Grassland and Environmental Research, Plas Gogerddan, Aberystwyth, Ceredigion SY23 3EB, UK*

[2]*Centre for Rural Research, University of Exeter, Lafrowda House, St German's Road, Exeter, Devon EX4 6TL, UK*

Introduction

The effectiveness and economic impact of alternative management strategies to control N losses from dairy farms have been examined in a desk study. This extended a previous systems study of losses from a hypothetical dairy farm (Jarvis et al., 1996) to six commercial farms in SW England.

Method

Six dairy farms, typical of the region and with at least 75% of the farm area under grass, were selected for study. Farms were between 59 and 110 ha in area with stocking rates of 1.7 - 2.3 livestock units ha^{-1} and milk outputs of 5790 - 6200 l per cow. All were predominantly slurry-based systems. Information was collected about the physical characteristics of the farms and their existing managements. This information was used to estimate N losses for each farm using the NCYCLE, SUNDIAL and MANNER models (Scholefield et al., 1991; Smith et al., 1996; Chambers et al., 1999) and typical emission factors for NH_3 volatilisation from housed cattle and manure stores (Misselbrook, 2001). Losses were then recalculated after adjusting the base data to simulate the farms under each of the following managements designed to reduce N losses: (a) grass fields changed to grass/clover swards without N fertiliser; (b) grass fields changed to grass/clover swards and managed to organic farming standards; (c) slurry applications restricted to the growing season and applied by trailing-shoe, N fertiliser matched to sward requirements; (d) forage maize grown to provide 50% of the silage requirement and grown with slurry instead of fertiliser; (e) combined clover + maize option; (f) combined slurry/fertiliser + maize option. The financial performance of the farms was assessed under their existing management and for each of the alternative managements.

Results

Total N losses from the farms under their existing managements were equivalent to 131 - 193 kg N ha^{-1} when averaged over the whole farm area. The effects of introducing the alternative managements on total N losses and financial performance are summarised in Table 1. Changing to clover-based swards was effective at controlling N losses, but reduced the output of milk. Financial margins could be maintained or increased by converting the farm to organic management but this was sensitive to the size of the organic milk premium. Improving slurry and fertiliser use was also effective at reducing losses and in this case, did not reduce milk output. However, the additional costs of the improved slurry application techniques were greater than the savings in fertiliser costs. These increased costs may have been overestimated, as it is likely that farmers could adopt a more flexible approach to slurry utilisation than was assumed for the simulation. The maize option resulted in modest improvements in margin but was the least effective at reducing N losses. Similarly, combining the maize option with the clover or improved slurry/fertiliser use options had little additional effect on the amount of N lost, when compared with these options alone.

Table 1. *Effect of alternative managements on the N loss and financial performance of the study farms expressed as the percentage change from the current management: range (mean).*

Management	Total N loss (%)		Financial margin (%)[1]	
Clover	-17 to -49	(-40)	-5 to +25	(+4)
Organic	-16 to -51	(-42)	+2 to +32	(+20)[2]
Slurry/fertiliser	-33 to -51	(-43)	-3 to -8	(-6)
Maize	-3 to -16	(-11)	+2 to +11	(+5)
Clover + maize	-25 to -53	(-44)	-4 to +26	(+6)
Slurry/fertiliser + maize	-33 to -48	(-41)	-4 to +11	(+2)

[1] excluding capital costs; [2] at 2002 milk price (22 pence per l: low organic premium)

Percentage reductions in leaching, NH_3 volatilisation and denitrification were broadly similar to those for the total N loss. The maize option slightly increased denitrification on four of the farms. Differences between the effectiveness of the managements on individual farms appeared to be due to differences in soil type, fertiliser rate, stocking rate, area of maize, type of slurry store and proportion of fertiliser applied as urea. Reductions in N loss were smallest on the least intensive farm, which also had the smallest losses under its base management.

Conclusions

The simulations indicated that changing to grass/clover swards, or improving the efficiency of slurry and fertiliser use, would reduce N losses from dairy farms. On some farms, the clover option was likely to improve financial performance, particularly if associated with a change to organic farming. Improving slurry use was likely to reduce margins slightly.

Acknowledgements

This study was funded by DEFRA (Contract NT1842). IGER is supported by the BBSRC.

References

Chambers, B.J., Lord, E.I., Nicholson, F.A. and Smith, K.A. (1999). Predicting nitrogen availability and losses following application of organic manure to arable land: MANNER. Soil Use and Management, 15, 137-143.

Jarvis, S.C., Wilkins, R.J. and Pain, B.F. (1996). Opportunities for reducing the environmental impact of dairy farming managements: a systems approach. Grass and Forage Science, 51, 21-31.

Misselbrook, T. (2001). Updating the Ammonia Emissions Inventory for the UK for 1999. Final Report to MAFF, London.

Scholefield, D., Lockyer, D.R., Whitehead, D.C. and Tyson, K.C. (1991). A model to predict transformations and losses of nitrogen in UK pastures grazed by beef cattle. Plant and Soil, 132, 165-177.

Smith, J.U., Bradbury, N.J. and Addiscott, T.M. (1996). SUNDIAL: A PC-based system for simulating nitrogen dynamics in agricultural land. Agronomy Journal, 88, 38-43.

Azofert: a new decision support tool for fertiliser N recommendations

P. Dubrulle[1], J.M. Machet[1] and N. Damay[2]
[1]INRA, Unité d'Agronomie de Laon-Reims-Mons, Rue Fernand Christ, 02007 Laon Cedex, France
[2]Laboratoire Départemental d'Analyse et de Recherche, Rue Fernand Christ, 02007 Laon Cedex, France

Introduction

The increasing demand for high quality crops and protection of the environment requires a more rigorous management of the N available for crops and an evaluation of environmental impacts. The Azobil software (Machet et al., 1990) based on a static predictive balance sheet method, has provided one means, most commonly used in France for predicting fertiliser N rates applied to annual crops (Meynard et al., 1997). Today a dynamic approach of the balance sheet method is possible, thanks to the development of a new decision support tool: Azofert® software (Machet et al., 2003). This is intended to be used by soil analytical laboratories, technical institutes, fertiliser companies and extension services giving N fertiliser advice to farmers. It can also be used for teaching. The present paper describes the software.

Model approaches

For the computer characteristics, Azofert® software:
- operates under Windows 9x and XP environments
- can have the parameters easily modified in order to adapt them to a large range of situations and pedoclimates. A Graphical User Interface allows modification of parameters in the different catalogues (soils, crops, organic amendments, crop residues, etc.) and grids defined in the software. Therefore, users become responsible for the N fertiliser advice given to the farmers
- does not have the functions of a Laboratory Information Management System, but is designed to be easily integrated into the data management system of a laboratory by using input/output files. The user's system constitutes the input file for Azofert and reads the output file from Azofert in order to publish a report of results, including interpretation and fertiliser N recommendations.

Azofert® software includes the following agronomic characteristics:
- all N inputs (soil inorganic N at the opening of the balance sheet, net mineralisation from humified organic matter, crop residues, catch crops and organic products, N in wet deposition, N in irrigation) and N outputs (N uptake by crop, fertiliser N immobilised and lost as gas, or by leaching) of the balance sheet are taken into account
- a dynamic approach, according to climatic conditions, is used to estimate N supplies from the soil, organic residues and mineral fertilisers (Figure 1)
- the fate of fertiliser N is included, particularly gaseous losses (NH_3 volatilisation) and microbial immobilisation

- the amount of fertiliser N is determined by adjusting the soil N supplies to the crop N requirements with recommendations on timing
- N advice is given for all the annual crops (cereals, industrial and vegetable crops) where the N requirements and the cycle of vegetation are known
- environmental risks (NO_3^- leaching, gas emissions) are assessed.

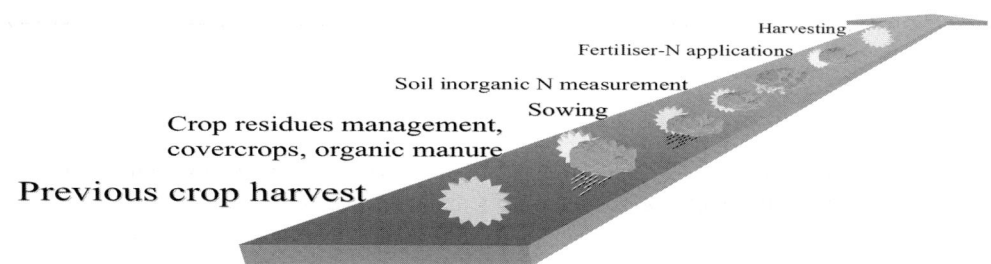

Figure 1. Different dates and cultural techniques are taken into account for the dynamic approach applied in Azofert®.

To make Azofert® software operational, the number of inputs are limited and easily obtained for farmers' fields. For any field, the main inputs are inorganic N profile at the opening of the balance sheet calculated from analytical results, soil characteristics (clay and lime contents, pH, total N content, bulk density, depth of rooting zone, thickness of the ploughed layer), crop description (cultivar, sowing date, crop density), cultural history, nature of the previous crop, organic amendments to the previous crop and crop residue management. The input of climatic data is for periods of ten days. To test Azofert® software, a set of samples with a large range of cultural situations and pedoclimates is constituted. The output of the Azobil and Azofert software are compared on the calculation of the rate of N fertiliser and the different items of the balance sheet method. Moreover, use of N experiments, where the optimal level of N fertilisation is determined for different crops, permits an evaluation of the accuracy of recommendations.

References

Machet, J.M., Dubrulle, P. and Louis, P. (1990). Azobil: a computer program for fertiliser N recommendations based on a predictive balance sheet method. Proceeding of 1st Congress of the European Society of Agronomy, S2 P21.

Machet, J.M., Recous, S., Jeuffroy, M.H., Mary, B., Nicolardot, B. and Parnaudeau, V. (2004). A dynamic version of the predictive balance sheet method for fertiliser N advice. (This volume).

Meynard, J.M., Justes, E., Machet, J.M. and Recous, S. (1997). Fertilisation azotée des cultures annuelles de plein champ. In : G. Lemaire and B. Nicolardot (eds) Maîtrise de l'azote dans les agrosystèmes. Les colloques de l'INRA, 83, 183-200.

Compost use in vegetable production: impact on gross N fluxes and implications for sustainable management practices

T.C. Flavel and D.V. Murphy
Centre for Land Rehabilitation, Faculty of Natural and Agricultural Sciences, The University of Western Australia, Crawley, WA, 6009 Australia

Introduction

Vegetable production in Western Australia occurs mainly on coarse textured sandy soils with low organic matter (OM) levels. A dry summer and wet winter (susceptible to large leaching events) typify the region. The inherent ability of these soils to store water and nutrients is low. In addition to rainfall (annual average 750 mm) crops are irrigated with 3 mm up to 3 times per day (soil water storage in 0-10 cm $\approx$10 mm). Carrot/lettuce rotations produce 2-3 crops per year and as synthetic fertilisers are applied at rates up to 400 kg N ha^{-1} per crop, the potential for N loss via leaching is high. Organic amendments have the potential to increase soil OM, soil water and nutrient holding capacity, and alter soil microbial populations and associated nutrient cycles. The aim of this research was therefore (i) to assess the timing and amount of soil N release from a selection of commercially available organic-based amendments and (ii) to supply growers with information they can use for N management on soils which have been treated with such amendments.

Materials and methods

Amendments assessed in this study included: a green-waste based compost (GWC), a straw based compost (SBC) and a vermi-compost (VC), a mixture of worm digested grape marc (grape skins, stalks and seeds) and composted manures. Amendment C:N ratios were 25.1 (GWC), 11.4 (SBC) and 12.2 (VC). Amendments were incorporated into a sandy soil (0.72% C, 0.05% N) at rates equivalent to 30 m^3 ha^{-1}. Amended soils were incubated in the dark for up to 142 days at 15°C (equivalent to 1 winter lettuce crop in Western Australia).

Microbial biomass-C (F-E method) and mineral N were extracted (1:4 soil: 0.5 M K$_2$SO$_4$) periodically over the incubation. On day 82, ^{15}N isotopic pool dilution techniques with a 3 µg N g^{-1} soil spike of 60 atom% (^{15}NH$_4$)$_2$SO$_4$ were extracted 2, 24 and 48 hours after spike addition (see Murphy et al., 2003 for methodologies). Gross N flux rates were determined numerically using FLUAZ (Mary et al., 1998).

Results

Microbial biomass levels remained relatively steady throughout the duration of the experiment. Average soil microbial biomass values for the 3 amendments were all similar and higher than the unamended control soil: VC, 171 mg C kg^{-1} (±3.2 se), GWC, 162 mg C kg^{-1} (±3.2 se), SBC, 151 mg C kg^{-1} (±2.8 se) and control 70 mg C kg^{-1} (±3.7se). NH$_4$$^+$-N remained below 1 mg N kg^{-1} in all 4 treatments for the duration of the incubation (data not shown). Both SBC and VC had higher levels of NO_3^--N than the control for the duration of the incubation, however, NO_3^--N levels were minimal in GWC and remained below levels of the control soil, indicating a 'N draw down effect' with this amendment (Figure 1). This was confirmed by ^{15}N pool dilution where immobilisation was greater than mineralisation in GWC, but not in other treatments (Figure 2).

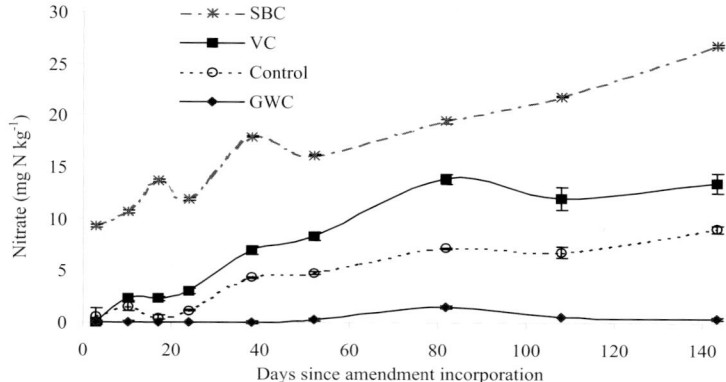

Figure 1. Changes in extractable NO_3^--N concentration with time. Error bars indicate S.E.

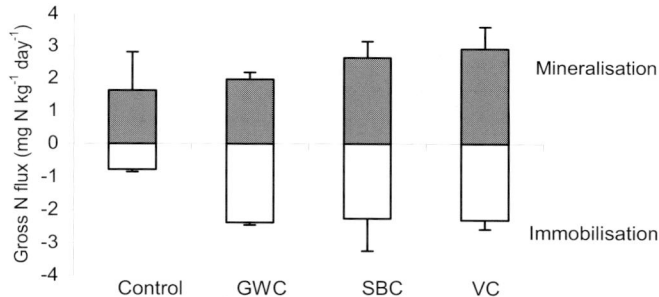

Figure 2. Nitrogen fluxes on day 82 of incubation. Error bars indicate C.I. at 95%.

Conclusions

Incorporation of organic amendments can increase soil microbial biomass and be beneficial in vegetable production systems. However, the quality of the amendment used should be matched to its purpose and each will have unique management requirements. Increasing microbial biomass will not necessarily increase plant available N. Amendments such as GWC should be used in conjunction with higher rates of inorganic N application than would be needed for unamended soils to prevent loss of plant available N.

References

Mary, B., Recous, S. and Robin, D. (1998). A model for calculating nitrogen fluxes in soil using ^{15}N tracing. Soil Biology and Biochemistry, 30, 1963-1979.

Murphy, D.V., Recous, S., Stockdale, E.A., Fillery, I.R.P., Jensen, L.S., Hatch, D.J. and Goulding, K.W.T. (2003). Gross nitrogen fluxes in soil: Theory, measurement and application of ^{15}N pool dilution techniques. Advances in Agronomy, 79, 69-118.

Ammonia volatilisation and soil nitrogen dynamics following application of pig deep-litter and pig slurry in different soil tillage systems

S.J. Giacomini[1], C. Aita[1], E. Guidini[1], E.B. Amaral[1] and A. Lunkes[1]
[1]Department of Soil, Federal University of Santa Maria, RS, Brazil

Introduction

Pig production in deep-litter systems is being introduced in South Brazil as an alternative to the traditional system (where liquid manure is produced) for economic and environmental reasons. Although the solid organic compost resulting from the deep-litter presents a potential N supply to plants, there are few studies involving its use as a fertiliser and impact on N dynamics in soil. The objective of this study was to compare NH_3 volatilisation and dynamics of N in soil with the use of pig deep-litter and pig slurry applied on the surface, or incorporated into the soil.

Materials and methods

The study was carried out from November 2002 to March 2003 at an experimental area of the Soil Science Department, Federal University of Santa Maria, State of Rio Grande do Sul, Brazil, in a Hapludalf soil. The treatments were application of pig deep-litter and pig slurry on the residues of oat, with and without incorporation into the soil. The deep-litter composition was pH 9.1; 10.6 kg Mg^{-1} total N; 1.5 kg Mg^{-1} total ammoniacal N (TAN=NH_3-N + NH_4^+-N); and 477 g kg^{-1} dry matter. Slurry composition was: pH 7.8; 2.15 kg m^{-3} total N; 0.94 kg m^{-3} TAN; and 46 g kg^{-1} dry matter. The deep litter and pig slurry were applied using a rate of, respectively, 65 m^3 ha^{-1} and 13 Mg ha^{-1}. The volatilisation of NH_3 was measured using a static half-open system (Nömmik, 1973) for a period of 80 h after manure application. The soil mineral N (NH_4^+ and NO_2^- + NO_3^-) was evaluated in 0-10, 10-30, 30-60 and 60-90 cm depth layers at 3, 9, 20, 37, 54 and 74 days after manure application.

Results

The NH_3 volatilisation was greater with the use of pig slurry than with deep litter (Table 1). Although the incorporation of pig slurry reduced the ammoniacal N losses by 78% (Table 1), soil mineral N was not different between treatments, with and without incorporation (Table 2). This could be explained by a larger microbial N immobilisation caused by the incorporation of the pig slurry mixture with oat straw. During the experiment, the treatments with deep-litter presented smaller mineral N in the soil than those with pig slurry.

The NO_3^--N in the soil after application of manures was greater for pig slurry than for pig deep-litter (Figure 1). The incorporation of pig slurry in the soil reduced the amount of NO_3^--N in the soil in the early phase after pig slurry application, reducing the potential of N losses by NO_3^--N leaching.

Table 1. Total NH$_3$ volatilisation as a percentage of total ammoniacal N (TAN) after deep-litter and pig slurry were either left on the soil surface or incorporated into the soil.

Soil tillage	NH$_3$ loss				
	kg ha^{-1}			% of applied TAN	
	Slurry	Deep-litter	Control	Slurry	Deep-litter
Surface	6.31Aa	0.89Ab	0.34Ab	9.30Aa	2.87Ab
Incorporated	1.39Ba	0.73Ab	0.36Ab	1.68Ba	1.94Ba

Means followed by the same upper case letter in a column and by the same lower case letter in a line are not significantly different at P=0.05.

Table 2. Temporal variation in soil mineral N (NH$_4^+$-N and NO$_2^-$ + NO$_3^-$) in the 0-90 cm layer at 3, 9, 20, 37, 54 and 74 days after deep-litter and pig slurry was either left on the soil surface or incorporated into the soil.

Soil tillage	Soil mineral N (kg ha^{-1})			Soil mineral N (kg ha^{-1})		
	Slurry	Deep-litter	Control	Slurry	Deep-litter	Control
	3 d			**9 d**		
Surface	48.9Aa	33.5Ab	20.3Bb	68.1Aa	39.2Ab	34.5Ab
Incorporated	43.9Aa	29.7Bb	29.7Ab	50.5Ba	33.4Ab	30.5Ab
	20 d			**37 d**		
Surface	61.6Aa	29.4Ab	22.3Ab	83.4Aa	53.5Ab	43.0Ac
Incorporated	52.7Aa	35.3Ab	23.1Ac	80.9Aa	48.2Ab	40.6Ac
	54 d			**74 d**		
Surface	45.3Aa	45.4Aa	34.5Ab	21.3Aa	20.0Aa	18.8Aa
Incorporated	46.8Aa	40.0Aab	38.1Aab	21.7Aa	19.7Aa	19.0Aa

Means followed by the same upper case letter in a column and by the same lower case letter in a line are not significantly different at P=0.05.

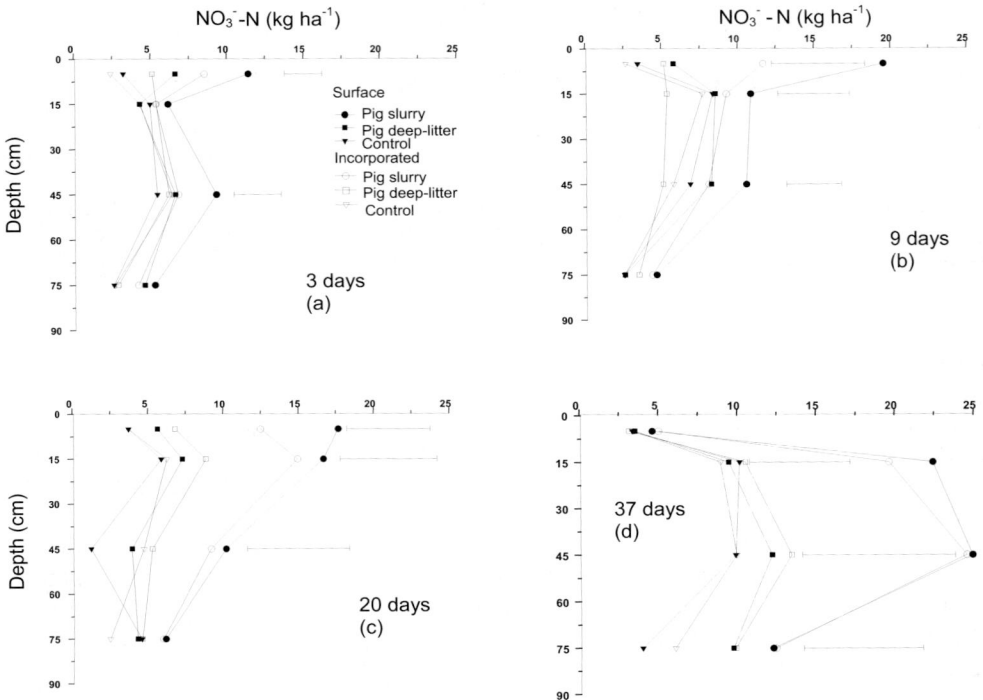

Figure 1. Temporal variation in soil NO$_3^-$-N at 3, 9, 20 and 37 days after deep-litter and pig slurry was either left on the soil surface or incorporated into the soil.

References

Nômmik, H. (1973). The effect of pellet size on the ammonia loss from urea applied to forest. Plant and Soil, 39, 309-318.

Estimation of nitrogen loading in Japanese prefectures and scenario testing of abatement strategies

M. Hojito[1], A. Ikeguchi[2], K. Kohyama[1], K. Shimada[1], A. Ogino[1], S. Mishima[3] and K. Kaku[1]
[1]*National Institute of Livestock and Grassland Science, Nishinasuno, Tochigi, 329-2793 Japan*
[2]*Headquarters, National Agricultural Research Organization, Tsukuba, Ibaraki, 305-8604 Japan*
[3]*National Institute for Agro-Environmental Sciences, Tsukuba, Ibaraki, 305-8604 Japan*

Introduction

Japan retains only a small agricultural area in relation to the human population. This, therefore, results in a very intensive agricultural production, which is heavily dependent upon chemical fertilisers and animal husbandry relying on imported feedstuffs. There is a substantial, positive N balance in the country and, consequentially, a high potential risk of N pollution of the environment. As a method of evaluating the N pollution risk, calculation of the likely N concentration in drainage water is effective (MacDonald, 2000), because it is easily compared with actual concentrations in surface or ground water. This paper reports the current situation of the country at the prefectural level and discusses possible scenario testing of abatement strategies to reduce N pollution.

Method

1. *Field level N balances*
Residual N at the surface of the field was calculated using statistical data according to the following equations:

$$Nres=Ninput-Noutput \tag{1}$$
$$Ninput=Nf+Nm+Nr+Nb \tag{2}$$
$$Noutput=Np+Nd+Ne \tag{3}$$

Nres: residual nitrogen, Ninput: nitrogen input, Noutput: nitrogen output, Nf: chemical fertiliser N, Nm: animal manure N, Nr: wet deposition N, Nb: symbiotic and non-symbiotic N fixation, Np: N removal with crop production, Nd: denitrification, Ne: NH_3 emission

2. *Modelled N concentration of drainage water*
Estimated N concentrations in the drainage water (assuming that surplus N in the soil will readily dissolve in water percolating through the soil) were derived using following equation;.

$$Nconc=Nres\times100/AP \tag{4}$$

Nconc: Estimated N concentration in drainage water mg N l^{-1}, Nres: soil residual N kg ha^{-1}, AP: annual percolating precipitation mm

3. *Abatement strategy scenarios*
Scenario 1: reduction of chemical fertilisers.
Scenario 2: manure N reduction using sewage treatment system, i.e. 'discounting' the N in the manures, by treatment in sewage treatment systems.
Scenario 3: utilisation of fallow fields for cropping.

Scenario 4: comprehensive, i.e. combination of fallow utilisation, 30% reduction of chemical fertiliser N, and removal of 30% of the manure by sewage treatment.

Results

1. Average N concentrations were 7.8, 8.8 and 2.9 mg N l^{-1}, for national, national excluding Hokkaido and Hokkaido, areas respectively. A wide variation in concentrations was observed, ranging from zero to >30mg N l^{-1}.
2. Many of the prefectures with high N concentrations coincided with high animal numbers and, hence, high manure N loading. The high concentrations could not be explained by the rate of chemical fertiliser application alone.
3. From abatement scenario 1, a reduction of 30% in chemical fertiliser N use resulted in a substantial reduction in N concentrations: i.e. from 7.8 - 5.4mg N l^{-1}(-31%), and 8.8 - 6.3mg N l^{-1}(-38%) according to national average and national average excluding Hokkaido, respectively.
4. From scenario 2, the effect of reducing the manure N loading, by sewage treatment systems, was not clear. It might, however, be worthwhile considering further, the extent of manure treatment that may be feasible through such systems.
5. From scenario 3, utilising all the fallow fields by cropping, decreased N concentrations markedly: 7.8 - 5.9 mg N l^{-1}(-24%), and 8.8 - 6.6 mg N l^{-1} (-25%) for the national average and national average excluding Hokkaido, respectively. Further significant reduction in the N concentration was estimated by combining the chemical fertiliser reduction with fallow field utilisation (Scenario 4).

Discussion

Results showed a clear risk of N pollution in drainage and surface waters arising from Japanese agriculture. Bearing in mind that, in some prefectures, the estimated N concentration in the drainage water was in excess of 20 mg l^{-1}, there is a considerable need for effective countermeasures against the high N loading. The results suggest that the most effective and practical abatement method is reduction of chemical fertiliser N use. Utilisation of fallow land for crop production is also effective for abatement. The calculation procedure, however, requires that a number of potential problems are addressed: i) Treatment of the whole prefecture as the unit of calculation; this means the N transfer such as trading as compost in the prefecture is completely free and also, all the N from different sources such as vegetation and manure type are treated equally in the prefecture ii) The use of annual mean N concentrations does not reflect the fluctuation associated with each precipitation event, nor the seasonal variation iii) The reduction of N loading, as a result of NH_3 emission, should be considered as another important source of (air) pollution. In Japan, the most popular animal manure treatment system is composting, which is known to result in substantial NH_3 emissions. These are important factors likely to contribute to a reduction in the accuracy of estimates and should be improved in the near future.

Conclusions

Even though the country average does not suggest a serious risk of N pollution, the risks are much higher within some prefectures. In such situations there is an urgent need for effective abatement measures such as fallow field utilisation, chemical fertiliser reduction or other strategies for more positive abatement of N loading.

References

MacDonald, K.B. (2000). Risk of water contamination by nitrogen. In: Environmental Sustainability of Canadian Agriculture. T. McRae et al., (eds) p117-123, Agriculture and Agri-Food Canada, Ottawa

Indicators for environmental and economic sustainability on UK dairy farms

E.C. Jewkes[1], D. Scholefield[1], M.M. Turner[2] and L. Brown[1]
[1]*Institute of Grassland and Environmental Research, North Wyke Research Station, Okehampton, Devon, UK*
[2]*Centre for Rural Research, Lafrowda House, University of Exeter, Exeter, UK*

Introduction

Sustainability has become a watchword in agriculture in recent years, but the term encompasses a broad spectrum of meanings. To the environmental and legislative bodies, a primary prerequisite of sustainable agriculture is for land use with minimised pollution and regard for the conservation and protection of natural resources. To the farmer, sustainability in the first instance is necessarily economic, in order to maintain livelihood. A major problem is the reconciliation of these two apparently disparate ends of the spectrum of the term; that is, to farm within environmentally acceptable limits whilst maintaining an economic level of production. This difficulty is highlighted by McInerny (1995), who suggests that environmental factors are increasingly incurring economic costs. Within dairy farming, reconciliation of environmental and economic factors is seen as an especially difficult task, as to maintain production, nutrient inputs (in the form of fertilisers and feeds) are often high, and guidelines pertaining to these inputs have, until recently, been based mainly on economic criteria. Such intensity of nutrient use has led to agricultural grassland becoming arguably, a significant source of diffuse N and P pollution, both to watercourses and the atmosphere.

Rationale

Using simple indicators of nutrient use and economic efficiency, it is envisaged that strategies for sustainability can be identified and some reconciliation of the environmental and economic disparities achieved. Variations in efficiencies will occur with differing management techniques and also according to site and prevailing conditions. By identifying and applying a range of indicators, it is hoped that the best combination of management factors can be delineated for a variety of sites and conditions. This approach uses a combination of analysis of real farm data and synthesis from models. Model synthesis is useful in providing information on environmental impacts, particularly nutrient losses, which are hard to obtain empirical data for, but is notably deficient in economic information. With respect to real farm data, the reverse is true. A broad range of indicators has been identified in a number of earlier projects (see Table 1 for examples), but those that link production with loss are likely to be the most useful in reconciling the two goals for sustainability.

Using indicators to assess performance

Using a combination of analysis and modelling, indicators for sustainable use of N and P are delineated and considered along with economic criteria for sustainability. An example is shown (Figure 1), where there is a general trend of increasing losses to the environment as margin over concentrates (MoC) increases, but exceptions where this is not the case, and a high MoC figure is obtained with a relatively small loss figure. The converse is also true, and poorer environmental performance can be seen with lower economic margins. Comparison of a number of indicators in this way will allow identification of management aspects that encourage sustainable nutrient use and economic viability, and consideration of the constraints or circumstances pertaining at individual sites or within certain managements.

Table 1. Examples of some indicators identified in previous studies.

Indicator	Category	Reference
Surplus N or P	Environmental	Aarts et al. (2000)
Surplus N per1000l milk	Environmental & economic	Jarvis (1999)
Product N perTotal N losses	Environmental & economic	Scholefield and Smith (1996)
Forage Support energy kg^{-1} DM	Economic	White et al. (1983)

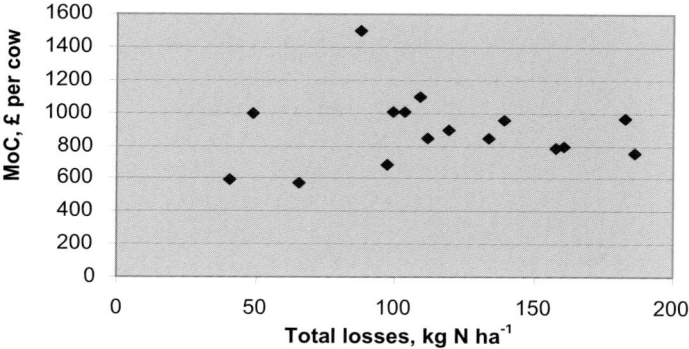

Figure 1. Margin over concentrates plotted against modelled total N losses.

Because of the variation in efficiencies between sites, or constraints at a given location, there is unlikely to be a management pattern that can be prescribed for all. However, the degree to which sustainable management attributes can be delineated may provide a measure of the ease by which environmental/economic sustainability can be reconciled in the current UK dairy farming climate.

Conclusions

Use of economic and environmental indicator values provides a method to identify management attributes and site specific factors that achieve some reconciliation of the two goals of sustainability. The degree to which this can be attained will also reflect how easily the requirement for sustainability can be achieved in the current dairy farming climate.

References

Aarts, H.F.M., Habekotte, B. and van Keulen, H. (2000). Nitrogen management in the "De Marke" dairy farming system. Nutrient Cycling in Agroecosystems, 56, 231-240.

Jarvis, S.C. (1999). Accounting for nutrients in grassland: challenges and needs. In: A.J. Corrall (ed), Accounting for Nutrients. BGS Occasional Symposium No. 33 pp3-13

McInerny, J.P. (1995). The economic context for grassland farming. In: G.E. Pollott (ed). BGS Occasional Symposium No. 29.

Scholefield, D. and Smith, J.U. (1996). Nitrogen flows in ley-arable systems. In: D. Younie (ed). BGS Occasional Symposium No. 30, pp 96-104

White, D.J., Wilkinson, J.M. and Wilkins, R.J. (1985). Support energy use in animal production from grassland. In: A.J. Corrall (ed). BGS Occasional Symposium No. 14.

Dairy production using an extended grazing management system - a preliminary assessment of nitrogen flows

E.C. Jewkes[1], D. Scholefield[1], M.R. Butler[1], J. Webb[2], T. Forrester[2], J. Lapworth[2], K. Russell[2], G. Bailey[3], A. Lathwood[4] and A. Clarke[4]

[1]*Institute of Grassland and Environmental Research, North Wyke Research Station, Okehampton, Devon, UK*
[2]*ADAS Research & Development, Woodthorne, Wergs Rd., Wolverhampton, UK*
[3]*ADAS Mamhead, Mamhead Castle, Mamhead, Exeter, Devon, UK*
[4]*ADAS Pwllpeiran, Cwmystwyth, Aberystwyth, Ceredigion, Wales*

Introduction

A poor economic climate has resulted in an increasing number of dairy farmers seeking to reduce costs by making greater use of grazed grass, whilst reducing reliance on bought concentrates and conserved forages. The resultant lengthening of the grazing season means that fertiliser is often applied earlier and later in the year than is typical in a conventional 180d housed/185d grazed system. As uptake of N is less efficient during these periods, it has been suggested that this practice may result in increased leaching and denitrification losses. However, the corollary to this would be that NH_3 losses may be reduced, as the volume of stored slurry requiring spreading to land will be smaller and the housing period is shorter. Ammonia emissions are known to be smaller from grazing compared with housing and other hard standings; Pain et al. (1998) calculated that, in a conventional system, 40% of dung and urine is deposited at grazing, but accounts for only 14% of the total NH_3 losses. An ADAS desk-study demonstrated that a 1-month extension of the grazing period could reduce NH_3 emissions by 9% and 7% for FYM and slurry-based systems, but suggested that, in some locations, NO_3^- leaching would increase.

Until recently, there have been no direct measurements of N losses from extended grazing systems in the UK. In the current study, four farms across a broad geographical range were selected in spring 2002 and a three-year programme of measurements of N flows has been implemented on a field at each. These include: herbage DM yield, herbage N yield, denitrification prior to and following fertiliser application, leaching losses and volatilisation from urea applications, slurry spreading and grazing. The selected farms encompass a range of management variables within the remit of the extended grazing principle, as shown in Table 1.

Table 1. Farm locations, soil types and management variables.

	Location	Soil type	Approx. Fert N use	Concs. use	Housing period
1	E. Devon	Sandy	Up to 250 kg N ha^{-1}	110 kg per cow	Dry cows only unless poor weather
2	Somerset	Sandy loam	Up to 350 kg N ha^{-1}	800 kg per cow	Housed Nov-Feb
3	SW Wales	Sandy clay loam	Approx. 350 kg N ha^{-1}	Zero	Housed Nov-Feb. max.
4	Staffordshire	Clay/clay loam	Approx. 250 kg N ha^{-1}	350-400 kg per cow	Housed Nov- Feb

Results

Recorded herbage DM yields from the first year of grazing were in the range 8.4 - 10t DM ha^{-1} (Table 2). This compares favourably with yields of 6 - 8t DM ha^{-1} recorded in trials on 'conventional' farms with a ca.185d grazing regime. Measurements from late spring 2002 to spring 2003 show that, despite late grazings and fertiliser applications, soil mineral N (SMN) did not accumulate to high levels (Table 2), particularly when compared with farm studies carried out in other projects. This concurs with Laidlaw et al. (2000), who found that SMN was not significantly affected by autumn grazing. However, the SMN measurements were not specifically timed to take grazing effects into account, and would be unlikely also to account for losses of excretal N by bypass flow.

Table 2. DM yields and N losses.

Farm	DM yield, kg ha^{-1}	SMN range, kg N ha^{-1}	Denit. range, kg N ha^{-1} d^{-1}	Peak NO$_3$-N conc. mg l^{-1}
1	8419	6.6 - 34.7	0.002 - 0.44	44.7
2	8940	9.6 - 30.9	0.003 - 0.43	42.3
3	10082	8.0 - 47.7	0.001 - 3.86	17.8
4	9991	5.3 - 14.3	0.003 - 0.65	N/A

Denitrification losses during 2002 were, in the main, modest and comparable to those found in other managements, with measured peak rates on farms 1, 2 and 4 below 0.7 kg N ha^{-1} d^{-1}. The exception to this was farm 3 in Wales, where denitrification rates were high, following fertiliser applications in late spring and autumn, with a peak measured loss of 3.86 N ha^{-1} d^{-1} (Table 2). This site has a high annual rainfall (local data 1466 mm in 2002) and the large N$_2$O fluxes were measured during wet spells, or in periods of high residual soil moisture. On farms 1-3, where suction cup samplers were installed, NO$_3^-$ concentrations were high during the first part of the winter (Table 2). Calculated total N losses, using meteorological data from close to the sites to estimate potential drainage, were also substantial, with 129, 152 and 95 kg N ha^{-1} leached, respectively. These losses are large given the modest SMN levels described earlier. Whilst grazing effects may account for some of this magnitude, the samplers were installed later in the year than ideal, and so disturbance of the soil may also have contributed to losses through enhanced mineralisation. It will become more apparent, if this was the case, during measurements in subsequent winters. The fourth site has tipping bucket assemblies installed, but data for N leaching from these are not yet available.

Conclusions

The first full year of study suggests that yields of utilisable herbage can be high in this system, whilst maintaining modest levels of SMN. Losses were high in some circumstances, but comprehensive conclusions as to the N budget of farms using the extended grazing management system should be deferred pending the remaining 2 years of study.

Acknowledgements

Many thanks to the farmers and herd managers involved for their effort and cooperation. This work is funded by Defra, London.

References

Laidlaw, A.S., Watson, C.J. and Mayne, C.S. (2000). Implications of nitrogen fertilizer applications and extended grazing for the N economy of grassland. Grass and Forage Science, 55, 37-46.

Pain, B.F., Misselbrook, T.M., Jarvis, S.J., Chambers, B.J., Smith, K.A., Phillips, V., Sneath, R.W. and Denmers, T.G.W. (1998). Inventory of ammonia emissions from UK agriculture 1996. Report to MAFF.

Decrease in the amount of residual nitrate in cultivated land

R. Lambert, V. Van Bol and A. Peeters
Laboratoire d'Ecologie des Prairies, Université catholique de Louvain, Place Croix du Sud, 5 bte1, B-1348 Louvain-la-Neuve, Belgium

Introduction

Nitrate remaining in the soil profile before winter is a good indicator of the risk of pollution. This indicator is used in some countries of the EU in the framework of the Nitrates Directive. A study carried out in the vulnerable area of the Brussels sands (Belgium) assessed the situation and its possible improvement (Lambert et al., 2002).

Materials and methods

Samples of soil were taken for NO_3^- profiles each year in November (from 1997 to 2000) in the each fields of 10 pilot farms to a depth of 150 cm. Each year, an average result was calculated for each farm and crop by weighting the data of each field by its surface area. During the study, the farmers were advised by the research team about fertilisation, catch crops and the management of animal manures. The results were presented and discussed annually in meetings with the farmers and researchers, and individual management targets were defined for the farmer.

Results and discussion

The measures recommended by the advisors quickly led to a reduction in the average value of the NO_3^- residue measured in autumn. When the study began in 1997, the average NO_3^- content amounted to 70 kg NO_3^--N ha^{-1} and remained at an average of 40 kg NO_3^--N ha^{-1} during the following years. Despite this marked improvement, the NO_3^- residue of some crops still remained high. This applied mainly to maize and potato. After these crops, the residues measured in the pilot farms in the year 2000 were, in general, similar to those obtained in experimental fields fertilised according to the best N fertiliser practices (Frankinet et al., 2001). The NO_3^- residues after flax and pea were also high, if they were not followed by a catch crop. However, these crops are of little importance, given the areas sown, with the exception of peas (7%).

Table 1 gives the average NO_3^- residues measured under each crop at the beginning and end of the study, as well as the relative importance of the crops in terms of the agricultural area in the year 2000. According to the relative area of each crop and the possible improvement, efforts must be concentrated on wheat and maize. The reduction of NO_3^- residues after wheat is responsible for more than 30% of the total reduction.

The effect of a catch crop on the reduction of the N residue was quite obvious: the residue after wheat amounted to 80 kg NO_3^--N ha^{-1}, on average, on bare soil and only 20 kg with a catch crop (*Sinapsis alba*). Soil cover with catch crops before spring crops went from 0% to

90% in the 10 pilot farms during the study. This is the most effective means to quickly reduce the NO_3^- residue before winter leaching. There was also a strong influence of the crop rotation. The rotation: peas - catch crop - spring crop is more efficient to maintain a low NO_3^- residue than the rotation: peas - winter cereal.

Table 1. Average NO_3^- residues (kg NO_3^--N ha^{-1}) in autumn after the harvest of several crops and the relative importance of these crops in the cropping pattern.

Previous crop	NO_3^- residues 1997	NO_3^- residues 2000	% area 2000
Wheat	68	45	30
Sugar beet	46	26	19
Grazed and mixed use grasslands	67	57	11
Chicory	37	52	8
Pea	131	58	7
Winter barley	55	19	6
Maize	110	89	3
Cut grassland	23	26	3
Potato	261	66	1
Flax	114	88	1

Conclusion

Nitrate residues in the soils before the leaching period can be reduced by ca. 50%. Improvement was mainly due to the reduction of NO_3^- residue after wheat and maize. Application of good agricultural practices and in particular, covering soil with catch crop, are effective means of reducing residual NO_3^-.

References

Frankinet, M., Renard, S., Dautrebande, S. and Casse, C. (2001). Gestion intégrée de l'azote en cultures arables et normes nitriques. Rapport final d'activités, 76 p.

Lambert, R., Van Bol, V., Maljean, J-F. and Peeters, A. (2002). Prop'eau-sable: Projet pilote pour la protection des eaux de la nappe des sables bruxelliens. Rapport final d'activités, 107 p.

Estimating nitrogen losses from animal manures using their phosphorus balance

R. Lambert, B. Toussaint and A. Peeters
Laboratoire d'Ecologie des Prairies, Université catholique de Louvain, Place Croix du Sud, 5 bte1, B-1348 Louvain-la-Neuve, Belgium

Introduction

Losses during storage, treatment and utilisation are an important aspect of the management of livestock manures. These losses mainly concern N (gaseous losses and leaching) and to a lesser extent K (leaching). The P contained in the manures is neither very mobile nor leachable (AFIDEL, 1987). It can be used as a reference for estimating the losses of other elements.

Materials and methods

The organic manures used in this study were produced on farms located in the south-east of Belgium. Manures with additives are not taken into account. It is considered that feeding and feedstuff efficiency are the same for animals producing each type of manure and that P does not suffer losses. It is also considered that N, P and K supplied by the straw bedding does not affect the ratio between elements in manure, because the ratios in straw are similar and inputs are negligible, compared with the inputs from animal urine and faeces (< 10%).

In the absence of losses, the ratios between the mineral elements should remain the same. If more N is lost in a particular type of manure, then the N:P ratio in that manure will be lower. The expected N content (or K) is compared with the average content measured. The manure expected N content (or K) is calculated according to the formula:

Expected N content (or K) = (N (or K) content of the slurry/ P content of the slurry) x analysed P content.

Losses are the difference between expected N (or K) and measured N (or K). The method was applied to two sets of data. The first one is a data base of organic manure composition from farms. The second one concerns the composition of farmyard manures (FYM) before and after composting.

Results and discussion

Table 1 gives the number of samples analysed for each kind of manure, their average content and the expected value based on the average content in P.

The K:P ratio reveals the amount of the K losses during storage and composting. When there are no K and P losses (slurry), the K:P ratio is 6 :1. This proportion is 5.8:1 in FYM and 5.5:1 in composted FYM. This small difference between the types of manure indicates that the losses of this element are small whatever the type of manure. The situation is very different for N.

Table 1. N, P and K contents of organic manures and N, K expected contents (kg t⁻¹).

	number	N	N*	K	K *	P
Slurry	176	4.5	-	4.2	-	0.70
Farmyard manure (FYM)	126	5.7	7.9	7.1	7.3	1.22
Composted FYM	235	6.0	11.0	9.4	10.2	1.70

* expected content of this element without losses

In cattle slurry, the N:P ratio is 6.4:1. It drops to 4.7:1 in FYM and is only 3.5:1 in compost. Nitrogen loss in FYM is thus greater than the loss in slurry. Per tonne of fresh FYM, about 2.2 kg N are lost compared with slurry. During composting, the N loss continues. The compost is, on average, 1.4 times more concentrated in P than the FYM. In other words, 1 t of manure gives 715 kg of compost. These results represent an average loss of 1.2 kg N t^{-1} of FYM during composting, or ca. 20% of the content. The loss occurred mainly in a gaseous form.

The method was also applied to heaps of FYM analysed before and after composting in 50 farms (Table 2). The average contents before composting are 5.4 kg N t^{-1} and 1.12 kg P t^{-1}. After aerating the heap and composting over a period of 2 to 3 months, the average contents are 6.45 kg N t^{-1} and 1.62 kg P t^{-1}. On the basis of these analyses and the N:P ratio, the average N loss during composting is shown to be 18% of the initial content, or about 1 kg N t^{-1} of manure.

Table 2. N and P contents of FYM before and after composting(kg t⁻¹).

	N	N*	P
Farmyard manure (FYM)	5.4		1.12
Composted FYM	6.45	7.81	1.62

* expected content of this element without losses

Conclusion

The N:P ratio can be used to estimate relative losses of N due to manure management. Nitrogen losses are more important in farmyard manure than in slurry. Composting leads to supplementary N losses.

References

A.F.I.D.E.L. (1987). Etudes statistiques des caractéristiques des lombricomposts. Association française interprofessionnelle de lombriculture. Document 2, Janvier 1987.

Encouraging farmers to utilise nitrogen more efficiently

K.A. Leach[1], J.S. Conway[1], J.P. Morgan[2], B.F. Pain[2] and D. Munday[2]
[1]*Royal Agricultural College, Cirencester, Glos. GL7 6JS, UK*
[2]*Creedy Associates, Shobrooke, Crediton, Devon, EX17 1AE, UK*

Introduction

Reducing losses of N from farms is an aim of policy makers wishing to reduce the environmental impact of farming systems. In order to achieve this, farmers need to be made more aware of the losses and wastage of N inputs, and inefficient use of N resources, which currently occur on their farms. An example of the difficulties encountered is the situation regarding manure utilisation. Although numerous publications have been produced recently, encouraging farmers to adopt 'Best Farming Practices', unfortunately the uptake of this information in relation to manure storage and utilisation has been limited, mainly due to farmers' lack of time to digest the quantity of published literature. This paper describes our experiences of two initiatives, encouraging farmers to improve their utilisation of N resources. These initiatives involved a discussion group focussing on farm gate N budgeting and a series of practical workshops on manure management.

Nitrogen budgeting discussion group

A discussion group was formed in spring 2002 by recruiting members through existing farmers' groups in Gloucestershire (viz. Farming and Wildlife Advisory Group (FWAG), Maize Growers' Association, Cotswold Arable Study Group and British Grassland Society). A group meeting was held to introduce the topic of nutrient budgeting. Ten members submitted farm data on inputs and outputs and a spreadsheet prepared by FWAG was used to calculate whole farm N budgets for the most recent complete cropping year. Deposition on all farms was assumed to be 25 kg N ha^{-1}. Nitrogen in sold outputs/N inputs (NUE) ranged from 10% to 57% and N surplus (N inputs - N in sold products) from 75 to 277 kg N ha^{-1} (Table 1). Farmers were surprised to learn how small a proportion of their purchased N inputs was sold in products. The individual farm gate N budget served as a valuable starting point for a wide ranging discussion of issues relating to N utilisation. Individual advisory visits were made to discuss management options that might improve NUE and reduce N surplus. Farmers' responses included reconsidering fertiliser application rates and livestock rationing, deciding to sample soil for mineral N in spring, and planning to improve utilisation of livestock manures. Reservations raised by farmers were a lack of confidence in relying on N from manures, and lack of storage preventing them saving slurry until the optimum application time. There were also fears about the yield penalty of reducing N applications to recommended levels for crops on thin soils, and objections to the cost of testing for soil mineral N. At a follow-up discussion meeting, farmers recognised the limitations of the farm-gate N budget and developed an interest in investigating component (crop/animal/manure) or field scale budgets, in cases where record keeping was good. All considered that it would be beneficial to repeat the budgeting exercise and to follow the consequences of management changes. A target of reducing N surplus by 10% was suggested.

Manure management workshops

Creedy Associates ran practical workshops on manure management on host farms during the winters of 2001-2 and 2002-3. The main aim was to improve utilisation of livestock manures (farmyard manure and slurry). Exercises included assessment of storage facilities and nutrient content of manures, demonstrations of manure spread at different rates, and calculations for balancing manure with bagged fertiliser to meet crop requirements. In the first winter, 95 farmers attended. Of those responding to a follow-up questionnaire, 94% said that the workshops had increased their awareness of the fertiliser value of manures. Over 60% of farmers subsequently attempted to estimate the nutrient content of manures; 80% claimed to have a better understanding of issues related to spreading manures and risks of water pollution. A significant proportion (35%) of farmers had reconsidered their historic spreading management and had reduced, or intended to reduce, manure application rates. An impressive 78% of farmers had reduced mineral fertiliser application rates to crops where manure had been applied. A further 10% intended to do this. Using manure more carefully has real financial benefits to farmers as well as environmental benefits. The practical approach of the workshops struck a chord with those attending and resulted in all-important action.

Table 1. N balances for 10 farms (kg N ha⁻¹).

System†	A	A	P/A	A/D	A/B	D	D	S/B	D	S
Inputs										
Fertiliser	103	158	137	148	55	194	204	0	177	81
Manures	36	28						37		
Feed + straw			291	28		74	53		116	21
Fixation	38	35	40	6	42		6	25		20
Livestock			9		1		2		2	
Seed			3	2	1	1	1	1		
Outputs										
Milk				15		48	48		37	
Crops	104	129	98	77	40			7		
Livestock/wool			111	2	9	6	3	6	6	16
Straw/silage	11		23			2				
N surplus	87	117	273	115	75	238	240	75	277	134
NUE (%)	56.8	52.5	45.9	44.9	39.5	19.1	17.5	15.1	13.4	10.5

†A - arable, B - beef, D - dairy, S - sheep, P - pigs

Conclusions

The response to these two approaches indicates that, with practically orientated personal attention targeted to individual farm situations, farmers can be made aware of the importance and benefits of giving greater consideration to improving the efficiency of use of N resources. The farm gate N budget is a good starting point for discussion of NUE. Both the N budget and the practical manure workshops encouraged farmers to consider N as a valuable resource for the farm. To be persuaded to take action to use N more efficiently, farmers must be convinced of a direct benefit to themselves.

Acknowledgements

These two projects were funded by Defra, London. We are grateful to FWAG for supporting the N budget group by providing advice and the N budget spreadsheet.

Conserving biologically fixed N to increase its utilisation and decrease gaseous losses

A.K. Løes[1], A.K. Bakken[2], T.A. Breland[3], R. Eltun[4] and H. Riley[5]
[1]*Norwegian Centre for Ecological Agriculture, NO-6630 Tingvoll, Norway*
[2]*Norwegian Crop Research Institute, Kvithamar Research Centre, NO-7500 Stjørdal, Norway*
[3]*Agricultural University of Norway, Department of Plant and Environmental Sciences, NO-1432 Aas, Norway*
[4]*Norwegian Crop Research Institute, Apelsvoll Research Centre, NO-2849 Kapp, Norway*
[5]*Norwegian Crop Research Inst., Apelsvoll Research Centre div. Kise, NO-2350 Nes, Norway*

Introduction

Biological fixation of N by legumes is essential in organic farming systems. As this farming method expands, an increasing number of Scandinavian farms are being converted to organic crop production systems with low or zero livestock density. Green manures are included in the rotation between cash crops such as cereals, or as herbage seed crops. A normal treatment with green manure is to mow the herbage 2-4 times during the growing season (partly to control weeds), leaving the cut material on the surface to fertilise the re-growth. Alternative use of the herbage, for instance as mulch in vegetables, is often restricted because such crops may be difficult to include in a cereal-based rotation. Potentially high gaseous losses of N may occur from mown green manure. This is undesirable because gases with a large negative environmental impact, such as NH_3 and N-oxides are produced. When green manure herbage is kept on the field during late autumn and winter, runoff and leaching losses of N, P and other nutrients may also occur. This adds to the negative environmental potential and reduces soil fertility.

It is possible that the utilisation of N may be increased if the green manure herbage is conserved and stored as hay or silage during winter and then applied as manure to the subsequent crop. Another possibility is to utilise the herbage as manure directly after harvest. Late-season green manure herbage could be conserved and used for fertilisation in the subsequent season.

Regardless of the state of the herbage (fresh, dried, ensiled etc.), a procedure is required to transform N into a form that may be taken up by plants. In herbage, 75-90% of the N is mainly present as protein, while smaller fractions are in nucleic acids and soluble amino compounds. Wilting causes hydrolysis of protein, and processes that are undesirable in the preparation of good quality roughage may, in fact, be desirable to further degrade herbage proteins. Such processes are heating of damp hay by thermophilic bacteria and silage fermentation dominated by *Clostridia* and *Enterobacteria*. Care must be taken that any NH_3 produced is conserved and that undesirable side effects such as unpleasant odours are avoided. Herbage (fresh, hay or silage) may be mixed into the soil before planting or used as mulch, but the possibility of extracting soluble N with water and using this water to fertilise growing cereal crops is also an interesting possibility. In organic systems, cereal growth is commonly restricted by N-availability early in the growing season, particularly in a cool climate and in years with a delayed spring.

Materials and methods

The topics outlined above are being studied in a field experiment (2003-05) with a crop rotation of green manure, barley and mixed oats/peas, located on two sites with different soil and climatic conditions. Kvithamar (Mid Norway) has excess precipitation and a clay soil, Apelsvoll (East Norway) has less precipitation and a loam soil. In 2004, degradation of protein N by various ways of handling the green manure herbage will be measured in the laboratory. The amounts of NH_4^+, amino acids and soluble plant protein in water extracts will be measured at various water/herbage ratios, temperatures etc. The influence of conditions that may affect the extraction (e.g. time and temperature) will also be recorded. In field trials, the effects of removing herbage from the surface, as compared with leaving it there, will be studied by determination of subsequent cereal yields.

In 2003, we conducted two preliminary studies on an experimental plot in Tingvoll, (West Norway) with a green manure composed of phacelia, vetches, red clover and ryegrass. The preliminary studies were a record of the N concentration in cut green manure after the 1st cut on July 11, and a record of the loss of dry matter (DM) in cut green manure after the 2nd cut on August 29. In the first cut, weeds, (notably *Spergula arvensis*), heavily infested the green manure. Plant residues were sampled daily from July 11 to 15, and for the last time on July 18. After the 2nd cut, 24 containers were filled with 40 g freshly cut and well mixed green manure (= 4.7 g DM) placed on a nylon net above a precipitation collector. Six large boxes, each holding four containers, were placed in the field. Leachate was collected regularly from September 1 to 17, on 8 sampling dates. The N concentration in the leachate will be measured. On each sampling date, decomposing green manure was sampled from 3 containers, and the DM content was measured.

Preliminary results and discussion

The period following the 1st cut of green manure was dry and windy. The plant material dried rapidly, and ryegrass started to grow through these plant residues after one week. During this period, the colour of the residues changed from green to brown. The N concentration did not change during this period, and the average value was 2.06% total N in plant DM. There seemed to be no relationship between the colour of cut green manure and N concentration.

The period following the 2nd cut was wet. Altogether, 41.2 mm rain was recorded between the date of cutting and the last sampling. Variable amounts of leachate were collected from each container. The accumulated values for the 3 containers that were sampled on the last day varied from ca. 200 to 500 ml of leachate, probably because of variation in wind exposure. The average DM content in each container decreased from 4.7 to 3.8 g during the study period. The green manure was brown and soft, but the structure of the cut leaves had not changed. Similar amounts of green manure that had been placed on the ground during the same period were almost completely decomposed by September 17. This shows that close contact to soil is important for the decomposition of organic matter.

Effect of animal treading intensity on the efficiency of N$_2$ fixation by clover in mixed grass/clover pasture, and potential implications for long-term soil N availability

J.C. Menneer[1], S.F. Ledgard[2], C.D.A McLay[3] and W.B. Silvester[1]
[1]*University of Waikato, Private Bag, Hamilton, New Zealand*
[2]*AgResearch Ruakura Research Centre, Private Bag 3123, Hamilton, New Zealand*
[3]*Environment Waikato, P.O. Box 4010, Hamilton, New Zealand*

Introduction

Current grass-clover pastoral systems often utilise intensive grazing practices to improve productivity and profitability. However, increased levels of animal treading in intensively grazed systems may impact on a range of soil and plant properties, including N$_2$ fixation. For example, treading in intensively grazed winter/spring swards has been shown to adversely affect soil properties and dry matter production of mixed pastures (Ledgard et al., 1996), but little is known about impacts on the clover component and N$_2$ fixation. Potentially, clover N$_2$ fixation could be reduced by animal treading directly, through plant tissue damage, or indirectly via soil structural deterioration. In the long-term, any reductions in N$_2$ fixation could drive soil N availability to a lower equilibrium status, leading to reduced farm productivity, and compromise the sustainability of the grass-clover system. This study examined the impact of animal treading on clover growth and N$_2$ fixation in an intensively-managed dairy pasture in the Waikato region of New Zealand.

Materials and methods

The experiment was conducted on an established dairy pasture on a Te Kowhai silt loam (Typic Orthic Gley Soil) with impeded subsoil drainage. Treatments consisted of three treading severities (nil, moderate and severe) replicated eight times in a randomised block design. Dairy cows were used to create moderate and severe treading treatments in early spring at a typical grazing intensity of 4.5 cows per 100 m^{-2}. Plots were then excluded from grazing and the pasture harvested every 20-40 days depending on the growth rate. Dry matter (DM) yield and pasture species composition were determined. Estimates of the total amount of N$_2$ fixed were determined using the ^{15}N isotope dilution method. A dynamic dairying N model was used to examine long-term impacts on soil N, pasture production and milk production.

Results and discussion

Annual pasture production was decreased under moderate and severe treading by 16% and 34%, respectively (Table 1). The corresponding decrease in clover DM yield was 9% and 52%, respectively. Under severe treading, the annual grass DM yield reduction was less than for clover (37% vs. 52%). This indicates that clover is potentially more susceptible to severe treading than ryegrass. The reduction in pasture growth was greatest during the first two harvests (48 days), but the effect was much less thereafter. Effects of treading on clover growth were

Table 1. The effect of treading severity on annual pasture production and N_2 fixation.

	Nil	Moderate	Severe	SED
Total pasture yield (kg DM ha^{-1})	10149	8562	6683	417***
Clover yield (kg DM ha^{-1})	2046	1854	989	479†
Total N fixed (kg N ha^{-1})	76	66	36	18†

*** and †significant at $P<0.001$, and $P<0.1$, respectively; 8 replicates.

prolonged and persisted for up to 220 days. The proportion of total clover N derived from N_2 fixation (%N fixed) was reduced to an average lower limit of 44% over the first three harvests (71 days) under severe treading.

Soil macroporosity decreased significantly from 21% in the control to an average of 15% in the moderate and severe treading treatments. This suggests that restricted soil aeration may have contributed to the initial reduction in N_2 fixation. The effect of severe treading on %N fixed and the associated reduction in clover yield across all treading treatments resulted in the total annual amount of fixed N in clover herbage decreasing from 76 kg N ha^{-1} in the control, to 66 kg N ha^{-1} (-13%), and 36 kg N ha^{-1} (-53%) in the moderate and severe treatments, respectively. The effect of decreased N_2 fixation on N availability, long-term pasture productivity and milk production was estimated using a dynamic dairying model of N cycling (DAMN) that has an interactive grass-clover component (Table 2). Using a 10-year scenario of annual moderate and severe treading damage, whereby annual N_2 fixation was reduced by 15% and 56%, grass yield was predicted to decrease by 14% and 24% under moderate and severe treading, respectively, after 10 years. Soil organic N was also predicted to decline to lower equilibrium levels by 3% and 7% under moderate and severe treading, respectively, after 10 years. The corresponding predicted loss in milk production was 11% and 35%. On a whole farm basis with moderate and severe treading on 50% and 10% of the farm (P. Singleton pers. comm.), respectively, this could represent a decrease in milk production of 9% (e.g. from a New Zealand average of 11,800 to 10,738 l ha^{-1} yr^{-1}).

Table 2. Predicted effects after 10 years of a single annual treading event (moderate or severe) on selected plant and soil variables using the DAMN model (Ledgard et al., unpublished). Data are relative to the 'default' simulation which was a dairy farm producing 11,800 l milk ha^{-1} yr^{-1} with nil N fertiliser. Clover data inputs for the treading treatments were varied to simulate the decrease in clover production measured in the field study.

Modelled Variable	Default	Moderate	Severe
Clover yield	100	87	40
Grass yield	100	86	76
N_2 fixation	100	85	44
Soil organic-N	100	97	93

Reference

Ledgard, S.F., Thom, E.R., Singleton, B.S., Thorrold, P.L. and Edmeades, D.C. (1996). Proceedings of the Ruakura Farmers' Conference, 48, 23-33.

Effect of type and dose of sewage sludge application in maize+ryegrass rotation in Galicia (NW Spain)

M.R. Mosquera-Losada, A. Amador-García and A. Rigueiro-Rodríguez
Crop Production Department, Escuela Politécnica Superior, Universidad de Santiago de Compostela, Lugo, Spain

Introduction

The increase in sewage sludge production is causing a disposal problem in developed countries. Therefore the EU is encouraging its use in agriculture with appropriate management, to prevent environmental damage and the transfer of heavy metals through the trophic chain. Sewage sludge comes from different water cleaning processes. Galicia (Spain) has an important dairy industry that generates a large quantity of sewage sludge and if no account of the N content is made, it can cause environmental problems. Better management of these residues will improve recycling in agriculture, the environment and human health. Maize ryegrass rotation are grown in order to make silage for feeding animals for one or two months in winter, and two or three months in autumn, which allows stocking rates to be increased compared with farms exclusively based on herbage (Mosquera & González, 1998).

Materials and methods

The experiment was carried out in two localities of Galicia differing in soil pH (Muimenta, pH 5.7 and Pol, pH 4.3), in a randomised block design with three replicates and four treatments: two doses of anaerobic sewage sludge, (200 and 400 kg total N ha^{-1}), applied as municipal (ML, low dose, and MH, high dose) and dairy (DL, low dose, and DH, high dose) sewage sludge. Municipal sewage sludge was applied in December (before *Lolium multiflorum* L. sowing) and May (before maize sowing) and the same doses of N were also applied as dairy sewage sludge in May. Pasture was harvested at the start of May and maize samples of four plants processed separately in September. Soil pH, total N and P content, were analysed and at both localities, maize was analysed for protein and P content.

Results and discussion

Dairy and municipal sewage sludge had similar pH, N, P, K, Ca, Mg and Na contents. Heavy metals were greater in municipal sewage sludge, but always under legal limits for agronomic use, (Rigueiro et al., 2002). Differences in composition can be explained by the different origin of the residues. Pasture production (Table 1) was double in those plots fertilised with a higher rate of municipal sewage sludge than the lower rate, and was higher in the locality with lower pH. Pasture production, was below average as it depended on soil fertility, and the lower pH limited the response to sewage sludge, despite a pH of ca.7. Maize production (Table 1) was higher at higher pH, as the low pH limited strongly the soil fertility. However, plots fertilised twice with municipal, (instead of once with dairy) sewage sludge, produced more, due to the residual effect caused by the previous application. Plots with municipal sewage sludge had a higher pH and P content (Figure 1). Sewage sludge applications did not modify the levels of these macronutrients in stem or corn cob. Protein concentration was higher at the highest doses of sewage sludge, and increased soil N content despite a possible dilution effect at the high rates of maize production. A similar response was found with P concentration, but only significant at Muimenta. Results showed the importance of pH on crop response to sewage sludge application and residual effect of sludge on crop production. Rates of sludge increased the P and crude protein content of maize.

Table 1. Pasture and maize production (kg dm ha⁻¹) in DL (low dose), DH (high dose) of dairy sewage sludge, ML (low dose), MH (high dose) of municipal sewage sludge treatments in Muimenta (MUI) and Pol (POL).

Treatment Production	DL	DH	ML	MH
Pasture-POL	-	-	1416.4[b]	3164.5[a]
Pasture-MUI	-	-	970.4	1637.7
Maize-POL	523[a]	2345[ab]	3957[a]	3021[ab]
Maize-MUI	4507[c]	5166[bc]	7882[ab]	10376[a]

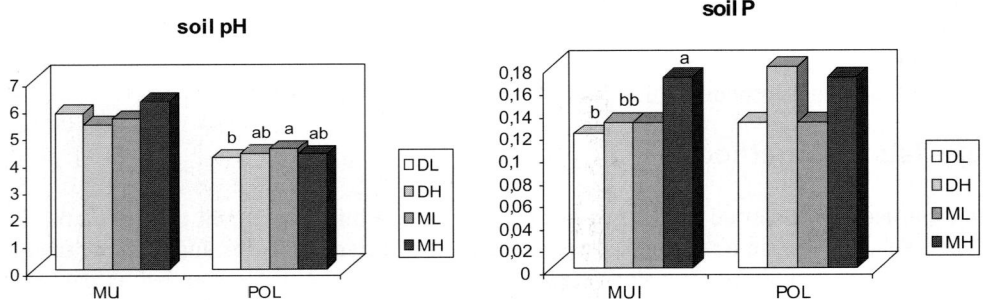

Figure 1. Soil pH and P concentration of soil at both localities and for each treatment. DL (low dose), DH (high dose) of dairy sewage sludge, ML (low dose), MH (high dose) of municipal sewage sludge. Different letters indicate significant differences between treatments in each locality.

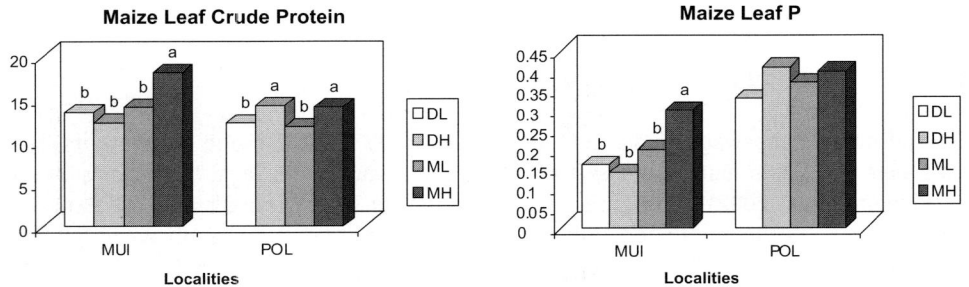

Figure 2. Crude protein and phosphorous concentration of maize leaves at both localities and for each treatment. DL (low dose), DH (high dose) of dairy sewage sludge, ML (low dose), MH (high dose) of municipal sewage sludge. Different letters indicate significant differences between treatments in each locality.

Acknowledgments

We thank AGROAMB-PRODALT for technical support.

References

Rigueiro-Rodríguez, A., Amador-García, A. and Mosquera-Losada, M.R. (2002). Characterisation of dairy and municipal sewage sludge as fertilisers. Proceedings of the VII European Society for Agronomy Congress.

Cattle slurry and vegetable, fruit and garden (VFG) waste compost in silage maize: fertiliser N no longer needed?

F. Nevens[1,2] and D. Reheul[1]
[1]*Ghent University, Plant Production Department, Ghent, Belgium*
[2]*Catholic University of Leuven, Centre for Agricultural and Resource Economics, Leuven, Belgium*

Introduction

The use of slurry on silage maize is, or will be, limited in the near future in order to comply with limits on N losses and NO_3^- leaching in particular. To maintain crop yields, this decrease in slurry use could possibly encourage relatively high applications of external mineral N input and also result in a decrease in soil organic matter contents. These assumptions were checked in the first experiment and a possible alternative (combining compost and slurry) was tested in a second experiment.

Materials and methods

In the first experiment (1983 - 2001), we compared the application of only mineral N with the use of a moderate amount of cattle slurry (on average 180 kg N ha^{-1}) on silage maize over 19 years. Additional mineral N rates (0, 50, 100, 150 and 200 kg N ha^{-1} yr^{-1}) were applied to establish quadratic yield response curves and to determine the mineral N fertiliser replacement value of the slurry-N (NFRV, Paré et al., 1993). We determined the efficiency of the slurry-N as the share of its NFRV in the total amount of applied slurry-N. The economic optimum of additional fertiliser N (N_{opt}) was determined as the N rate above which the extra yield value did not pay for the cost of extra fertiliser N (Neeteson & Wadman, 1987) (see Nevens, 2003). In a second experiment on the same experimental site (1997 - 2000), we combined the use of slurry N (on average 42 Mg ha^{-1} yr^{-1}, 140 kg N ha^{-1} yr^{-1}), vegetable fruit and garden waste (VFG) compost (22.5 Mg ha^{-1} yr^{-1}, 334 kg N ha^{-1} yr^{-1}) and additional mineral N rates (0, 100 and 200 kg N ha^{-1} yr^{-1}). Again, we determined N_{opt} (see Nevens & Reheul, 2003).

The risk of excessive N-leaching was assessed with the amount of residual soil NO_3^--N: according to Flemish legislation, this amount should not exceed 90 kg ha^{-1} (0-90 cm, measured between 1 October and 15 November).

Results

Experiment 1 revealed that the slurry-N use efficiency increased from ± 20% in 1983 to ± 60% in 2000 (Figure 1). The average of the economic optima, for additional mineral N during the most recent 5 years, was 89 kg N ha^{-1}. Obviously, the amount of slurry-N applied was insufficient to keep a good soil quality: we observed that the soil organic matter content decreased during the experiment, down to sub-optimal values in 2000. The contribution of slurry to stable soil organic matter is limited.

Combining cattle slurry and compost resulted in a decreasing N_{opt}: 94, 43, 22 and 12 kg N ha^{-1} during the first, second, third and fourth application years, respectively. The slurry + compost combination seems to offer possibilities to exclude the use of fertiliser N in silage maize.

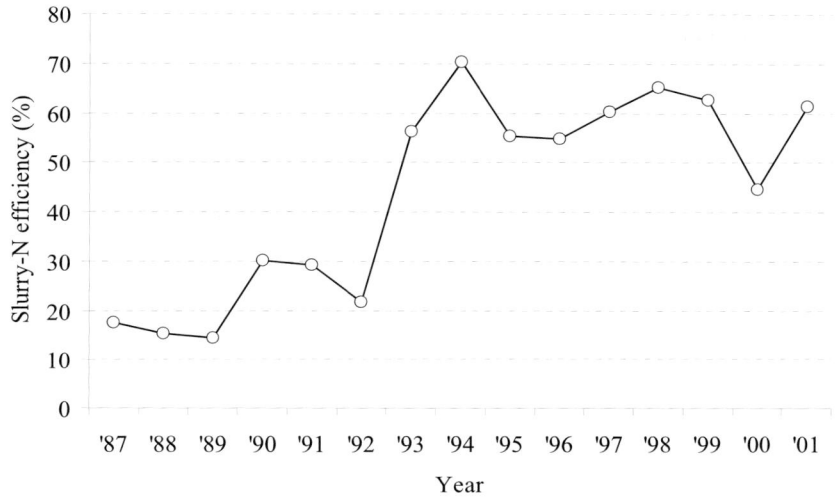

Figure 1. Evolution of slurry-N efficiency (Neff = 100* NFRV/total slurry-N applied).

Already after 4 application years, we have observed an increased soil organic matter content on the plots with compost application. In spite of the high amounts of applied N and the low N-output : N-input ratio, we did not observe excessive amounts of residual mineral soil N, provided that N_{opt} was not exceeded: the ecological and the economic optimum of N application coincided well.

Conclusions

There is evidence that the use of cattle slurry and vegetable, fruit and garden waste compost could result in a silage maize management system in which fertiliser-N is no longer needed and the soil is maintained in good condition.

References

Neeteson, J.J. and Wadman, W.P. (1987). Assessment of economically optimum application rates of fertilizer N on the basis of response curves. Fertilizer Research, 12, 37 - 52.

Paré, T., Chalifour, F.-P., Bourassa, J. and Antoun, H. (1993). Forage-corn production and N-fertilizer replacement values following 1 or 2 years of legumes. Canadian Journal of Plant Science, 73, 477 - 493.

Nevens, F. (2003). Nitrogen use and nitrogen use efficiency in grassland, silage maize and ley/arable rotations. Ph.D. Thesis, Ghent University, Belgium, 231pp.

Nevens, F. and Reheul, D. (2003). The application of vegetable, fruit and garden waste (VFG) compost in addition to cattle slurry in a silage maize monoculture: nitrogen availability and use. European Journal of Agronomy (in press).

Applying a ley/arable rotation to reduce N input in forages

F. Nevens[1,2] and D. Reheul[1]
[1]*Ghent University, Plant Production Department, Ghent, Belgium*
[2]*Catholic University of Leuven, Centre for Agricultural and Resource Economics, Leuven, Belgium*

Introduction

Grasslands and forages comprise a significant part of the utilised agricultural area in many European regions; the reduction of N losses and the increase of N use efficiency is a major and justified concern when considering the sustainability of agriculture. Hence, working on grassland and forage systems with enhanced N use efficiency is very effective to increase the sustainability of the agricultural sector as a whole. The soil-crop component of a cattle farm N cycle offers the best prospects for improving N use efficiency and decreasing N losses (Jarvis & Aarts, 2000). Reducing the use of mineral fertiliser N is an important way to reduce the soil N balance surplus (Stedula, 2003). In a long-term experiment, we studied the possibilities of a ley/arable rotation to contribute to a cattle farm system with less fertiliser input, keeping up grass and forage yields and limiting environmental risks.

Materials and methods

In 1966, an experiment was established on a sandy loam soil of the experimental site of Ghent University in Melle (Belgium). The four major treatments were permanent arable land (PA), permanent grazed grassland (PG, sown in 1966 and never reseeded since), 3 years of arable land alternating with 3 years of grazed grassland (LA 1) and 3 years of grazed grassland alternating with 3 years of arable land (LA 2) (see Nevens & Reheul, 2002, 2003). Three mineral N rates (0, 75 and 180 kg ha^{-1}) on the arable plots (on which silage maize was grown) enabled us to determine the economic optimum of N fertiliser use (N_{opt}) during the last decade. This optimum was determined as the N rate above which the extra yield value did not pay for the cost of extra fertiliser N (Neeteson & Wadman, 1987).

The risk of excessive N-leaching, particularly following the cultivation of temporary grasslands, was assessed from the amount of residual soil NO_3^--N: according to Flemish legislation this amount should not exceed 90 kg ha^{-1} (0-90 cm, measured between 1 October and 15 November). The grassland yields were determined indirectly: the live weight based energy demands of the grazing heifers were calculated from the stocking rates and the length of the subsequent grazing periods (Nevens & Reheul, 2003).

Results

The 9-year average of economically optimal N fertiliser applications in silage maize on permanent arable land was 175 kg ha^{-1}. On arable land in the first, second and third season (following three year old grazed grassland) this was 2, 139 and 154 kg ha^{-1}, respectively. On both types of plots, the same average yield was observed: 19.8 Mg DM ha^{-1}. So, during the three year arable period of the ley/arable rotation, 231 kg N ha^{-1} could be saved, compared with monoculture maize on permanent arable land. Taking into account the application of slurry in actual agricultural practice, we concluded that about 30 kg N ha^{-1} yr^{-1} could be saved on arable plots in ley/arable rotation (Nevens, 2003). This is a significant saving when we consider that a reduction of N input to the soil of 90 kg N ha^{-1} is necessary to reach a soil-crop N efficiency of 70% (arbitrary, 10% below the technically feasible maximum, Jarvis & Aarts, 2000). Applying 30 kg less N on 25% of the Flemish silage maize land would mean a yearly saving of ca. 1.2 million kg of fertiliser-N; applied on 50% of the silage maize land we could save some 2.4 million kg of fertiliser-N.

The 90 kg residual soil NO$_3^-$-N ha^{-1} (0 - 90 cm) threshold was exceeded when the N$_{opt}$ was exceeded (we conclude that the ecological and the economic optimum of N rate coincided well). During the first season after ploughing the three year old grazed grassland, the risk of excessive amounts of residual soil NO$_3^-$ -N following silage maize was high, even without N application. Growing fodder beet provided a perfect solution: with their long growing season, high yields and deep rooting system, the beet scavenged the soil and left almost no unused N in the soil profile. The yields of the permanent grassland (after 33 years still in fairly good botanical condition) and the temporary grasslands were not significantly different; either on average over a 33 year period, or during the most recent years.

Conclusions

The (re-)introduction of ley arable rotations could be an effective tool to further enhance the N use efficiency on Flemish dairy farms.

References

Jarvis, S.C. and Aarts, H.F.M. (2000). Nutrient management from a farming systems perspective. In: K. Soegaard, C. Ohlson, J. Sehested, N.J. Hutchings. and T. Kristensen (eds). Grassland farming. Balancing environmental and economic demands. Grassland Science in Europe, Volume 5. British Grassland Society, Reading, UK, pp. 363 - 373.

Neeteson, J.J. and Wadman, W.P. (1987). Assessment of economically optimum application rates of fertiliser N on the basis of response curves. Fertiliser Research, 12, 37 - 52.

Nevens, F. 2003. Nitrogen use and nitrogen use efficiency in grassland, silage maize and ley/arable rotations. Ph.D. Thesis., Ghent University, Belgium, 231pp.

Nevens, F. and Reheul, D. (2002). The nitrogen- and non-nitrogen contribution effect of ploughed grass leys on the following arable forage crops: determination and optimum use. European Journal of Agronomy, 16, 57-74.

Nevens, F. and Reheul, D. (2003). Permanent grassland and three-year leys alternating with three years of arable land: 31 years of comparison. European Journal of Agronomy, 19, 77 - 90.

Stedula (Steunpunt Duurzame Landbouw) (2003). Jaarverslag 2002. (Annual report of the Flemish Policy Research Centre for Sustainable Agriculture). Stedula, Gontrode, Belgium, 272pp. (in Dutch).

Farm N budgets with estimated nitrogen losses by use of soil N modelling

A.H. Nielsen, B.M. Petersen and I.S. Kristensen
Danish Institute of Agricultural Sciences, Department of Agroecology, P.O. Box 50, DK-8830, Tjele, Denmark

Introduction

Accounting for the different N losses in the N balance for a defined farming system involves certain difficulties. Losses occur along with several of the flows of N in the observed system (e. g. the farm). The static nature of element balances is contrary to the dynamic nature of the loss processes for which relatively simple assumptions must be made. When looking at field or farm scale N balances, it is tempting to make the simplifying assumption that the N surplus equals emissions to the environment. This may not be true since a nutrient surplus may be temporarily stored within the system e.g. in the soil. For Danish soils, Heidmann et al. (2002) showed a considerable annual change in the soil N pools. This study used an approach with three major elements: 1) calculation of a surface balance, 2) estimation of NH_3 and denitrification losses and 3) simulation of soil N developments, thereby allowing an estimated NO_3^- leaching calculated by difference.

Materials and methods

An inventory of 28 farm types representing the Danish farming sector was constructed on the basis of 2239 farm accounts from 1999 (Dalgaard & Halberg, 2003). Within each farm type, a coherent farm model was established by use of the accounts data on crop rotation, crop yields etc. Partial N balances for the herd and for crop production were established. Symbiotic N_2-fixation was predicted from modelled crop-specific standard values. Standard values were used to estimate the total atmospheric deposition, the NH_3 losses from stables, manure storage facilities and manure excreted on or applied to fields and losses by denitrification. An extended version of C-TOOL (Petersen et al., 2002) was utilised for the simulations of soil N developments, thereby allowing an estimated NO_3^- leaching calculated by difference. The model was initiated with soil data from the Danish National Square Grid System (Heidmann et al., 2002). Where possible, the calculated N turnover and emissions of N were checked against national level statistical data. For this paper, two major farm types with animal husbandry on sandy soil (<10% clay) were selected: 1) mixed dairy farming with medium livestock rate (1.7 LSU ha^{-1}) and 2) pig farms with low-medium rates (1.4 LSU ha^{-1}).

Results

The mixed dairy farm had the largest N surplus (197 kg N ha^{-1}) compared with a lower level at the pig farm (160 kg N ha^{-1}). The pig farm had large inputs and outputs at farm level, primarily because of N in purchased feed, sold meat and sold crops. The mixed dairy farm, in contrast, had large internal recycling of N between the herd and the fields. In spite of a lower livestock rate at the pig farm, there was a larger NH_3 loss per hectare compared with the dairy farm. This is primarily to be explained by higher standard N emissions from pig stables than from dairy stables. The simulations of soil N development showed a relatively moderate increase of 5 kg N ha^{-1} at the mixed dairy farm, compared with the large annual internal turnover of N. The farming practice at the pig farm tended to deplete soil N pools (Δ-N soil: -10 kg N ha^{-1}). The leaching estimate calculated by the difference method resulted in an average level of the order of 100 kg N ha^{-1} at the pig farm and ca.10% more at the mixed dairy farm.

Discussion

Calculated N surpluses were in good agreement with values measured earlier at private Danish farms (Nielsen & Kristensen, 2002). Differences in soil N development at the dairy and pig farms, as estimated in this study, have two main causes: 1) high initial level of C (equivalent to N) in the soil at dairy farms and 2) larger annual input of C to the soil deriving from manure and crops at dairy farms. Nitrate leaching, as calculated by difference, only increased slightly in consequence of increasing the livestocking rate, whereas farm N surplus increased markedly (in agreement with earlier findings at Danish farms). The farm types in this study are not representative of individual farm cases parallel to private farmers. Formation of coherent farm models and detailed cross validations at different scale levels (farm and national) were used to ensure a consistent analysis. Estimating leaching by the difference method may introduce additional uncertainty to the results.

Conclusions

Farm N budgets with estimated turnover, losses and changes in soil N pools at 28 Danish major farm types representative of the Danish agricultural sector were established. The use of dynamic soil N modelling allowed the estimation of NO$_3^-$ leaching by the difference method. Increase in loss of NO$_3^-$ as a result of increased stocking rate may be partly counterbalanced by C-driven changes in soil N pools and by increased gaseous losses. Structural differences of farm types with respect of external and internal C and N flow, land use and previous history of soil are related to differences in environmental impact between farm types.

References

Dalgaard, R. and Halberg, N. (2003). An LC inventory based on representative and coherent farm types. In: N. Halberg and B. Weidema (eds) 4th International conference on: Life cycle assessment in the agri-food sector, Bygholm Park Hotel, October 6-8, 2003, Working papers, Danish Institute of Agricultural Sciences, Department of Agroecology, 114-122.

Heidmann, T., Christensen, B.T. and Olesen, S.E. (2002). Changes in soil C and N content in different cropping systems and soil types. In: S. O. Petersen and J.E. Olesen (eds) Greenhouse Gas Inventories for Agriculture in the Nordic Countries. DIAS report - Plant Production, 81, 77-86.

Nielsen, A.H. and Kristensen, T. (2002). N-overskuddet og kvægbedriftens tilpasningsmuligheder (N-surplus and possibilities for adjustment at dairy farms). In: Kvælstofbalancer på landbrugsbedriften - status og perspektiv. Temadag arrangeret af Afd. for Jordbrugssystemer 24. april 2002. Forskningscenter Foulum. Intern rapport, 157, 67-82.

Petersen, B.M., Olesen, J.E. and Heidmann, T. (2002). A flexible tool for simulation of soil carbon turnover. Ecological Modelling, 151, 1-14.

Political transformations and nitrogen balances of dairy farms in Poland

S. Pietrzak[1] and O. Oenema[2]
[1]Institute for Land Reclamation and Grassland Farming at Falenty, 05-090 Raszyn, Poland
[2]Wageningen UR, Alterra, P.O. Box 47, NL-6700 AA Wageningen, The Netherlands

Introduction

The transformation from a political system with central planning into parliament democracy with market economy in Poland in 1989 had dramatic effects on Polish agriculture. The economic conditions for agriculture deteriorated strongly, in part also because of the decrease in export markets after the collapse of the former Soviet Union. The contribution of agriculture to the gross domestic production declined from 8.5% in the beginning of the 1990s to less than 4% in 2000, while employment in agriculture decreased from about 25% to 18% in the same period. Numbers of dairy cows decreased from 5 million in 1990, to 3 million in 2000, i.e. by 40%. Total milk production decreased by 25%, from 15.4 billion l in 1990 to 11.5 billion l in 2000 (Rural Poland 2000, Rural Development Report, FDPA, Warsaw). During the last few years, Poland has prepared itself for membership of the European Union (EU). A complex element in the preparation is the adjustment of Polish agriculture to fit the requirements of the EU. There will be a further decrease in animal stock and milk production, because of the limit of 8.9 billion l of milk agreed recently in Copenhagen. Also important is the introduction of standards for food quality and regulations for environmental protection. These new standards and regulations require expensive investments (e.g. for storage and spreading of animal manure and fertilisers). Many farmers have or will have to abandon production. Other farmers try to remain in the market and improve the economic profitability of farming by specialisation, intensification and enlargement of the farm. We are particularly interested in the changes in farm structure and management during the agricultural transition, and in the consequences of these changes for N use efficiency and N losses. We also explored possibilities for improvements in N use efficiency of farms that remain in the market.

Methods

Four farms, representative of those that want to remain in the market, were selected in Gródek in north-eastern Poland. Farmers' interviews were made regularly and statistical data of farms were collected from 2000 onwards. Detailed assessments were made of all N inputs and N outputs. Changes in some farm characteristics are shown in Table 1 and in Figure 1. Milk production per farm increased, on average by about 6000 kg per farm per year, through an increase in both number of dairy cows and milk production per cow.

Results and discussion

In the period 1999 to 2002, N surpluses ranged from 56 to 166 kg N ha^{-1}. There were no clear relationships between changes in dairy cow number, livestock density and N surpluses. In two farms, N surpluses tended to increase, while in the other two there was a decreasing trend.

Table 1. Changes in farm structure of four dairy farms in Gródek between 2000 and 2003.

Specification	Farm							
	KT		WP		KS		CS	
	2000	2003	2000	2003	2000	2003	2000	2003
Farm size, ha	23.4	29.4	26.8	33.8	23.4	32.4	15.1	16.6
Grassland area, %	40.2	44.9	11.0	33.6	54.7	48.8	45.8	52.1
Number of dairy cows	16	20	11	21	19	23	15	15

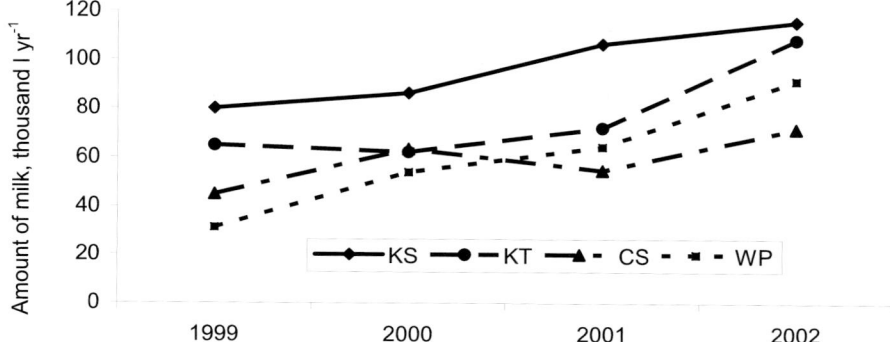

Figure 1. Changes in the amount of milk sold per farm in the period 1999 to 2002.

The efficiency of N utilisation (the ratio of total N output to total N input) ranged between 13 and 40% (Table 2). The farm with the highest N use efficiency had the lowest percentage grassland area and dairy cow numbers. A low efficiency of N utilisation coincided with a high milk production per ha and with high usages of N fertiliser and feed concentrates and fixation by legumes. The wide range suggests that there is scope for improvement in at least some farms. The surroundings of Gródek have high natural value, as shown for example, by pristine waters, the abundance of fish in the nearby river Nurzec, and by the presence of beaver lodges. It is important that the intensification of animal production in this area results in satisfactory incomes to farmers, but not at the expense of natural values.

Table 2. Changes in N balance in four dairy farms in Gródek over the years 1999 to 2002.

Farm	N surplus, kg N ha^{-1}				N use efficiency, %			
	1999	2000	2001	2002	1999	2000	2001	2002
KT	145.5	146.7	132.5	123.2	16.6	12.8	14.1	17.7
WP	98.2	73.3	56.5	85.9	28.5	36.8	40.3	32.4
KS	81.4	132.3	139.4	166.5	22.7	14.0	19.0	13.6
CS	140.5	143.6	117.3	114.7	14.9	14.5	19.8	24.2

Conclusion

There were no clear trends in N surpluses, because farmers had different strategies to adjust to the changes in the market position, and these had contrasting effects on N surpluses. The time period was probably also too short to detect trends. There is scope for improving the efficiency of N use, through improved (i) animal feeding, (ii) use of manure, and (ii) crop rotations.

Acknowledgements

This project is funded by Polish State Committee for Scientific Research (No 3P06S 053 22).

Dirty water - a valuable source of nitrogen on dairy farms

K. Richards[1], M. Ryan[1] and C.E. Coxon[2]
[1]Teagasc, Johnstown Castle, Wexford, Ireland
[2]Department of Geology, Trinity College Dublin, Dublin 2, Ireland

Introduction

It is estimated that the Irish dairy sector generates 19.6 M tonnes of dirty water (DW) annually (Stapleton et al., 2000). Dirty water is dilute farm yard effluent, comprising rainwater contaminated with livestock excreta, silage effluent, milk bulk tank rinsings, milk parlour washings and leachate from stored manures. This study quantified the daily N content of DW and the impact of DW irrigation on NO_3^- leaching.

Materials and methods

The experiment was conducted on an intensive grassland dairy farm in the south of Ireland on a free draining sandy loam soil with a stocking rate of 2.6 livestock units ha^{-1}. DW was sampled daily from the storage tank and was analysed for Total Kjeldhal N (TKN) and Total Available N (TAN = NH_4^+-N + NO_3^--N). Daily N loads were calculated by multiplying N concentration by daily volumes of DW irrigated. Leaching of NO_3^--N was quantified by soil and soil solution sampling from two grazed 1 ha plots receiving annual DW irrigation loads of 140 mm (low) and 444 mm (high), equivalent to 437 and 756 kg TKN ha^{-1} yr^{-1}. Inorganic soil N (autumn 1995) was determined using KCl extraction of depth-specific bulked soil samples (20 cores per plot); the results were then summed for 0-90 cm. Fortnightly soil solution samples were taken between October 1995 and February 1996 using 12 ceramic cups per plot clustered in 4 blocks per plot (3 cups per block) at 0.5 m.

Results

The DW contained 1349 kg TKN yr^{-1} including 623 kg TAN yr^{-1}. TKN concentrations ranged from 8 to 1360 mg l^{-1} (average of 188 mg l^{-1}). TAN, on average, accounted for 98% of the inorganic N in DW and concentrations ranged from 1.3 to 284 mg l^{-1} (mean of 79 mg l^{-1}). The highest DW concentrations and nutrient loads applied were observed during summer months. Daily mean DW volumes generated were 25 m^3 day^{-1} and the average volume per cow was 88.5 m^3 yr^{-1}. Autumn inorganic N contents (0 to 90 cm) in the low and high DW plots were 168 and 320 kg N ha^{-1}, respectively, compared with an average of 103 kg N ha^{-1} observed in five non DW plots on the same farm. The mean soil solution NO_3^--N concentrations in the low and high DW plots were 49 and 59 mg l^{-1}, respectively: maximum mean NO_3^--N concentrations were 90 mg l^{-1} and 124 mg l^{-1}, respectively. During the drainage period, 210 and 253 kg N ha^{-1} was leached from the low and high DW plots, respectively, with considerable in-plot variation (Table 1).

Discussion

In England and Wales, Cumby et al. (1999) reported similar DW, TKN and TAN concentrations of 950 ($\pm$770) and 580 ($\pm$487) mg l^{-1} and lower DW irrigation volumes of 2.7 to 60.8 m^3 yr^{-1} per cow compared with the mean 88.5 m^3 yr^{-1} per cow reported in this study.

Table 1. *The volume and duration of dirty water irrigation and the effect on soil solution nitrate-N loads (kg N ha*$^{-1}$*) leaching from irrigated plots.*

Plot	DW volume applied (m^3 ha^{-1})	Irrigation duration (d)	Mean block NO_3^--N loads leached (kg ha^{-1})				Mean plot NO_3^--N (kg ha^{-1})	Standard Error ($\pm$)
			1	2	3	4		
Low	1404	83	264	199	164	212	210	21
High	4435	185	323	288	307	95	253	53

The temporal variation of dirty water quality observed is related to the spring calving of the herd which resulted in peak milk production in summer. In winter milking herds, there is a potential that high N loads would be generated in winter at a time of low grass growth and storage of DW should be considered. In this study, the high and low rates of DW irrigation, combined with dairy cow grazing, resulted in rates of NO_3^- leaching considerably greater than the limits stated in the Nitrates Directive (EEC, 1991). When DW is irrigated at high rates on restricted areas of land, the NO_3^- leaching potential can be significantly increased. In this study DW irrigation was especially important during summer periods when the TKN load is highest and the hydraulic load may leach nutrients beyond the grass rooting zone.

Conclusions

Dirty water irrigation has the potential to contribute significantly to farm N budgets and it also poses a risk to water quality because of the nutrient and hydraulic loading when irrigated onto land. High hydraulic loading rates can leach nutrients beyond the rooting zone greatly increasing the risk of NO_3^- leaching. It is important that when irrigation plans are developed they should ensure that the total N application does not exceed plant requirements and that the hydraulic loading is low.

Acknowledgements

This project was funded by Teagasc under the Walsh Fellowship programme.

References

Cumby, T.R., Brewer, A.J. and Dimmock, S.J. (1999). Dirty water from diary farms, I: biochemical characteristics. Bioresource Technology, 67, 155-160.

EEC (1991) Council Directive 91/676/EEC of 12 December 1991 concerning the protection of waters against pollution caused by nitrates from agricultural sources, OJ L 375, 31.12.1991, 8pp.

Stapleton, L., Lehane, M. and Toner, P. (2000). Irelands environment - A millennium report, Environmental Protection Agency, Ireland, 270pp.

Effect of date and dose of sewage sludge application in grasslands production and nitrogen soil concentration in Galicia (NW Spain)

A. Rigueiro-Rodríguez, A. Casanova-Vigo and M.R. Mosquera-Losada
Crop Production Department, Escuela Politécnica Superior, Universidad de Santiago de Compostela, Lugo, Spain

Introduction

An important environmental problem of European Union (EU) will arise from the increase in sewage sludge production (40% from 1999 to 2005, COM (1999) 752)). The EU Directives (Council Directive 91/271/EEC) specify a need to increase disposal of urban sludge into agriculture, based on previous practical application experiences, in order to increase the value of this residue, prevent environmental damage, and increase confidence in agronomic sewage sludge use. The use of sewage sludge as fertiliser has to be based on N content, provided heavy metal limits are not exceeded, and taking account of plant N availability, application date and dose. Assays made in Galicia, regarding timing and doses of N application have been made with mineral N and indicated that the use of N earlier, will not modify pasture production in warm areas, as clover was favoured. However, in colder areas, pasture production was reduced because of leaching (Mosquera & González, 1999). Conclusions from experiments testing different dates of sludge application in Galicia mountain areas (cold and low pH) indicated that with low doses (between 50 and 100 kg total-N ha^{-1}), there were no effects on pasture production (Mosquera et al., 2002). Comparisons between date of sowing (spring or autumn) indicated that spring application of sludge without incorporation by ploughing in, is a more sustainable management practice that introduce this residue in the autumn at sowing (Mosquera et al., 2001).

Materials and methods

The experiment was carried out on flat land in Galicia (annual precipitation, 1350 mm, mean temperature 28 °C), following a randomised block design with four replicates in small plots (2x4 m^2). Previous soil characteristics were defined by pH of 6.5, low organic matter content (1.55%), high P-Olsen content (28 mg kg^{-1}), medium K concentration (86 mg kg^{-1}) and a low CEC (6.7 cmol(+) 100 g soil^{-1}) and previously had grass and potatoes crops. Municipal sewage sludge had a pH of 7.0, an N content of 2.5% and a dry matter of 25%. Heavy metal contents of sewage sludge were under the allowed limits indicated by Spanish legislation. Treatments consisted of four dates (December, January, February and March) and two doses (160, N$_1$, and 320, N$_2$, kg total N ha$^-$) of application in 2000 and 2001. A no fertiliser treatment (NF) was also established. Plots were sown with 12.5 kg ha^{-1} cooksfoot, 12.5 kg ha^{-1} ryegrass and 5 kg ha^{-1} white clover. Four harvests were made (April, May, July and November) from areas of 4 x 1.1 m. Total N was estimated after micro-kjeldahl digestion. At the end of the experiment, soils were sampled to a depth of 25 cm, following legislation (RD 1310/90). Organic matter was determined by potassium dichromate digestion.

Results and discussion

Pasture production was significantly affected by date harvest and dose harvest interactions. There was a positive response of pasture production to dose of sewage sludge; low and high sewage sludge doses were significantly different from NF until the last harvest, when the low dose was not significantly different from NF treatment. The positive effect of the sewage sludge application on pasture production has been previously described (Smith, 1996). Dates of application also modified significantly DM yield, the last date of application reduced yield with respect to other dates, but this effect reverted in the second and fourth harvest, and meant that the annual yield did not differ between treatments. However, the delay in pasture production at the last date, will mean an economic cost increment for the farm, as it implies the use of concentrates for feeding animals when animal needs are higher. Total N-soil was reduced if sewage sludge was applied as mineralisation was promoted because of C introduction and removal in crops. C/N relationship was not modified by dates of application, differences being explained by doses. Total pasture production was only affected by doses, but not by date of application.

Sewage sludge recommendations must take into account the local characteristics, and the type of soil: its nutrient concentration is important. Optimisation of the use of sewage sludge must also take into account the use of the pasture; if silage is being made, the date of application is not important. In Galician conditions, herbage is grazed and a period of ca. two months should be left between sludge application and grazing.

Table 1. Pasture production (t dm ha^{-1}, mean of two years) for different fertiliser applications (NF: not fertilisation; N_1: low dose of sewage sludge; N_2: high dose of sewage sludge) and application dates (D: December; J: January; F: February; M: March), N soil concentration (%) and C/N ratios. Different letters indicated significant differences between treatments (doses or dates).

Harvest	Dose			DateNF				
	NF	N_1	N_2	NF	D	J	F	M
Pasture								
April	3.16c	3.98b	4.68a	3.16b	4.93a	4.96a	4.48a	2.95b
May	0.98b	1.14a	1.26a	0.98c	0.92c	1.06bc	1.18b	1.65a
July	0.20b	0.23ab	0.31a	0.2	0.31	0.22	0.27	0.29
November	1.0b	1.68b	2.05a	1.6b	1.93ab	1.72b	1.62b	2.14a
Total	5.94b	7.03ab	8.3a	5.94b	8.09a	7.96a	7.55ab	7.03ab
Soil								
N-soil	0.37a	0.27b	0.31b	0.37a	0.3abc	0,26bc	0.24c	0.36ab
C/N	2.44b	5.06a	5.05a	2.44b	5.20a	5.23a	5.13a	4.62a

Acknowledgments

We thank J. Javier Santiago Freijanes, M.T. Piñeiro-López, Divina Vázquez-Varela and José Alberto Lamas-Díaz for laboratory and field work.

References

Mosquera, M.R., Fernández, C. and Rigueiro, A. (2002). Nitrogen Mineralisation, pasture production and botanical composition in silvopastoral systems installed in very acid soils. Proceedings of the VII meeting of the European Society of Agronomy, 352-357.

Mosquera, M.R. and González, A. (1999). Use of first nitrogen in South Europe template grassland. Proceedings of 10th Nitrogen Workshop, vol. 2.,IV.32

Mosquera, M.R, López, M.L. and Rigueiro, A. (2001). Effect of date of application of sewage sludge on nitrogen and pasture production. Proceedings of First World Congress on Conservation Agriculture: A worldwide challenge, 409-411

Smith, S.R. (1996). Agricultural recycling of sewage sludge and the environment. CAB International. 225 pp. WRC Marlow Buckinghamshire (UK).

Nitrate losses in forestry nurseries using municipal sewage sludge

A. Rigueiro-Rodríguez, J. Rasche-Castillo and M.R. Mosquera-Losada
Crop Production Department, Escuela Politécnica Superior, Universidad de Santiago de Compostela, Lugo, Spain

Introduction

The employment of sewage sludge as part of the substrate for growing plants is a way of recycling this residue, because it will reduce costs of plant cultivation and allow a suitable means of diposal. On the other hand, the input of heavy metals and NO_3^- to soil after afforestation will be low, especially as saplings are planted at a low density. Nowadays, the use of sewage sludge in a responsible way in agriculture is encouraged by the EU, because of the growing production of this residue. The use of sewage sludge as fertiliser on forest plantations will permit cultivation that will not be eaten by humans, avoiding direct health risks. The use of this organic residue has a different behaviour compared with mineral fertiliser, (NO_3^-) which is more readily available. To avoid potential health problems, the Drinking Water Directive (CEC,1980) has a limit on potable water supplies of 11.3 mg NO_3^--N l^-.

Materials and methods

The study was carried out in spring in a greenhouse in Galicia (NW Spain), where four treatments were established in a randomised block design with four replicates, that consisted of 27 plants of *Betula alba* L., each placed in a superleach container. Treatments consisted of four combinations of perlite (P), pine bark (B), sewage sludge (S) and a control (C): 75% P + 25% B; low dose of sludge (L): 25% P + 25% S + 50% B; medium dose of sludge (M): 25% P + 37.5 % S + 37.5 % B and high dose of sludge (H): 25% P + 50% S + 25% B. The control treatment was fertilised with compound fertiliser (14N:7P $_2$ O$_5$:16 K$_2$O) at a dose of 2.7 kg m^3). Plants were harvested in September and divided into root, stem and leaf fractions that were weighed after drying. Height and diameter were also measured. From the start of the experiment, water samples were taken from each replicate every two days and placed in a freezer. NO_3^--NO_3 was analysed from 28 April to 9 August.

Results and discussion

All the production criteria (Table 1) were higher in H treatment than the control, but similar to those found for L and M treatment, with the exception of height and leaf weight that were significantly lower in L and M treatments. The positive effect of the inclusion of sewage sludge as a component of the substrate will improve the growth of different forest species, but this response depended on the species cultivated. Studies developed with *Pinus pinaster* (Rigueiro et al., 2001b) indicated that the best production is obtained with a L treatment, for *Pseudotsuga menziesii* (Mosquera et al., 2001a) and *Pinus sylvestris* (Rigueiro et al., 2001a) it was M treatment and for *Euclayptus nitens* (Mosquera et al., 2001b) it was between L and M treatment. Mean NO_3^- leaching was under the EU limit (CEC, 1980) and was significantly higher in H (7.9 mg

NO_3 1^{-1}) and C (7.8 mg NO_3 1^{-1}) than in L (6.6 mg NO_3 1^{-1}) and M (5.92 mg NO_3 1^{-1}) treatments. An interaction of date of sampling and treatments was found, indicating that initially, NO_3^--N leaching was above the EU limit in H and C treatments (27 and 30 mg NO_3^- 1^{-1}). However, a month later, all treatments with sludge as a part of the substrate were above the limit (30 mg NO_3^- 1^{-1}), (Figure 1). Nitrate leaching was different between treatments depending on easily mineralisable N content and, it was also increased between 16 and 35 days after sowing.

Table 1. Tree height (h, cm), diameter (d, mm), root biomass (r, g dm per plant), leaf biomass (le, g dm per plant) and stem biomass (st, g dm per plant) in each treatment: high (H), medium (M), low (L) sewage sludge dose and control (C) at harvesting.

Treatment	H	M	L	C
h	25.89[a]	22.47[b]	19.13[c]	15.36[d]
d	0.25[a]	0.26[a]	0.24[a]	0.20[b]
r	0.58[a]	0.65[a]	0.58[a]	0.38[b]
le	0.37[a]	0.27[b]	0.27[b]	0.15[c]
st	0.56[a]	0.46[a]	0.47[a]	0.21[b]

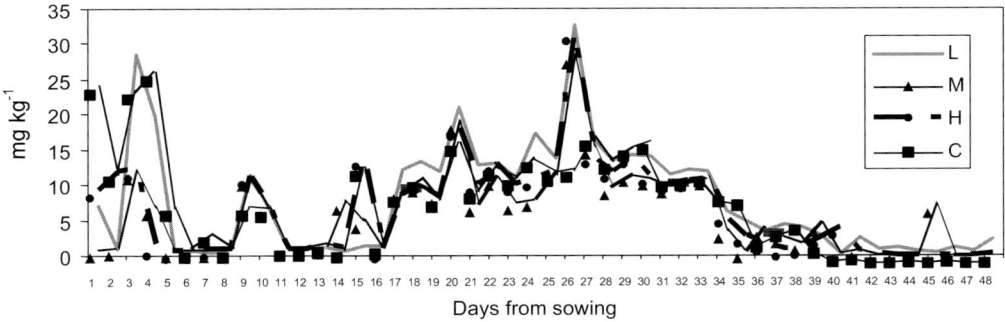

Figure 1. Nitrate leaching over time. L:low dose, M: Medium dose; H: High dose; C:control.

Sewage sludge can substitute as a part of the substrate for nursery plant production in containers, as it is cheaper and avoids the use of mineral fertilisers and at optimised concentrations it will reduce NO_3^- leaching. This will reduce cost of forest plant production.

Acknowledgments

We thank GESTAGUA, S.A. and J. Javier Santiago Freijanes, M. T. Piñeiro-López, Divina Vázquez-Varela and José Alberto Lamas-Díaz for field and laboratory work.

References

CEC (1980). Council Directive of 15 July 1980 relating to the quality of water intended for human consumption (80/778/EEC). Official Journal of the European Community. No. L 229/11-29.

Mosquera, A., Rigueiro, A. and Quintela, V. (2001). Efecto de la proporción de distintos componentes del sustrato en el crecimiento de *Pseudotsuga menziesii* Douglas en vivero. III Congreso Forestal Nacional, 3, 416-422.

Mosquera, A., Rigueiro A. and Vila, T. (2001b). Producción y contenido de nutrietnes en raíz tallo y hojas deplanta de vivero de *Eucalyptus nitens* (Deane & Maiden) cultivada en envase con sustrtos en losqueseincluyen distintas proporciones de lodo de depuradora urbana III Congreso Forestal Nacional, 3, 422-429.

Rigueiro, A., Mosquera, A. and Quintela, V. (2001a). Contenido de metales en planta de *Pinus sylvestris* L. cultivada en vivero sobre sustratos con lodos de depuradora urbana. III Congreso Forestal Nacional, 3, 430-434.

Rigueiro, A., Mosquera, A. and Vila, T. (2001b). Efecto de la proporción de distintos componentes del sustrato en el crecimiento de Pinus pinaster Aiton en envase en vivero, II Congreso Forestal Nacional, 3, 410-415.

Assessment of measures to reduce nitrogen losses from dairy farms

J. Scheringer[1] and J. Isselstein[2]
[1]Centre for Agriculture and the Environment, University of Göttingen, Am Vogelsang 6, 37075 Göttingen, Germany
[2]Institute for Agronomy and Plant Breeding, University of Göttingen, Germany

Introduction

During the last decade a range of management options have been suggested to reduce N losses (Bussink & Oenema, 1998) and these have been disseminated via extension services to farmers. However, as current N-losses show, the situation is far from satisfactory. A better understanding of the degree of implementation of measures and their effectiveness on practical farms is necessary in order to improve the efficiency of the use of N and further reduce losses.

Materials and methods

In a research project carried out on 46 dairy farms in north west Germany, the degree of implementation of improvement measures and their effectiveness were investigated. The investigated farms were both conventional and organic farms. The average N surplus of the whole farm balance for all farms was 132 kg N ha^{-1} and was taken as the indicator of N losses (Scheringer, 2002). Measures which were introduced on these farms to some degree, and therefore assessed more closely, were: 1) the integration of low protein forages such as maize into the diet of the cows (Jarvis & Aarts, 2000), 2) increasing the milk per lactation to allow for a more efficient use of dietary N and reduced stocking levels (Bussink & Oenema, 1998), and 3) handling excreta more efficiently to reduce N losses (e.g. Bussink & Oenema, 1998).

Results and discussion

1. The predominantly fed low-protein feed was maize silage. Of the 36 farms which had arable land, 32 grew maize. However, growing more maize did not improve the whole farm surplus: there was no significant correlation found between the proportion of land cultivated with maize and the N surplus of the whole farm balance (r = 0.261, n.s., p < 0.05). The farms which grew maize tended to have higher stocking rates (Figure1). This defeated the positive effects of maize, as stocking rate is the driving force of intensity, and higher intensity in dairy farming leads to higher N surpluses (Scheringer, 2002).
2. Farms with a higher milk yield per cow had higher N surpluses. These farms also had higher stocking rates (Figure 1). Moreover, there is evidence that farmers achieve higher milk yields by increasing the purchase of concentrates (Figure 2).
3. Efficient excreta handling: 40 farms housed their cows in cowsheds with slurry systems. Only 12 of these 40 farms covered the slurry pits and proportionally, more of these were organic, rather than conventional farms. Slurry additives were hardly used and stirring the slurry was rarely practised. Broadcasting with a splash plate was the main application technology. Trailing hoses were used in only a few cases and injection only on one farm.

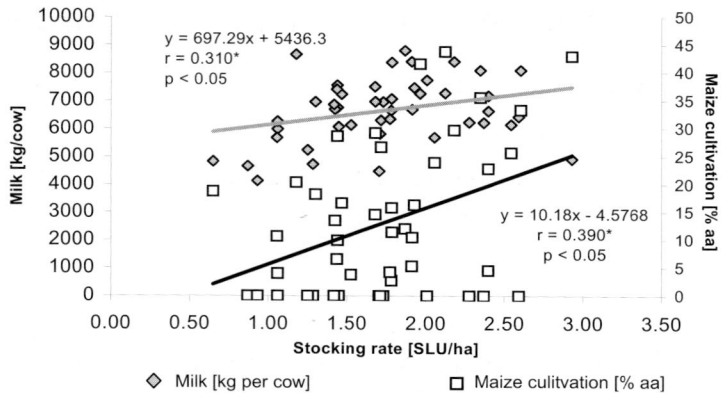

Figure 1. Stocking rates, maize cultivation area (aa) and milk yields.

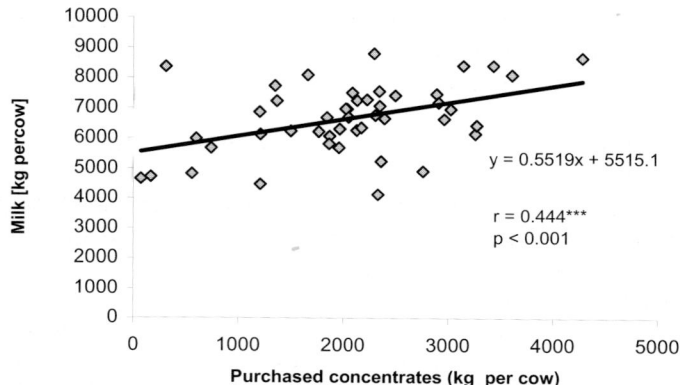

Figure 2. Purchased concentrates and milk yields.

Conclusion

The assessment of options to reduce N losses from dairy farms showed that important measures to reduce N losses from dairy farms have not yet been implemented to a satisfactory degree. Moreover, because of framework conditions and farmers' management practices (which aim to increase income by increasing intensity), some measures implemented do not result in reduced N-surpluses and, therefore, reduced N-losses. A stronger control needs to be put on the conditions and practices of implementation of such measures on working farms.

References

Bussink, D. W. and Oenema, O. (1998). Ammonia volatilisation from dairy farming systems in temperate areas: a review. Nutrient Cycling in Agroecosystems, 51, 19-33.

Jarvis, S. C. and Aarts, H. F. M. (2000). Nutrient management from a farming system perspective. Grassland Science in Europe, 5, 363-373.

Scheringer, J. (2002). Nitrogen on dairy farms: balances and efficiency. Göttinger Agrarwissenschaftliche Beiträge 10. excelsior p.s.

Nitrogen use efficiency in a water saving ground cover rice production system in Beijing, North China

H. Tao[1,2], K. Dittert[1], S. Lin[2], C. Kreye[1] and B. Sattelmacher[1]

[1]Institute of Plant Nutrition and Soil Science, Christian Albrechts University, D-24118, Kiel, Germany
[2]Department of Plant Nutrition, China Agricultural University, 100094, Beijing, P.R. China

Introduction

Typically, the N use efficiency (NUE) of paddy rice is in a range of 30-40% (Cassman, et al., 1993), which is low compared with other crops. In our experiment, a new approach, the ground cover rice production system (GCRPS) was evaluated. In GCRPS, soil is irrigated to approximately 90% water holding capacity (WHC) before rice seedlings are transplanted or seeded. Thereafter, the soil is covered by plastic film or plant mulch, and the soil is kept at approximately 90% WHC without a standing water layer during the whole growing period (Lin et al., 2002). The aims of the experiment were to monitor N dynamics in the soil and to quantity the NUE for identification of opportunities to improve NUE.

Materials and methods

Two-year field experiments were conducted on a light loamy soil at the China Agricultural University in the North-west of Beijing (39.95°N and 116.3°E). The difference and ^{15}N isotope methods were used for the measurement of NUE in 3 treatments: paddy control, GCRPS-film and GCRPS-straw. Different N fertilisation methods were tested in 2 years. In 2001, urea was applied as basal fertiliser at 225 kg N ha^{-1} in GCRPS; the same amount of urea was applied in the paddy, but split and applied at a ratio of 4:3:3. In 2002, urea-N (180 kg N ha^{-1}) was split and applied on each plot at a ratio of 3:4:3. Additionally, in 2002, organic manure equivalent to 45 kg N ha^{-1} was applied as basal fertiliser.

Results

In 2001, rice yield was substantially reduced in all GCRPS with single N dressing; in 2002, when N fertiliser was given in 3 splits, yield of GCRPS-film was only 8% lower than in paddy control (Table 1). Nitrogen use efficiency was lower in GCRPS than in paddy control over two

Table 1. Yield[1] and NUE[2] of GCRPS compared with paddy rice production system.

Treatments	2001				2002			
	Yield (kg ha^{-1})	NUE (%)	FP (kg grain kg N^{-1})	NHI	Yield (kg ha^{-1})	NUE (%)	FP (kg grain kg N^{-1})	NHI
Paddy control	8515 a	47	38	0.48	7386 a	42	41	0.50
GCRPS-film	5863 b	2.4	26	0.56	6772 b	19	37	0.55
GCRPS-straw	3579 c	1.3	16	0.44	3345 c	13	18	0.41

[1]data in each column followed by the same letters are not significantly different, P< 0.05, n=3; [2]measured by ^{15}N isotope method

years. In comparison with 2001, NUE and fertiliser productivity (FP) in GCRPS were improved with split N application in 2002. The N harvest index in GCRPS-film was slightly higher than in paddy control over 2 years (Table 1.).

Sixty days after fertilisation, there was little NO_3^- -N left in the rooting depth in 2001 (Figure 1). When in 2002 the N application method was changed to 3 splits, the soil NO_3^- -N remained at a higher level to the end of the reproductive stage, which contributed to a higher grain yield, compared with 2001.

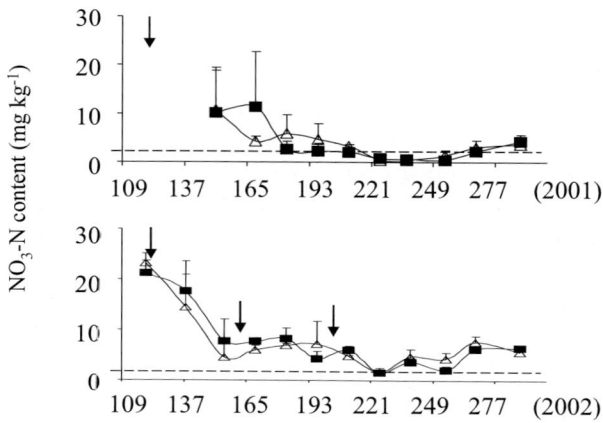

Figure 1. Soil NO_3^--N content in 20cm soil depth of GCRPS (vertical arrows indicate fertiliser applications ; n=3).

Discussion

In 2002, when fertiliser management in GCRPS was improved and an acceptable grain yield was obtained, the NUE of GCRPS was still low. We attribute the reduction from 42%, which is in a typical range for paddy rice (Norman, et al., 1989), to 19% or less in GCRPS to N losses to the atmosphere (see Kreye et al., this volume) and to a large extent to NO_3^- leaching.

Acknowledgements

This work was supported by the Natural Science Foundation of China (NSFC) and the German Research Council (DFG).

References

Cassman, K.G., Kropff, M.J., Gaunt, J. and Peng, S. (1993). Nitrogen use efficiency of rice reconsidered: what are the key constraints? Plant Soil, 155-156, 359-362.

Lin, S., Dittert, K., Tao, H., Kreye, C., Xu, Y.C., Shen, Q.R. and Fan, X.L. (2002). The ground-cover rice production system (GCRPS): a successful new approach to save water and increase nitrogen fertilizer efficiency? In. B.A.M. Bouman, et al.(eds). Water- Wise Rice Production. IRRI. pp187-195.

Norman, R.J., Wells, B.R. and Moldenhauser, K.A.K. (1989). Effect of application method and dicyandiamide on urea-nitrogen-15 recovery in rice. Soil Science Society of American Journal, 53, 1269-1274.

Effect of herbicide, maize variety precocity and sowing date of three winter cover crops on forage yield and on herbage N removal from an intensive double-cropping system

H. Trindade, J. Coutinho and N. Moreira
Department of Plant Science and Agricultural Engineering, UTAD, Ap. 1013, 5001-911 Vila Real, Portugal

Introduction

Important amounts of residual mineral N may be present in soil after maize cropping. This soil mineral N (mainly as NO_3^--N) is highly vulnerable to leaching when the drainage period starts in autumn (Trindade et al., 1997). In double-cropping forage systems, the early establishment of winter crops achieved by the use of earlier-maturing maize varieties may reduce NO_3^- leaching losses by allowing uptake of the soluble N pool. Meanwhile, early sowing of winter crops may lead to the use of less persistent herbicides in the maize crop. The aim of the work reported here was to evaluate the effect of these cropping modifications, together with the use of three different winter crops on forage yield and forage N content and removal.

Materials and methods

The trial was conducted over a 3-year period (May 1997-May 2000) on the same experimental plot on a commercial farm in the NW region of Portugal. The soil was a deep well-drained sandy loam dystric cambisol derived from granite, with pH_{water} value of 6.0, OM content of 25 g kg^{-1} and 4.9 $cmol_c$ kg^{-1} of CEC. Two winter crop sowing dates were tested, mid-October (D1) and 3 weeks later (D2), achieved by cropping two different maize varieties: Action (FAO 400), an earlier-maturing variety for the region, and Dedra (FAO 600), respectively. These treatments were factorially combined with two types of herbicide applied to the maize crop, (alachlor + atrazine) (Hr) and (bromoxynil + nicosulfuron) (Hc), and with three winter crops: Italian ryegrass (It.r) and mixtures of oats with Italian ryegrass (O+It.r) and oats with rye (O+rye). Hc herbicide has a contact plus residual action but is less persistent in soil than Hr. The maize was fertilised with cattle slurry (ca. 60 m^3 ha^{-1}, surface applied) and mineral N (40 kg ha^{-1} at sowing plus 100 kg ha^{-1} top-dressed at the 6-7th leaf stage). The winter crops were fertilised with cattle-slurry (ca. 45 m^3 ha^{-1}) and 75 kg ha^{-1} of mineral N top-dressed in February. Growth data for winter crops was based on two cuts in 1997/8 and 1998/9 and on a single cut in 1999/2000. The trial was laid out as a randomised complete block design with 3 replications and two factors (herbicide and sowing date) arranged in strips and the other factor (winter crop) in split plots.

Results and discussion

Three-year average DM yields of early (D1) and late (D2) maize maturing varieties reached, respectively, 20.3 and 22.0 t DM ha^{-1}. Winter crop DM yield results revealed an interaction effect among winter crop type, sowing date and year. The highest DM yields in the first two years were obtained with the mixtures containing oats sown at D1 (9.7 and 11.3 t DM ha^{-1}, respectively, for O+It.r in 1997/8 and for O+rye in 1998/9). In 1999/2000, the highest winter crop DM yields were obtained on treatments It.r×D1 and O+rye×D2 with values of, respectively, 8.5 and 8.6 t DM ha^{-1}. These results seem to indicate an advantage of oat mixtures in dry years, or when forage was harvested by a single cut and an advantage of Italian ryegrass in wet years or when forage was harvested by two cuts. No differences among treatments were observed for yearly total DM yields (maize + winter crop), suggesting that the lower maize yield at D1 was compensated by increased yield of the winter crops.

Forage N content results showed a combined effect of sowing date and preceding winter crop over maize N content (Table 1). Winter crop N contents were affected by sowing date and herbicide treatment. The values obtained on treatment combinations D1×Hc, D2×Hc, D1×Hr and D2×Hr were, respectively, 18.4, 17.9, 16.9 and 16.9 g N kg^{-1} DM. Table 2 shows the interaction effect of sowing date × year on winter crop and on yearly total N removal. The data indicate that crop N removal on treatment D1 was higher, in particular in years with a very wet autumn as occurred in 1997/8. Results of forage total N removal per year were also affected by sowing date × herbicide type. Herbicide treatment Hc allowed a higher yearly total N removal on sowing date D1 (399 kg N ha^{-1}) than on D2 (375 kg N ha^{-1}), while Hr treatments showed similar values on both sowing dates (381 kg N ha^{-1}).

Table 1. Effect of winter crop sowing date (or maize variety precocity) and type of winter crop on maize N content (g N kg^{-1} DM).

Winter crop sowing date	Type of winter crop		
	It.r	O+It.r	O+rye
D1	11.2 abc	11.3 ab	11.7 a
D2	10.8 ab	10.9 abc	10.4 c

Data within a group of interaction effect followed by the same letters do not differ at p<0.05 level, Tuckey test.

Table 2. Effect of winter crop sowing date and year on winter crop and yearly total (maize + winter crop) forage N removal (kg N ha^{-1}).

Sowing date	Year		
	97/98	98/99	99/00
Winter crop			
D1	127 b	230 a	117 b
D2	90 c	205 a	136 b
Maize + Winter crop			
D1	399 b	462 a	310 d
D2	348 c	449 a	337 cd

See footnote on Table 1.

Conclusions

The results obtained suggest that it is possible to use earlier-maturing maize varieties together with less persistent herbicides without harmful effects on forage production and possibly with beneficial environmental effects. The lower maize DM yield obtained when an earlier-maturing variety was used may be compensated by increased yield of the early established winter crop. The sowing of a earlier-maturing maize variety and the earlier establishment of the winter crop allowed the production of forage with higher N content and markedly increased the N removal by the winter crop in years with a wet autumn.

Acknowledgements

This project was funded by INIA - Project PAMAF no. 3005.

References

Trindade, H., Coutinho, J., Van Beusichem, M.L., Scholefield, D. and Moreira, N. (1997). Nitrate leaching from sandy loam soils under a double-cropping forage system estimated from suction-probe measurements. Plant and Soil, 195, 247-256.

Effects of grassland renovation on herbage yields and nitrogen losses

G.L. Velthof[1] and I.E. Hoving[2]
[1]*Alterra, P.O. Box 47, 6700 AA Wageningen, The Netherlands*
[2]*Animal Sciences Group, P.O. Box 2176, 8203 AD Lelystad, The Netherlands*

Introduction

Most grasslands in dairy farming systems in NW Europe are occasionally ploughed up and reseeded in order to maintain or increase the sward productivity and quality (Conijn et al., 2002). However, grassland renovation also increases the risk of N losses to ground and surface waters and the atmosphere. In 2002, field experiments were started in the Netherlands in order to obtain a quantitative insight into the effects of timing and method of grassland renovation on herbage yield and quality and on N losses. The results will be used for development of measures and tools for farmers to achieve environmentally, and agriculturally, sound grassland systems.

Materials and methods

Three identical experiments were started in 2002 on fields with a poor sward quality: on dry sand, moist sand and heavy clay. The treatments (in duplicate) consisted of five renovation treatments and four N application rates (0, 150, 300, and 450 kg N ha^{-1} yr^{-1} as mineral fertiliser). The renovation treatments were: T1) undisturbed grassland (control), T2) grassland chemically killed, ploughed to 25 cm and reseeded in April 2002, T3) grassland chemically killed, ploughed to 25 cm and reseeded in September 2002, T4) grassland chemically killed, ploughed to 25 cm and not reseeded in 2002 (to be reseeded in spring 2003) and T5) grassland chemically killed and reseeded without soil cultivation in September 2002. The measurements included herbage dry matter (DM) yields and the contents of mineral N and soluble organic N (0.01M CaCl$_2$ extraction) in the soil up to 90 cm at 7 sampling times in 2002.

Results and discussion

DM yields were higher in undisturbed than reseeded grassland (Table 1). Renovation in April resulted in higher DM yields than renovation in September. The relatively low DM yields after renovation in September were caused by the dry conditions in the period after reseeding.

Renovation in April enhanced soil mineral N contents for about two months (not shown). This indicates that the soil N supply was higher than the N uptake of the new sward in this period. No N leached to deeper soil layers during the growing season. Soil mineral N contents in November were similar for undisturbed grassland and for grassland renovated in spring, at all sites (see Figure 1 for the dry sandy soil). However, mineral N contents were much higher when grassland was renovated in autumn, especially for the dry sandy soil (Figure 1). The results suggest that the risk of N losses is higher when grassland is renovated in autumn than in spring. Soil mineral N contents after renovation without soil cultivation were similar to those after ploughing and reseeding grassland (Figure 1). Apparently, soil cultivation did not increase

Table 1. *Dry matter yields in 2002 (tonne ha^{-1}) at 0, 150, 300, and 450 kg N ha^{-1} yr^{-1}.*

Soil	Renovation treatment	0 N	150 N	300 N	450 N
Dry sand	T1	9.9	14.8	15.1	16.6
	T2	8.7	12.2	13.1	13.7
	T3, T4, T5	8.3	11.8	11.6	12.8
Moist sand	T1	6.5	10.3	12.2	16.5
	T2	5.8	10.0	10.1	10.0
	T3, T4, T5	3.7	7.5	9.5	14.4
Clay	T1	5.0	10.4	12.7	14.4
	T2	4.5	8.2	9.5	10.2
	T3, T4, T5	3.7	8.7	10.2	10.9

N mineralisation in the old sward. Leaching of soluble organic N may be a pathway of N pollution of ground and surface waters. None of the grassland renovation treatments affected the contents of soluble organic N in the soil, at all sites (data not shown).

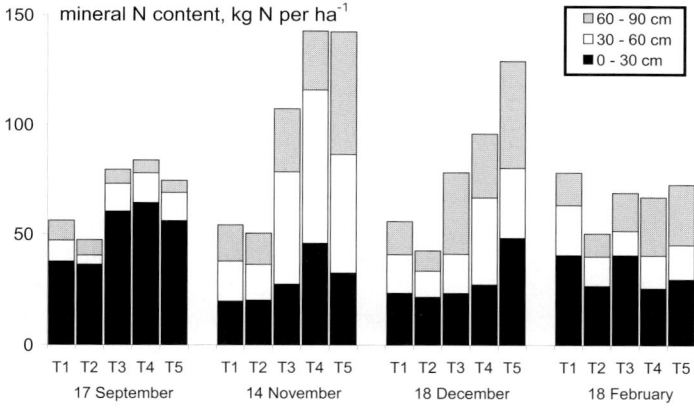

Figure 1. *Soil mineral N in autumn and winter in the dry sandy soil at 300 kg N ha^{-1}.*

Conclusions

The results show that in the year of grassland renovation, dry matter yields decrease and risk of N loss increases. The risk of N loss is smaller for renovation in spring than in autumn. A good evaluation of the agronomic and environmental consequences of grassland renovation can only be made when herbage yields and quality, and N losses are quantified over several years. The experiments are continuing.

Acknowledgements

This research project was funded by the Dutch ministry of Agriculture, Nature and Food Quality (LNV programme 398-II).

References

Conijn, J.G., Velthof, G.L. and Taube, F. (eds.). (2002). Grassland resowing and grass-arable crop rotations. International Workshop on Agricultural and Environmental Issues, Wageningen, The Netherlands, 18 & 19 April 2002. Wageningen, Plant Research International, report 47, 128 pp.

Nitrogen budgets of Flemish specialised dairy farms during 1990-2000

I. Verbruggen[1], F. Nevens[1], A. Mulier[1] and G. Hofman[2]
[1]*Flemish Policy Research Centre for Sustainable Agriculture, Ghent University and Catholic University of Leuven, Potaardestraat 20, 9090 Gontrode, Belgium*
[2]*Departement Soil Management and Soil Care, Ghent University, Coupure links 653, 9000 Ghent, Belgium*

Introduction

Farm nutrient budgets are a useful tool to evaluate the environmental impact of nutrient use by agriculture. There is no Flemish legal obligation to calculate farm N budgets but some farmers use them for their management. Since the early nineties there has been Flemish manure decree which limits the use of manure and fertiliser (Anonymous, 1991 and Anonymous, 2000). This regulation has important effects on the dairy farmers' fertiliser strategies and N budgets.

Materials and methods

We calculated farm and soil N-budgets of specialised dairy farms monitored in the Farm Accountancy Data Network (FADN) of the Flemish Centre for Agricultural Economics for the period 1990 - 2000. The N-inputs were mineral fertiliser, concentrates and by products (e.g. sugar beet pulp), straw, purchased roughage and imported manure (e.g. manure from pig farms), deposition and N-fixation. The outputs were milk, sold animals and also a small amount of sold arable crops. We calculated the N-efficiency as:

N-efficiency = (N-outputs/N-inputs) x 100. (1)

Results and discussion

The average *farm N surplus* decreased from 359 kg N ha^{-1} in 1990 to 248 kg N ha^{-1} in 2000, the corresponding N-efficiency increased from 16.1 to 22.9% (Table1). This was the result of a strong reduction in mineral fertiliser use: on grassland it decreased from 287 kg N ha^{-1} in 1990 to 213 kg N ha^{-1} in 2000 and on arable land from 94 to 57 kg N ha^{-1}, respectively. Also the import of N in concentrates decreased, because of a better milk production/concentrate intake ratio and protein content of the concentrates also decreased. Large variations in N-surplus between farms were found with the lowest N-surplus at ± 100 kg N ha^{-1}.

Also *soil N surplus* showed a decreasing tendency (from 318 kg N ha^{-1} in 1990 to 209 kg N ha^{-1} in 2000) with corresponding increasing soil-N-efficiencies (44.1 to 56.8%). The most efficient farms had a soil N-surplus of about 80 kg N ha^{-1} with a corresponding soil N-efficiency of 75%, which is close to the technically achievable value of 77% of Jarvis & Aarts (2000). A further optimisation is possible for many farms. An improved nutrient management is very important in reaching the same targets as the most efficient farms.

Table 1. Farm characteristics of monitored specialized dairy farms during the period 1990-2000.

		'90	'92	'94	'96	'98	'00
Number of farms		165	129	117	102	94	79
Farm area	ha	27.8	27.3	29.6	31.6	32.6	32.1
Share of grassland	%	70	67	65	61	61	64
Concentrates	kg cow^{-1}year^{-1}	1169	1172	1289	1244	1282	1158
Byproducts	kg DM cow^{-1}year^{-1}	480	372	439	621	402	361
Mineral Fertilizer	kg N ha^{-1}	225	203	171	171	172	150
Milk production	litre cow^{-1}year^{-1}	5365	5501	5684	5920	6126	6023
Milk production	litre ha^{-1}year^{-1}	9567	9648	10094	10090	10271	10020
Farm N-surplus	kg N ha^{-1}	359	319	299	298	284	248
Farm N-efficiency	%	16.1	18.1	19.4	19.4	20.2	22.9
Soil N-surplus	kg N ha^{-1}	318	280	257	255	244	209
Soil N-efficiency	%	44.1	48.2	50.3	50.7	52.3	56.8

Conclusions

A large reduction in N-surpluses without a reduction in productivity (milk per ha) was realised. The results of the best farms provide a new goal for the other farms: a farm N-surplus of ± 100 kg N ha^{-1} and a soil surplus of ± 80 kg N ha^{-1}. These figures may not be always obtainable for all farmers because of economic constraints.

References

Anonymous (1991). Decree concerning the protection of the environment against pollution by fertilisers (in Dutch and French). Belgian Law Gazette, 28.02.1991, N91-535, 3829-3838 (with amendments).

Anonymous (2000). Decision of the Flemish government of 30th November 2000 to set a renumeration standard concerning the protection of the environment against pollution by fertilisers and in order to change the decision of the Flemish government of 26th May 2000 concerning the execution of certain articles of the same decree (in Dutch and French). Belgian Law Gazette, 22.11.2000, N2000-2957, 38813-38824.

Jarvis, S.C. and Aarts, H.F.M. (2000). Nutrient management from a farming systems perspective. In: K. Søegaard, C. Ohlsson, J. Sehested, N.J. Hutchings and T. Kristensen (eds) Grassland farming: Balancing environmental and economic demands, Proceedings of the 18th General Meeting of the European Grassland Federation Aalborg, Denmark, 2000, pp. 363-373.

Substitution of N fertiliser supply for maize with lupin as a winter crop in rotations under zero and conventional tillage in southern Brazil

L. Zotarelli[1], B.J.R. Alves[1], E. Torres[2], S. Urquiaga[1] and R.M. Boddey[1]
[1]Embrapa Agrobiologia, Caixa Postal 74.505, Seropédica, 23890-000, RJ, Brazil
[2]Embrapa Soja, Caixa Postal 1061, Londrina, 86001-970, PR, Brazil

Introduction

It is estimated that in 2003, Brazil will produce 49 million tonnes of soybean and today over 50% of this crop is produced under zero tillage (ZT). Since 1992, the area under ZT in Brazil has increased rapidly from approximately 1 M ha to over 17 M ha. In southern Brazil, the favourable rainfall regime permits year-round cropping and the use of ZT has encouraged farmers to diversify their rotation systems to avoid pest and disease problems such that in at least one year in three, a summer crop other than soybean is planted, most frequently maize. Wheat, oats, oil radish and temperate-region legumes, such as vetch and lupin, are most favoured as winter crops but, except for wheat, these crops are used generally only for maintaining soil cover and green manure. In this study, we examined the potential of lupin as a N_2-fixing green manure crop to substitute N fertiliser for maize under both conventional tillage (CT) and ZT.

Materials and methods

The study was performed in the 4th and 5th years of an experiment on 3 crop rotations with soybean or maize as summer crops and wheat, oats and lupin in winter, managed under either CT (disc plough followed by disc grader) or ZT. The experiment was conducted on a clayey Oxisol at the field station of the Embrapa Soybean Centre (Londrina, Paraná State). In both years, the effect of the winter lupin crop on the yield of the following maize crop was compared with that of a prior crop of oats. In 2000, the maize was fertilised with 80 kg N ha[-1] as ammonium sulphate, and in 2001, N fertiliser was added only to the maize after oats. Total dry matter and N accumulation of lupin and oats (winter 2000 and 2001), and maize (summer 2000/01 and 2001/02) were evaluated as well as maize grain production and N content. The contribution of N_2 fixation (BNF) to the lupin crop was evaluated using the ^{15}N natural abundance technique (Shearer & Kohl, 1986) which allowed the computation of a simple N balance for each system using the data for crop N accumulation, BNF contribution and N fertiliser inputs.

Results and discussion

Tillage practice did not affect the final accumulation of total plant dry mass or N. The mean grain yield of soybean, considering both tillage systems, was 5.6 t ha[-1] and 4.4 t ha[-1] for the harvests 2000 and 2001, respectively. Grain N accumulation ranged from 322 kg N ha[-1] under ZT to 330 under CT, for the first harvest, and from 278 kg N ha[-1] under ZT to 289 under CT, for the second. For the 2000 crop, soybean reliance on BNF was estimated using the ureide relative abundance technique (Herridge et al., 1988) and in 2001, using the ^{15}N natural abundance technique. In the former, BNF contribution to soybean under ZT and CT was found to be 81 and 74%, respectively and for the 2001 crop it was 69 and 84%, respectively. For the N balance of the soybean crop for the first harvest, the difference between the BNF input and the N exported in grain gave a positive value estimated at 6 kg for ZT and -36 kg N/ha for CT (Table 1). For the second harvest, the N balances for this crop were negative whichever tillage system had been used.

In both years under both tillage systems, the lupin crop accumulated approximately 10 Mg ha^{-1} of dry matter and 300 kg ha^{-1} of N, there being no significant difference between tillage systems or years. The ^{15}N abundance data indicated that approximately 200 kg N ha^{-1} (70% of crop N) was added to the system from BNF through lupins in accumulated biomass. Oats cropped after soybean in the winter of 2000 accumulated on average 6.5 t ha^{-1} of plant dry matter and in 2001, 10.1 t ha^{-1}. There were no differences between tillage systems in either year. Mean total plant N accumulation was 100 and 160 kg N ha^{-1} for 2000 and 2001 winter seasons, respectively. As oats were used as green manure, all accumulated N was left on the cropped area and the N balance for this crop was considered nil. In the summer of 2000/01, maize following the lupin crop yielded 7.3 to 8.0 Mg ha^{-1}, while maize following oats and fertilised with 80 kg N ha^{-1} yielded significantly less at 6.0 to 7.5 Mg ha^{-1}. In the following summer, maize following lupin yielded 8.3 to 10.0 Mg ha^{-1} and non-N fertilised maize following oats yielded 4.5 to 4.6 Mg ha^{-1}. As shown in Table 1, the N balance for maize cropped on lupins residues was always higher than that following oats (as maize after oats was supplemented with 80 kg ha^{-1} of N fertiliser). Also, the soil N accumulation promoted by lupins allowed higher maize yield and higher N exportation in the grain compared with maize after oats.

The overall N balances (not including gaseous or leaching losses) for the sequence soybean/lupin/maize under CT and ZT were, respectively, +61 and +86 kg N ha^{-1} in the 1999 to 2000 period (Table 1). This same sequence, but using oats instead of lupins, resulted in overall negative N balances. These results indicate that under either tillage system, the utilisation of a winter lupin crop in this region can substitute well in excess of 80 kg N ha^{-1} of N fertiliser.

Table 1. N balance (kg N ha^{-1}) for the cropping sequences soybean (S)/lupin(L)/maize(M) or soybean(S)/ oats(O)/ maize(M).

Years/tillage	System	Soybean	Maize after:		Total crop sequence balance	
			Lupin	Oats	S/L/M	S/O/M
1999 - 2001	CT	- 35.7	- 119.5	-32.5	+ 60.8	- 68.2
	ZT	+ 6.2	- 112.5	- 10.1	+ 85.6	- 3.9
2000 - 2002	CT	- 29.9	- 124.5	- 67.5	+ 32.6	- 97.4
	ZT	- 2.2	-150.1	- 69.0	+ 89.7	- 71.2

Acknowledgements

The project was funded by IAEA/FAO through the research contract 10 953/RO. The authors thank CNPq and CAPES for the fellowship support.

References

Herridge, D.F., O'Connell, P. and Donnelly, K. (1988) The xylem ureide assay of nitrogen fixation: Sampling procedures and sources of error. Journal of Experimental Botany, 39, 12-22.

Shearer, G. and Kohl, D.H. (1986). N_2-fixation in field settings: Estimations based on natural ^{15}N abundance. Australian Journal of Plant Physiology, 13, 699-756.

SECTION 6
MODELS AND DECISION SUPPORT SYSTEMS

NGAUGE: A decision support system to optimise N fertilisation of UK grassland for economic and/or environmental goals

L. Brown, D. Scholefield, E.C. Jewkes, A. del Prado and D.R. Lockyer
Institute of Grassland and Environmental Research, North Wyke Research Station, Okehampton, Devon, EX20 2SB, UK

Model development

The existing NCYCLE model was taken as the basis for the new model. The submodels within NGAUGE calculate N cycling through components and processes on a monthly basis. NGAUGE extends the capabilities of the original NCYCLE model by including increased detail of average weather, sensitivity to within-year ongoing weather and simulation of losses of NH_3 from, and mineralisation of, organic manures. The major development within NGAUGE is the inclusion of an optimisation process, which produces a monthly fertiliser recommendation. This enables a user-specified target of outputs to be met while maximising efficiency of N use.

Model components

Plant uptake
Data from experiments were used to derive a set of plant uptake curves, which describe the relationship between inorganic N and plant N (including N in roots) for each month, using predictions from the mineralisation submodel for N supply from this source. In order to calculate total N in plant, assumptions were made about the 'u' factor (proportion of N in the whole plant that is harvested by the animal or by cutting).

Mineralisation
Mineralised N is considered to be influenced by previous land use, the herbage production in the current year, dung and applied manures. The total mineralised N is calculated for different months according to relationships describing the effect of soil moisture and temperature on mineralisation. Mineralisation from the current year's residues is calculated using empirically derived functions, which relate monthly plant N to observed or estimated mineralisation.

Denitrification
Denitrification is modelled as a function of soil inorganic N, water-filled pore space (%WFPS) and temperature. WFPS is a function of soil texture and rainfall zone, and is related to monthly denitrification using a relationship derived from controlled laboratory experiments. Denitrification rate was assumed to increase linearly with temperature from 2 to 20°C. Losses of N oxides, via denitrification and nitrification, have recently been implemented using relationships with the main controlling variables of %WFPS, temperature and N in soil, derived from published data.

Nitrate leaching

From the total soil inorganic N, uptake of N by the plant, volatilisation of urine and N lost by denitrification are subtracted. The remaining 'leachable N' contributes to the total annual leaching. For most months, leaching is assumed not to occur, and 'leachable N' is passed to the succeeding month as a component of the inorganic N pool.

Manure management

Two slurry types and 2 farmyard manure types are available for selection in NGAUGE, each associated with default values of ammoniacal N, organic N and total N, following selection of dry matter content by the user. Following application of manure, volatilisation of NH_3 is simulated.

Optimisation

The optimisation procedure is the means by which the best fertiliser distribution is calculated. It is based on the set of monthly plant uptake relationships. A field-specific target can be set by the user, and may be herbage N, N loss, or fertiliser N. For each of these, the user selects the value desired. The end point of the optimisation is achieved when the model reaches the target value, satisfying the optimisation criteria (Figure 1).

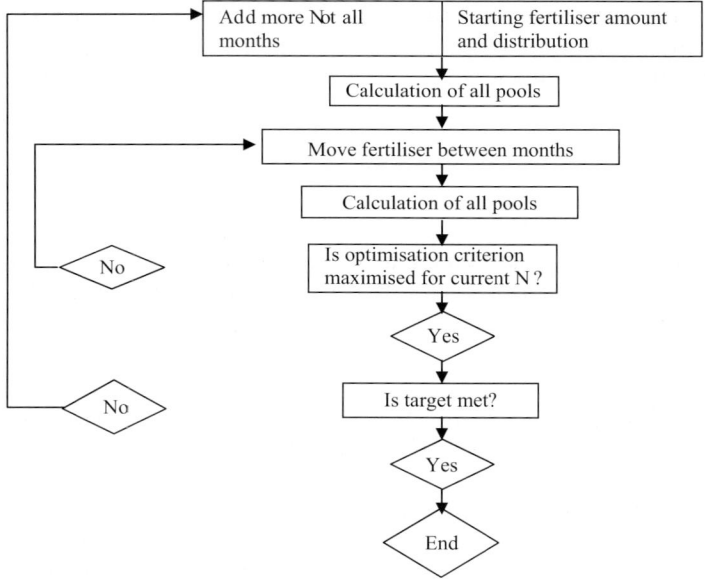

Figure 1. Operation of the optimisation procedure in NGAUGE DSS.

The DSS output includes a monthly fertiliser recommendation which is field and target specific. The recommendation, therefore, takes account of soil type, site history, manure applications, climate and on-going weather to produce the best distribution and amount of N fertiliser to meet the target of the user, while maximising the efficiency with which N is used in the grassland system.

Acknowledgements

The authors would like to thank to DEFRA for funding the development of this work.

UK-DNDC: a mechanistic model to estimate N_2O fluxes in the UK

L. Brown[1], B. Syed[2], S.C. Jarvis[1], R.W. Sneath[3], V.R. Phillips[3], K.W.T. Goulding[4] and C. Li[5]

[1]*Institute of Grassland and Environmental Research, North Wyke Research Station, Okehampton, Devon, EX20 2SB, UK*
[2]*Soil Survey and Land Research Centre, Cranfield University, Silsoe, Bedford, UK*
[3]*Silsoe Research Institute, Silsoe, Bedford, UK*
[4]*Institute of Arable Crops Research, Rothamsted, Harpenden, Herts., UK*
[5]*Institute for the Study of Oceans and Space, University of New Hampshire, USA*

Introduction

The DeNitrification-DeComposition model (DNDC) was originally developed for the USA (Li et al., 1992). It is driven by climate, soil characteristics, crop type and fertiliser application and considers the detailed mechanisms involved in the biogeochemistry of C and N in agricultural systems. The model simulates emissions of N_2O, NO, N_2, NH_3 and CO_2 from agricultural soils. Our UK consortium used this model as a base from which to produce UK-DNDC (Brown et al., 2002) in order to apply it to the UK. The original objective of the development of UK-DNDC was to provide an improved estimate of N_2O emission from the UK, for comparison with the IPCC (1997) methodology.

Model description

DNDC comprises 4 main sub-models:
- climate: calculates hourly and daily soil temperature and moisture.
- crop growth: simulates crop biomass accumulation and partitioning based on thermal degree days and daily N and water uptake.
- decomposition: calculates decomposition, nitrification, NH_3 volatilisation and CO_2 production on a daily timestep.
- denitrification: follows the sequential reduction from NO_3^- to N_2, based on soil redox potential and dissolved organic carbon (DOC) concentration.

The components of the database for UK-DNDC are shown in Figure 1. Information on crop areas (for 18 crop types) for each of 72 UK counties is linked to the crop library in which data on farming practice associated with that crop is held. This includes information such as planting and harvesting date and timing and rate of fertiliser application. Animal numbers for each county (for 6 categories of animal) are used to calculate excretion of N and C as dung and urine. These returns are divided between different categories of grazed grassland and animal waste management systems, i.e. slurry or farmyard manure (FYM). Slurry and FYM are then applied to grassland and arable areas on different dates in the year.

The soil file contains data for the three dominant soil types in each county. For each UK simulation, the model is run three times, for each soil type, and for each, the minimum and maximum soil organic C values are used.

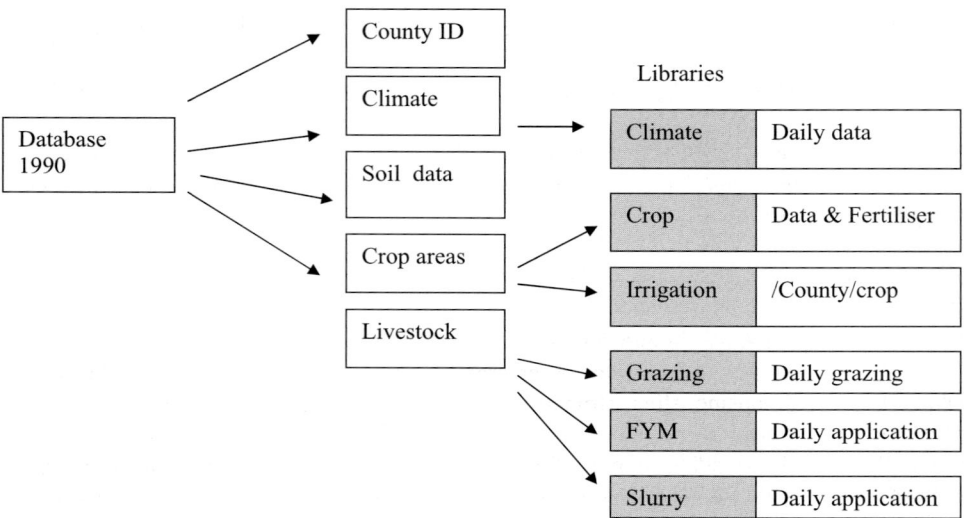

Figure 1. Structure of UK-DNDC database.

The model was validated using 17 datasets from the UK, of contrasting N source, soil and crop type and was used to estimate the total emissions of N_2O from agriculture in the UK, and the results were compared to the IPCC estimates (IPCC, 1997). Emissions were converted into emission factors (EF) to maintain consistency with the IPCC approach. Individual EFs were therefore, calculated for each soil/crop/county/N source combination. The model facilitates the identification of key geographical areas of emission and key sources in terms of crop types and farming practice. We are currently using UK-DNDC to explore options for mitigation of N_2O emission (Brown et al., 2003).

Acknowledgements

The development of UK-DNDC was funded by DEFRA, London under OC9601.

References

Brown, L, Cardenas, L., Bellamy, P, Jarvis, S.C., Hollis, J., Sneath, R.W., Yamulki, S. and Goulding K.W.T. (2003). A model-based evaluation of options for the mitigation of agricultural nitrous oxide emission. (This volume).

Brown, L., Syed, B., Jarvis, S.C., Sneath, R.W., Phillips, V.R., Goulding, K.W.T. and Li, C. (2002). Development of a mechanistic model to estimate emission of nitrous oxide from UK agriculture. Atmospheric Environment, 36: 917-928.

IPCC, (1997). IPCC Revised 1996 Guidelines for National Greenhouse Gas Inventories, IPCC WGI Technical Support Unit, Hadley Centre, Meteorological Office, Bracknell, UK.

Li, C., Frolking, S. and Frolking, T.A. (1992). A model of nitrous oxide evolution from soil driven by rainfall events. 1. Model structure and sensitivity. Journal of Geophysical Research, 97, 9759-9776.

MAST - a model of ammonia volatilisation with an examination of abatement strategies for a dairy farm

E.C. Jewkes, C.A. Ross, D. Scholefield and S.C. Jarvis
Institute of Grassland and Environmental Research, North Wyke Research Station, Okehampton, Devon, EX20 2SB, UK

Introduction

Agriculture is responsible for approximately 80% of the total emissions of NH_3 to the atmosphere in the UK (Pain et al., 1999). The losses from agricultural operations occur during slurry application, animal housing, slurry storage, grazing, fertiliser application and from crops. Animal excretions are the primary source in the first four of these loss pathways. Recent estimates suggest that 74 Mt of cattle manure, 11Mt of pig manure and 4 Mt of poultry manure are produced annually in the UK, with a total N content of about 450,000 t (Chambers et al., 1999). Losses of NH_3 present a threat to the environment after deposition, through acidification, eutrophication and direct toxic affects. However, it also represents a loss from the plant-soil system, where more effective utilisation of manures could allow reduced fertiliser applications whilst maintaining dry matter (DM) yields.

MAST (Model for Ammonia System Transfers at the farm scale) was developed to estimate losses and transfers of NH_3 from livestock agriculture (Ross et al., 2002). It is a mass-flow model with a yearly time-step, and calculates NH_3 losses on the basis of emission factors from the NH_3 emissions inventory (Misselbrook et al., 2000). It has modules for dairy, beef, pig, sheep and poultry production. Here, it is used to calculate losses from, and abatement strategies for, a hypothetical 'typical' dairy farm in SW England (based on Jarvis, 1993), which uses 250 kg N fertiliser ha^{-1} to produce sufficient dry matter yield for 131 dairy livestock units (LU).

Model structure

MAST consists of five separate sub-models that calculate emission from grazing, housing, manure storage, manure application and fertilisers, for a livestock or mixed farm. Each sub-model requires an input by the user of easily obtainable farm information.

Grazing
Emissions from grazing are calculated on LU per day basis, and therefore an accurate estimate of animal numbers at different ages/weights is required on the input page. MAST accounts for milking time in the summer months by reducing the grazing period by 3 h^{-1}day^{-1} for the dairy cows and increasing the housing time by the same amount.

Housing
The required inputs include type and duration of housing and whether it is slurry or FYM-based. This is important because of a large difference in emission factors between the two systems. Again, the emission factor is based on livestock units and so animal numbers are important. MAST allows the option of having both systems (i.e. slurry and FYM) in use on one farm.

Storage

Manure storage considers both whether the system is slurry or FYM based and also the form of storage used. The emission factors are calculated as loss per surface area exposed per day. Crusting of the slurry surface is accounted for in the model. Slurry stores are assumed to always contain some manure and therefore be a potential NH_3 source all year round. FYM, however, varies in the time it is stored, and an average figure of six months storage has been used. The volume of slurry entering the stores is calculated for the housing period as 0.057 m^3 cow^{-1} day^{-1} and 0.012 m^3 cow^{-1} day^{-1} for cattle <1 years old. For FYM systems, a factor of 1.5 is used to account for the increased volume of manure entering storage. MAST also accounts for storage of dirty water.

Manure application

Required inputs to the manure application sub-model include: amounts spread, percentage spread during summer and winter and slurry DM content. For the case study farm, surface spreading is assumed to be the 'baseline' for emissions.

Fertiliser application

Inputs to the fertiliser module include the type of fertiliser used and the application rate. Three fertiliser types are included at present, namely: urea, ammonium nitrate and 'other'.

Losses and abatement strategies

Using the case-study farm described above, annual NH_3 emissions were calculated for both slurry and FYM based dairy systems. These were based on a range of scenarios, from best- to worst-case, for each system. Abatement strategies considered to achieve improvements in losses included: extending the grazing period by 10d, changing from a slurry-based housing to straw-based, covering slurry stores, altering slurry application timing, using low-emission application techniques and using ammonium nitrate fertiliser instead of urea. Some of these options increase losses by other pathways, which cannot be accounted for in MAST. This is discussed in more detail by Ross et al., (2002). Emissions from the slurry-based system ranged from 27 kg NH_3-N ha^{-1} yr^{-1}, achieved with a 'best-case' combination of abatement strategies, and 107 kg NH_3-N ha^{-1} yr^{-1}, calculated for a 'worst case' scenario. For the FYM system, the range of emissions was between 33 and 86 kg NH_3-N ha^{-1} yr^{-1}. The greatest reductions were achieved by manipulating options linked to fertiliser usage and manure application.

Conclusions

MAST provides a tool to examine losses of NH_3 from farming systems. A range of abatement options can be investigated, which can produce substantial reductions in losses.

References

Jarvis, S.C. (1993). Nitrogen cycling and losses from dairy farms. Soil Use and Management, 9, 99-105.

Misselbrook, T.H., Van der Weerden, T.J., Pain, B.F., Jarvis, S.C., Chambers, B.J., Smith, K.A., Phillips, V.R. and Demmers, T.G.M. (2000). Ammonia emission factors from UK agriculture. Atmospheric Environment 34, 871-880.

Pain, B. F., Misselbrook, T.H. and Chadwick, D.R. (1999). Controlling losses of nitrogen as ammonia from manures. In: A.J. Corrall, (ed). BGS Occasional Symposium No. 33. Accounting for nutrients, Great Malvern, UK

Ross, C.A., Scholefield, D. and Jarvis, S.C. (2002). A model of ammonia volatilisation from a dairy farm: an examination of abatement strategies. Nutrient Cycling in Agroecosystems, 64, 273-281

Development of the OVERSEER® nutrient budget model to examine implications of pastoral management practices on nitrogen flows and losses

S.F. Ledgard[1], D.M. Wheeler[1], C.A.M. de Klein[2], R.M. Monaghan[2] and K. Johns[1]
[1]*AgResearch Ruakura Research Centre, Private Bag 3123, Hamilton, New Zealand*
[2]*AgResearch Invermay Research Centre, P.O. Box 50034, Mosgiel, New Zealand*

Introduction

Decision support models for N offer the potential for users to examine impacts of specific farm systems on N flows and losses, and to assess the effectiveness of mitigation strategies in reducing losses. In 1998, the OVERSEER® nutrient budget model was first developed as a decision support model (covering N, P, K and S; annual time-step) primarily for farmers and their support specialists (e.g. farm consultants, fertiliser industry staff) (Ledgard et al., 1999). It is New Zealand's (NZ) most widely used nutrient model and covers all pastoral farming systems, and some arable and horticultural crops, although this paper will describe pastoral systems only. Recently, a series of meetings and interviews were used to define limitations of the model as perceived by the main end-users.

Development of model upgrade

End-user meetings with farmers and consultants defined the need for integration of other key nutrients, and in particular, the requirement for information on interpretation of model outputs and the ability to do scenario analyses. Government policy bodies have become strong advocates for use of the model and they requested that greenhouse gases be incorporated, as a way of educating farmers on the farm factors that contribute to losses.

The new upgrade of the model includes calculated budgets for N, P, K, S, Ca, Mg, Na, H^+ and greenhouse gases (CH_4, N_2O and CO_2). Key aspects of the N sub-model are:
- Estimation of the amount of N excreted by grazing animals. This is the main driver of N cycling and losses. It is calculated from the difference between N intake (dry matter intake derived from animal productivity, and dietary N concentration) and N incorporated in products (milk, meat, fibre). Excreta-N is partitioned into urine and dung forms based on diet N concentration.
- Internal databases of the N concentrations of animal products, supplementary feeds and fertilisers.
- Calculation of N transfers in animal excreta. This includes transfer of excreta-N to animal camp sites in hill country and transfer of excreta-N to farm lanes and the milking shed on dairy farms (and the associated fate of effluent).
- Estimation of N loss by leaching, denitrification and NH_3 volatilisation based on empirical relationships accounting for effects of soil texture and drainage, rainfall, animal type, rate of N excreted, N fertiliser type, timing and rate, and period of grazing during the year.

- Calculation of N_2O emission from N fertiliser rate, estimates of the amounts of N excretion, leaching and NH_3 loss, and IPCC-based emission factors (NZ-modified).

Model outputs and practical application

After entering farm specific information, the user can examine a range of model outputs. The outputs include a nutrient budget page itemising all nutrient inputs and outputs, many dialogue boxes whereby specific information on best management practices can be obtained, and an environmental N page. The latter has estimates of N leaching, the NO_3^--N concentration in drainage and specific components of N loss (e.g. direct leaching loss of fertiliser N and N loss from effluent). These are compared with data for 'average' NZ farms to provide a benchmark.

The user can apply the model for scenario evaluation, with up to four scenarios able to be compared 'on-screen'. Scenarios could include altering input information such as animal stocking rate, N fertiliser rate and timing, and level of supplementary feeding. A number of key mitigation management practices are specifically identified, such as winter management of animals (e.g. all winter grazing compared with use of feed-pad systems) and diet manipulation (e.g. substitution of N-boosted pasture with low protein supplements). One benefit of the model covering a range of nutrients is that interactions and multiple benefits of management changes can be examined. For example, practices to increase N use efficiency can result in decreased N leaching to groundwater, reduced soil acidification (and lime requirements) and decreased N_2O emissions.

In NZ, the local government authorities and the fertiliser industry encourage farmers to use nutrient budgets as a tool to improve nutrient use efficiency and reduce nutrient losses to the environment. The main dairy company, which collects milk from about 95% of all dairy farms, has recently developed a 'Clean Streams Accord' with local and central government agencies, whereby they will require all dairy farmers to be using nutrient budgeting to manage nutrient use on farms by 2007. Thus, it is envisaged that many farmers and almost all dairy farmers will have nutrient budgets done for their farms. In some cases they will do this themselves, and in others, it will be done for them by one of their advisors. Some local authorities are assessing how N budgets can be used to benchmark farms in sensitive catchments and to achieve defined reductions in N emissions to waterways.

Reference

Ledgard S.F., Williams, P.H., Broom, F.D., Thorrold, B.S., Wheeler, D.M. and Willis, V.J. (1999). OVERSEER™ - a nutrient budgeting model for pastoral farming, wheat, potatoes, apples and kiwifruit. In: L.D. Currie, M.J. Hedley, D.J. Horne and P. Loganathan (eds). Best Soil Management Practices for Production. Occasional report No. 12. FLRC, Massey University, Palmerston North. pp. 143-152.

The MANNER model: predicting the crop available N supply from farm manure applications

F.A. Nicholson[1], B.J. Chambers[1], E.I. Lord[2], K.A. Smith[2], S.A. Anthony[2] and M.M. Gibbons[2]
[1]*ADAS Gleadthorpe Research Centre, Meden Vale, Mansfield, Notts. NG20 9PF, UK*
[2]*ADAS Wolverhampton, Woodthorne, Wergs Road, Wolverhampton, WV6 8TQ, UK*

Introduction

Recent research has led to significant advances in understanding the N supply characteristics of farm manures, which is reflected in current UK advice (e.g. Anon., 2000). However, annual fertiliser use statistics in Britain indicate that farmers generally make little or no allowance for the N contribution of manures to crop requirements: lack of confidence in estimating manure N availability is thought to be one of the major reasons (Smith & Chambers, 1995). Simple models, or decision support systems, have an important role to play in the practical application of research information and improving manure N utilisation on farms.

Approaches

A WINDOWS compatible (ActiveX control) decision support system to predict the fate of N following organic manure applications to land was developed, drawing together the latest UK and other European research information on factors affecting manure N availability. The ADAS MANure Nitrogen Evaluation Routine (MANNER) takes into account the following factors :
- manure N analysis (total N, NH_4^+-N and uric acid-N)
- NH_3 volatilisation (depending on speed of incorporation and, for slurries, dry matter content)
- NO_3^- leaching (depending on effective rainfall and soil texture)
- mineralisation of manure organic N.

Only a few simple inputs are required, which should all be readily provided by a farmer/consultant; however, default information is available if these are not known. A single screen printable output is produced, summarising data inputs and predictions of the N volatilised or leached and the fertiliser N replacement value for the next crop grown.

Validation

Predictions from MANNER have been evaluated by comparison with independently collected experimental data. Good agreement ($p<0.001$) was found between predicted and actual fertiliser N replacement values for poultry manure, pig slurry and cattle slurry applied to arable crops (Figure 1). This confirmed that MANNER can provide a simple, quick and accurate estimate of the fertiliser N value of different farm manures spread under a range of circumstances.

User take-up and future developments

By the end of March 2003, MANNER 3.0 had been distributed on disk or CD to more than 7500 farmers, advisors and researchers throughout the UK and abroad. An enhanced version of the software (MANNER 4.0) has recently been released. This includes the integration of a climatic

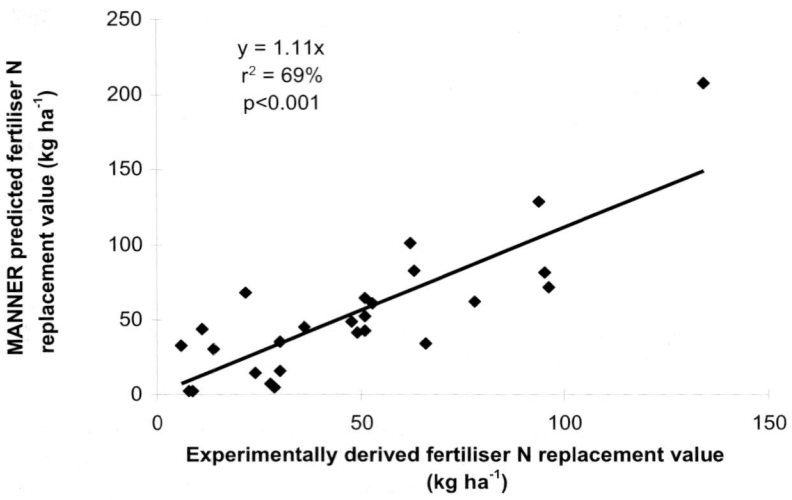

Figure 1. Comparison between experimentally-derived and MANNER-predicted fertiliser N replacement values for cattle and pig slurries applied to cereals.

database, which calculates monthly rainfall and actual evapotranspiration rates at a 10 km^2 spatial resolution, so that calculations of manure crop available N supply can be performed using climatic data specific to the selected part of Britain. To aid the identification of the location of a farm, a telephone directory has been added to the program so that the user can enter the standard area code, and the appropriate climatic data are automatically selected.

Further versions of MANNER are currently under development, which will synthesise the results of recent (post-1996), Defra-funded research on NH_3 volatilisation, N_2O emissions and NO_3^- leaching losses following manure applications, and the mineralisation of manure organic N. MANNER-PSM will provide a tool for policy makers to assess the effects of different manure management practices on N losses via NH_3 volatilisation, NO_3^- leaching and N_2O emissions (i.e. 'pollution swapping'), and the fertiliser N value of organic manures. MANNER-NPK will be an enhanced decision support system for farmers and advisors to estimate the P, K, Mg and S supply from organic manures (in addition to N) and to calculate the economic value of manure applications.

Acknowledgements

This work was funded by Defra.

References

Anon. (2000). Fertiliser Recommendations for Agricultural and Horticultural Crops (RB 209), 7th edition. The Stationery Office, Norwich, 176pp.

Smith, K.A. and Chambers, B.J. (1995). Muck : from waste to resource. Utilisation : the impacts and implications. Proceedings of the MUCK'95 Conference. Agricultural Engineer 1995, pp33-35.

The nutrient leaching model ANIMO

J. Roelsma, P. Groenendijk and O.F. Schoumans
Alterra Green World Research, Wageningen University and Research Centre, Wageningen, The Netherlands

Introduction

The ANIMO model aims to quantify the relationships between fertilisation level, soil management and the leaching of nutrients to groundwater and surface water systems for a wide range of soil types and different hydrological conditions (Groenendijk & Kroes, 1999; Rijtema et al., 1999; Kroes & Roelsma, 1998). The model is a functional model incorporating simplified formulations of processes. The organic matter cycle plays an important role for quantifying the long term effects of land use changes and fertilisation strategies. Attention has been paid to the most relevant processes governing the organic cycle. Currently, the ANIMO model serves as one of the parent models for the development of the Dutch consensus leaching model STONE (Wolf et al., 2003)

Carbon, nitrogen and phosphorus cycles

The simulated transformation processes are all part of the C, N and P cycles. In the organic C cycle, the following processes are described: (a) application of organic materials (e.g. manure); (b) decomposition of root materials; (c) decomposition of fresh organic materials and transformation to humus; (d) turnover of humus. For each kind of organic material, fractions are specified which have different decomposition rates, and N and P contents. The organic part of both the N and P cycle in the soil runs largely parallel to the organic C cycle.

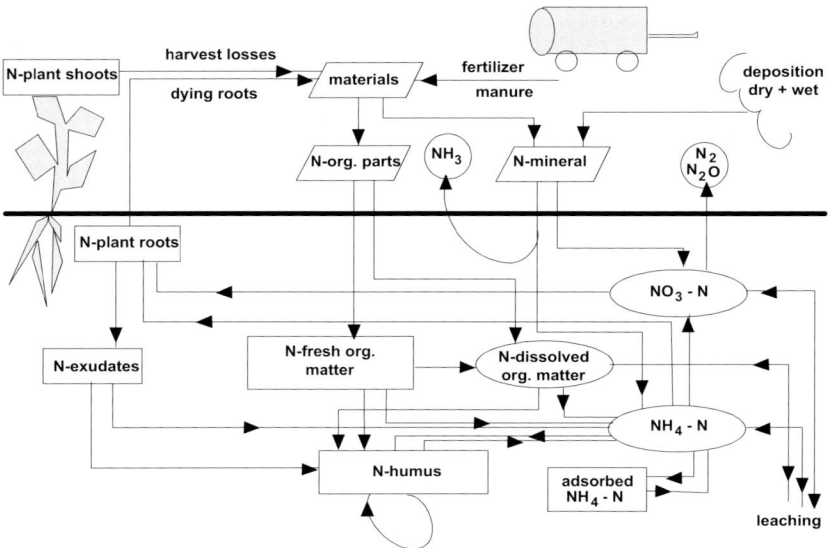

Figure 1. Relationship diagram of the nitrogen cycle in the ANIMO model.

In the inorganic part of the N cycle the following processes are described: (a) addition of mineral N (e.g. fertilisers); (b) NH_3 volatilisation; (c) NH_4^+ sorption; (d) nitrification; (e) denitrification; (f) N uptake by crop (Figure 1). In the inorganic part of the P cycle the following processes are described: (a) addition of mineral P; (b) P sorption; (c) precipitation; (d) P uptake by crop. In the ANIMO model the rate variables for organic matter transformation are corrected for the influences of temperature, moisture, pH and O_2 demand. The nitrification rate is corrected for the effects of temperature, moisture and pH.

Hydrological schematisation

The ANIMO model requires hydrological data supplied by an external field plot model (e.g. SWAP) or regional groundwater flow model (e.g. SIMGRO). The vertical schematisation resulting from the spatial discrimination, as applied in the water quantity model, forms part of the input of the model. A water quantity model should simulate all relevant terms of the water balance.

Transport and transformations

The substances that can be transported with water fluxes are: NH_4^+, NO_3^-, PO_4^{2-} and the dissolved organic matter fractions. For this transport combined with production or consumption, a transport- and conservation-equation is used. The calculation procedure follows the flow direction in the schematic column. For the first soil compartment, the boundary condition for the incoming flux from above is the precipitation with a concentration of the precipitation flux. For the last compartment, the boundary condition of the incoming flux is the seepage flux with a concentration of the soil solution below the described profile. Physical dispersion is simulated by the thickness of the model compartments and the length of the timestep. For the additions to the soil system and for the runoff to surface waters, the model has an extra reservoir on top of the compartment-division. The additions can be added to this reservoir and infiltrate into the soil system with the precipitation-flux. The runoff to surface water will take place from this reservoir. The reduction factor for crop uptake is determined on the basis of the summarised crop uptake during previous timesteps. For grassland the uptake includes diffusion.

References

Groenendijk, P. and Kroes, J.G. (1999). Modelling the nitrogen and phosphorus leaching to groundwater and surface water. ANIMO 3.5. Report 144. DLO Winand Staring Centre, Wageningen.

Kroes, J.G. and Roelsma, J. (1998). ANIMO 3.5. User's guide for the ANIMO version 3.5 nutrient leaching model. Technical Document 46. DLO Winand Staring Centre, Wageningen.

Rijtema, P.E., Groenendijk, P. and Kroes, J.G. (1999). Environmental impact of land use in rural regions. The development, validation and application of model tools for management and policy analysis. Series on environmental science and management, Vol. 1. Imperial College Press, London.

Wolf, J., Beusen, A.H.W., Groenendijk, P., Kroon, T., Rötter R. and Van Zeijts, H. (2003). The integrated modeling system STONE for calculating nutrient emissions from agriculture in the Netherlands. Environmental Modelling and Software, 18, 597-617.

The SUNDIAL Model

J.U. Smith[1], P. Smith[1], A.G. Dailey[2], M.J. Glendining [2], G. Tuck[2] and P.K. Leech[2]
[1]School of Biological Science (Plant & Soil Science), Aberdeen University, Cruickshank Building, St. Machar Drive, Aberdeen, AB24 3UU, UK
[2]Rothamsted Research, Harpenden, Herts, AL5 2JQ, UK

Introduction

SUNDIAL is a dynamic simulation model of N and organic matter turnover in the soil / plant system (Bradbury et al., 1993; Smith et al., 1996a). It includes simple descriptions of all major processes, providing detailed results using only simple inputs that are readily available to farmers and growers (Figure 1). It has been well developed and tested for a selection of arable and horticultural crops using both inorganic fertilisers and organic manures, and a new version has been developed to describe grassland.

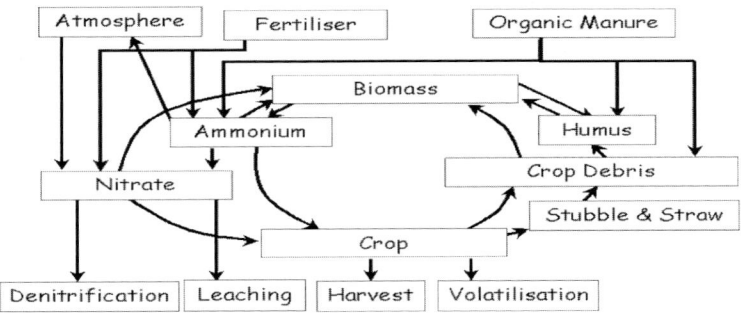

Figure 1. The structure of the SUNDIAL model.

Applications

SUNDIAL has been used for fertiliser recommendation, and to simulate whole farm rotations and catchments. The fertiliser recommendation system, SUNDIAL-FRS (Smith et al., 1996b) includes a graphical user interface for data entry, and is fully supported by defaults (Figure 2a). The dynamic results are presented so that farmers can understand the fate of N applied to the crop (Figure 2b). Recommendations are given for fertiliser amount and timing.

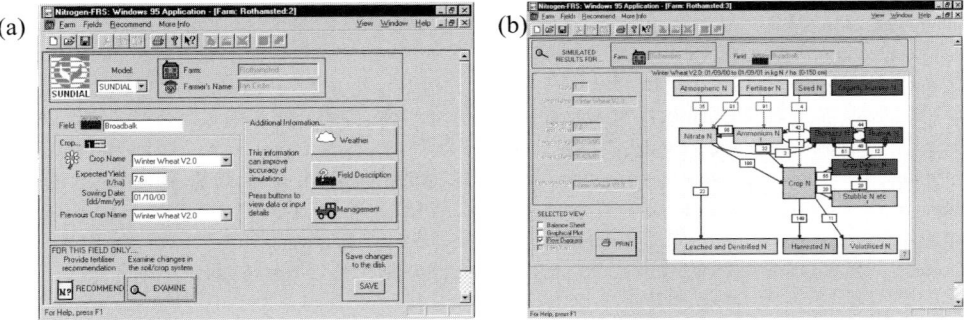

Figure 2. The SUNDIAL-FRS graphical user interface (a) example input screen (b) example results screen.

Controlling nitrogen flows and losses

Simulation of whole farm rotations uses the farm status and imposed rules and regulations to draw up a list of all allowed rotations (Smith et al., 1997). The model is run for each rotation, with outputs showing the rotations giving the best and worst-case N losses. The catchment modelling system, SUNDIAL-CAT calculates N turnover in the catchment, using only information that is generally available at the catchment scale. The uncertainty associated with the calculation is included using weighted distributions (Figure 3).

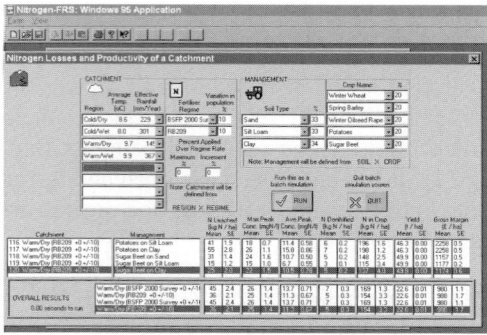

Figure 3. The SUNDIAL-CAT graphical user interface.

Conclusions

Dynamic models of N turnover in the soil/plant system have great potential for site and season specificity. Their potential to solve real-world problems can be realised in user-friendly decision support systems. Development of such systems enables full use to be made of scientific advances in the understanding of soil organic matter and N turnover.

Acknowledgements

Rothamsted Research receives grant-aided support from the Biotechnology and Biological Sciences Research Council of the United Kingdom

References

Bradbury, N.J., Whitmore, A.P., Hart, P.B.S. and Jenkinson, D.S. (1993). Modelling the fate of nitrogen in crop and soil in the years following application of ^{15}N-labelled fertilizer to winter wheat. Journal of Agricultural Science, Cambridge, 121, 363-379.

Smith, J.U., Bradbury, N.J. and Addiscott, T.M. (1996a). SUNDIAL: A PC-based system for simulating nitrogen dynamics in arable land. Agronomy Journal, 88, 38-43.

Smith, J.U., Dailey, A.J., Glendining, M.J., Bradbury, N.J., Addiscott, T.M., Smith, P., Bide, A., Brown, E., Chorley, R., Cook, S., Cousins, S., Draper, S., Dunn, S., Fisher, A., Griffiths, P., Hayes, C., Locke, A., Malone, C., McKay, J., Nettleton, D., Overman, H., Rickwood, S., Scholey, A. and Taylor P. (1996b). Constructing a nitrogen fertilizer recommendation system: What do farmers want? Soil Use and Management, 13, 225-228.

Smith, J.U., Glendining, M.J. and Smith, P. (1997). The use of computer simulation models to optimise the use of nitrogen in whole farm systems. Aspects of Applied Biology, 50, 147-154.

SECTION 7
THEMED WORKING GROUPS

Organic matter: does it matter, or can technology overcome most problems related to soil fertility?

Report by S. Recous[1] and G. Cadisch[2]

[1]*INRA, Unité d'Agronomie LRM, rue Fernand Christ, 02007 Laon Cedex, France*
[2]*Department of Agricultural Sciences, Imperial College London, Wye Campus, Wye, Ashford, Kent, UK*

This thematic group facilitated a critical evaluation of the potential and limitations of technology options to replace some of the functions of soil organic matter and helped to identify knowledge gaps and future research needs and priorities.

Is organic matter a problem?

The organic matter, its use, recycling and accumulation in soils is often supposed to provide much benefit to agro- and ecosystems, as it is associated with the concept of quality in production and intrinsic benefits for the environment. However, in many situations the accumulation of OM in soil either in the short term (e.g. decomposing crop residues), or in the long term (accumulation of humified OM) can translate into greater NO_3^- pollution or enhanced gas emissions. This might be due to the fact that we have little control over N release from soil OM, with potentially a lack of synchrony between the nutrient release and the crop N demand, and/or with a potential excess in the nutrient balance, since the mineralisable SOM pool usually increases with higher OM returns to the soil. Drastic changes in management of soils with a high OM content (e.g. ploughing grassland) can also increase dramatically NO_3^- losses, as the release of N is fast and substantial and cannot be controlled and evaluated. Soil organic matter can also contribute to the stabilisation of contaminants, or be a reservoir of pathogens, so that its accumulation in soil can also be a problem in the long term. Because of the uncertainty and difficulty in controlling its dynamics, one might consider that OM is a problem rather than an asset and hence we would be better off without it. This proposition actually came out of the discussion during the oral presentation and due to its radical nature continued to spark fierce discussions within the working group.

Technology replacement options for organic matter functions

In order to examine whether OM matters, we need to understand its functioning in ecosystems and which of these functions can be replaced safely and economically by technology. Current experiences show that nutrient supply functions of OM can be substituted by appropriate fertiliser use, precision agriculture, etc. The question therefore arises, can we replace other functions of OM such as physical effects (soil structure, water holding capacity, etc.), buffering capacity (nutrient retention, pH, pollutants), detoxification of contaminants (size and diversity of soil microflora) as well? During the session, the working group put together a range of options (Table 1) which showed that potentially, most of OM functions could be replaced by technological innovations, probably with the exception of biological functions, (for example in the case of detoxification), but even here there exist potential options.

Table 1. Technological options to replace the roles of organic matter in soil.

Role of organic matter	Technology replacement option
• Nutrient supply	Fertilisers, precision agriculture, ...
• Soil physical properties (structure, WHC,..)	Irrigation, hydroponic systems, artificial support, soil conditioners, ...
• Buffering capacity and retention (pH, nutrients, pollutants,...)	Active charcoal, liming, resins
• Detoxification and soil health (degradation of pollutants, pest control, ..)	Inoculation, enzymes, pesticides,...
• Source of energy and habitat for organisms (preservation of biodiversity)	Synthetic C source, porous materials, ...
• Landscape aesthetics	Artificial landscaping, gardening

It could be argued, that in some cases, such technologies could even be safer for the environment than relying on, or building up soil OM, e.g. we would have more control over nutrient release and supply characteristics and timing of fertilisers than from OM. Should we therefore, put more effort into developing appropriate technologies, rather than investing in soil OM research? Opinion in the forum on this issue was obviously divided. While most people recognised the potential benefits that alternative technologies could provide, they did not think that technology could provide the complete solution. The danger with technology driven approaches is that interrelations and implications for whole systems are often overlooked. Additionally, the cost of many of these technologies to date is not competitive, though that might change in the future.

It also became clear that even if we achieve a widespread replacement of soil OM by technology, the new system would continue to produce organic plant and animal residues. These organic residues would need to be removed from the agroecosystem (e.g. excess animal manures in Netherlands) by either export, burning or landfill, otherwise they would enhance soil OM and increase uncertainty (Figure 1). However, these removal options themselves create environmental problems rather than solutions.

At best, the consensus was that we should use technology to derive better control and enhance functioning of OM and integrated agroecosystems.

What are the alternative options to control some of the adverse effects of SOM?

Current changes in agricultural practices and land use to increase sustainability, bio-diversity and environmental functions are leading to the use of more organic resource-based systems (green manures, catch crops, ley rotations, organic farming, etc.), and even in intensive agricultural systems, which require a better understanding of OM functions and the scope for its management.

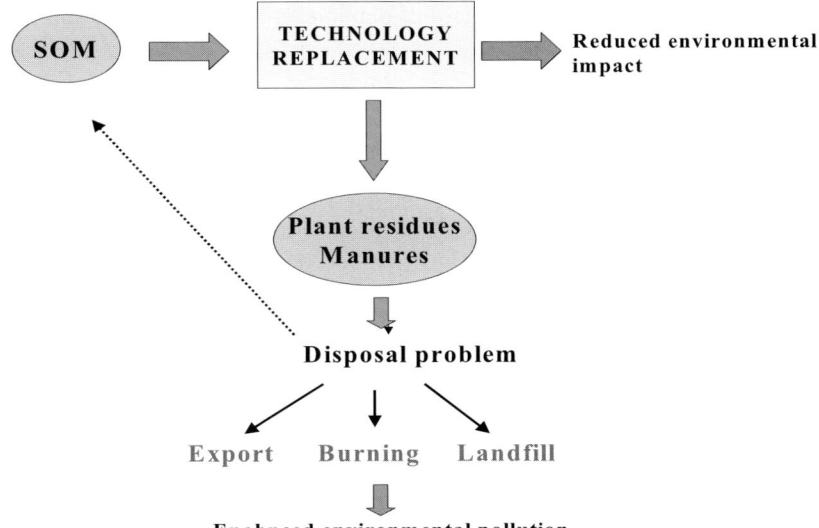

Figure 1. Potential feedback mechanisms of technology options for organic matter replacement.

Control of excess nutrients and uncontrolled release

Potential of transgenic plants

With the advances in molecular biology and genetically modified organisms and plants, new approaches (and potential dangers) to control nutrient flows emerge. Three ways by which modified plants could influence (directly or indirectly) N cycling in soils were identified: i) manipulating plant quality by changes in sugar/protein composition, increasing or modifying polyphenols or changing in lignin structure; ii) specific enzymes, e.g. protease inhibitors and iii) manipulating root growth pattern and timing of N uptake.

Modifications to plant quality attributes are mostly associated with plant breeding for improved animal nutrition. Alteration of lignin structure is potentially also of interest to the pulp industry and the effect on soil respiration has been investigated (Hopkins et al., 2001). Kumar et al. (2003) suggest that incorporation of specific protease inhibitors into rice or tobacco could also slow down the decomposition of proteins in soils. Research is still at an early stage and preliminary studies show some effect. During discussion of the working group, the potential problems with this approach were discussed, such as expression timing, duration and location. Indeed some of the alternative effects of such modifications could also influence crop resistance against insect attack and cause a slower N supply in the animal digestive tract and hence these need be considered as well. Perhaps more promising approaches to regulate N cycling are changes in root growth and timing of N uptake, but these are likely to be associated with multiple traits and as such, more difficult to genetically modify.

In conclusion, there needs to be a balance between research which evaluates such new potential approaches and concurrent efforts which assess their impact on other soil functions and the wider environment. In the current climate of public anti-GMO feelings, it will be challenging to develop these ideas.

Manipulation of OM quality

The possibility of manipulating the quality of SOM by mixing organic residues of various biochemical qualities was discussed   (e.g. mixing manure and fertiliser, or mixing easily degradable and recalcitrant crop residues). The discussion developed on whether the effects are 'additive' (e.g. enhanced decomposition of low C:N residues due to the release of N from high C:N residues) or a real positive (or negative) 'interaction' (e.g. growth of microorganisms that promote later decomposition of other substrates). It was concluded that it is difficult to assess the interactions, and it is still unknown which conditions give a positive interaction.

Does the quality of inputs affect the quality of SOM? It was emphasised by the convenors that little is known about the mechanisms that link the initial quality of organic inputs and the quality of the humified organic matter (Figure 2). Is the quality of SOM affected by the initial quality of inputs, and the pattern of decomposition, or is the residual SOM virtually the same when the short-term decomposition is over? It was felt that this may be dependent on the fractions considered, and that is is probably true for *labile fractions* (with which quality might be more closely linked to the initial inputs). The opinions were divided for stable fractions. Many people thought that the quality of SOM is more soil dependent than residue dependent.

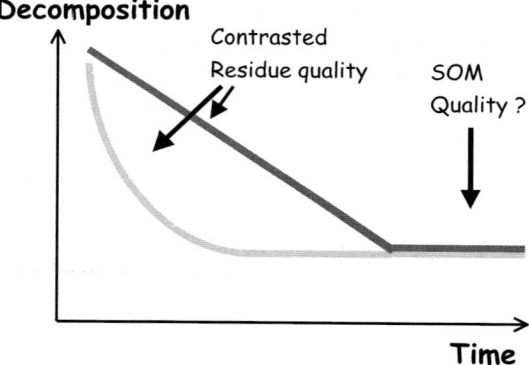

Figure 2. *Potential pathways of decomposition of crop residues of different initial quality and their potential impact on the quality of soil organic matter.*

What is the financial cost of soil organic matter?

Most participants were unfamiliar with the concept of giving a value to soil C and had little idea about the actual monetary value of soil OM or even current C trading prices under the CDM.

Estimates of the financial value of C come mainly from the C trading debate. Optimistic estimates of the theoretical potential to sequester C in agricultural soils (restoration, set-aside, minimal tillage) are (worldwide) ca. 24-43 Gt C over 50 years . This assumes restoration of large parts of degraded tropical soils and C lost historically from soils. Externalities of C have been calculated to be ca. US$95 per tonne emitted as CO_2 (Pretty & Ball, 2001). Existing trading schemes around the world pay between US$1-38 per tonne of C equivalent to about 1-65 US$ per tonne SOM (Pretty & Ball, 2001). These values are probably considerably less than the optimistic wishes of some farmers. In the UK, market current prices are ca. £12 per tonne CO_2. This amounts to ca. £3.2 t^{-1} C or £5.6 t^{-1} OM.

The current interest in C trading ignores potential effects on N cycling. If we assume that soil organic matter has a C:N ratio of approximately 12:1, then for every tonne of C sequestered, there is a N sink of ca. 83 kg N. This may be desirable in intensive agricultural systems, where we often have an excess of mineral N in soils, but for small scale farmers on N poor soils in the tropics, this is potentially a substantial cost, which has to be offset by increased fertiliser inputs. Some of the degradation problems in tropical pastures are associated with increased C stocks and immobilisation of mineral N. Yet such costs are not considered explicitly in current discussion on C trading issues. In order to convince policy makers of the importance of management of soil OM, we, as scientists, need to be able more readily to put a value on natural resource commodities, such as soil OM.

References

Hopkins, D.W., Webster, E.A., Chudek, J.A. and Halpin, C. (2001). Decomposition in soil of tobacco plants with genetic modifications to lignin biosynthesis. Soil Biology and Biochemistry, 33, 1455-1462.

Kumar, K., Rosen, C.J. and Russelle, M.P. (2003). A novel approach to regulate soil nitrogen mineralization. In: 12th N workshop: Controlling N flows and losses. Book of Abstracts. IGER, Exeter, Devon, UK.

Pretty, J. and Ball, A. (2001). Agricultural influences on carbon emissions and sequestration: A review of evidence and the emerging trading options. Centre for Environment and Society, University of Essex, UK, 31pp.

Optimising N additions: can we integrate fertiliser and manure use?

Report by J.J. Schröder[1] and R.J. Stevens[2]
[1]*Agrosystems Research, Plant Research International, Wageningen University and Research Centre, P.O. Box 16, 6700 AA Wageningen, The Netherlands*
[2]*Department of Agriculture and Rural Development, Agriculture, Food and Environmental Science Division, Newforge Lane, Belfast BT9 5PX, UK*

Abstract

Mineral fertiliser N is nothing but a supplement to the N provided by recycled by-products such as manure. To reduce mineral fertiliser rates, the N fertiliser value (NFV) of manures should hence be correctly asssessed and exploited as fully as possible. Ideas of experts involved in this working group are summarised in the present paper. It was concluded that, when proper attention is given to the composition of manure and decisions on rates, timing and placement are made according to best management practices, the relative NFV of manure can be enhanced from the current 10-50% to 40-80% in the near future. This could lead to a substantial reduction of mineral fertiliser use, nutrient surpluses and pollution.

Introduction

Environmental and economic reality has stimulated farmers to re-think the value of manures. Nowadays, most of them, at least roughly, estimate the value of nutrients available via manure before deciding on mineral N rates. However, manure is still seen as an unreliable source of N rather than a valuable asset. Inexpensive mineral fertilisers have disrupted the delicate local balance between crop production and animal production and made the recycling of 'wastes' no longer self-evident (Schröder, 2004). Manure is not an easy to use source of N. Extensive research confirms that the relative N fertiliser value (NFV) of manure ('N equivalency'), depends on numerous factors such as the manure composition, the rate, timing and method of its application, the crop type, the field history, soil and weather conditions. Researchers and extension services in Europe have different opinions about the N equivalency of manure, even when controlling factors seem more or less equal. Consequently, recommendations on the necessary mineral supplements vary greatly, too. The apparent controversies justify further analysis to which one of the Themed Working Groups (TWG) of the 12th Nitrogen Workshop (22-24 September 2003, Exeter, UK) has been dedicated. The discussions were started with three oral presentations by Frank Nevens, Hans-Werner Olff and Jaap Schröder, after which Jim Stevens chaired a discussion which was triggered by twelve challenging propositions. This paper summarises some of the considerations given to each of the three questions.

How can we correctly assess the N fertiliser value of manures?

Manuring does not always exactly deliver the N one expects it to provide. A commonly expressed view often attributes this variability to the 'unpredictable mineralisation' of manures. Such a view may incorrectly distract attention from other relevant explanations, such as the unknown composition, uneven spreading and the occurrence of gaseous or water-borne N losses. Anyhow, the level of uncertainty causes farmers to apply unnecessarily high amounts of mineral fertiliser N. Consequently, losses to the environment are augmented. Therefore, we need methods to correctly assess the NFV of manures.

Proper definitions are a prerequisite for the assessment of the NFV. Confusion may arise from different ways to characterise the N supplying ability of manures, as NFV is sometimes confused with the N recovery value, or with the fraction of the manure-N input that is mineralised over a certain period of time. These definitions of N availability yield different numbers. NFV in terms of 'N equivalents' can only be properly assessed in experiments including mineral fertiliser N treatments (in which the other elements are sufficiently supplied) next to manured treatments. Response curves, constructed from data for the mineral N fertiliser treatments, provide the basis for the assessment of the NFV (e.g. Schröder et al., 1997a; Nevens & Reheul, 2002). The mathematical analysis of the curves requires attention too, as it may strongly affect the calculated NFV (e.g. Schröder et al., 1998).

The relative NFV of manure equals the ratio of the recovery of manure-N and the recovery of mineral fertiliser-N. Recoveries can be determined not only by the difference method, but also by istotope dilution techniques (Diekmann et al., 1993; Thomson, 2001; Chantigny et al., 2003; Powell & Kelling, 2003). When applying N, a so-called added N interaction (ANI) may occur. Apparent ANIs refer to the substitution of ^{15}N for ^{14}N, causing the recovery to be underestimated, when based on isotope dilution. Real ANIs refer to phenomena such as priming, better N availability via enhanced rooting depth and a better utilisation of indigenous N, due to the simultaneous application of deficient elements other than N (Giller, 2002). Real ANIs generally result in an overestimation of the N recovery (Harmsen & Moraghan, 1988; Rao et al., 1993). When N application rates applied are high, relative to crop requirements, recoveries based on isotope dilution appear to be in the order of 10% higher than those based on the difference method. When N rates applied are low, relative to the amounts of soil N potentially involved in mineralisation-immobilisation turnover processes, recoveries based on isotope dilution tend to be in the order of 10% lower than those based on the difference method (Varvel & Peterson, 1990; Timmons & Baker, 1991; Powlson et al., 1992; Torbert et al., 1992; Diekmann et al., 1993; Jokela & Randall, 1997). So, both assessment methods for recoveries have artefacts. As the relative NFV is by definition the ratio of recoveries, one could reason that the ratio itself is not affected by the assessment method of recovery. This, however, is only true if the ANI per unit manure N, equals the ANI per unit mineral fertiliser N, times the ratio of the real recoveries of manure N and mineral fertiliser N. As manures often contain NH_4^+-N (which is prefered by the soil microbial biomass over NO_3^-) and interact with a microbially active environment, ^{15}N-based NFV assessements of manure deserve just as much suspicion, as NFV's based on the difference method (Harmsen & Moraghan, 1988; Rao et al., 1993).

It is demanding of research resources to assess the NFV for each manure type, and practically impossible to assess the NFV for each batch of manure. Therefore, attempts have been made to better characterise the composition of the manure and make predictions based on the characteristics. Distinguishing between the NH_4^+-N and the organic N fractions in manure is certainly a first step (Smith & Chambers, 1993; Chambers et al., 1999). A more detailed analysis of the organic fraction has had little practical application so far (Chadwick et al., 2000; Morvan & Nicolardot, 2003).

Organic fertilisers show a residual N effect beyond the year of their application, as decomposition of organic material usually takes more than one year. When organic fertilisers are used repeatedly, residual effects cumulate and significantly increase the availability of N (Wolf et al., 1989; Wolf & Van Keulen, 1989; Dilz et al., 1990; Schröder & Van Keulen, 1997). However, farmers find it difficult to account for the N which mineralises from previous organic inputs. Advisers tell them to credit manure N before deciding on mineral fertiliser supplementation. Most estimates of the N contribution from manure are generally conservative, as they commonly only account for the available N in the first season, following manuring (Table 1). Reliable a priori estimates of residual N effects, based on soil analysis, are difficult to make (Jarvis et al., 1996; Powlson, 1997). However, implicit accounts for residual N are made in recommendation systems that include soil organic N (e.g. Machet et al., 2003).

Manures with a large C to N ratio ($\approx$ small NH_4^+-N to total N ratio) showing relatively little mineralisation in the first year(s) after application are often seen as 'environmentally benign' (Nevens & Reheul, 2003). However, in the long term, an equilibrium will be established between the yearly organic inputs and the yearly mineralisation. It may take decades before such an equilibrium is attained, depending on the ratio of (reactive) C and (reactive) N (Sims, 1995). The larger the C : N ratio of a manure, the greater the difference between NFV in the first year after application and the long term NFV. This implies that the risk of underestimating both the NFV and the environmental impact is more likely with FYM and composts than with slurries. Table 1 confirms that it does make a difference whether recommendations take residual effects into account or not. The presented data also show that differences are larger for solid manures than for slurries. For spring-applied solid manures, NFV's range from 0.20 to 0.75 per kg N applied, whereas NFV's of slurries range from 0.45 to 0.75 kg per kg N applied.

How can we maximise the fertiliser value of manures?

Nitrogen should be water-soluble during the growing season to be available to plants, whereas it should be bound from late summer to spring to avoid losses. Manures differ significantly in the ratio of mineral N to organically bound N. Hence, preference for a certain type of manure should be influenced also by the time windows for spreading as imposed by the soil type. Manures with a large organic N fraction (e.g. FYM) are generally better suited for autumn applications than manures with a small organic N fraction (e.g. slurries) (e.g. Smith & Chambers, 1993). When applied in spring, FYMs may not be able to meet the crop requirements in time, especially if the associated bedding material (e.g. cereal straw) is not yet fully decomposed and hence tends to immobilise N (Thomson, 2001; Berry et al., 2002). From slurries applied in late summer, on the other hand, considerable amounts of mineral N can be lost, unless the N is taken up by cover crops and successfully immobilised until the subsequent spring (Schröder et al., 1997a). When the soil type permits field traffic in early spring, the NFV of slurry is maximised by postponement of applications until spring (Schröder et al., 1993). Synchronisation may be further improved when the application of manure is postponed until after crop emergence. Such a deliberate postponement is not without risks in annual crops, however. The equipment involved may damage the soil structure and/or crop, especially as incorporation is needed to avoid NH_3 volatilisation.

Table 1. Composition (kg per 1000 kg fresh material) and N fertiliser value of manure (i.e. kg of fertiliser N saved per 100 kg manure N applied) as communicated to farmers by national extension services when incorporation has taken place within 5 hours.

Type[1]	Country[2]	kg per 1000 kg				NFV when applied on					
		N total	NH$_4$-N	P$_2$O$_5$	DM	1 March			1 September		
						1st year[3]	2nd year[3]	Long term[3]	1st year[3]	2nd year[3]	Long term[3]
FYM	EW	6.0	1.5	3.5	250	0.25	-	-	0.10	-	-
	F	5.5	0.5	3.5	200	0.25	-	-	0.10	-	-
	Dk[5]	-	-	-	-	(0.25)	(0.15)	-	0.25	0.15	-
	Sw	5.1	1.1	2.7	200	0.20-0.25	-	0.25-0.50	0.20-0.25	-	0.20-0.25
	BF	8.3	2.4	2.9	210	0.31	-	-	0.16-0.18	-	-
	NL	6.9	1.6	3.8	235	0.45	-	0.65	0.25	-	0.50
	P	5.0	-	3.0	210	0.20-0.40	0.35-0.40				
LCS	EW	3.0	1.5	1.2	60	0.45	-	-	0.20	-	-
	F	2.5	1.3	1.4	75	0.50	-	-	n.a[4]	-	-
	Dk[5]	-	-	-	-	0.55	0.10	-	(0.55)	(0.10)	-
	Sw	2.2	1.2	0.9	45	0.47-0.56	-	0.53-0.70	0.47-0.56	-	0.53-0.70
	BF	4.6	2.1	1.4	84	0.54	-	-	0.22-0.28	-	-
	NL	4.9	2.6	1.8	90	0.60	-	0.75	0.25	-	0.40
LPS	EW	4.0	2.4	2.0	40	0.55	-	-	0.20	-	-
	F	5.5	3.5	6.0	70	0.70	-	-	n.a[4]	-	-
	Dk[5]	-	-	-	-	0.60	0.10	-	(0.60)	(0.10)	-
	Sw	3.0	2.1	1.9	25	0.50-0.60	-	0.50-0.70	0.50-0.60	-	0.50-0.70
	BF	7.8	4.5	4.5	83	0.62	-	-	0.19-0.27	-	-
	NL	7.2	4.2	4.2	90	0.70	-	0.75	0.25	-	0.35
PM	EW	16	8	13	300	0.50	-	-	0.25	-	-
	F	28	7	23	680	0.65	-	-	n.a[4]	-	-
	Dk[5]	-	-	-	-	-	-	-	-	-	-
	Sw	30	8	19	650	0.35	-	0.40	0.35	-	0.40
	BF	33	5	28	690	0.52	-	-	0.36-0.40	-	-
	NL	19	9	24	640	0.60	-	0.75	0.30	-	0.40

[1] FYM, LCS, LPS, PM indicating farm yard manure from cattle, slurry from dairy cows, slurry from fattening pigs and poultry manure, respectively.

[2] EW = England and Wales, F = France, DK = Denmark, Sw = Switzerland, BF = Belgian Flanders, NL = Netherlands, P = Poland.

[3] NFV's in the first main crop following application, the second main crop following application (i.e. residual effect in the second year after application) and in the long term (i.e. cumulative apparent NFV in any year after repeated applications).

[4] n.a. not allowed.

[5] Numbers refer to NFV's farmers are expected to achieve when calculating their DK-annual fertiliser budget; these numbers are meanwhile raised by 0.10 points to further reduce N loss.

Lack of synchronisation between supply and demand can also occur at the end of the season. N mineralisation from manure continues naturally beyond the period in which crops take up N, but is redundant and can be lost if the crop is followed by a long fallow period. The timely establishment of a catch crop can help to save N and transfer it to the next growing season, thus improving the apparent NFV (Schröder et al., 1996). The NFV of manures is not only determined by proper timing of application, but also by the place of its application, since the reduction of NH_3 losses is a major consideration. This is especially relevant for manures, rich in mineral N (slurries, urine); NH_3 losses can be substantially decreased by injection. Too deep a placement, however, is not conducive to maximise the NFV. Proper attention must also be given to spreading techniques. Irregular, patchy spreading patterns increase the heterogeneity of the soil fertility. Consequently, some parts of the field become over-fertilised, whereas other parts may become deficient. Even spreading is not synonymous with uniform spreading, however. For example, in crops with a wide distance between rows, yields may benefit from techniques that apply manure close to the anticipated position of rows (Sawyer et al., 1991; Schröder et al., 1997b).

As indicated in the previous sections, the composition of manure has a strong impact on the actual and potential NFV. In turn, composition is the result of off-field measures such as diets, use of bedding material and manure additions and treatment techniques. The answer to the question: 'which manure type is preferable?' should therefore, not just be inspired by consideration of how to maximise the NFV under given pedoclimatic circumstances, but should also take account of the implications for N utilisation at higher scales than just the field and cropping season. Whether to go for composted FYM's or slurries is a typical example of a decision with (scale-related) contrasting implications for N utilisation (Neeteson et al., 2003).

How should we assess the need for mineral N fertiliser?

Utilisation of manure N is not just determined by proper timing and placement, but, also by avoiding excessive application rates. It is generally impossible to meet the N requirements of the crop rotation as a whole with manure only, if excessive applications of phosphate are to be avoided. The reason for this is that the N : P requirement in most crops exceeds the ratio of effectively available N : P in manures. This discrepancy is larger with solid composted manures (because of their inherent lower N : P ratio) than with, for example, cattle slurry (Schröder, 2004). Even slurry-based systems, however, are always short of N. In organic farming systems this relative N deficiency must be met by the presence of legumes in the sward or rotation, whereas in conventional systems, mineral fertiliser N must be supplemented.

From this point of view, mineral N fertiliser should be seen as a supplement rather than the basis for fertiliser plans. To adjust such mineral N supplements to the actual mineralisation and crop requirements, indicator-based, split-application strategies can be useful. Nowadays, numerous indicators for a more precise assessment of N requirements are available. Examples of these indicators are the mineral N supply of the soil, the NO_3^- content of plant tissues, or the colour of the leaf or crop canopy (Schröder et al., 2000; Link et al., 2003). Tissue tests are of limited value in supporting decisions on N supplementation, compared with indicators that are directly related to the soil or to the measurement of greenness. Greenness tests and tissue tests are less suitable to quantify excessive availability of N at early crop stages, as opposed to soil related indicators. Moreover, crop tests often need an on-site calibration against reference plots. Although indicator-based, site specific N rates may save money, the indicators themselves carry a price too. So far, surprisingly little attention has been paid to a cost-benefit analysis of indicator-based, site-specific N management in general (Schröder et al., 2000).

Conclusions

When proper attention is given to the composition of manure, and decisions on rates, timing and placement are made according to best management practices, the N fertiliser value of manure can be enhanced. The current conservative estimates of the relative N fertiliser value of 10-50%, could be raised to values of 40-80% in the near future. This could lead to a drastic reduction of mineral fertiliser use, nutrient surpluses and pollution. Under such conditions, 'manure', 'precision farming' and 'environment' are no longer contradictions in terms.

Acknowledgements

We would like to thank our colleagues H. Menzi (Switzerland), B. Chambers and K. Smith (England and Wales), F. Nevens (Flanders), F. Vertes and T. Morvan (France), S. Samborski (Poland) and the persons who they in turn have consulted, for the data presented in Table 1.

References

Berry, P.M., Sylvester-Bradley, R., Philipps, L., Hatch, D.J., Cuttle, S.P., Rayns, F.W. and Gosling, P. (2002). Is the productivity of organic farms restricted by the supply of available nitrogen? Soil Use and Management, 18, 248-255.

Chadwick, D.R., John, F., Pain, B.F., Chambers, B.J. and Williams, J. (2000). Plant uptake of nitrogen from the organic fraction of animal manures: a laboratory experiment. Journal of Agricultural Science, Cambridge, 134, 159-168.

Chambers, B.J., Lord, E.I., Nicholson, F.A. and Smith, K.A. (1999). Predicting nitrogen availability and losses following application of organic manures to arable land: MANNER. Soil Use and Management, 1, 137-143.

Chantigny, M.H., Angers, D.A., Morvan, T. and Pomar, C. (2003). Distribution of ^{15}N labeled pig slurry in sandy soils cropped to maize. Abstracts 12th N Workshop, Exeter, Devon, UK.

Diekmann, K.H., De Datta, S.K. and Ottow, J.C.G. (1993) Nitrogen uptake and recovery from urea and green manure in lowland rice measured by ^{15}N and non-isotope techniques. Plant and Soil, 148, 91-99.

Dilz, K., Postmus, J. and Prins, W.H. (1990). Residual effect of long term applications of farmyard manure to silage maize. Fertilizer Research, 26, 249-252.

Giller, K.E. (2002) Targetting management of organic resources and mineral fertilizers: can we match scientists' fantasies with farmers' realities. In: B. Vanlauwe, J. Diels, N. Sanginga, and R. Merckx (eds). Integrated Plant Nutrient Management in Sub-Saharan Africa. CAB International, 155-171.

Harmsen, K. and Moraghan, J.T. (1988) A comparison of the isotope recovery and difference method for determining nitrogen fertilizer efficiency. Plant and Soil, 105, 55-67.

Jarvis, S.C., Stockdale, E.A., Shepherd, M.A. and Powlson, D.S. (1996). Nitrogen mineralisation in temperate agricultural soil: processes and measurement. Advances in Agronomy, 57, 187-235.

Jokela, W.E. and Randall, G.W. (1997) Fate of fertilizer nitrogen as affected by time and rate of application to corn. Soil Science Society of America Journal, 61, 1695-1703.

Link, A., Jasper, J. and Olfs, H.W. (2003). Variable nitrogen fertilization by tractor-mounted remote sensing. (This volume)..

Machet, J.M., Recous, S., Jeuffroy, M.H., Mary, B., Nicolardot, B. and Parnaudeau, V. (2003). A dynamic version of the predictive balance sheet method for fertilizer N advice. (This volume).

Morvan, T. and Nicolardot, B. (2003). Decomposition of soluble compounds obtained after fractionation of different animal wastes. (This volume).

Neeteson, J.J., Schröder, J.J. and ten Berge, H.F.M. (2003). A multi-scale system approach to nutrient management research in The Netherlands. Netherlands Journal of Agricultural Science 50, 141-151.

Nevens, F. and Reheul, D. (2002). The nitrogen and non-nitrogen contribution effect of ploughed grass leys on the following arable forage crops: determination and optimum use. European Journal of Agronomy, 16, 57-64.

Nevens, F. and Reheul, D. (2003). The application of vegetable, fruit and garden waste in addition to cattle slurry in a silage maize monoculture: nitrogen availability and use. European Journal of Agronomy, 19, 189-203.

Powell, J.M. and Kelling, K.A. (2003) Differential ^{15}N labeling and use of dairy manure components for N cycling studies. Abstracts 12th N Workshop, Exeter, Devon, UK.

Powlson, D.S. (1997). Integrating agicultural nutrient management with environmental objectives - current state and future prospects. Proceedings Fertilizer Society 402, London, 42 pp.

Powlson, D.S., Hart, P.B.S., Poulton, P.R., Johnston, A.E. and Jenkinson, D.S. (1992) Influence of soil type, crop management and weather on the recovery of ^{15}N-labelled fertiliser applied to winter wheat in spring. Journal of Agricultural Science, Cambridge, 188, 83-100.

Rao, A.C.S., Smith, J.L., Papendick, R.I. and Parr, J.F. (1993) Influence of added nitrogen interactions in estimating recovery efficiency of labeled nitrogen. Soil Science Society of America Journal, 55, 1616-1621.

Sawyer, J.E., Schmitt, M.A., Hoeft, R.G., Siemens, J.C. and Vanderholm, D.H. (1991). Corn production associated with liquid beef manure application methods. Journal of Production Agriculture, 3, 335-344.

Schröder, J.J. (2004). Revisiting the agronomic benefits of manure: correct assessment and exploitation of its fertilizer value spares the environment. Bioresource Technology (accepted for publication).

Schröder, J.J. and van Keulen, H. (1997). Modelling the residual N effect of slurry applied to maize land on dairy farms in The Netherlands. Netherlands Journal of Agricultural Science, 45, 477-494.

Schröder, J.J., ten Holte, L., van Keulen, H. and Steenvoorden, J.H.A.M. (1993). Effects of nitrification inhibitors and time and rate of slurry and fertilizer N application on silage maize yield and losses to the environment. Fertilizer Research, 34, 267-277.

Schröder, J.J., van Dijk, W. and de Groot, W.J.M. (1996). Nitrogen fluxes in maize cropping systems as affected by cover crops. Netherlands Journal of Agricultural Science, 44, 293-315.

Schröder, J.J., ten Holte, L. and Janssen, B.H. (1997a). Non-overwintering cover crops: a significant source of N. Netherlands Journal of Agricultural Science, 45, 231-248.

Schröder, J.J., ten Holte, L. and Brouwer, G. (1997b). Effects of slurry placement on silage maize yields. Netherlands Journal of Agricultural Science, 45, 249-261.

Schröder, J.J., Neeteson, J.J., Withagen, J.C.M. and Noij, I.G.A.M. (1998). Effects of N application on agronomic and environmental parameters in silage maize production on sandy soils. Field Crops Research, 58, 55-67.

Schröder, J.J., Neeteson, J.J., Oenema, O. and Struik, P.C. (2000). Does the crop or the soil indicate how to save nitrogen in maize production? Field Crops Research, 62, 151-164.

Sims, J.T. (1995). Organic wastes as alternative nitrogen sources. In: P.E. Bacon, (ed.) Nitrogen fertilization in the Environment. Marcel Dekker, New York, pp. 487-535.

Smith, K.A. and Chambers, B.J. (1993). Utilizing the nitrogen content of organic manures on farms: problems and practical solutions. Soil Use and Management, 9, 105-112.

Thomsen, I.K. (2001) Recovery of nitrogen from composted and anaerobically stored manure labelled with [15]N. European Journal of Agronomy, 15, 31-41.

Timmons, D.R. and Baker, J.L. (1991) Recovery of point-injected labeled nitrogen by corn as affected by timing, rate and tillage. Agronomy Journal, 83, 850-857.

Torbert, H.A., Mulvaney, R.L.,Vanden Heuvel, R.M. and Hoeft, R.G. (1992) Soil type and moisture regime effects on fertilizer efficiency calculation methods in a nitrogen-15 tracer study. Agronomy Journal, 84, 66-70.

Varvel, G.E. and Peterson, T.A. (1990) Nitrogen fertilizer recovery by corn in monoculture and rotation systems. Agronomy Journal, 82, 935-938.

Wolf, J. and van Keulen, H. (1989). Modeling long-term crop response to fertilizer and soil nitrogen. II. Comparison with field results. Plant and Soil, 120, 23-38.

Wolf, J., de Wit, C.T. and van Keulen, H. (1989). Modeling long-term crop response to fertilizer and soil nitrogen. I. The model. Plant and Soil, 120, 11-22.

Controlling gaseous N emissions - what is achievable?

Report by R. Harrison[1] and E.M. Baggs[2]
[1]ADAS Boxworth, Cambridge, CB3 8NN, UK
[2]Department of Agricultural Sciences, Imperial College London, Wye Campus, Wye, Ashford, Kent TN25 5AH, UK

Introduction

Agricultural systems are an important source of emissions of NH_3, which contributes to acidification and eutrophication of soils and waters, and N_2O, a significant greenhouse gas and implicated in the destruction of stratospheric O_3 (Badr & Probert 1993). In Europe in 1990, agriculture was responsible for ca. 98% of total NH_3 emissions and ca. 7% of total greenhouse gas emissions (Brink et al., 2001). These emissions predominantly result from application of inorganic N fertilisers, animal manures and plant biomass to soils, resulting in reduced N-use efficiency, and from animal housing and manure stores. It has been estimated that $1.25 \pm 1\%$ of inorganic N applied to soil is lost as N_2O during denitrification or nitrification (Bouwman, 1996), but this has since been shown to vary greatly depending on system and management practice (e.g. Smith et al., 1997). Large quantities of animal manures (slurry, farmyard manure and compost) are produced and applied as a valuable N-fertiliser to crops, thereby, reducing the requirement for mineral N-fertiliser. Manure N is composed of NH_4^+ and organic N, but it has been shown that the crop uptake of N applied in animal manure in the first year can be calculated on basis of the NH_4^+ content. Ammonium in manure may be transformed and emitted as NH_3, N_2 or N_2O from animal houses, manure stores, or from applied manure. In consequence, the fertiliser value of manure, in the first growing season, is lower and more variable than that of commercial fertilisers, and mineral fertilisers are often applied to 'manured' crops to compensate for the emission. Reduction in gaseous emissions of oxidised and reduced N will reduce the need for fertiliser N and the cost for crop production, and reduce the detrimental impact of agriculture on the environment.

To date, research on gaseous N emissions has tended to concentrate on quantifying emissions from different agricultural systems under different management practices. It is now necessary to synthesise this knowledge, to propose appropriate mitigation strategies for NH_3 and N_2O, and to identify key areas of research focus required to refine these mitigation strategies and to improve prediction of emissions. As part of the 12th Nitrogen Workshop, Themed Working Group 3 discussed the challenges associated with the proposal of mitigation strategies, with the objective of specifying limitations in our knowledge about the processes controlling gaseous N emission and, in consequence, defining research needed either to improve prediction of emissions, or to improve measures (management, technological) that reduce emissions. Research required was considered in terms of the N cycle, the form of N applied, the growth season or crop rotation, and farm management.

Research required with respect to the nitrogen cycle

There were 21 papers in this workshop (see Anon, 2003) concerned with gaseous N emissions and aspects of the N cycle (Table 1). These were divided between:

Table 1. Topics addressed in relation to gaseous N emission and aspects of the N cycle (from Anon, 2003)

Authors	Current uncertainty	Opportunity to control emissions
Baggs et al.	Effect of increasing pCO_2 denitrification, nitrification & CH_4 oxidation, hence on gaseous N losses	Manipulation of soil and plant biology
Wrage et al.	Differentiation between sources of N_2O; nitrification, nitrifier denitrification, denitrification & other sources	Understanding of processes contributing to emissions
Bateman et al.	Relative contribution of heterotrophic bacteria & fungi to nitrification, hence N_2O under different conditions (WFPS)	
Kesik et al.	Functional relationships between environmental factors and emissions	Integration of knowledge in models
Cárdenas et al. Heinen	Applicability and robustness of process models of denitrification	
Chatskikh et al. Werner et al.	Application of mechanistic models to inventory estimates	
Le Cadre et al.	Mechanistic understanding of effect of soil, climate and technical parameters on NH_3 emission	
Le Cadre et al.	Heterogeneity resulting from dissolution of ammonium nitrate granules	DSS for fertiliser practice (to abate NH_3)
Cannavo et al.	Variations in denitrifying activity below root zone due to environmental parameters (seasonal)	Land & landscape management
Zwart	Quantitative estimates of nitrate removal by denitrification in deeper soil layers	
Cors & Tychon	Denitrification in riparian buffer strips	
Murray et al.	Effect of horizontal transfers at sub-surface impermeable layer on N transformations	
Reay et al. Thomas et al.	Emission factors of N_2O for leached N	
Murray et al. Syväsalo et al.	Factors (including organic C) which regulate denirification in soils	
Koponen et al.	Low-temperature dependence of denitrification	
Master et al.	Effect of irrigation of effluent on gaseous N losses in arid & semi-arid areas	
Petersen et al.	Effect of urea concentration on N_2O emission from grassland	

- studies developing a deeper understanding of the microorganisms responsible for N_2O and N_2 production and the physical chemistry associated with NH_3 volatilisation
- modelling (ultimately for policy and management)
- studies of the role of denitrification in reducing NO_3^- before it enters ground- and surface-waters
- investigations of particular factors of relevance to management (organic C, temperature, irrigation)

Many of these topics were highlighted in the working group discussion, although there was a recognition that, in considering gaseous emissions in the context of the N cycle, it is necessary to take a wide view of the ultimate objectives. This is, of course, implicit in much of the work presented, e.g. where denitrification is a method of reducing potential NO_3^- pollution of ground- and surface-waters. Although there continues to be a significant effort to improve the scope of mechanistic models by incorporating current scientific understanding of the key processes, there is still a need to continue to work towards a more co-ordinated approach to research in this area. It will never be possible to completely reconcile competing objectives by manipulation of the N cycle, but there will always be new opportunities to improve the functioning of the system as our understanding and societal demands evolve. It is worth noting that, in general terms, the economic effect of gaseous N losses in intensive agriculture on producers is often low and this represents a significant challenge to scientists in proposing management strategies to mitigate emissions.

A number of key limitations and opportunities were identified for the prediction of emissions and for practice to reduce emissions. Spatial and temporal variability in processes and resulting emissions still need to be adequately understood and accounted for, and improved techniques are required to quantify the contributions of individual processes within a heterogeneous environment. Manipulation of soil biology was considered a possibility to achieve specific goals. To date, plant breeding has mostly been directed towards increasing yields, but in the future, greater consideration may be given to plant breeding (and potentially GM) programmes that increase the efficiency of resource-use, particularly N. Our knowledge of the dynamics of C and N in soils is reasonably well-advanced, but most proposed interventions tend to relate the use of C-rich substrates to immobilise excess labile N, thereby indirectly reducing gaseous N emissions, but there are still uncertainties associated with interaction between C and N cycles.

Research required with respect to the form of N applied

There were only six posters considering the effect of fertiliser form on gaseous emissions (Table 2), despite the obvious potential this has for manipulating N cycling. Two studies compared the effect of nitrification inhibitors and two studies contrasted emissions from inorganic vis-à-vis organic materials. Only one paper dealt with the interaction between practices that sequester C in soil (e.g. through the use of organic manures) and mitigate N_2O emission. This would seem to be an important topic for the future, if we are to adopt a holistic approach to global warming.

Despite the limited number of posters presented, the working group considered that there was scope for further exploiting the known opportunities in this area. These include: further

Table 2. Topics addressed in relation to gaseous emission and aspects of the form of N (from Anon, 2003).

Authors	Current uncertainty	Opportunity to reduce emissions
Jones et al.	Effects of differing fertiliser materials on NO & N_2O (& CH_4)	Variations in characteristics of range of organic and inorganic N sources
Vallejo et al.	Effect of differing fertiliser materials on N_2O emission	
Lampe et al.	Temporally integrated (1-year) N_2O emissions as affected by form and amount	
Báez et al. Merino et al.	Comparative effect of nitrification inhibitors on N_2O emission with organic and inorganic fertilisers	Fertiliser formulation to modify behaviour in soil
Beheydt et al.	Effect of measures to sequester C in soils on N_2O emission	Multiple environmental goals

promotion of the economic value of animal manures, (see also report from Themed Working Group 2) the use of inhibitors and other formulations to modify transformations of N inputs in soil, and combining organic and inorganic fertilisers to improve the synchrony between availability of N and crop demand. However, the variable composition of organic N inputs (manure, sewage sludge, etc.), uncertainty of the mechanisms involved in any interactions between organic and inorganic inputs and a lack of consensus on the relative importance of NH_3, N_2O and NO emissions, were seen as significant limitations to progress in this area.

Research required with respect to crop growth, the crop environment and crop rotation

There were seven contributions dealing with gaseous emissions and aspects of crop growth, the crop environment or crop rotation in the workshop (Table 3). These posters consider interactions between the crop, soil and atmosphere that influence emissions, and therefore, point the way to further research that may be useful in managing processes at the field scale.

Table 3. Topics addressed relating to gaseous emission and crop growth, the crop environment and crop rotation (From Anon, 2003).

Authors	Current uncertainty	Opportunity to reduce emissions
Génermont et al.	Interaction of crop canopy and physiology effects with soil & environmental controls on emission	Responses at crop-system scale
Laville et al.	Influence of arable crop management on NO emission	
Mori et al.	Effect of vegetation on N_2O fluxes in grassland	Interactions between plant and soil
Thyme & Ambus	Emission factor of N_2O for biologically fixed N	
Regina et al.	Temporally integrated (1-year) N_2O emissions in boreal region	
Kanerva et al.	Ecosystem response to elevated CO_2 and/or O_3	
DiMarco et al.	High temporal resolution of N_2O & NH_3 fluxes from grassland	

A significant point raised during the discussion was that, whereas crops and crop rotations are amenable to management, the weather often has a major influence on emissions, but cannot be managed or even adequately predicted. An improvement in prediction of rainfall might lead to 'real-time' management advice systems, e.g. for timing of fertiliser application. There is still great uncertainty as to emissions resulting from crop residue management, and the potential for emissions to be reduced by manipulation of residue quality, use of catch crops, intercropping or bi-cropping.

Research required with respect to farm management

There were 20 posters dealing with gaseous emissions and aspects of farm management in the workshop (Table 4). Eight of these reported findings directly relevant to international commitments entered into by Governments (e.g. Gothenburg protocol, Kyoto protocol) and nine investigated potential measures that might contribute to meeting such commitments.

Research opportunities identified to help reduce emissions, fell into two main areas. On the technological side, further progress was anticipated as a result of increased 'precision', e.g. in the management of soil water, in terms of spatial and temporal co-location of C and N, and in terms of more precise application of N fertilisers in space (precision agriculture) and time (in relation to weather, see above). However, the role of management is also important, and in this context, issues such as farmer education and the need to combine biophysical and socio-economic knowledge and considerations must be addressed.

Conclusions

In the keynote paper introducing this theme, Davidson and Mosier (2003) discussed a number of management options to reduce gaseous emissions from agricultural systems. They concluded that the challenge is to develop technologies, management practices and incentives for farmers and ranchers to adopt N management plans that address both the goals of society and the needs of the farmer.

In this working group, research opportunities were identified which will either improve the precision of current interventions, or utilise current knowledge in the context of the whole system, including socio-economic aspects. We identified potential roles of plant breeding, nitrification inhibitors, tillage practice, combined organic and inorganic inputs, manipulation of residue quality, and precision agriculture in reducing gaseous N emissions. However, advances in the proposal of sound and appropriate mitigation strategies and in our predictive capability are currently hindered by our need to understand spatial and temporal variability, the contributions and importance of particular processes, the nature of interactions between C and N, and between organic and inorganic inputs, and by adequate prediction of rainfall.

In order to predict and reduce global warming, it is necessary to devise research programmes that consider CO_2 and CH_4, as well as N_2O, and provide a global warming potential for systems and management practices investigated. Strategies proposed to reduce greenhouse gas emissions should also consider the reduction of NH_3 emission. This is a particular challenge as there can be interaction between the processes of CH_4 oxidation and nitrification that

Table 4. Topics addressed in abstracts relating to gaseous emission and farm management (From Anon, 2003).

Authors	Current uncertainty	Opportunity to reduce emissions
Davidson & Mosier	Designing technologies, management practices and incentives for N goals	Using science to drive policy development
Brown et al. Kuikman et al. Webb et al.	Evaluating a range of mitigation strategies (for N_2O & NH_3)	
Hyde et al. Reidy et al.	National NH_3 inventory	
Kelliher et al.	National N_2O inventory	
Thomas et al.	National inventory of N leached	
Thomas et al.	Effects of management factors in vegetable production on N_2O emission	Evaluation of specific measures
Kreye et al.	Evaluating rice production systems re. N_2O emissions	
Hyde et al.	Relative effect of grassland management and soil type *cf.* N application on N_2O	
Webster et al. Rochette et al.	Effect of tillage practice on N_2O (& CH_4 + CO_2) emission	
Demmers et al.	Evaluating animal housing systems re. NH_3 emissions	
Thorman et al. Hansen & Sommer	Evaluating manure-handling strategies re. NH_3 & N_2O (& CH_4) emissions	
Morrissey et al.	Effect of NH_3 capture by woodland on N_2O emission	
Bol et al.	Urine-N content on N_2O emission from soil	Modifying system inputs and characteristics
Petersen et al.	Urea concentration on N_2O emission from soil	
Menneer et al.	Soil structure effects on denitrification	

determine N_2O and net CH_4 emissions from soils, and strategies most appropriate for mitigation of one greenhouse gas, or for reduction in NH_3 emission, may not necessarily be appropriate for a reduction in Global Warming Potential.

References

Anon. (2003). 12th N Workshop: Controlling N flows and losses. Book of Abstracts, IGER, Exeter, Devon, UK.

Badr, O. and Probert, S.D. (1993). Environmental impacts of atmospheric nitrous oxide. Applied Energy, 44, 197-231.

Bouwman, A.F. (1996). Direct emission of nitrous oxide from agricultural soils. Nutrient Cycling in Agroecosystems, 46, 53-70.

Brink, C., Kroeze, C. and Klimont, Z. (2001). Ammonia abatement and its impact on emissions of nitrous oxide and methane - Part 2: application for Europe. Atmospheric Environment, 35, 6313-6325.

Smith K.A., McTaggart I.P. and Tsuruta H. (1997). Emissions of N_2O and NO associated with nitrogen fertilization in intensive agriculture, and the potential for mitigation. Soil Use and Management, 13, 296-304.

Missing N: is the solution in dissolved N?

Report by A.J. Macdonald[1] and D.L. Jones[2]

[1]*Rothamsted Research, Agriculture and the Environment Division, Harpenden, AL5 2JQ, UK*
[2]*University of Wales, School of Agricultural and Forest Sciences, Bangor, LL57 2UW, UK*

Abstract

A themed working group was convened as part of the 12th Nitrogen Workshop in September of 2003. The working group discussions focused primarily on the extent to which transformations and losses of soluble N can account for the missing N in balance studies. The main points arising from these discussions indicated that:

1. Of the N inputs to agricultural systems, between 20 and 70% may be unaccounted for.
2. Nitrate leaching and gaseous N losses account for a significant proportion of the missing N. In some systems, increases in soil organic N may account for some of the missing N.
3. Losses of Dissolved Organic N (DON) account for 4-29% of the N leached in intensive agricultural systems and up to 90% in some low intensity agricultural systems.
4. Some N loss processes may be operating that we are not currently monitoring, e.g. gaseous emissions of NH_3 and volatile organic N (VON) from plants, and N_2O losses in super-saturated drainage water.
5. Some participants considered that, although there may be other small N loss processes operating, our focus should still remain on managing the NO_3^- problem effectively.
6. Models of soil N dynamics require further modifications to accurately integrate processes occurring at different scales.
7. A critical examination of models together with a comprehensive review of the literature is required to help prioritise future areas of research on the fate of N in the environment.

Keywords: N enrichment, N balance, leaching, denitrification, dissolved organic N (DON)

Introduction

A themed working group of ca. thirty people was convened as part of the 12th Nitrogen Workshop held at Exeter University in September of 2003. Five of the workshop participants (K. Leach, J. Williams, P. Murray, S. Thomas and K. Zwart) were invited to present briefly some of their work and highlight relevant discussion points. Each presentation was followed by some discussion and there was further dialogue at the end of the meeting. One of the main objectives of this working group was to evaluate the extent to which transformations and losses of dissolved N could account for the N lost from terrestrial ecosystems. The purpose of this paper is to provide a summary of these discussions and the work that was presented together with other relevant information from the literature.

Missing N: is it a problem?

In a recent review of the impact of human activity on the global N cycle, Jenkinson (2001) reported that in 1996-97 some 160 million tonnes of N per year was fixed as a result of global industrial and agricultural activity. Of this, 98 million tonnes was fixed by the Haber-Bosch process, 40 million tonnes was a result of the cultivation of legumes (soybeans, groundnuts, forage clovers etc.) and another 22 million tonnes was fixed during combustion processes, mainly for the generation of energy. This is substantially more than the 90-130 million tonnes N yr^{-1} fixed by biological processes in pre-industrial times. The increased amounts of N fixed by human activity have, in turn, increased both the amounts of chemically combined N entering ground and surface waters and the atmospheric concentrations of NH_3, N_2O, NO and NO_2. Much of the N fixed by the Haber-Bosch process is used as agricultural fertiliser in developed countries and industrial regions. However, it is becoming increasingly apparent that to sustain the projected population increases in the developing world, especially Asia, the use of synthetic fertiliser N on a global scale will almost certainly increase (Jenkinson, 2001). This, together with the continuing industrialisation of the developing world, indicates that global N enrichment may well continue to increase for the foreseeable future. Increasing N inputs to natural, semi-natural and agro-ecosystems are of concern, because amongst other things, they have been shown to have significant effects on the pH and nutrient status of both soil and water (Hornung & Langan, 1999). In the longer-term, they can also have significant effects on the abundance and diversity of plant and animal species (Yeo & Blackstock, 2002). In addition, N_2O is a potent greenhouse gas and NO_x (NO and NO_2) can contribute to enhanced tropospheric ozone (Dalal et al., 2003).

N Balance Studies

Much research has been devoted to the construction of N balances to identify the major N inputs and losses for both agricultural and extensively managed ecosystems. Jenkinson (1990) reported that of the annual N inputs in fertiliser (189 kg N ha^{-1}) and other external N inputs (about 50 kg N ha^{-1}) to winter wheat on the Broadbalk Wheat Experiment at Rothamsted, ca. 29% was lost from the crop-soil system. Five main potential pathways of N loss were identified, including losses in drainage (mainly NO_3^- leaching), losses of organically bound N (by erosion), NH_3 volatilisation, gaseous N losses of N_2O and N_2 during bacterial denitrification, and NO, NO_2 and N_2O during bacterial nitrification. Losses of fertiliser N applied in spring to winter wheat on Broadbalk were thought to be largely a result of gaseous N loss.

Leach et al. (2003) calculated annual farm gate N balances over 7 years on a mixed farm (arable, sheep and dairy) in the Cotswolds using estimates of N inputs and outputs from farm records. Leaching losses of mineral N and dissolved organic N were estimated using porous cups. Gaseous losses were modelled using the UK ammonia inventory and the DNDC model. On average, N use efficiency was 45.6% and only 6.3% of N inputs were unaccounted for. In all years, leaching accounted for most of the N losses. There was an indication of an inverse relationship between missing N and winter rainfall, indicating that N may accumulate in soil in a dry year and may be lost in a wet year. However, the relationship between inputs and outputs was not significantly different from 1:1 (i.e. all N was accounted for).

N Loss pathways

Nitrate leaching
Much is already known about the factors which contribute to NO_3^- losses from agricultural land. These include excessive fertiliser and manure applications to arable crops and grasslands, leaving

soil bare over winter when drainage (and hence leaching) is greatest, taking insufficient account of the N supply from manures when calculating fertiliser application rates and the applications of slurries and poultry manures in autumn (Shepherd, 2003). Whilst the excessive use of mineral fertiliser in arable cropping systems may contribute to NO_3^- leaching, especially where crop growth and N uptake is limited by drought (Glendining et al., 1992) or disease, Macdonald et al. (1997) reported that the majority of the NO_3^- at risk to leaching in the winter (following harvest of winter wheat), was derived from the mineralisation of soil organic N. Consequently, even with the best agronomic practices, it may be difficult to reverse the trend of increasing NO_3^- concentrations in the drier arable areas of Eastern England (Shepherd, 2003). The use of animal manures (especially slurries) in arable cropping systems, provides another potential source of NO_3^- which may be leached. Williams et al. (2003) reported that N losses in drainage were lower for winter applications of pig slurry compared with autumn applications, largely because of the smaller amount of drainage after the later applications. In addition, N losses from autumn ploughed land in the winter of 1999/2000 were 75% less than from uncultivated stubble. Autumn ploughing may sever cracks and fissures in the soil profile which permit drainage to pass from the surface into the sub-soil (Goss et al., 1990). Consequently, NO_3^- leaching, following autumn applications of slurry to crops established by direct drilling, may be greater than those under cultivated land.

Loss of DON - Natural, semi-natural and agro-ecosystems

The importance of soluble organic N (SON) and its contribution to DON in soil solution and losses in drainage has received relatively little attention, compared with studies on the effects of land use and management on the soil mineral N pool. This may, in part, be because of its limited importance for plant nutrition, particularly in high input agro-ecosystems (Owen & Jones, 2001). However, Macdonald et al., (2003) reported that 2 M KCl extractable SON in soils (0-15cm) under seven different land use classes in Britain (including tilled land, managed grass, rough grass, heathland, wetland, broad-leaved and coniferous woodland), ranged from 10 to 37 kg N ha^{-1}. Arable soils contained less SON compared with that found under grassland and woodland systems, presumably because of the greater accumulation of soil organic matter under less physically disturbed systems and the poorer quality of plant residue inputs (i.e. high C:N ratio and lignin/tannin content) under these systems. However, the amount of SON measured was dependent on the methodology used. Substantially more SON was extracted from soil with 2 M KCl compared with water, perhaps indicating that a significant proportion of the SON was physically and chemically adsorbed. Similarly, Jones et al., (2002) acknowledged the importance of methodology when assessing DON in soil solution.

Macdonald et al., (2003) reported that losses of DON in drainage under continuous winter wheat accounted for 4-28% of total soluble N leached. Both mineral N (NO_3^- and NH_4^+) and DON losses were greatest following long-term inputs of farmyard manure. Similarly, Williams et al., (2003) reported that DON accounted for 10-29% of the soluble N lost in drainage from arable land receiving pig slurry. Leach et al., (2003) reported that losses of DON from arable land and grassland accounted for 6 and 10% of the total N leached, respectively. Therefore, leaching losses of DON may contribute significantly to N losses from arable land, especially following the application of slurry or repeated applications of organic manures. It was concluded that measurement of SON and DON should be included more frequently in N balance studies. Although it is clear that DON is an important loss pathway in some agricultural systems, the nature of this DON is still poorly understood. To date, there have been few studies investigating the chemical nature of DON. While this may not be important from an N budget perspective, it does have implications on the downstream impacts of DON losses. In particular, whether DON contributes to eutrophication and whether it should be viewed as a pollutant remains unknown. While it has been speculated that DON is relatively recalcitrant in soil, exposure to light and subsequent photo-oxidation may induce the release of highly bio-available inorganic N (Anesio & Graneli, 2003). Even if DON does not contribute to pollution, it can be expected to have a significant impact on freshwater stream ecology (Biggs, 2000).

Gaseous N losses

Missing N in balance studies is often attributed to gaseous N losses during bacterial denitrification and nitrification. Denitrification occurs in soil in the presence of NO_3^- under anaerobic conditions, whilst nitrification is largely an aerobic process. Both processes are highly spatially and temporally variable, but it is thought that under some conditions they may occur simultaneously in different micro-sites within the soil (Jenkinson, 2001). Thomas et al. (2003) investigated subsoil denitrification of ^{15}N-labelled NO_3^- from poorly drained soil cores after the application of simulated rainfall. Gaseous losses of ^{15}N were monitored, as were changes in soil solution mineral N and water soluble C. Of the ^{15}N applied, 10-20% was lost as N_2 and N_2O from the soil surface after 300 days. In addition, 0-14% of the labelled N leached was lost as N_2O. It was concluded that significant N losses could occur in poorly drained soils as a result of subsoil denitrification, but that it was limited mainly by available C. Murray et al., (2003) also reported significant increases in gaseous loss as $^{15}N_2O$ and ^{15}NO following applications of ^{15}N-labelled urea and $K^{15}NO_3$ to hydrologically isolated field plots. However, Zwart (2003) reported that potential denitrification decreased sharply below 0.6 and 0.4 m for grassland and arable land, respectively. He concluded that denitrification below 0.4-0.6 m contributed little to N losses unless biodegradable soil organic matter was present in addition to NO_3^-.

The potential for new pathways of N_2O loss were also highlighted. Recent work has shown that significant amounts of N_2O may be lost in drainage which feed directly to freshwater (Reay et al., 2003). The N_2O contained within this often super-saturated drainage water is quickly lost once it enters the stream (within a few metres) and is consequently rarely measured. The impact of this loss pathway on overall N budgets remains unknown, but warrants further research. In addition, although the adsorption of inorganic N from the atmosphere by plants is well documented (Herrmann et al., 2002) losses of both inorganic and organic N by plants is poorly understood. The potential for the bi-directional exchange of volatile organic N (VON) between the atmosphere and the earth's surface has been illustrated recently by a study of amino acids found in dew deposited on a PTFE sheet mounted horizontally above a grass field (Scheller, 2001). However, there have been few quantitative field studies to measure losses of (VON) from agricultural fields. Further work is required to assess this flux at different times of the growing season.

Problems of Scale

Of increasing importance is the development of mitigation measures to prevent the negative impacts of excessive N loading within the environment. To achieve this, however, requires that we understand N flows at a range of spatial scales. While our process level understanding of

N dynamics at the soil profile level is good, our knowledge of N flows at the landscape level is relatively poor. For example, while N leaching within a soil profile can be described reasonably well with current mathematical models, predicting the flow of NO_3^- or DON at a large landscape level remains difficult, particularly in complex land use environments (Ren et al., 2003). Further work is therefore required to gain a better understanding of N behaviour during the flow of groundwater to surface waters.

Conclusions

It was concluded that it is still not possible to quantify all N fluxes with complete accuracy, largely because of the great spatial and temporal variability associated with some of these processes (e.g. denitrification). Future technological developments which can help overcome these limitations, such as continuous flux monitoring systems, may help to improve our N balances. Whilst denitrification is often thought to account for much of the missing N in N balance studies, it does not appear to be able to account for all of it. Therefore, other N loss pathways/processes may be operating that we are either unaware of, or are not measuring accurately, e.g. N_2O losses in super saturated drainage water, or gaseous N losses from plants. A critical examination of current models to identify knowledge gaps, together with a comprehensive literature review of N loss processes, would help prioritise future areas of research on terrestrial N cycling.

Acknowledgements

Rothamsted Research receives grant-aided support from the UK Biotechnology and Biological Sciences Research Council. This work was funded in part by the UK Natural Environment Research Council and the Department for Environment Food and Rural Affairs.

References

Anesio, A.M. and Graneli, W. (2003). Increased photoreactivity of DOC by acidification: Implications for the carbon cycle in humic lakes. Limnology and Oceanography, 48, 735-744.

Biggs, B.J.F. (2000). Eutrophication of streams and rivers: dissolved nutrient-chlorophyll relationships for benthic algae. Journal of the North American Benthological Society, 19, 17-31.

Dalal, R.C., Wang, W.J., Robertson, G.P. and Parton, W.J. (2003). Nitrous oxide emission from Australian agricultural lands and mitigation options: a review. Australian Journal of Soil Research, 41, 165-195.

Glendining, M.J., Poulton, P.R. and Powlson, D.S. (1992). The relationship between inorganic N in soil and the rate of fertilizer N applied on the Broadbalk Wheat Experiment. Aspects of Applied Biology, 30, 95-102.

Goss, M.J., Howse, K.R., Harris, G.L. and Colbourn, P. (1990). The leaching of nitrates after spring fertilizer application and the influence of tillage. In: R. Merckx, H. Vereecken and K. Vlassak (eds). Fertilisation and the Environment. pp 20-25. Leuven University Press, Leuven, Belgium.

Herrmann, B., Mattsson, M., Fuhrer, J. and Schjoerring, J.K. (2002). Leaf-atmosphere NH_3 exchange of white clover (*Trifolium repens* L.) in relation to mineral N nutrition and symbiotic N_2 fixation. Journal of Experimental Botany, 53, 139-146.

Hornung, M. and Langan, S.J. (1999). Nitrogen Deposition: Sources, impacts and responses in natural and semi-natural ecosystems. In: S.J. Langan (ed) The Impact of Nitrogen Deposition on Natural and Semi-Natural Ecosystems., Kluwer Academic Publishers. Dordrecht, The Netherlands.

Jenkinson, D.S. (1990). Leaks in the Nitrogen Cycle. In: R. Merckx, H. Vereecken and K. Vlassak (eds). Fertilisation and the environment, 35-49. Leuven University Press, Leuven, Belgium.

Jenkinson, D.S. (2001). The impact of humans on the nitrogen cycle, with focus on temperate arable agriculture. Plant and Soil, 228: 3-15.

Jones, D.L., Owen, A.G. and Farrar, J.F. (2002). Simple method to enable the high resolution determination of total free amino acids in soil solutions and soil extracts. Soil Biology and Biochemistry, 34, 1893-1902.

Leach, K.A., Goulding, K.W.T., Hatch, D.J., Conway, J.S. and Allingham, K. (2004). Nitrogen balances over 7 years on a mixed farm in the Cotswolds. (This volume).

Macdonald, A.J., Poulton, P.R., Powlson, D.S. and Jenkinson, D.S. (1997). Effects of season, soil type and cropping on recoveries, residues and losses of ^{15}N-labelled fertilizer applied to arable crops in spring. Journal of Agricultural Science, Cambridge, 129, 125-154.

Macdonald, A.J., Francis, S.M.J., Stockdale, E.A., Goulding, K.W.T., Willett, V.B., Jones, D.L., Baddeley, J.A., the late Green, J.J., Watson, C.A., Saunders, G. and Cadisch, G. (2003). Effects of land use and management on soluble organic N in UK soils. In: Abstracts of the 12th N Workshop, 21st-24th September, 2003, Exeter, Devon, UK.

Murray, P.J, Dixon, E.R.., Hatch, D.J., Granger, S.J., Laughlin, R.J. and Stevens, R.J. (2004).Using a system of undisturbed, in situ lysimeters to determine nitrogen transformations in a sub-surface clay interface. (This volume).

Owen, A.G. and Jones, D.L. (2001). Competition for amino acids between wheat roots and rhizosphere microorganisms and the role of amino acids in plant N acquisition. Soil Biology and Biochemistry, 33, 651-657.

Reay, D.S., Smith, K.A. and Edwards, A.C. (2003). Nitrous oxide emission from agricultural drainage waters. Global Change Biology, 9, 195-203.

Ren, L., Ma, J.H. and Zhang, R.D. (2003). Estimating nitrate leaching with a transfer function model incorporating net mineralization and uptake of nitrogen. Journal of Environmental Quality, 32, 1455-1463.

Scheller, E. (2001). Amino acids in dew-origin and seasonal variation. Atmospheric Environment, 35, 2179-2192.

Shepherd, M.A. (2004). Controlling nitrogen losses to water - where next? (This volume).

Thomas, S., Clough, T.J., Francis, G.S., Sherlock, R.R. and Beare, M.H. (2004). Seasonal subsoil denitrification of leached ^{15}N-labelled nitrate. (This volume).

Williams, J.R., Chambers, B.J., Cross, R.B. and Hodgkinson, R.A. (2004). Nitrogen losses in drainage water following pig slurry application to an arable clay soil (2003). (This volume).

Yeo, M.J.M. and Blackstock, T.H. (2002). A vegetation analysis of the pastoral landscapes of upland Wales, UK. Journal of Vegetation Science, 13, 803-816.

Zwart, K. (2004). Denitrification in top soil and sub soil, data and model results. (This volume).

Pollution problems: mitigation or are we swapping one form of pollution for another?

Report by B.J. Chambers[1] and O. Oenema[2]
[1]*ADAS Gleadthorpe Research Centre, Meden Vale, Mansfield, Notts. NG20 9PF*
[2]*Alterra, Wageningen University and Research Centre, P.P. Box 47, 6700 AA Wageningen, The Netherlands*

Introduction

Fertiliser and organic manure N additions, biological N fixation, atmospheric deposition and the mineralisation of soil organic N reserves can all lead to N emissions to the air and water environments (Figure 1). This working group debated the potential for creating alternative problems when mitigations options are employed for one particular emission issue, i.e. pollution swapping.

Figure 1. Nitrogen emissions to the air and water environments.

Nitrogen loss mitigation

The Working Group members were asked to assess the N loss form(s) that was their country's main target(s) for mitigation, viz: score 1: high priority; score 2: secondary priority and score 3: low priority.

The priorities of the Working Group members are summarised in Table 1 and provided the following clear messages:
- Nitrate loss mitigation was a high priority for every country (overall score 1.0), reflecting the worldwide importance of the 50 mg NO_3^- l^{-1} limit in legislation (e.g. EU Nitrates Directive) and for drinking water quality (e.g. New Zealand and Japan)
- Ammonia emission mitigation (overall score 1.7) was seen overall as the second most important N loss priority, particularly in central and northern Europe (score 1 for all these countries)
- All other N loss forms had an overall score of >2.0

Table 1. Summary of main N mitigation targets for countries represented in Working Group

Country/Region	Air				Water			
	Ammonia	Nitrous oxide	Oxides of nitrogen	Di-nitrogen	Nitrate	Ammonium	Nitrite	Dissolved organic N
	(NH_3)	(N_2O)	(NO_x)	(N_2)	(NO_3^-)	(NH_4^+)	(NO_2^-)	(DON)
Belgium	2	2	3	3	1	3	3	3
Denmark	1	1	2	3	1	2	2	2
England & Wales	1	2	2	3	1	1	2	3
Finland	2	3	3	3	1	3	3	3
France	2	3	3	3	1	3	3	3
Germany	1	2	3	3	1	3	3	2
Ireland	2	2	3	3	1	3	3	3
Japan	3	2	2	2	1	1	1	1
New Zealand	3	2	3	3	1	3	3	3
Norway	1	1	2	3	1	3	2	2
Pakistan	1	2	3	3	1	3	3	3
Portugal	2	2	3	3	1	3	3	3
Scotland	2	1	3	3	1	2	3	3
Spain	2	3	3	3	1	3	3	3
Sweden	1	3	3	3	1	2	3	3
Average	**1.7**	**2.1**	**2.7**	**2.9**	**1.0**	**2.5**	**2.7**	**2.7**

Statements

The Working Group members were asked to discuss and debate the following statements:

1. *All N loss pathways are equally damaging?*

The Working Group felt that all N loss pathways represented a loss of N resources and that some N loss forms were more damaging than others, but there was no universal agreement on the ranking of damage from each N loss form. The importance placed on any particular N loss form depended upon the perspective of individuals and their research interests, and the environmental priorities and values of individual countries. However, it was possible to draw out three main N loss forms that were of concern, although it was not possible to rank these in any order of priority:
- N_2O, because of its powerful effect as a greenhouse gas. However, it was recognised that mitigation measures would take a long time to influence the rate of climate change.
- N deposition (wet and dry) on sensitive ecosystems, as a result of N losses for agriculture (e.g. NH_3) and human activities (e.g. power generation, cars, etc.)
- Water quality, in particular as a result of NO_3^- leaching losses and NH_4^+- N transfers to freshwater systems.

2. *'Pollution swapping' is largely a response to government policies that focus on one N loss form (e.g. NO_3^-, NH_3).*

The view of the Working Group was that N 'pollution swapping', as a result of N mitigation measures, was a common occurrence e.g.

- policies that require the ploughing of manure rapidly into the soil to minimise NH_3 emissions may exacerbate N_2O emissions and NO_3^- leaching losses
- policies that have closed spreading periods in the autumn to minimise NO_3^- leaching losses and promote spring application to growing crops may exacerbate NH_3 emissions.
- the use of no till systems to encourage C sequestration in arable soils (e.g. in the United States) may exacerbate N_2O and CH_4 emissions.
- the group felt that many of the potential 'pollution swapping' issues were a result of policy makers and researchers working separately on air and water emissions. The group felt that a holistic approach should be taken in N emissions research, that tackled air and water emissions together, and where appropriate, other loss forms (e.g. P, pathogens, etc). This would improve the scientific robustness of N loss studies and would allow the identification of 'win-win' situations and where 'trade-offs' had to be made. This ultimately would provide more integrated and informed policy advice for Government and Regulators.

3. *What is the best way to assess the success of mitigation measures?*

The group felt that it was important to quantify the success of any mitigation measures in terms of improvements in water quality (e.g. reduced NO_3^-- N concentrations) or air quality (e.g. N deposition to sensitive ecosystems).

At the farm gate and national level, the group felt that the N surplus approach had a great deal of merit. This indicator was sensitive to decreases in N inputs (e.g. fertiliser use, manure inputs etc.) and to increases in N utilisation efficiency per unit of output (e.g. N outputs in milk and meat products etc.)

Conclusions

The Working Group considered that 'pollution swapping' was a real issue that occurred largely as a result of narrow-focused policies and research work that considered air and water emissions separately, such that the possibilities for 'pollution swapping' were not always fully recognised. The group felt that there was a need for further integrated research to assess the effects of mitigation measures on the target N loss form alongside other N loss routes (plus other losses of concern), to ensure that potential 'pollution swapping' issues were addressed adequately. This would increase the scientific robustness of experimental work and would provide better and more informed advice to policy makers, so that they could more clearly identify and quantify 'win-win' and pollution 'trade-off' situations.

Systems studies; do we need them, or can they be replaced by desktop studies?

Report by H.F.M. Aarts[1] and A. Pflimlin[2]
[1] *Agrosystems Research, Plant Research International, Wageningen University and Research Centre, P.O. Box 16, 6700 AA Wageningen, The Netherlands*
[2] *Institut de l'Elevage, 149 rue de Bercy, 75595 Paris cedex 12, France*

Introduction

There is much that is known, but also much that is not, about the internal functioning of agricultural systems and their interactions with the environment. As agronomists, we are faced with two main challenges: 1) to illuminate hidden structures (a scientific challenge), and 2) to make use of what we already know, to synthesise proper systems or to manage systems in a way that performance is improved (an engineering challenge; Passioura, 1996). In all scientific disciplines, growth has been stimulated by the evolution and testing of hypotheses, sometimes driven by the development of a theory that stimulated new experimental work and sometimes by unexpected sets of measurements that provided a basis for a new theory. Until the 1960s, agricultural research almost completely relied upon experimental and empirical work, combined with statistical analyses. Though progress had been impressive, constraints and limitations to this type of research became more and more evident: location and time-specific results were difficult to generalise and extrapolate, and processes were often described (rather than explained) in terms of underlying processes, i.e. research was analytical rather than synthetic. Since the 1960s a wide range of scientists became involved in agro-ecological modelling. First, the goal was to understand the behaviour of a system (analysis), later also, the possibility to help to improve and to manage a system was recognised (synthesis).

Because modelling in the form of computer simulation developed into such a powerful and relatively cheap tool to analyse and to design, there is a danger that it will weaken the link between theory and real (as opposed to 'imaginary') experiments (Monteith, 1996). The cost of 'imaginary' experiments (desk top research) will further decrease, because of technological improvements, while the costs of real experiments are expected to grow, because of increased costs of labour. Excessive reliance on computer-experiments means that the check on reality is lost and that the absence of unexpected sets of measurements prevents finding new theories. On the other hand, too much effort on experiments will lead to high costs and low profits from results, because possibilities to analyse systems and to extrapolate results are very limited.

During the 12th N Workshop in Exeter, the participants of this working group focussed attention on research methodology by answering the question 'System studies: do we need them, or can they be replaced by desktop studies?' For a practical reason, the question was divided into sub-questions. First, the advantages and limitations of experimental systems and desktop studies (modelling) were gathered and discussed. Next, participants tried to indicate situations in which desk top studies or experimental studies were most desirable, should be avoided or should be combined. Subsequently, the working group discussed profitable ways of combining experimental and desktop activities as part of research methodology. Finally, attention was paid to the desirability and feasibility to build a tool to help researchers to define the most suitable research approach, including the role of experimental systems and desk top studies. This paper summarises the main opinions.

What are the main advantages and limitations of experimental system studies and desk top studies?

Desk top studies may be the only way to integrate the many processes that have to be studied in complex systems, as agricultural systems generally are. Models can help researchers to combine disciplinary research results and to test hypotheses regarding interactions at a system level. Modelling can also help to provide managers with a tool that assists in managing a crop, a farm, a watershed or even a whole country. Models are most important for high system levels (mostly large scale systems), for instance a watershed, because the number of processes involved is high and studying all relevant processes in such a system (in an experimental way) is impossible, e.g. from a financial point of view. For only a low cost, a high number of imaginary situations can be calculated. A problem is that models are based on assumptions that should be derived from previous experiments. If the underlying processes of a system are misunderstood, model results will not reflect reality. If a model is complex, relations between inputs and results are difficult to understand. For this reason, and because results cannot be demonstrated in practice, acceptance by some categories of 'customers' can be restricted.

Experimental system research is very costly, and can only be used to study small systems, such as fields or farmlets. Weather conditions have a high influence on outcomes. If a system is large, only parts (sub-systems or selected processes) can be studied extensively. On the other hand, no important aspect can be kept out of research, because the results of experiments will always reflect reality. Reality of data can make it difficult to explain system results and therefore, unexpected results can force us to adapt theories about involved processes.

Table 1. Main advantages and limitations of experimental system studies and desk top system studies (modelling)

	Strong points	Weak points
Desk top system research	• Low costs • Low risks (controlled conditions) • Not limited by system scale or local constraints	• Reality of results • Acceptance of results
Experimental system research	• Reality of results • Integration of all aspects • Possibility of communicating with farmers and other users of results, by demonstrating the function of the system in practice	• High costs • Only small and simple systems or parts of larger systems can be studied • Results depend on weather conditions • Biosecurity and management risks • Interpratation of data

At what system level and under what conditions should experimental system research and desk top system research be preferred or combined?

The field or crop (rotation)

Typically, an agro-ecological system is an independent unit, influenced by farmers (as part of the system) to improve performance by the increase in benefits or the reduction of damage. Different system scales can be distinguished. The smallest scale is the field, with a crop rotation. Crops can be managed in different ways and results also depend on specific soil and climate conditions. Research on field level systems will be dominated by experimental work. It is easy to make adaptations, to study effects on performance, such as changes in fertilisation or grazing intensity. It is important that effects on performance can be demonstrated to the people that are thought to profit from the results: 'It is impossible to convince farmers only with books and journals. The truth should be forced into their heads by clear visible examples' (Van den Elsen, 1918) (father Gerlachus van den Elsen founded the biggest Dutch farmers union, the Rabo-bank and an infrastructure to disperse agricultural knowledge). For a farmer, the field and barn are the preferred venues for exchange of management information. However, that might change in the future, because the new generation will grow up with computers, maybe even more than with crops or cows. At the field scales, models can be used to explore ways of improving management, and in that way they are very useful to plan experiments and to forecast experimental results. Models can help to analyse actual results and, therefore, can draw attention to gaps in understanding and thereby stimulate new experimental or theoretical work.

In principle, models can be used to capture knowledge in a management tool. However, there are only a few examples of successful applications of models as a management tool at the field scale, even when the models have been specially tailored for use by farmers or extension personnel. Maybe, such crop models would be more acceptable if modellers make them more transparent (as well as more robust) by making more effort to simplify their models by removing components that contribute noise rather than numerical precision to the final output (Monteith, 1996). Most crop management models were primarily designed for detailed analytical purposes, to adsorb knowledge, rather than to disperse knowledge.

The farm

Compared with the field or crop scale, at the farm scale it is more difficult to restrict research to experiments only. A farm is unique (therefore replications are not possible) and, especially if livestock are involved, a lot of processes are relevant and important interactions between farm components exist. The farmer, with his personal interests and skills, is an important part of the system and, therefore, also has a high influence on farm performance. The difficulty to replicate and the complexity of each farming system forces us to increase the role of modelling in research. However, the acceptance of research results by farmers, relies to a large extent on the possibility to demonstrate the functioning of the system in real practice, which forces us to do experimental research.

A combination of the strong point of modelling and experimenting, in a methodology called 'prototyping', can be useful if sustainable livestock farming systems are to be developed and introduced in practice (Aarts, 2000). Simple, transparent models have to be used to perform calculations for a wide range of situations with respect to the individual farm components (livestock, manure, soils and crops) and their management. Outcomes are used for the design of farming systems that, in theory, meet the quantitative objectives of the farmer and of governmental rules. From this set of systems, the system judged as most interesting from a research point of view, is implemented on an experimental farm. Its performance can be monitored and the system further developed, by comparing results with the desired targets. With the results, underlying models are improved and used to extrapolate, by assuming certain conditions. Pilot commercial farms play an important role in demonstrating the reliability of the models to practical farmers and they provide more insight in the role of the farmer (skills, goals) and farming conditions (soil type, milk quota) on farm performance. Therefore, pilot farms also have to be monitored and analysed, but less intensively compared with the experimental farm.

Advantages of this research method are that attention has to be paid to the whole chain, from raw materials to end products and that connections among the various components of the farm cannot be neglected. By bridging the gap between theory and experimental testing on an experimental farm, it is possible to avoid desk work that is devoted to elements with little practical relevance, while insufficient attention is paid to the more important aspects. Transfer of knowledge is promoted by allowing people to visit the experimental farm and pilot commercial farms and to discuss the performance of the systems with the farm staff. The prototype system on the experimental farm should not be copied by the visiting farmers, but must stimulate farmers into thinking about the functioning of their own farm in a more scientific way. With the prototype system as an example and with the help of its scientific staff, this should lead to farm specific ideas for improvement. Next, the farmer can visit pilot commercial farms in his region, to discuss his/her experiences for improvements, on a soil type and at a farm intensity that are comparable to his/her own. For the farmer this is a safe way for farm development. He/she understands the background and the effects of each step taken. The number and size of these steps, depend on personal feelings about needs and risks. The farmer can always find help (by returning to the experimental farm or to one of the pilot commercial farms) for further improvements, or if expected improvements are not realised. The distrust of farmers on the outcome of models can be explained by the fact that a wrong decision in farm development can be catastrophic. The effects of decisions and their driving forces therefore should be very clear; in general, more clarity is needed than models currently provide.

Prototyping has its drawbacks, however. The experimental system, implemented on an experimental farm, is selected from various options, and there is no possibility of comparing different systems: only a single system is put into practice and that system continuously develops. Therefore, results cannot usually be tested statistically. These disadvantages can be overcome, in part, by very extensive and specific measurements, by involving experts to explain the results, and if necessary, through comparative research of farm elements with comparable elements on farms elsewhere. Besides, key questions relating to farm components can be answered by comparing farm components on a smaller scale in a conventional way, making it possible to analyse results statistically. For instance, the ratio of grass and maize area of the experimental farming system can be compared on a small scale with alternatives (farmlets), taking effects on cattle diet and related manure production into account by simulating (modelling), or experimenting (groups of cows fed with diets depending on grass: maize ratio).

The watershed and country level
At the watershed scale, the system has a size and complexity that means that modelling will dominate research activities. The manager of a watershed has to take into account the performance of the farming systems that are part of the watershed. Performance depends on the goals and skills of farmers and on management restrictions (mainly regulations to protect environment). The system can be modelled technically by using the knowledge gathered at field and farm scales, but also, other relevant knowledge is needed, such as information about regional water flows or processes in the subsoil. Analyses to find the weakest points in the model can help to define the most profitable experiments, to improve descriptions of processes, or to test model behaviour (effects of regulations).

Much more than for field or farm scale, model results are likely to be accepted by the manager of the system: (mainly the government or an organisation that is strongly related to the government) who will accept a model as the best tool to explore long-term consequences of policy objectives for rural areas. Such managers are used to basing decisions mainly on written information, not on 'pictures', 'conversations' or 'feelings', as do farmers. Therefore at this level, models can strongly support decision making. At country scales, models are indispensable for forecasting and evaluating the effects of regulations. Experimental research, in general, is restricted to monitoring to provide the model with data to validate.

Table 2. Intensity of the use of experimental research and of desk top research, at different system levels (+ = low intensity, ++++ = high intensity).

System level	Experimental system research	Desk top system research
Field	++++	+
Farm	+++	++
Watershed	++	+++
Country/planet	+	++++

Is it desirable to prescribe a research method?

For a researcher involved in agro-ecological system research, it is most important to be familiar with the system, not only from a technical point of view, but also with the social aspects, if people are also part of the system. It is necessary to detect the main technical and economical problems that should be solved, and to define potential solutions, taking cultural aspects into account (including traditions, acceptance of risks, importance of short term versus long term profits). Furthermore, the researcher is usually part of a group with its own culture and history, with individuals that differ in knowledge, skills and interest. In fact he/she is part of a research system with common and individual interests.

The optimal research procedure depends on the system that has to be studied, the questions, available or unavailable tools and the characteristics of the customer and of the group of scientists. For such complex situations, no standard can be prescribed. However, before starting system research, a few points should be discussed: what is the scale of the system? what are the key questions to be answered?, who is the user of the results?, what expertise and tools do we need?, what expertise and tools do we have and how can we deal with shortcomings? At farm system scales, the combination of farm models, experimental farms and pilot farms can be optimal, but how to act if one of these tools is missing? For instance, models can be applied directly on pilot farms, without preliminary research on an experimental farm, if people are aware of the increased risks of failures. Scientists can learn from experiences (success and failures) of groups of colleagues in other countries, with comparable objectives. For that purpose, a Farming System Group was recently formed as a working group within the European Grassland Federation.

References

Aarts, H.F.M. (2000). Resource management in a 'De Marke' dairy farming system. PhD Thesis, Wageningen University, The Netherlands, 222 p.

Monteith, J.L. (1996). The quest for balance in crop modelling. Agronomy Journal, 88, 695-697.

Passioura, J.B. (1996). Simulation models: science, snake oil, education or engineering? Agronomy Journal, 88, 690-694

Van den Elsen, G. (1918). Sociology of farmers, Abbey Berne, Heeswijk, The Netherlands. 631 p.

Model answers: can we improve their level of confidence and applicability?

Report by S.F. Ledgard[1] and N.J. Hutchings[2]
[1]*AgResearch Ruakura Research Centre, Private Bag 3123, Hamilton, New Zealand*
[2]*Danish Institute of Agricultural Sciences, P.O. Box 50, DK-8830 Tjele, Denmark*

Introduction

There is a wide range of models of N cycling and loss from agricultural systems. Table 1 summarises some of the models used by researchers in papers at the 12th N Workshop. They cover a range of scales from soil core through to national inventories. Some are for specific N processes, while others attempt to cover the whole N cycle with estimates of all N losses. The models fall into distinct categories of mechanistic 'science' models and decision support system (DSS) models and were discussed at this themed working group session at the Workshop.

The mechanistic models improve understanding through integration of detailed components of processes as affected by estimates and soil conditions. The main application of these science models is to assess the interactions between a range of variables and the effects of different management practices. The DSS models generally target improving the management of N on the farm, particularly fertiliser or manure use, to reduce environmental impacts.

Decision Support System N models for farmers

Key aspects in the development of DSS models include:
1. Having a clearly defined purpose.
2. Involvement of farmers and advisors at all stages.
3. Being as simple as possible.
4. All input requirements should be based on readily-obtainable and reasonably reliable information.
5. Outputs should be useful and easily understandable.
6. Information should be provided on use of the DSS (e.g. help system) and interpretation of outputs.
7. There should be the capacity to assess a range of N management options in models that have an environmental focus.

The involvement of farmer and/or advisors throughout the development of the DSS will help to ensure that points 4 and 5 are fulfilled and that the resulting tool is well focussed on the problem that it aims to address. These conclusions back up those found by Parker & Sinclair (2001).

The complexity of any model or models that lie within the DSS is a balance. The basis for more complexity is the need to reflect the underlying biological and physical processes, whilst that for simplicity is the need to avoid demanding inputs that are numerous, difficult to obtain or subject to large errors, difficulties in obtaining reliable parameter values and the increase in coding errors that occurs with complex programmes.

Table 1. Range of nitrogen models used in papers at the 12th Nitrogen Workshop.

SCALE	N process				Model application		Decision Support for	
	Leaching	N$_2$O, denitrification	NH$_3$	Mineralisation	Science or Inventory	Decision Support	Mitigation	Fert Advice
Core/plot/lab	SOIL- SOIL N LIXIM			LIXIM	SOIL- SOIL N LIXIM			
Paddock	DAISY STICS AZOFERT SOIL N MANNER ANIMO SUNDIAL	AZOFERT ANIMO SUNDIAL	AZOFERT SOIL N MANNER ANIMO SUNDIAL	ICBM/N MANNER SUNDIAL	DAISY STICS ICBM/N SOIL N ANIMO SUNDIAL	AZOFERT MANNER		AZOFERT MANNER
Farm	FASSET Your farm & NVZ OVERSEER NGAUGE	DNDC FASSET OVERSEER NGAUGE	NGAUGE MAST	NDICEA FASSET	NDICEA DNDC FASSET	Your farm & NVZ OVERSEER NGAUGE MAST	Your farm & NVZ OVERSEER NGAUGE MAST	
Catchment	CANDY ARLAS MAGPIE SUNDIAL	SUNDIAL	ARLAS	SUNDIAL	CANDY ARLAS SUNDIAL	MAGPIE	MAGPIE	
District/country	DNDC (UK)	DNDC (UK)	DYNAMO NARSES		DNDC (UK) DYNAMO NARSES			

Ideally, the DSS should be linked to an existing farm recording system to reduce duplication and increase ease of use of the model. Similarly, the use of fertiliser advice models can be enhanced by linkage to other soil testing alternatives (e.g. AZOFERT in France) or other tools used by the agricultural advisors.

The main dangers for scientific organisations who undertake development of DSSs are;
- The failure to understand the needs of the farmers, either because the farmers are never asked or because they are not understood. The latter can sometimes be avoided by including a moderator in the user group (i.e. someone trained in interpreting verbal communication).
- The tendency to add more complexity than is necessary or desirable, sometimes as a result of over-enthusiasm, but often from a failure by scientists to recognise the relative (un)importance of their particular field of research to farmers.
- The failure to recognise that the widespread uptake of DSSs requires their promotion and practical demonstration of their benefits (financial and/or environmental). Examples of DSSs that have achieved successful uptake include the UK's model MANNER, which targets efficient manure use, and NZ's OVERSEER nutrient budget model, which targets efficient whole-farm nutrient use for reduced environmental impacts.
- The failure to anticipate the demand for training and user support after release of the DSS.
- The failure to allow time for adequate documentation (especially when development uses staff on short-term contracts).
- The failure to undertake follow-up research in factors influencing effective use of the DSS by end users.

To be fair to scientific organisations, a number of the pitfalls associated with producing a successful DSS is often created by a lack of continuity or foresight by funding bodies; the focus of scientific funding bodies is usually on the quantity and quality of the science involved and not on the demands or needs of users. The latter may be left to another funding body, to commercial organisations or, not infrequently, to no one.

Decision support system N models for policy makers

Many points raised in the previous section apply to models developed for policy application, but the latter often target a scale larger than the farm. One consequence of moving to larger scales is that fewer data are usually available or data may be aggregated (e.g. to protect the privacy of individuals or companies). In systems dominated by linear relationships, the 'averaging' effect can remove the need for detail in the underlying models, leading to more simple models that demand fewer or more aggregated data as inputs. Unfortunately, the relationships that drive losses of N to the environment are frequently non-linear so such simplification has to be treated with care (Dalgaard et al., 2003).

Another consequence of moving to higher scales is that other processes begin to operate. For example, in areas designated as Nitrate Vulnerable Zones, manure may be exported or imported from farms. Changes to manure production, or in the capacity to utilise manure in one area may, therefore affect nutrient inputs in other areas (Hutchings et al., 2003). There is also increasing demand by policymakers for a qualitative or quantitative assessment of changes in N management on other nutrients or other policy areas, e.g. economic and social aspects.

Outputs from policy models should also include a distinction function (e.g. through use of sensitivity analyses) to identify uncertainty. Such functions can reduce 'over-interpretation' of model outputs.

References

Dalgaard, T., Hutchings, N.J. and Porter, J.R. (2004). Agroecology, scaling and interdisciplinarity. Agricultural Ecosystems and Environment (in press).

Hutchings, N.J., Dalgaard, T., Rasmussen, B.R., Hansen, J.F., Dahl, M., Rasmussen, P., Jørgensen, L.F., Ernstsen, V., von Platen-Hallermund, F. and Pedersen, S.S. (2004). Watershed Nitrogen Modelling (This volume).

Parker, C. and Sinclair, M. (2001). User-centred design does make a difference. The case of decision support systems in crop production. Behaviour & Information Technology, 20, 449-460.

Author index

Aarts, H.F.M. – 609
Abbaspour, K.C. – 417
Accoe, F. – 131
Aizpurua, A. – 212
Akhonzada, N.A. – 133
Alexandre, M. – 444
Alfaro, M.A. – 136
Allingham, K.D. – 39
Alonso, A. – 212
Alves, B.J.R. – 559
Amador-García, A. – 527
Amaral, E.B. – 504
Ambus, P. – 365
Anderson, M. – 309
Andrada, M.P. – 145
Anthony, S.A. – 573
Argyropoulos, G. – 437, 447
Arroyo-Sanz, J.M. – 229, 231
Artetxe, A. – 420
Askegaard, M. – 389
Aubrion, G. – 487
Baeta, J. – 145
Báez, D. – 285, 484
Baggs, E.M. – 73, 78, 90, 288, 290, 594
Bailey, G. – 513
Bailey, J.S. – 133
Bakken, A.K. – 523
Balík, J. – 148
Bálint, Á. – 138
Ball, B. – 268
Bañuls, J. – 222
Barlow, H.E. – 359
Bateman, E.J. – 290
Beare, M.H. – 357
Beaudoin, N. – 487
Beheydt, D. – 293
Bellamy, P. – 298
Beretta, G.P. – 449
Berntsen, J. – 389
Bhogal, A. – 63
Bioteau, T. – 54
Bittman, S. – 490
Björnsson, H. – 140

Blagodatsky, S. – 326
Blum, H. – 288
Boddey, R.M. – 559
Boeckx, P. – 131, 143, 151, 293, 428
Boels, D. – 397
Bol, R. – 164, 295, 351
Boldreghini, P. – 239
Bommelé, L.M. – 493
Bontemps, P.Y. – 415
Bontems, P. – 54
Bot, J. – 143
Bounaix, F. – 490
Breland, T.A. – 224, 523
Brown, F.A. – 70
Brown, L. – 298, 423, 510, 565, 567
Brussaard, L. – 101
Buchan, M. – 323
Buegger, F. – 197
Burgers, S.L.G.E. – 397
Butler, M.R. – 513
Butterbach-Bahl, K. – 326, 372
Černý, J. – 148
Cabral, F. – 156, 158, 236
Cadisch, G. – 78, 90, 205, 288, 290, 406, 581
Cameron, K.C. – 431
Cannavo, P. – 300
Cardenas, L. – 298
Cárdenas, L.M. – 303
Careaga, L. – 340
Carranca, C. – 145
Carton, O.T. – 316, 318
Casanova-Vigo, A. – 541
Castellón, A. – 212
Cerri, C.C. – 65
Chadwick, D. – 303
Chadwick, D.R. – 345, 363
Chambers, B.J. – 63, 363, 463, 573, 606
Chaves, B. – 151
Chen, N.C. – 496
Chisholm, E. – 461
Christensen, B.T. – 234
Christofides, C. – 351
Ciavatta, C. – 239
Clark, H. – 323

Clarke, A. – 513
Clough, T.J. – 357
Convertini, G. – 171, 194
Conway, J.S. – 39, 520
Cookson, W.R. – 101
Coppens, F. – 153
Cordovil, C.M.d.S. – 156, 158
Costa, M. – 145
Coutinho, J. – 156, 158, 168, 236, 285, 484, 551
Coxon, C.E. – 439, 539
Cross, R.B. – 463
Cuttle, S. – 182
Cuttle, S.P. – 498
Dahl, M. – 47
Dahlin, S. – 160
Dailey, A.G. – 85, 577
Dalgaard, T. – 47
Damay, N. – 500
Dampney, P.M.R. – 70
Davidson, E.A. – 251
De Blander, H. – 425
De Gruijter, J.J. – 397
De Klein, C.A.M. – 571
De Neve, S. – 151
De Ruijter, F.J. – 397
De Varennes, A. – 145
Decamps, C. – 425
Decrem, M. – 417
Dekker, P. – 459
Del Hierro, O. – 420
Del Prado, A. – 340, 420, 423, 565
Dela-Cruz, J. – 81
Delin, S. – 162
Demmers, T.G.M. – 306
Dendooven, L. – 244
Deprez, B. – 425
Devillers, C. – 415
Dhondt, K. – 428
Di, H.J. – 431
Díez, J.A. – 367
Dise, N.B. – 88
Dittert, K. – 164, 295, 331, 336, 549
Dixon, E.R. – 347, 349
Doledec, A.F. – 209
Dorsainvil, F. – 444
Dubrulle, P. – 500

Edwards, A.C. – 354
Eltun, R. – 523
Engström, L. – 166
Ericsson, L. – 166
Eriksen, J. – 434
Ernstsen, V. – 47
Esala, M. – 184, 224
Estavillo, J.M – 212, 340
Fanning, A. – 316
Farrar, J.F. – 441
Farthofer, R. – 67
Fawcett, C.P. – 70
Fernandes, A. – 168
Ferri, D. – 171
Fillery, I.R.P. – 101
Finlayson, J.D. – 76
Flavel, T.C. – 502
Flura, D. – 311
Forge, T. – 490
Fornaro, F. – 194
Forrester, T. – 513
Francis, G.S. – 357, 359, 361
Freyer, B. – 67
Friedel, J.K. – 67
Friesen, A. – 490
Fumagalli, L. – 449
Gallardo, M. – 454
García-Torres, L. – 367
Garnier, P. – 153
Garot, T. – 93
Geers, L. – 293
Génermont, S. – 311
Georgiou, T. – 437, 447
Giacomini, S.J. – 504
Gibbons, M.M. – 70, 573
Gimenez, C. – 454
Gioacchini, P. – 239
Glegg, G. – 81
Glendining, M.J. – 85, 577
Godoy, R. – 143
González-Murua, C. – 340
Goodlass, G. – 182
Goossens, A. – 293
Goulding, K.W.T. – 39, 298, 369, 567
Gowing, D.J. – 88
Granger, S.J. – 349

Green, E.R. – 73
Groenendijk, P. – 575
Guidini, E. – 504
Haberle, J. – 173
Hackett, R. – 439
Hack-Ten Broeke, M.J.D. – 397
Hansen, J.F. – 47
Hansen, M.N. – 295
Haroulis, A. – 437, 447
Harrison, R. – 363, 594
Hartwig, U.A. – 288
Hatch, D.J. – 39, 182, 347, 349
Hawkins, J.M.B. – 303
Hawkins, M. – 316
Haygarth, P. – 81
Hedderley, D.I. – 357, 359
Heinen, M. – 314
Heltai, Gy. – 138
Hénault, C. – 311
Hernáiz, P. – 367
Herre, C. – 209, 214
Herrmann, A. – 113
Herzog, F. – 417
Heß, J. – 197
Hodgkinson, R.A. – 463
Hofman, G. – 131, 151, 203, 428, 557
Hojito, M. – 342, 507
Hollis, J. – 298
Hooker, K.V. – 439
Hoving, I.E. – 397, 554
Hoyle, F.C. – 101
Hunt, D.E. – 490
Hutchings, N.J. – 47, 615
Hyde, B. – 316
Hyde, B.P. – 318
Ikeguchi, A. – 507
Inanaga, S. – 496
Ineson, P. – 345
Isselstein, J. – 547
Jadczyszyn, T. – 176
Jakobsson, C. – 29
Jarvis, S.C. – 298, 347, 349, 567, 569
Jasper, J. – 188
Jensen, A. – 179
Jensen, E.S. – 197
Jensen, L.S. – 179, 224

Jeuffroy, M.H. – 191, 311
Jewkes, E.C. – 510, 513, 565, 569
Johns, K. – 571
Jollands, N.A. – 76
Jones, D.L. – 101, 441, 600
Jones, S. – 268
Jørgensen, L.F. – 47
Joynes, A. – 182
Jung, K. – 138
Justes, E. – 444
Justus, E. – 122
Kaku, K. – 507
Kanerva, T. – 321
Karhu, K. – 321
Karyotis, T. – 437, 447
Katsilouli, I. – 437, 447
Kätterer, T. – 113
Kay, R. – 306
Kelliher, F.M. – 323
Kelling, K.A. – 217
Kersebaum, K.C. – 207
Kesik, M. – 326, 372
Knoden, D. – 425
Kohyama, K. – 507
Koivisto, K. – 321
Kokkonen, A. – 184
Kondo, H. – 342
Koponen, H.T. – 329
Kowalenko, C.G. – 490
Krejčová, J. – 173
Kreye, C. – 331, 549
Kristensen, I.S. – 534
Kubiak, R. – 207
Kuikman, P.J. – 333
Kumar, K. – 186
Laanbroek, H.L. – 260
Lafolie, F. – 300
Lagoa, R. – 236
Lambert, R. – 415, 425, 516, 518
Lampe, C. – 336
Laplana, R. – 54
Lapworth, J. – 513
Lathwood, A. – 513
Lau, A.K. – 490
Laughlin, R.J. – 338, 347, 349
Laurent, F. – 487

Laville, P. – 311
Leach, K.A. – 39, 520
Lean, I.J. – 90
Ledgard, S.F. – 76, 323, 361, 525, 571, 615
Leech, P.K. – 577
Legaz, F. – 222
Li, C. – 372, 567
Limpinuntana, V. – 78
Lin, S. – 331, 549
Lindén, B. – 166
Link, A. – 188
Lockyer, D.R. – 565
Løes, A.K. – 523
Loiskandl, W. – 67
López-Fernández, S. – 367
Lord, E.I. – 381, 573
Lundström, C. – 224
Lunkes, A. – 504
Macdonald, A.J. – 600
Machet, J.M. – 191, 500
Macleod, C. – 81
Maiorana, M. – 194
Manninen, S. – 321
Mantovi, P. – 449
Marco, C.Di – 309
Marcoen, J.M. – 93
Martikainen, P.J. – 329
Mary, B. – 122, 191, 209, 214, 487
Master, Y. – 338
Matsunami, H. – 342
Matthews, R. – 363
Mayer, J. – 197
McGinn, S. – 490
McLaughlin, N. – 490
McLay, C.D.A – 525
McTaggart, I. – 268
Mee, L. – 81
Melillo, J.M. – 65
Menéndez, S. – 340
Menneer, J.C. – 525
Menzi, H. – 277
Merckx, R. – 153
Merino, P. – 340
Mészáros, Cs. – 138
Milford, C. – 309
Millon, F. – 214

Milton, N. – 101
Mishima, S. – 507
Misslebrook, T. – 318
Mitsimponas, Th. – 447
Mohimont, A.C. – 93
Monaghan, R.M. – 571
Montemurro, F. – 171, 194
Moreira, N. – 168, 285, 484, 551
Morgan, J.P. – 520
Mori, A. – 342
Morrissey, T. – 345
Morvan, T. – 200
Mosier, A.R. – 251
Mosquera-Losada, M.R. – 83, 527, 541, 544
Mountford, J.O. – 88
Mulier, A. – 203, 557
Munday, D. – 520
Murphy, D.V. – 101, 502
Murray, P.J. – 347. 349
Mutabaruka, R. – 205
Neeteson, J.J. – 29
Neill, C. – 65
Nendell, C. – 207
Neto, C. – 145
Nevens, F. – 493, 530, 532, 557
Nicholson, F.A. – 63, 363, 573
Nicolardot, B. – 191, 200, 209, 214, 300
Nieder, R. – 207
Nielsen, A.H. – 534
Nievergelt, J. – 417
Nótás, E. – 138
O'Prey, K. – 347
Oenema, J. – 397
Oenema, O. – 96, 260, 333, 457, 536, 606
Ogino, A. – 507
Olesen, J.E. – 389
Olfs, H.-W. – 188
Oliveira, A. – 145
Ortuzar, M.A. – 212
Oyarzún, C. – 143
Pain, B.F. – 520
Palmason, F. – 224
Pampulha, M. – 145
Panagopoulos, A. – 447
Panoras, A. – 447
Papen, H. – 326, 372

Pärnä, A. – 329
Parnaudeau, V. – 191, 214
Pateras, D. – 447
Patni, N. – 490
Pedersen, A. – 179, 224
Pedersen, S.S. – 47
Peeters, A. – 415, 425, 516, 518
Perraud, A. – 209
Petersen, B.M. – 534
Petersen, S.O. – 295, 351
Pfefferli, S. – 277
Pflimlin, A. – 609
Phillips, V.R. – 567
Piccolo, M.C. – 65
Pietrzak, S. – 536
Pietsch, G. – 67
Pinto, M. – 340, 420
Plume, H. – 323
Powell, J.M. – 217
Power, S.A. – 73
Prazeres, A. – 145
Primo-Millo, E. – 222
Promsakha Na Sakonnakhon, S. – 78
Quemada, M. – 219
Quiñones, A. – 222
Radersma, S. – 397
Rasche-Castillo, J. – 544
Rasmussen, B.M. – 47
Rasmussen, P. – 47
Reay, D.S. – 354
Recous, S. – 153, 191, 581
Rees, R. – 268
Regina, K. – 321
Reheul, D. – 493, 530, 532
Reidy, B. – 277
Renault, P. – 300
Reuter, S. – 207
Richards, K. – 439, 539
Richardson, S.J. – 70
Richaume, A. – 300
Richner, W. – 417
Richter, M. – 288
Rigueiro-Rodríguez, A. – 83, 527, 541, 544
Riley, H. – 523
Robert, P. – 214
Roderick, S. – 182

Rodríguez-Barreira, S. – 83
Roelsma, J. – 575
Rosen, C.J. – 186
Ross, C.A. – 569
Rotillon, G. – 54
Rowe, E. – 406
Rozbicki, J. – 226
Russell, K. – 513
Russelle, M.P. – 186
Ryan, M. – 316, 539
Rys, G. – 323
Saad, J.K. – 487
Salazar, F.J. – 136
Salo, T. – 224
Samborski, S. – 226
Sattelmacher, B. – 331, 336, 549
Scheringer, J. – 547
Schloter, M. – 197
Scholefield, D. – 81, 303, 342, 423, 469, 510, 513, 565, 569
Schoumans, O.F. – 575
Schröder, J.J. – 29, 397, 586
Scott, T.S. – 369
Shavit, U. – 338
Shaviv, A. – 338
Shepherd, M.A. – 381
Sherlock, R.R. – 323, 357
Shimada, K. – 507
Silvennoinen, H. – 329
Silvester, W.B. – 525
Skiba, U. – 268, 309
Smit, A. – 397, 452
Smith, J.U. – 85, 101, 577
Smith, K.A. – 354, 573
Smith, P. – 85, 577
Smith, R.V. – 461
Sneath, R.W. – 298, 567
Soler-Rovira, J. – 229, 231
Solimando, D. – 239
Sørensen, P. – 234
Sousa, J.R. – 236
Sprosen, M.S. – 76
Stamatiadis, S. – 351
Stenberg, B. – 224
Stevens, C.J. – 88
Stevens, R.J. – 338, 347, 349, 586

The Mechanics' Magazine, Museum, Register, Journal And Gazette Vol. Xxx October 6th, 1838-march 30th, 1839